Handbook of Research on Developments and Trends in Industrial and Materials Engineering

Prasanta Sahoo
Jadavpur University, India

IGI Global
DISSEMINATOR OF KNOWLEDGE

A volume in the Advances in Chemical and Materials Engineering (ACME) Book Series

Published in the United States of America by
IGI Global
Engineering Science Reference (an imprint of IGI Global)
701 E. Chocolate Avenue
Hershey PA, USA 17033
Tel: 717-533-8845
Fax: 717-533-8661
E-mail: cust@igi-global.com
Web site: http://www.igi-global.com

Library of Congress Cataloging-in-Publication Data

Names: Sahoo, Prasanta, editor.
Title: Handbook of research on developments and trends in industrial and materials engineering / Prasanta Sahoo, editor.
Description: Hershey, PA : Engineering Science Reference, 2019. | Includes bibliographical references and index. | Summary: “”This book explores applied research in the areas of industrial and materials engineering”--Provided by publisher”-- Provided by publisher.
Identifiers: LCCN 2019034100 (print) | LCCN 2019034101 (ebook) | ISBN 9781799818311 (h/c) | ISBN 9781799818335 (eISBN)
Subjects: LCSH: Industrial engineering--Research. | Materials--Research.
Classification: LCC T56.42 .H36 2019 (print) | LCC T56.42 (ebook) | DDC 620--dc23
LC record available at https://lccn.loc.gov/2019034100
LC ebook record available at https://lccn.loc.gov/2019034101

This book is published in the IGI Global book series Advances in Chemical and Materials Engineering (ACME) (ISSN: 2327-5448; eISSN: 2327-5456)

British Cataloguing in Publication Data
A Cataloguing in Publication record for this book is available from the British Library.

For electronic access to this publication, please contact: eresources@igi-global.com.

Advances in Chemical and Materials Engineering (ACME) Book Series

J. Paulo Davim
University of Aveiro, Portugal

ISSN:2327-5448
EISSN:2327-5456

Mission

The cross disciplinary approach of chemical and materials engineering is rapidly growing as it applies to the study of educational, scientific and industrial research activities by solving complex chemical problems using computational techniques and statistical methods.

The **Advances in Chemical and Materials Engineering (ACME) Book Series** provides research on the recent advances throughout computational and statistical methods of analysis and modeling. This series brings together collaboration between chemists, engineers, statisticians, and computer scientists and offers a wealth of knowledge and useful tools to academics, practitioners, and professionals through high quality publications.

Coverage

- Materials to Renewable Energies
- Coatings and surface treatments
- Sustainable Materials
- Computational methods
- Heat Treatments
- Composites
- Polymers Engineering
- Nanomaterials
- Metallic Alloys
- Electrochemical and Corrosion

IGI Global is currently accepting manuscripts for publication within this series. To submit a proposal for a volume in this series, please contact our Acquisition Editors at Acquisitions@igi-global.com or visit: http://www.igi-global.com/publish/.

The Advances in Chemical and Materials Engineering (ACME) Book Series (ISSN 2327-5448) is published by IGI Global, 701 E. Chocolate Avenue, Hershey, PA 17033-1240, USA, www.igi-global.com. This series is composed of titles available for purchase individually; each title is edited to be contextually exclusive from any other title within the series. For pricing and ordering information please visit http://www.igi-global.com/book-series/advances-chemical-materials-engineering/73687. Postmaster: Send all address changes to above address.

Titles in this Series

For a list of additional titles in this series, please visit:
https://www.igi-global.com/book-series/advances-chemical-materials-engineering/73687

Applications and Techniques for Experimental Stress Analysis
Karthik Selva Kumar Karuppasamy (Indian Institute of Technology, Guwahati, India) and Balaji P.S. (National Institute of Technology, Rourkela, India)
Engineering Science Reference • © 2020 • 350pp • H/C (ISBN: 9781799816904) • US $245.00

Design of Experiments for Chemical, Pharmaceutical, Food, and Industrial Applications
Eugenia Gabriela Carrillo-Cedillo (Universidad Autónoma de Baja California, Mexico) José Antonio Rodríguez-Avila (Universidad Autónoma del Estado de Hidalgo, Mexico) Karina Cecilia Arredondo-Soto (Universidad Autónoma de Baja California, Mexico) and José Manuel Cornejo-Bravo (Universidad Autónoma de Baja California Mexico)
Engineering Science Reference • © 2020 • 429pp • H/C (ISBN: 9781799815181) • US $255.00

Nanocomposites for the Desulfurization of Fuels
Tawfik Abdo Saleh (King Fahd University of Petroleum and Minerals, Saudi Arabia)
Engineering Science Reference • © 2020 • 300pp • H/C (ISBN: 9781799821465) • US $225.00

Enhanced Heat Transfer Mechanism of Nanofluid MQL Cooling Grinding
Changhe Li (Qingdao University of Technology, China) and Hafiz Muhammad Ali (University of Engineering and Technology, Taxila, Pakistan)
Engineering Science Reference • © 2020 • 350pp • H/C (ISBN: 9781799815464) • US $235.00

Advanced Catalysis Processes in Petrochemicals and Petroleum Refining Emerging Research and Opportunities
Mohammed C. Al-Kinany (King Abdulaziz City for Science and Technology, Saudi Arabia) and Saud A. Aldrees (King Abdulaziz City for Science and Technology, Saudi Arabia)
Engineering Science Reference • © 2020 • 257pp • H/C (ISBN: 9781522580331) • US $195.00

Nanotechnology in Aerospace and Structural Mechanics
Noureddine Ramdani (Research and Development Institute of Industry and Defense Technologies, Algeria & Harbin Engineering University, China)
Engineering Science Reference • © 2019 • 356pp • H/C (ISBN: 9781522579212) • US $265.00

Polymer Nanocomposites for Advanced Engineering and Military Applications
Noureddine Ramdani (Research and Development Institute of Industry and Defense Technologies, Algeria & Harbin Engineering University, China)
Engineering Science Reference • © 2019 • 435pp • H/C (ISBN: 9781522578383) • US $225.00

IGI Global
DISSEMINATOR OF KNOWLEDGE

701 East Chocolate Avenue, Hershey, PA 17033, USA
Tel: 717-533-8845 x100 • Fax: 717-533-8661
E-Mail: cust@igi-global.com • www.igi-global.com

List of Contributors

Table of Contents

Detailed Table of Contents

The general goal of warehouse management is to ensure the correct supply of materials to the production line. In this context, this document presents a supposed instance for a general problem of a raw materials warehouse with poor control of inventories and poor internal traceability. An internal material requirement planning (MRP) and storage location-allocation problem (SLAP) for sixty materials was used to supply the restocking store that provides for the production line, to see the materials with the highest turnover and to reassign the materials in the general warehouse to minimize the internal long-term logistics costs. As a result of the methodology, the company determines the products with the highest turnover in a variable demand environment. Also, the company reallocates these products in the warehouse of raw materials reducing logistics costs and supply in 100% on-time the stock store that, in turn, supplies the production line.

The advent of Radio Frequency Identification (RFID) technique, low-cost wireless sensor devices, and web-based technologies are ushering new ways to take control of the plethora of data created by the Internet of Things (IoT) approach has gained popularity in the modern manufacturing industry. This mechanism provides an opportunity of remotely storing and retrieving data from RFID-tags. Products labelled with such tags can be scanned efficiently using RFID-readers that do not require line-of-sight. This type of identification and access to information on tags are used by manufacturing business operations, can lead to improving logistics, and better customer service. However, the widespread use of RFID technology also introduces serious security and privacy risks since the information stored in tags can easily be retrieved by hidden RFID-readers, eventually leading to the violation of genuine business operation privacy. This chapter discusses the open research challenges of RFID applications in the manufacturing supply chain operations.

Manufacturing communities around the globe are eagerly witnessing the recent developments in semantic web technology (SWT). This technology combines a set of new mechanisms with grounded knowledge representation techniques to address the needs of formal information modelling and reasoning for web-based services. This chapter provides a high-level summary of SWT to help better understand the impact that this technology will have on wider enterprise information architectures. In many cases it also reuses familiar concepts with a new twist. For example, "ontologies" for "data dictionaries" and "semantic models" for "data models." This chapter presents the usefulness of a proposed architecture by applying a theory to integrating data from multiple heterogeneous sources which entails dealing

requirements and then into material selection decision. In the proposed approach, weights of the material selection criteria are obtained through Quality Function Deployment (QFD) and then the TOPSIS (Technique for Order Preference by Similarity to Ideal Solution) methodology is employed to achieve the optimum selection of material under a multi-attribute decision-making environment. An illustrative example is taken from the literature and presented in this chapter to demonstrate the proposed methodology.

This study deals with a plant layout where there were ninety predefined locations which have to be inspected by using three multiple robots in such a way that there would not be any collisions between the robots. A heuristic integrated multiobjective particle swarm optimization algorithm (HPSO) is developed for allocating tasks to each robot and planning of path while moving from one task location to another. For optimal path planning of each robot the research utilized A* algorithm. The task allocation for each robot is carried out using a modified multiobjective particle swarm optimization algorithm where the earliest completion time (ECT) inspired technique is used to make the algorithm applicable in multi robot task allocation problems. At the later stage of this study, in order to test the capability of HPSO an instance is solved by the algorithm and is compared with the existing solutions of a genetic algorithm with the A* algorithm. The computational results showed the superiority of the proposed algorithm over existing algorithms.

This chapter describes how corrosion is the degradation of a material due to a reaction with its environment. Wear is another deteriorating phenomenon which erodes the contacting surfaces of any mechanical component gradually. Although corrosion and wear are two different phenomenon they are often interdependent and in most practical situations take place simultaneously. Fortunately, several surface modification methods exist which can protect the surface of metal against corrosion and wear. Among them, electroless nickel coatings is a technology where a coating is applied without the use of external current source. The primary advantages of electroless coatings include uniform coating and ability to coat non-conductive materials. Electroless nickel coatings possess excellent properties such as high hardness, good wear resistance and corrosion resistance. The corrosion and wear behavior of these coatings mainly depend on bath ingredients, deposition conditions, heat treatment temperature, etc.

In the present work, the mechanical, tribological behavior of the machinability using electro discharge machining on a hybrid reinforced with different compositions of SiC and glass particulates are fabricated through a stir casting route. The effect of composition with machine parameters on the responses was observed. The Taguchi L9 orthogonal array was used for conducting the experiments, and the optimal EDM machining condition was obtained by performing MULTI-MOORA method integrated with PCA method. However, the effect of load, speed, and composition was investigated for tribological properties. It has been observed that SiC and glass particulates, significantly affect the compressive strength and hardness. However, the impact energy reduces an increase in the brittle nature of the composite. The glass particles do not contribute to the tensile strength but significantly improve the tribological property. The composition and pulse current found to be the most influencing parameters, the composition (Al-2014+2wt% glass+3 wt% SiC) has led to the best combination for machinability.

The laser bending process, also called the laser forming process, consists of irradiating the surface of a sheet or a plate by means of a moving laser beam with a predefined scanning strategy to generate the desired shape through thermally induced residual stress. This chapter presents the mechanisms of a laser bending process and the technological aspects concerning laser v-bending of rectangular AISI 304 plates for full plates and plates with a central cutout at its middle to highlight the process fundamentals and how processing affects the final bending angle of the workpieces. Laser bending involving plates with a cutout will have numerous applications for car bodies, such as front and rear panels where bending is required to be performed on panels with cutout geometries. To investigate the effects of shape and size of the cutout on temperature distribution, stress distribution, and final bending angle, different shapes such as circular, ellipse, rectangular, and square, as well as different dimensions of cutouts have been chosen.

In last few decades, continuous efforts have been put by the researchers on developing the field of structural systems considering the applicability and enhancing material properties of the structural elements. In this regard functionally graded materials (FGMs) have gained tremendous popularity and have been the active area of research interest for researchers in recent decades. In this chapter, a detailed method of analysis for FGM cracked beam subjected to static and dynamic loadings is presented. Method based on Timoshenko beam theory and Ritz method is successfully implemented to get the displacement and natural frequency of cracked beam subjected to static as well as dynamic loadings.

and biomedical applications. This chapter outlines the underlying physics of field-fluid interactions on one hand; on the other, it cites novel techniques in ferrofluid-actuated electronics cooling, hydrostatic & hydrodynamic bearings, extreme-boundary lubrication, damping systems, biomimetic locomotion, and medical diagnostics (e.g. magnetic drug targeting & magnetic cell sorting).

Fluidized bed technology is a well-established environment friendly technology, by which energy can be generated through combustion and gasification techniques. It is widely prevalent today owing to its excellent heat transfer, mixing characteristics and compactness. The design and scale-up of the fluidized beds are vital to the enhancement of heat transfer and mixing characteristics. However, heat transfer characteristics play a key role in determining the combustion and gasification characteristics. CFD is a technique which helps to optimize the design and operation of fluidized bed combustor and gasifiers. Enhancement of computing speed and numerical techniques has led to CFD being used as a widely implemented tool to provide a bridge between laboratory scale and industrial study. In this chapter, a comprehensive review of CFD modelling and experimental study on the conical fluidized bed has been carried out. Primarily this chapter demonstrates probable future accomplishments and identifies trends and regions where further research is required.

System dynamics is a method that allows analysts to separate complex social, and behavioral systems into components, to visualize them by reconstructing them as a whole again, and to develop a simulation model. In this chapter, a system dynamics model is developed for a flexible manufacturing system. The case study is developed using Vensim package program. To manage a flexible manufacturing system it is important to determine production schedule rules and examine results. System dynamics is a tool to test these rules without any production loss.

Total quality management (TQM) is a set of the systematic management approaches for the continuous improvement of quality standards of products, services, and business relations with employees and consumers. This chapter mainly focusses on the eight key principles of TQM, the involvement of workers, leadership, process approach, strategic approach, continuous improvement, together for a factual approach to decision-making and communication. This chapter also discusses a four-part management model that implements continuous quality improvements and process control in different stages of an organization based on the Deming cycle or the Shewhart cycle. Quality engineering encompasses a broad range of methodologies and tools, which include quality management systems, advanced product quality planning (APQP), tools like quality function development (QFD), failure modes, and effects analysis (FMEA), statistical process control (SPC), and are widely accepted methodologies used in industries.

Preface

Handbook of Research on Recent Developments and Trends in Industrial and Materials Engineering provides a domain for professionals, academicians, researchers, policy makers, and practitioners working in the fields of industrial and materials engineering, reliability engineering, and management to present their research findings, propose new methodologies, disseminate the latest findings, and to learn from each other's research directions. It focuses on applied research in the areas of industrial and materials engineering with an emphasis on inter and intra disciplinary research. This book contains full-length research manuscripts, insightful research, and practice notes, as well as case studies.

Handbook of Research on Recent Developments and Trends in Industrial and Materials Engineering provides a forum for industrial and materials engineering educators, researchers, and practitioners to advance the practice and understanding of applied and theoretical aspects of industrial and materials engineering and related areas. The book contains practical, empirical and theoretical research on the development, improvement, implementation, and evaluation of integrated systems in engineering. In addition, the book will especially be interested in those research studies that show a significant contribution to the area by way of intra and inter disciplinary approaches in industrial and materials engineering. The target audience of this book is composed of professionals and researchers working in the field of industrial engineering including administrative sciences, management and education. Moreover, the book provides insights and support executives concerned with the management of expertise, knowledge, information and organizational development in different types of work communities and industries. The book contains nineteen chapters contributed by academic and industrial researchers from around the globe.

"Demand Forecasting: A Challenge for Furniture SMEs" discusses a proposal for demand management in furniture SMEs located in the city of Puebla, México. The historical production data is reviewed, and the literature of SMEs was analyzed as well as the current situation and statistical information is sought. Additionally, it presented an overview of the models to forecast demand and applied to the most relevant articles. The ANN model is developed with TANSIGMOID transfer function using MATLAB software. The appropriate forecasting techniques are selected according to the variability of the demand of the articles taking a short-term planning horizon. This research helps reduce uncertainty, forecasting sales and achieving better production planning through ANNs.

"Methodology Proposal for Logistics Management in an Automobile Company" presents a study case where the authors propose a methodology for logistics management that encloses the layout redesign and the proposal of indicators to improve the area and the company through the gathering of previous research and the use of good practices in warehouse management. Logistics impacts the costs and effectiveness of the client service system directly; therefore, it is of paramount importance to adequately design and operate warehouses in order to achieve lower costs and higher performance.

The general goal of warehouse management is to ensure the correct supply of materials to the production line. In this context, "Methodology for Storage Location Allocation Based on the Planning of Material Requirements" presents a supposed instance for a general problem of a raw materials warehouse with poor control of inventories and poor internal traceability. An internal material requirement planning (MRP) and storage location-allocation problem (SLAP) for sixty materials are used to supply the restocking store that provides the production line, to know the materials with the highest turnover and to reassign the materials in the general warehouse to minimize the internal long-term logistics costs. As a result of the methodology, the company determines the products with the highest turnover in a variable demand environment.

"Radio Frequency Identification Technology and Its Security Concerns in Manufacturing Industry" classifies the existing security and privacy threats into those, which target the radio frequency identification (RFID) components such as the tag, the communication channel, and the overall system threats. The advent of RFID technique, low-cost wireless sensor devices, and web technologies are ushering new hope to take control of plethora of data created by the Internet of Things (IoT) approach has gained popularity in modern manufacturing company. While IoT offers the capability to connect and communicate with both digital and physical world of manufacturing supply chain partners – however managing privacy and security in IoT data presents a significant challenge. Finally, the chapter discusses the open research challenges that need further investigation, especially with the rapid introduction of diverse RFID applications in manufacturing supply chain management (SCM).

"Ontology-Assisted Enterprise Information Systems Integration in Manufacturing Supply Chain" presents the usefulness of a proposed architecture by applying theory to integrating data from multiple heterogeneous sources entail dealing with semantic mapping between source schema and resource description framework (RDF) ontology are described declaratively using a specific query language (i.e. SPARQL) queries. The manufacturing community around the globe is eagerly witnessing the recent developments in semantic web technology (SWT). This technology combines a set of new mechanisms with grounded knowledge representation techniques to address the needs of formal information modelling and reasoning for web-based services. This chapter provides a high-level summary of SWT to help better understand the impact that this technology will have on wider enterprise information architectures. In many cases it also reuses familiar concepts with a new twist. The semantic of query rewriting is further discussed and a query rewriting algorithm is presented.

"Manufacturing, Control, and Automation" identifies the common needs for process controls and automation that include methodologies to enable in-situ-level process controls, optimization at the plant or industry level, open-architecture software tools, adaptive control systems, methods and diagnostic tools for condition-based maintenance of process equipment in a manufacturing industry. Manufacturing is concerned with the process of making useful products from raw materials. These raw materials are subjected to pass through a sequence of processes to make individual products of a certain value. Conventionally, manufacturing can be divided into three major categories viz., mass manufacturing, batch manufacturing, and job-shop manufacturing. In this era of global competitiveness, it is extremely important for the manufacturing industry to meet their customer's demands more effectively. In this course, all manufacturing processes continuously strive to improve its product quality, identifying methods to cut down costs, fast throughput and incorporating production flexibility.

"Performance Evaluation of Cloud Systems by Switching the Virtual Machines Power Mode Between the Sleep Mode and Active Mode" studies energy savings of data centers by consolidation and switching off those virtual machines which are not in use. Data centers are cost-effective infrastructures for storing large volumes of data and hosting large-scale service applications. Cloud computing service providers are rapidly deploying data centers across the world with huge number of servers and switches. These data centers consume significant amounts of energy, contributing to high operational costs. According to this policy, c virtual machines continue serving the customer until the number of idle servers attains the threshold level *d*; then *d* idle servers take synchronous vacation simultaneously, otherwise these servers begin serving the customers. Numerical results are provided to demonstrate the applicability of the proposed model for the data center management in particular, to quantify theoretically the tradeoff between the conflicting aims of energy efficiency and Quality of Service (QoS) requirements specified by cloud tenants.

"QFD-Based TOPSIS Methodology for Material Selection" presents user-centered material selection methodology which paves a path for reflecting user preferences into design requirements and then into material selection decision. In the proposed approach, weightages of the material selection criteria are obtained through quality function deployment (QFD) and then the technique for order preference by similarity to ideal solution (TOPSIS) methodology is employed to achieve optimum selection of materials under multi-attribute decision making environment. An illustrative example is taken from the literature and presented in this chapter to demonstrate the proposed methodology.

"Task Allocation and Path Planning of a Multi-Robot System Using Heuristic Coupled Particle Swarm Optimization Algorithm" deals with a plant layout where there were ninety predefined locations which have to be inspected by using three multiple robots in such a way that there would not be any collision between the robots. A heuristic integrated multi objective particle swarm optimization algorithm is developed for allocating tasks to each robot and planning of path while moving from one task location to another. For optimal path planning of each robot the research utilized A* algorithm. The task allocation for each robot is carried out using modified multi-objective particle swarm optimization algorithm where the earliest completion time (ECT) inspired technique is used to make the algorithm applicable in a multi-robot task allocation problem.

"Corrosion and Wear Behavior of Electroless Nickel Coatings" deals with two important material degradation phenomena. Corrosion is the degradation of a material due to a reaction with its environment. Wear is another deteriorating phenomenon which erodes the contacting surfaces of any mechanical component gradually. Although corrosion and wear are two different phenomena, they are often interdependent and in most practical situations take place simultaneously. Fortunately, several surface modification methods exist which can protect the surface of metal against corrosion and wear. Among them, electroless nickel coatings is matured technology where a coating is applied without the use of external current source. The primary advantages of electroless coatings include uniform coating and the ability to coat non-conductive materials. Electroless nickel coatings possess excellent properties such as high hardness, good wear resistance, and corrosion resistance. The corrosion and wear behavior of these coatings mainly depend on bath ingredients, deposition conditions, heat treatment temperature, etc.

"Tribological Behavior, Machinability, and Optimization of EDM of AA-2014 Hybrid Composite Reinforced With SiC and Glass Particulates" considers mechanical, tribological behavior with the machinability using Electro Discharge Machining on the hybrid reinforced with different composition of SiC and Glass particulates fabricated through stir casting route. The effect of composition with machine parameters on the responses was observed. Taguchi L9 orthogonal array was used for conducting the experiments, and the optimal EDM machining condition was obtained by performing MULTI-MOORA method integrated with PCA method. However, the effect of load, speed and composition was investigated for tribological properties. It has been observed that SiC and glass particulates, significantly affect the compressive strength and hardness. However, impact energy reduces an increase in the brittle nature of the composite. The glass particles do not contribute to the tensile strength but significantly improve the Tribological property. The composition and pulse current found to be the most influencing parameters, the composition (Al- 2014+2wt% glass+3 wt% SiC) has led to the best combination for machinability.

The laser bending process, also called the laser forming process, consists in irradiating the surface of a sheet or a plate by means of a moving laser beam with a predefined scanning strategy to generate the desired shape through thermally induced residual stress. "Temperature Gradient-Based Laser Bending of Full Plates and Plates With Cutout" presents the mechanism of laser bending process and technological aspects concerning laser V-bending of rectangular AISI 304 plates for full plates, and plates with a central cutout in the middle that highlights the process fundamentals and how processing affects the final bending angle of the workpieces. Laser bending involving plates with a cutout will have numerous scopes for applications in car bodies, front and rear panels, where bending is required on panels with cutout geometries. To investigate the effects of shape and size of the cutout on temperature distribution, stress distribution and final bending angle, different shapes such as circular, ellipse, rectangular and square, and different dimensions of cutouts are chosen.

In the last few decades, continuous effort has put by the researchers on developing the field of structural systems considering the applicability and enhancing material properties of structural elements. In this regard, functionally graded materials (FGMs) have gained tremendous popularity and have been the active area of research interest for researchers in recent decades. In the chapter "Modelling and Analysis of Functionally-Graded Cracked Beams Subjected to Static and Dynamic Loadings," a detailed method of analysis for a FGM cracked beam subjected to static and dynamic loadings is presented. The method is based on the Timoshenko beam theory and the Ritz method is successfully implemented to get the displacement and natural frequency of cracked beam subjected to static as well as dynamic loadings.

The chapter "Corrosion Protection of Construction Steel" highlights and summarizes the different grades of steel for their high corrosion resistance. Reinforced concrete is one of the most versatile materials for constructions, but the performance is limited by corrosion, cracking and spalling of steel rebars. The steel embedded in concrete is protected by a passive film from a corrosive attack from chlorides, carbon dioxide, and sulphates. As the concentration of chlorides, carbon dioxide or sulphates increases above a certain threshold value at the concrete rebar interface, the passive film breaks and leads to a severe increase in corrosion rate. Further, the dynamic loading and temperature of the surroundings also affect the durability of the reinforcements. The rebars may be protected from such corrosion attack by the suitable selection of material, improving the concrete quality and tailoring its composition or application of protective coatings. Further, surface engineering and the application of corrosion resistance coatings for the prevention of corrosion of construction steel rebars has been also discussed.

"Non-Destructive Testing of Carbon Fibre Reinforced Polymer (CFRP) Composite Using Thermosonic Technique" deals with failure detection of composite materials. Composite materials are often subjected low velocity impacts which leads to delamination in subsequent layers. Linear ultrasound-based approaches are not accurate enough to detect it properly. The local defect resonance (LDR) based has proven to be an efficient candidate for detection of such defects. LDR frequency excitation leads to high amplitude vibration which raises defect temperature drastically, detectable by an infrared camera. Here, a numerical investigation of LDR frequency excited ultrasound thermography is carried out on a delaminated carbon fibre reinforced polymer (CFRP) plate. The location and size of the delamination can be easily seen from the thermal signature. The temperature gradient variation is found to be high at first and then it decreases due to higher heat conduction rate. The delamination in CFRP plate is detected by standard phased array ultrasound testing (PAUT) using flat bottom hole in aluminium plate as a case study. Delamination detection by PAUT is found to be very time-consuming process compared with thermosonic technique.

"Basics and Applications of Ferrofluids" outlines the underlying physics of field-fluid interactions on one hand; on the other, it cites novel techniques in ferrofluid-actuated electronics cooling, hydrostatic & hydrodynamic bearings, extreme-boundary lubrication, damping systems, bio-mimetic locomotion, and medical diagnostics (e.g. magnetic drug targeting & magnetic cell sorting). Engineered 'smart' fluids, called ferrofluids, which permit static and dynamic control by external magnetic fields of low or moderate strengths present a novel and challenging domain for scientists and technologists. The introduction of a 'controllable' or 'tunable' external force into the momentum transport equations opens up new fields of physical phenomena. In situations where the external field influence becomes strong enough to compete with gravitational forces or acts as a sole external agency to drive flows in the hypo-gravity environment, a new class of hydrodynamic phenomena becomes accessible. Ferrofluids exhibit special rheological properties that make them suited for a number of technological and biomedical applications.

The chapter "A Future Trend on Research Scope of Numerical Simulation on Conical Fluidized Bed" presents a comprehensive review of CFD modelling and experimental study on the conical fluidized bed. Fluidized bed technology is a well-established environment friendly technology, by which energy can be generated through combustion and gasification techniques. It is widely prevalent today owing to its excellent heat transfer, mixing characteristics and compactness. The design and scale-up of the fluidized beds are vital to the enhancement of heat transfer and mixing characteristics. However, heat transfer characteristics play a key role in determining the combustion and gasification characteristics. CFD is a technique which helps to optimize the design and operation of fluidized bed combustor and gasifiers. Enhancement of computing speed and numerical techniques has led to CFD being used as a widely implemented tool to provide a bridge between laboratory scale and industrial study. Primarily this chapter demonstrates probable future accomplishments and identifies trends and regions where further research is required.

System dynamics is a method that allows analysts to separate complex social, and behavioral systems into components, to visualize them by reconstructing them as a whole again, and to develop a simulation model. In the chapter, "Supply Chain Management via System Dynamics in Flexible Manufacturing System," a system dynamics model is developed for flexible manufacturing system. The case study is developed using Vensim package program. To manage the flexible manufacturing system, it is important to determine production schedule rule and examine results. System dynamics is a power tool to test these rules without any production lots.

Total quality management (TQM) is a set of the systematic management approach for the continuous improvement of quality standards of products, services, and business relations with the employees and consumers. The chapter "Total Quality Management and Quality Engineering" mainly focusses on the eight key principles of TQM, the involvement of workers, leadership, process approach, strategic approach, continuous improvement, factual approach to decision-making and communication. This chapter also discusses a four-part management model that implements continuous quality improvement and process control in different stages of an organization based on the Deming cycle or Shewhart cycle. Quality engineering encompasses a broad range of methodologies and tools, which includes quality management systems, advanced product quality planning (APQP) tools like quality function development (QFD), failure modes, effects analysis (FMEA), and statistical process controls (SPC), are widely accepted methodologies used in industries.

The editor would like to thank all the authors for contributing their valuable research work as chapters for this book. The authors submitted the chapters and responded for the revisions on time which could make it possible to publish this book on time. The editor is thankful to all the reviewers who painstakingly reviewed all the submissions and the revisions, wherever needed. Finally, the editor is thankful to IGI Global, for allowing publication of the book, and also to the book development editor and other team members at IGI Global for their motivation, encouragement, advice and support provided during organization of this edited volume that led to the publication of this book.

Prasanta Sahoo
Jadavpur University, India

Chapter 1
Demand Forecasting: A Challenge for Furniture SMEs

Natalie Nicole Mar Hernández
UPAEP University, Mexico

Patricia Cano-Olivos
UPAEP University, Mexico

Diana Sánchez-Partida
UPAEP University, Mexico

José-Luis Martínez-Flores
UPAEP University, Mexico

Santiago Omar Caballero-Morales
https://orcid.org/0000-0002-9986-7768
UPAEP University, Mexico

ABSTRACT

This chapter provides a proposal for demand management in furniture SMEs located in the city of Puebla, México. The historical production data reviewed, and the classification of the most critical articles was made using the ABC classification methodology. Subsequently, the literature of SMEs was analyzed as well as the current situation and statistical information was sought. Additionally, it presented an overview of the models to forecast demand and applied to the most relevant articles. Due to the results and previous study, it was decided to implement a forecasting technique which is modelled by artificial neural networks. The ANN model developed with the TANSIGMOID transfer function by using MATLAB software. The appropriate forecasting techniques selected according to the variability of the demand of the articles takes a short-term planning horizon. This research will help the company reduce uncertainty, forecasting sales, and achieve better production planning through ANNs.

DOI: 10.4018/978-1-7998-1831-1.ch001

INTRODUCTION

Currently, small and medium-sized enterprises (SMEs) have faced new challenges; then, the companies require strategies for making-decisions, which allow obtaining a competitive advantage. Demand planning is one strategic area since the majority of activities depend on it. Therefore, forecast techniques are vital for making-decisions from historical information, enabling minimizing errors and eliminating the uncertainty in the planning. However, the problem is that most of the SMEs do not make use of it, either because of its lack of knowledge or not knowing how to choose the right forecast technique. In Mexico, SMEs represent 1.15% gross domestic product (GDP), employ to 92,107 Mexicans (National Institute of Statistics and Geography [INEGI], 2018), according to data, are having up to 250 employees (Organization for Economic Co-operation and Development [OECD], 2013).

SMEs must begin to know techniques that allow them to have a competitive advantage. Methods of forecasts have been a powerful tool which reduces the uncertainty of suffering erroneous planning and in this way, achieving an improvement in the performance of the supply chain. Forecasts predict the future based on historical data and support decision making in the organization. That is why it is part of the administration of the supply chain. Through the application of traditional forecasts, it can observe the limitations that affect the accuracy of them. However, artificial neural network algorithms showed a better response predicting the demand due to the ability to accommodate nonlinear data (Chang & Wang, 2006). "Artificial intelligence forecasting techniques have been receiving much attention lately to solve problems that are hardly solved using traditional methods. ANNs can learn like humans, by accumulating knowledge through repetitive learning activities. Animal brain's cognitive learning process simulated in ANNs" (Kumar, Herbert & Rao, 2014).

This research was applied to a Mexican enterprise of furniture, which is a company of family tradition and artisanal, the owners manage the internal operations, and they are not specialists in the area. This company sells its products in different regions of the country, offering their customers products for home, garden and offices. Therefore, the main objective is to apply the right forecasting technique that will estimate the demand with minimal error.

BACKGROUND

SMEs in the current economy are of great importance due to the generation of employment at the national and regional level that they generate. In recent years these companies are linking with larger industries and provide a regional balance through the distribution of more equitable investments (Taiwo, Ayodeji M, & Yusuf A, 2012). It should consider that the conditions under which SMEs operate are not precisely favourable. Its low level of technological acceptance, insufficiently qualified workers and/or owners, inefficient administration, low productivity, among others (Zeballos, 2003), to minimize competitive advantages. In Mexico, SMEs represent about 4% of the total of economic units, employ 32% of the population, and produce 36% of the GDP (Ministry of Economic, 2012). The classification of the companies is according to the people occupied and the amount of income per year. The criteria recommended by the European Union and OECD are for small companies of 10 to 49 people and annual sales less than 10 million of euros, for medium-sized companies from 50 to 249 people and yearly transactions of less than 50 million euros (National Institute of Statistics and Geography [INEGI], 2014). SMEs means a vital component of the productive network in the region: employ around 67% of the total number of workers and represents approximately 99% of the total number of companies. It should be noted that the contribution to GDP is relatively low, compared to large companies (Economic Commission for Latin America and the Caribbean [CEPAL], 2018).

In recent years, SMEs are vital to a country's economic growth, which strengthens competitiveness and entrepreneurship (Zeballos, 2003), hence the government has given them more considerable attention, especially in third world countries, because they use more local raw materials, generate more employment, encourage rural development and entrepreneurship, mobilize domestic savings, link to more substantial industries, provide regional equilibrium through the distribution of investments more provide self-employment and provide opportunities to train managers and semiskilled workers (Taiwo, Ayodeji & Yusuf, 2012). The people in charge of SMEs decision-making faced with a number of day-to-day challenges, the skills and capacities for decision-making that they possess will be the difference between success and failure (Schreier, Udomkit & Macchi, 2016), within the companies, the decision always taken through two approaches, the rational decision-making, which based on the theory and the empirical data; and intuitive decision making, which is based on intuition and non-rational data (Wiratmadja, Rumanti, Reynaldo & Kurniawati, 2016).

To make decisions plays a role in topics such as marketing, human resources development, research and development, public relations and investment (Schreier et al., 2016). These companies have in the domestic markets is successful in terms of growth in the economy and redistribution of income. Therefore, one of the essential tasks for public and private sector institutions in the region is to take advantage of the potential of these at the local level (Zeballos, 2003). SMEs, when faced with several obstacles, must seek ways to save money or be more productive, using empirical tools. The adaptation and creativity of SMEs are of great importance to stay in a globalized environment (Soto & Dolan, 2003). This phenomenon affects both negatively and positively all companies regardless of size, turn or location (Lopez, Acosta, Espinoza & Delfín, 2007). Due to globalization, SMEs face various challenges for their development, which include quality, flexibility, costs, training, among others that are of great importance for the growth of companies (Gunasekaran, Rai & Griffin 2011).

At the ministerial-level meeting of the OECD Council (2017), SMEs were described as essential for achieving globalization and present growth. Unfortunately, some countries face the challenge of low growth, weak trade and investment, and persistently high risk or inequality (OECD, 2017). Because of this, it is necessary to create conditions that benefit the companies giving a more significant benefit to the economy and society. Inter-American Development Bank (IDB) (2010) mentions that the opening of international trade entails the globalization of supply chains. Depending on the characteristics, SMEs of logistic costs are usually between two and three times higher than those of large companies. For SMEs, logistics costs are approximately 40% of the value of the product (IDB, 2011), where internal logistical costs can be 42% of total sales, compared to levels from 15% to 18% in the case of large companies (World Trade Organization [WTO], 2016).

Logistic performance of the supply chain of SMEs in Mexico is 13.3% (Ministry of Economic, 2011). Rubio (2017) mentions that of 20 organizations studied; only 55% use technique of forecasting and 45% do not use it. A forecast is a prediction which is an extension of the past. Although the presence of any random phenomenon makes them uncertain, is a process in which quantitative and qualitative methods can be used to perform them. The prognosis is fundamental in strategic management and is a crucial element in a total quality program of any company (Alonso & Villegas, 2005). In this sense, the main objective of the research is to apply forecasting methods for planning and optimize the resources.

A forecast is necessary for the areas of administrative planning; it is forecast which helps to reduce the uncertainty. Due to the variability of the demand, SMEs make decisions for intuition and most of the time are unaware of the importance. By using these techniques, you should know how and where to use them in a way that contributes and be beneficial to the company (Hanke & Wichern, 2010). Sales forecasts are essential regardless of the activity or size. Forecast allows to improve customer service, reduces costs and takes advantage of the opportunities that the market offers for obtaining good results. It is evaluated systematically and rationally, giving a future vision of demand (Morlidge & Player, 2010). Acosta, Diaz & Anaya (2009) mention that the application of individual and combined forecasts for 263 data series of the same company achieved significant savings due to the right prediction.

A forecast is necessary for the areas of administrative planning; it is forecast which helps to reduce the uncertainty. Due to the variability of the demand, SMEs make decisions for intuition and most of the time are unaware of the importance. By using these techniques, you should know how and where to use them in a way that contributes and be beneficial to the company (Hanke & Wichern, 2010). Sales forecasts are essential regardless of the activity or size. The forecast allows to improve customer service, reduces costs and takes advantage of the opportunities that the market offers for obtaining good results. It is evaluated systematically and rationally, giving a future vision of demand (Morlidge & Player, 2010). Acosta, Diaz & Anaya (2009) mention that the application of individual and combined forecasts for 263 data series of the same company achieved significant savings due to the right prediction.

Corporate management has been affected by volatility, but SMEs are more affected than large companies. Most SMEs do not use quantitative and qualitative methods, while larger companies use sophisticated methods or software. However, it does not have a difference between the accuracy of the forecasts (Hofer, Eisl & Mayr, 2015). Austrian SMEs often use the extrapolation of tendencies through different mathematical methods. In large companies, qualitative methods are more widely used, such as the opinion of an expert or a group of experts and market indicators (Hofer et al., 2015). Flores, Stading & Klassen (2007) mention that SMEs apply judgmentally, counting, or survey for forecasting. As observed a serious problem for SMEs is the lack of research and development, in addition to high operating costs, making it difficult to have access to large loans from banks. These companies are affected by high taxes, the costs of social security and the protection of employment affect the incentives for the creation, operation and development of companies, slowing the development of the formal sector (OCDE, 2013).

Hanke and Wicher (2010) mention that in most of the regional businesses do forecast by intuition, so it is necessary to reduce this degree of concern in companies. Smith III et al. (1996) mention that forecasting plays an essential role in every major functional area of management. Forecast methods can involve mathematical model, subjective or intuitive prediction, or it can include a combination of these (Mendez & Lopez 2014). Short-Term forecasts are necessary for an organization. However, these are used more frequently in the areas of marketing, production and administration. Moving averages, exponential smoothing and trend extrapolation, are typical in short-term projections (Heizer & Render, 2010). Any forecast aims to be useful from the information available and apply the technique to the most appropriate for the different patterns that demand (Krajewski, Ritzman & Malhotra, 2008). The model that each company chooses depends on the horizon to be forecast, the availability of the data, and the accuracy required, the budget for the estimates and the qualified personnel for the calculation (Chase, Jacobs & Aquilano, 2014).

Some of the forecasting techniques are the time-series analysis that involves the use of past data to predict the future. These methods analyzed cycles, seasonal sales and random factors which might affect the forecast. Other techniques using historical data are correlation methods. Regression analysis is used to correlate historical data to one or more variables. Any estimate aims to be useful from the information available and apply the technique to the most appropriate for the different patterns that demand (Krajewsky et al. 2008). According to Chase et, al. (2014), the model that each company chooses depends on the horizon to be forecast, the availability of the data, the accuracy required, the budget for the estimates and the qualified personnel for the calculation.

Some examples of these forecast methods can be simple average where the horizontal pattern of a series of time is based on the average of the demands, simple moving averages where the application of this method requires merely calculating the average demand for the most recent periods, in order to use it as a forecast for the next period, weighted mobile average where each one of the historical demands that intervene in the average can have its weighting and the sum of the weights should be equal to 1.0, the average is obtained by multiplying the weighting by the value of said period and adding the products, Brown`s exponential smoothing is a method that uses a weighted average of the past values of a series of time to obtain a short-term forecast, Winter`s model is the indicated method when you have series of time with a clearly defined seasonal component, each of the components of time, the smoothed average, the smoothed trend and the seasonal factor are combined separately to obtain the forecast (Krajewski et al. 2008), Holt`s exponential smoothing is a very refined weighted moving average method that allows calculating the average of a series of time, assigning to the recent demands higher weight than to the previous requirements (Zuluaga, Molina & Guisao, 2011).

Some authors identify the behaviour and conditions of the products through the application of methods which provides the most critical items. Due to most studies in the area of inventory management aim to reduce the money or the space tied up in inventories and therefore use volume and product characteristics (e.g. space needs or unit cost) (Van Kampen, Akkerman & van Donk, 2012). The ABC method determines which are the most critical items (Flores & Whybark, 1988). This technique classifies the products based on the demand value or demand volume (Van Kampen et al., 2012), so that, the category A represents 80% of the accumulated revenue value and are typically 20% of the items. Category B contribute 15% of the accumulated revenue value and are 80%-90% of the items and C items are the last 5% of the revenue (Dhoka & Choudary, 2013). However, textbooks of operations management and supply chain management have not followed their lead but continue to discuss ABC analysis based on the idea of annual dollar volume. The objective of the technique is to focus on a few products that represent most of the income, obtaining significant reductions in inventory costs (Dhoka & Choudary, 2013). Inventories large amounts of finished products, spare parts and raw materials complicate material handling, so answering at minimum two questions: how much to order and when to order (Ravinder & Misra, 2014).

Zuluaga, et al. (2011) mentions that the advantages offered by demand forecasts are operational and strategic, and minimizing excess inventory. Forecasts improve the level of customer service due to the balance between supply and demand (Efendigil, Önüt & Kahraman 2009). The accuracy of the demand forecast significantly affects inventory levels, costs and customer satisfaction levels. To set up all of the logistic steps required for producing and selling a product, apparel managers must rely on efficient and accurate forecasting systems (Thomassey, Happiette & Castelain, 2002). Also improved demand forecasting figures play a major role in planning for short shelf-lived products (Raju, Kang, Moroz, Clement, Hopwell & Duffy, 2015). Having a clear vision of the future demand allows companies to reduce their costs, improve and adapt the service offered to their customers (Slimani, Farissi & Achchab, 2015). Frank, Garg, Raheja & Sztandera (2003) forecasted demand by using three different methods: exponential smoothing, Winter's method and NN using the same historical data. They stated that the forecasting accuracy of the NN method is higher than those of the other two techniques. Escoda, Ortega,

Sanz & Herms (1997) compared the NN, fuzzy NN and Winter's method for demand forecasting. They had 73 historical data points and used 59 of them in the training stage and 14 in the validation stage.

Artificial neural networks (ANNs) are inspired by the behaviour of the human brain, especially because of the neurons and their connections. It tries to create artificial models to solve problems that are difficult to solve using conventional algorithmic techniques. A mathematical definition of a neural network is "a directed graph consisting of nodes with interconnecting synaptic and activation links" (Haykin, 2009). A neural network has a set of inputs (x), of n components, a set of synapses that are characterized by a weight or strengths of its own (W), which represent the interaction between the input of synapse (j) connected to neuron (k) is multiplied by the synaptic weight (W_{kj}), an adder for summing the input signals weighted by the respective synaptic strengths of the neuron, an activation function for limiting the amplitude of the output of a neuron and output which is provided by the output function (Y_k), a graphic representation of it can be observed in the Figure 1, which forms the basis for designing a large family of neural networks (Haykin, 2009).

ANNs was applied the first time in 1964; when an adaptable linear ANN was used for climate forecasting due to the absence of the training algorithm for multilayer ANN. Forecasts of multiple periods is a single model that simultaneously generates the forecasts of multiple periods, that is, an ANN with multiple outputs (Figure 1). The combination of non-linear modelling and learning from the data makes ANNs flexible tools of general modelling attractive for their application in making forecasts. Even though there are numerous applications of ANNs for forecasting with satisfactory results, these do not have a standardized methodology that guarantees good accuracy of the estimates (Salazar & Cabrera, 2007). ANNs are flexible non-linear data-driven models that have attractive properties for forecasting. Statistical methods are efficient for data with seasonal or trend patterns while ANNs accommodate the data influenced by the special case, like a promotion or extreme crisis demand fluctuation (Kourentzes, 2013).

A neural network takes several binary inputs where x1, x2, x3 . . . are independent variables and contains one or more layers that are hidden from input to output nodes as can be observed in figure 2. ANNs are proved to be efficient in modelling complex and poorly understood problems for which sufficient data collected (Dhar & Stein, 1997). ANN is a technology which has been mainly used for predic-

Figure 1. Nonlinear model of a neuron
Source: Haykin, 2009

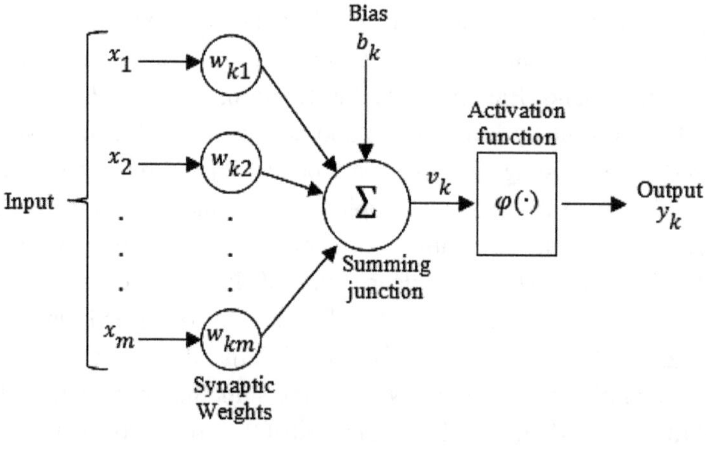

tion, clustering, classification and alerting of abnormal patterns. The capability of learning examples is probably the vital property of neural networks in applications and can be used to train network with the records of the past response of a complex system (Wei, Zhang & Li, 1997).

Many ANN models have been proposed, a multi-layer perceptron is a feed-forward network which is composed of several layers of neurons between the inputs and output what allows to decide on areas more complex than a single-level sensor (Hilera & Martinez, 1995). Artificial neural networks can learn from experience, to generalize from previous cases to new cases, to essential abstract characteristics from entries that represent irrelevant information, etc. Backpropagation is a type of neural network that has been applied to resolve many problems due to the ease of learning, for example, robot control, prediction, pattern recognition. This algorithm applied with more than two layers which consists of learning a predefined set of input-output pairs given as an example, using a propagation-adaptation cycle of two phases: first an entry pattern is applied; Xp: xp1, xp2, …, xpN as a stimulus for the first layer of the neurons of the network, is propagated through all the upper layers until generated the desired output that the network must make: d1, d2, …, dM, the result obtained in the output neurons is compared with the output to obtain, and an error value calculated for each output neuron (Hilera & Martinez,1995). The activation function is also called the transfer function, and it determines the relationship between inputs and outputs of a node and a network. These can be (Zhang, Patuwo, & Hu, 1998):

The sigmoid (logistic) function:

$$f(x) = \left(1 + \exp\left(-x\right)^{-1}\right) \tag{1}$$

The hyperbolic tangent (tanh) function:

$$f(x) = \left(\exp\exp(x) - \exp(-x)\right) / \exp(x) + \exp(-x) \tag{2}$$

The sin or cosine function:

Figure 2. Multilayer Neural Network with multiple outputs trained
Source: Haykin, 2009.

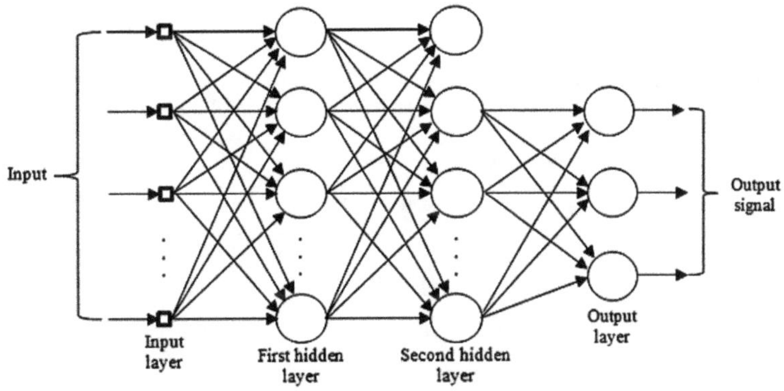

$$f(x) = \sin\sin(x)\, or\, f(x) = \cos(x) \tag{3}$$

The linear function:

$$f(x) = x \tag{4}$$

Zhang et al. (1998) also mention that "performance measures should be calculated based on the rescaled outputs, that a training and a test sample typically required for building an ANN forecaster and these may affect the performance of ANN, there are many performance measures for an ANN forecaster like the modelling time and training time, the ultimate and the most critical measure of performance is the prediction accuracy it can achieve beyond the training data and an accuracy measure often defined in terms of the forecasting error which is the difference between the actual (desire) and the predicted value. The most frequently used are:

The mean absolute deviation:

$$(MAD) = \frac{\Sigma|e_t|}{N} \tag{5}$$

The sum of squared error:

$$(SSE) = \Sigma(e_t)^2 \tag{6}$$

The mean squared error:

$$(MSE) = \frac{\Sigma(e_t)^2}{N} \tag{7}$$

The root mean squared error:

$$(RMSE) = \sqrt{MSE} \tag{8}$$

The mean absolute percentage error:

$$(MAPE) = \frac{1}{N}\Sigma\frac{|e_t|}{y_t} * 100 \tag{9}$$

METHODS USED IN RESEARCH

To ensure the selection of the right forecasting technique to reach the demand forecasting with the minimum error, it was necessary to go through several phases. First, there was collected the demand history of the SME. When more information becomes available, the level and trend factors can be updated. In this study, the methods used were in first instance the ABC classification which allowed dividing the items into three categories A, B, C: being the items A the most valuables which represent 70% of the income and B and C being the least valuables items. In this way, the project could be focused on the products that represent a greater monetary value for the company. Once that information was obtained the traditional forecast techniques were applied just for the items A. Time series methods are used to make detailed analyses of the historical patterns of a variable over time and project them toward the future.

The method of time series that was applied was the simple average, which is used for stabilized series and the environment is generally unchaining. The second method was simple moving average; the application of this method requires calculating the average demand for the most recent n periods, to use it as a forecast for the next period. The third method is weighted moving average; each of the historical demands involved in the average can have their own weight, and the sum of the weights must be equal to 1.0, the average is obtained by multiplying the weight by the value of said period and adding the products (Krajewski et al., 2008). The next traditional method of time series that was applied was the Holt´s Model; which is optimal for demand patterns that present a tendency and a constant seasonal pattern, in which it is intended to eliminate the impact of irregular historical elements by focusing on periods of recent demand adding a beta (β) smoothing constant, whose function is to reduce the error that occurs between the actual demand and the forecast (Hanke & Wichern, 2010).

Later Brown´s method was applied to the historical data, which consist of two exponential smoothings from which the estimated value or the wanted forecast will be obtained calculated with a simple expression. First applied to the values observed in the time series and the second to the attenuated series obtained by the first attenuation. And the last traditional method applied was Winters´s method, which uses three weights or smoothing parameters (α y β) to update the components in each period. It makes use of the level, trend and seasonal components to generate forecasts. After the application of the mentioned methods, there was calculated the average error over the n periods, a common average error measurement is the Mean Absolute Percentage Error (MAPE) which expresses the error relative to the magnitude of the demand (Hanke & Wichern, 2010). However, the error presented by the forecast applied were high; it was decided to make use of the artificial neural networks (ANNs) to observe if the error rate decreased. Due to there are many different ways to construct and implement a neural network for forecasting through the literature review, it was observed that most studies use back-propagation algorithm applied to the multi-layer perceptrons (MLP). However, there are few references of works developed with the application of this technique.

ANNs could reduce the error rate if we compared the rate with the ordinary forecast techniques. There are three basic points of multilayer perceptrons: one is that each neuron in the network includes a nonlinear activation function that is differentiable, the network contains one or more layers that are hidden from both the input and output nodes and the network exhibits a high degree of connectivity, the extent of which is determined by synaptic weights of the network. Before an ANNs can use any kind of task, first it has to be trained to do so. A common training "for multilayer perceptrons is the back-propagation algorithm. The training proceeds in two phases: in the forward phase, the synaptic weights of the network are fixed, and the input signal is propagated through the network, layer by layer until it reaches the output. In the backward phase, an error signal is produced by comparing the output of the network with the desired response. The resulting error signal is propagated through the network, again layer by layer, but this time the propagation is performed in the backward direction." Implementing the method, the demand values were represented as input vectors to generate the learning capability and compare the difference between the desired output data with the obtained output to this way calculate the error (Haykin, S. (2009).

The training algorithm used to find the weights that minimize the error was back-propagation, which is represented mathematically as:

$$vj\left(n\right) = \sum_{i=0}^{m} wji\left(n\right) yi\left(n\right) \tag{10}$$

Where $vj\left(n\right)$ produced at the input of the activation function associated with neuron j is therefore where m the total number of inputs (excluding the bias) is applied to neuron j. The synaptic weight $wj0$ equals the bias bj applied to neuron j. Hence, the function signal $yi\left(n\right)$ appearing at the output of neuron j at iteration n is

$$yi\left(n\right) = \varphi j\left(vj\left(n\right)\right) \tag{11}$$

RESEARCH METHODOLOGY

For the elaboration of the proposal it was made the following methodology:

1. Problem approach: The selected company dedicated to the manufacture of contemporary furniture with a traditional carpentry process, it is quality characteristic distinguishes them from the competition. This organization started operations in the city of Puebla in the year 2002. It currently has 40 operative workers and eight administrative workers. This company was representative of Mexican talent and was selected by PROMÉXICO to participate in the furniture exhibition called Wanted Design in New York City in 2017. It participates in different months of the year in the most crucial furniture exhibitions in Mexico such as International Furniture Expo (Guadalajara, Jalisco), Decoestylo (CDMX) and Expo Habitat (CDMX).

It has a store, and its primary customers are in the city of Puebla. The rest of the clients are in CDMX, Morelos, State of Mexico, Jalisco, Michoacán, Chihuahua and Coahuila. Its portfolio includes eight different collections of furniture in which each one includes: Bedrooms, Dining Rooms, TV Furniture, Coffee Tables and accessories. The company divided into four main areas: Commercial, which is responsible for generating and receiving orders, making the production plan, collection and deliveries to customers; Logistics, which includes the departments of warehouse of raw Material, purchases, inventories, warehouse of finished product, packaging and transport to final clients; Production, which consists of the areas of design, carpentry, funding, finishing and quality control, administration, which is responsible for the management of capital and human assets of the company.

2. Data: The company makes the purchase of raw materials from demand planning with a horizon of 15 days. There is no demand forecast, and there is no limited storage capacity for the finished product. For the present case of study, the authors gave the task of obtaining the historical data of the demand for the products that the company manufactures. They provided us with information on the months between December 2016 and November 2017.

In this study to reach the most effective forecast with the minimum error it was include the following phases:

Phase One: It was necessary to carry out the ABC classification to only concentrate on the “A” products that are the most produced and demanded during the year according to the historical data provided.

The historical data of the products that fall into the A classification were collected and concentrated in a table as can be observed in table 2 and table 3.

Phase Two: The implementation of the traditional forecast techniques were applied to the 4 principal products of the enterprise starting with the time series method simple average, to later apply the simple moving average to next make use of the weighted moving average and proceed with Holt ´s method, Brown´s method and finally Winters´s method. This allowed us to make a comparative

Table 1. ABC classification results

Item	Consumption	Accumulated Percentage	ABC
Lamp	1338	28%	A
Chair	1322	56%	
Table	318	62%	
Bureau	221	70%	
Headboard	176	72%	B
Buffet	151	74%	
Backrest	135	76%	
Mirror	129	79%	
Bed base	127	82%	
Chest of drawers	102	84%	
Console	59	85%	
Bench	54	86%	
Shelf	54	87%	
Armchair	51	88%	
Modules	48	89%	
Others	507	100%	C
Total	4792		

Source: (Own elaboration)

Table 2. Historical data of the demand for products A (2017)

2017	Dec	Jan	Feb	Mar	Apr	May	Jun	Jul	Aug	Sep	Oct	Nov
Bureau	27	25	20	12	13	14	10	18	20	21	18	17
Lamp	52	51	152	131	166	177	125	99	104	100	93	90
Table	13	19	40	23	30	20	26	23	43	42	22	20
Chair	65	82	105	157	134	70	117	124	194	139	86	64

Source: (Own elaboration)

Table 3. Historical data of the demand for products A (2016)

2016	Jan	Feb	Mar	Apr	May	Jun	Jul	Aug	Sep	Oct	Nov	Dec
Bureau	19	28	14	5	12	27	33	14	16	20	16	27
Lamp	9	11	9	3	12	12	89	9	10	11	8	10
Table	30	25	60	65	60	80	76	80	95	85	90	52
Chair	17	34	11	4	11	12	20	24	30	32	20	13

Source: (Own elaboration)

analysis of the error percentages. It could be observed the high error that the use of these techniques reflected. This can be due to the tendency of the demand of the company. Which bring us to opt for another forecast technique.

Phase Three: The historical data was analysis to define the best way to calculate a forecast demand through the use of a neural network. For this research there was decided to develop a multilayer network with a training method of back-propagation algorithm which assign certain forms of behavior to all of its neurons through a specification of the error correction learning algorithm. Due to the company has a low monthly sales record only the collection of two-year historical data was used. The elaboration of the neural network was carried out in the tool MATLAB 18. First, the 2016 sales were taken as inputs and 2017 sales were taken as the desired outputs to train the network and obtain a forecast for 2017. The adaptive learning function was Learngdm which calculates the weight change and bias (errors) learning function and the training algorithm was Trainlm which is the fastest backpropagation algorithm in the toolbox MATLAB. After the experiments, the number of hidden layers was varied between 2 and 20 looking for each product as the best option. The number of hidden layers that were established varied between 2 and 20 looking for each product the best option. As goal it was established a margin of 1E-2 to 1E-5 which gave the best result. The more trained the network, the less chance of error it was. After the information obtained from that trained network the results were collocated as inputs to predict the demand of 2018.

Phase Four: There was compared the error (MAPE) between the forecast methods applied.

RESULTS

According to the historical data of the year 2017 provided by the company, six quantitative methods were evaluated, analyzed, and applied: Simple Average, Simple Mobile Average, Weighted Moving Average,

Brown Exponential Smoothing, Winters Model, Holt Model, to determine the future demand that the company will have for the next three months. Forecasts made with the methods as mentioned earlier, and the error calculated in each of them, which helps us to define which the best technique to forecast the demand is. In the following table are reflected the percentages of the errors resulted from each of the forecasts applied as can be observed in the Table 4.

The term error refers to the difference between the forecast values and real data. It is possible to observe than in the Brown and Holt exponential smoothing models, the errors are lower, being for chair 32%, table 26%, lamp 17% and bureau 21%. Due to the methods of double exponential smoothing are the most suitable for this type of products whose variability in demand does not present a trend or seasonality. In this method, the most recent events are indicators of the future than those of the more distant past. For the bureau and lamp products used Brown method as shown in the following table 5.

For the case of the other two products classification A, which are: Tables and Chairs, the forecast was made using the Holt Exponential Smoothing Method and gave us forecast results for December, January and February shown in the table 6.

After several attempts, through the training neural networks the results were reflecting a lower forecast error compared with the high error reflected with the traditional forecasting techniques as can be observed in table 7, table 8, table 9 and table 10 and they are represented graphically in the graph 3, graph 4, graph 5 and graph 6.

Table 4. Errors of applied forecasting methods

Item	Simple Average	Simple Moving Average	Weighted Moving Average	Holt	Brown	Winters	ANNs
Chair	36%	46%	44%	32%	34%	41%	11%
Table	32%	33%	33%	26%	30%	33%	10%
Lamp	29%	29%	26%	29%	17%	43%	10%
Bureau	35%	33%	29%	22%	21%	26%	11%

Source: (Own elaboration)

Table 5. Predicted demand brown method

	Dec	Jan	Feb	Mar	Apr	May	Jun	Jul	Ago	Sep	Oct	Nov	Dec
Bureau	27	25	20	12	13	14	10	18	20	21	18	17	18
Lamp	52	51	152	131	166	177	125	99	104	100	93	90	92

Source: (Own elaboration)

Table 6. Predicted demand holt method

	Dec	Jan	Feb	Mar	Apr	May	Jun	Jul	Ago	Sep	Oct	Nov	Dec	Jan	Feb
Table	13	19	40	23	30	20	26	23	43	42	22	20	31	33	34
Chair	65	82	105	157	134	70	117	124	194	139	86	64	124	123	123

Source: (Own elaboration)

Table 7. Predicted demand of bureau with 10 layers and 80 epochs

Bureau					
Input	**Output**	**Forecasting 2017**	**Error**	**MAPE**	**Forecasting 2018**
19	27	26.7489	0.2511	0.0093	14.2865
28	25	24.6055	0.3945	0.01578	10
14	20	17.9782	2.0218	0.10109	21.261
5	12	12.7453	0.7453	0.06210833	24.2886
12	13	24.2886	11.2886	0.86835385	19.1297
27	14	14.2865	0.2865	0.02046429	17.9782
33	10	10	0	0	10.9221
14	18	17.9782	0.0218	0.00121111	26.2371
16	20	18.3632	1.6368	0.08184	21.261
20	21	21.261	0.261	0.01242857	10.1145
16	18	18.3632	0.3632	0.02017778	26.2371
16	17	18.3632	1.3632	0.08018824	21.3678
				11%	

Source: (Own elaboration)

Table 8. Predicted demand of Lamp with 10 layers and 10 epochs

Lamp					
Input	**Output**	**Forecasting 2017**	**Error**	**MAPE**	**Forecasting 2018**
30	51	55.4776	4.4776	0.08779608	90.2045
25	152	151.2512	0.7488	0.00492632	104.3559
60	131	132.9034	-1.9034	0.01452977	104.3559
65	166	135.2566	30.7434	0.1852012	104.3559
60	177	132.9034	44.0966	0.24913333	104.3559
80	125	87.4972	37.5028	0.3000224	104.3559
76	99	90.0965	8.9035	0.08993434	104.1347
80	104	87.4972	16.5028	0.15868077	104.3515
95	100	102.0693	-2.0693	0.020693	104.2527
85	93	88.3022	4.6978	0.05051398	100.8532
90	90	98.6995	8.6995	0.09666111	98.6995
52	96	95.9408	0.0592	0.00061667	102.8273
				10%	

Source: (Own elaboration)

Table 9. Predicted demand of Table with 10 layers and 22 epochs

Table					
Input	Output	Forecasting 2017	Error	MAPE	Forecasting 2018
17	19	19	0	0	19.7278
34	40	23.2858	16.7142	0.417855	26.6892
11	23	21.5	1.5	0.065217391	43
4	30	32.7491	-2.7491	0.091636667	41.9973
11	20	21.5	1.5	0.075	21.4987
12	26	22.3521	3.6479	0.140303846	43
20	23	21.4987	1.5013	0.065273913	43
24	43	43	0	0	26.6894
30	42	41.9973	0.0027	6.42857E-05	26.6894
32	22	21.9999	0.0001	4.54545E-06	42.9975
20	20	21.4987	-1.4987	0.074935	21.4987
13	17	21.9521	-4.9521	0.2913	19
				10%	

Source: (Own elaboration)

Table 10. Predicted demand of Chair with 10 layers and 123 epochs

Chair					
INPUT	OUTPUT	Forecasting 2017	ERROR	MAPE	Forecasting 2018
83	82	81.9986	0.0014	1.70732E-05	137.1065
36	105	95.16446	9.8354	0.093670476	64.0001
51	157	193.9988	36.9988	0.235661146	64.0001
33	134	135.4149	1.4149	0.010558955	64.0001
70	70	69.9986	0.0011	1.57143E-05	69.9989
35	117	96.9942	20.0058	0.170989744	64.0001
37	124	94.685	29.315	0.23641129	64.0001
32	194	190.4745	3.5255	0.01817268	64.0001
45	139	138.9486	0.0514	0.000369784	64.0001
38	86	94.5729	8.5729	0.099684884	64.0007
40	64	94.7709	30.7709	0.480795313	127.4447
65	84	84.0001	0.0001	1.19048E-06	64.5327
				11%	

Source: (Own elaboration)

Figure 3. Graphical comparison of real and predicted demand for bureau
Source: (Own elaboration)

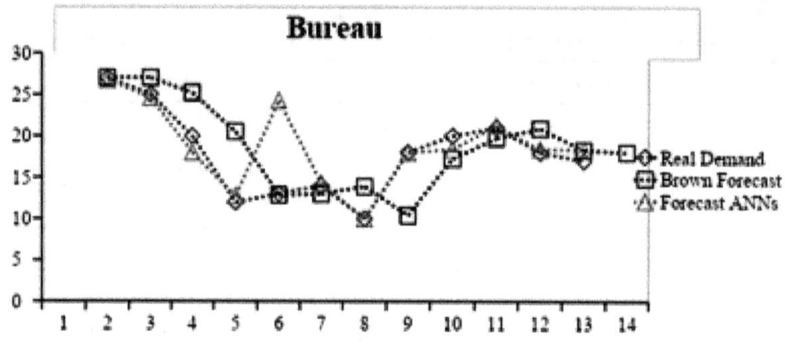

Figure 4. Graphical comparison of real and predicted demand for lamp
Source: (Own elaboration)

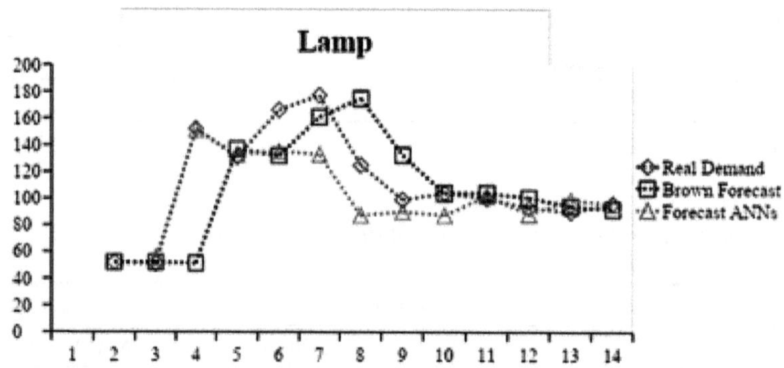

Figure 5. Graphical comparison of real and predicted demand for table
Source: (Own elaboration)

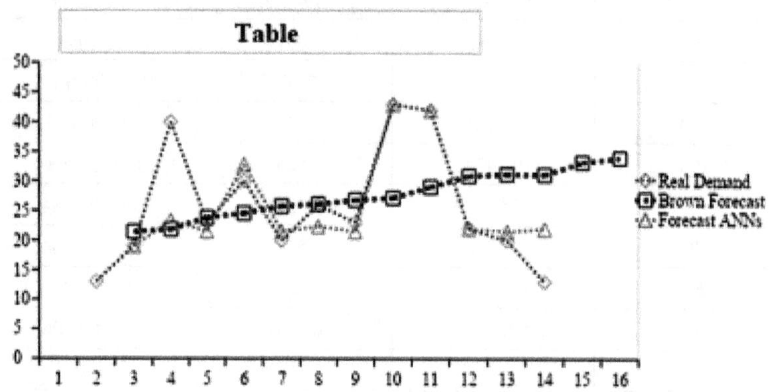

Figure 6. Graphical Comparison of real and predicted demand for chair
Source: (Own elaboration)

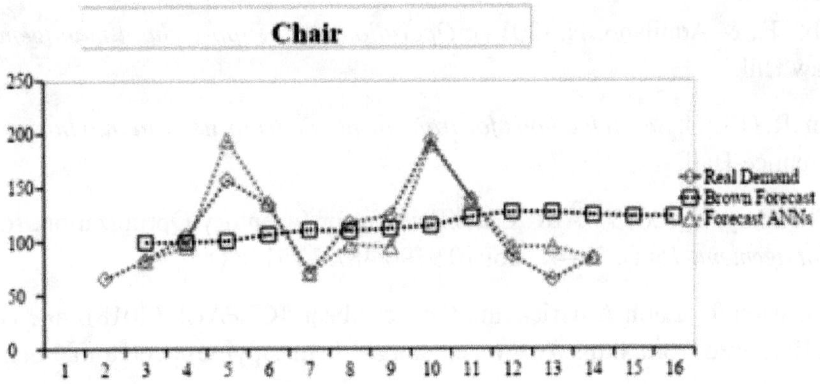

CONCLUSION

A forecast is a vital element for any organization such as SMEs, thanks to these forecasts, important decisions can be made in management. An estimate must be the basis of production planning and supply of demand. With a good forecast, the production, logistics and operations staff of an SME will plan their capacities, distribute their facilities and personnel, and schedule their daily production and inventories. It is possible to observe than some simple tools, and without much investment, entrepreneurs can support SMEs and achieve their production goals and customer service.

An important step in selecting an appropriate forecasting technique is to identify data patterns that exist within a time series. Once they are identified the most suitable forecasting method can be used to each case.

Forecasts are specifics for each item; it is vital to consider the variability of the demand for applying the appropriate method due to could guarantee a better estimate. In this case, were double exponential smoothing, which will allow the company to improve the acquired resources and eliminate the uncertainty of the orders. It is important to keep in mind that the data must be analyzed and updated over time, to take into account unplanned events due to the constant change in demand.

Due to the inefficiency of the historical data, the errors can result in a bad forecast, but it could be possible to minimize the mistake through ANNs. When an artificial neuronal network trained with more historical data, the neuron processes the information provided to obtain an optimal result.

REFERENCES

Acosta, M., Diaz, R., & Anaya, A. (2009). Review of techniques of multiple criteria decision analysis (MCDA) as a support to complex problems: Demand predictions. *Revista Científica Guillermo de Ockham*, *7*(2), 91–110.

Alonso, P., & Villegas, E. (2005). La importancia de los criterios cualitativos de los pronósticos en los agronegocios. *Revista Mexicana de Agronegocios*, *9*(16), 464–477.

Chang, P., & Wang, Y. (2006). Fuzzy Delphi and back-propagation model for sales forecasting in PCB industry. *Expert Systems with Applications*, *30*(4), 715–726. doi:10.1016/j.eswa.2005.07.031

Chase, R., Jacobs, F., & Aquilano, N. (2014). *Operations and supply chain management* (13th ed.). Mexico: Mc Graw Hill.

Dhar, V., & Stein, R. (1997). *Seven methods for transforming corporate data into business intelligence, Nueva Jersey*. Prentice-Hall.

Dhoka, D., & Choudary, Y. (2013). ABC Classification for Inventory Optimization. *IOSR Journal of Business and Management*, *15*(1), 38–41. doi:10.9790/487X-1513841

Economic Commission for Latin America and the Caribbean [CEPAL]. (2018). *Acerca de Microempresas y Pymes*. Retrieved from: https://www.cepal.org/es/temas/pymes/acerca-microempresas-pymes

Efendigil, T., Önüt, S., & Kahraman, C. (2009). A decision support system for demand forecasting with artificial neural networks and neuro-fuzzy models: A comparative analysis. *Expert Systems with Applications*, *36*(3), 6697–6707. doi:10.1016/j.eswa.2008.08.058

Escoda, I., Ortega, A., Sanz, A., & Herms, A. (1997). Demand forecast by neuro-fuzzy techniques. *Proceedings of the Sixth IEEE International Conference*, 3, 1381-1386.

Flores, B., Stading, G., & Klassen, R. (2007). The business forecasting process: A comparison of differences between small and large Canadian manufacturing and service firms. *International Journal of Management and Enterprise Development*, *4*(4), 387–402. doi:10.1504/IJMED.2007.013152

Flores, B., & Whybark, D. (1988). Know Your ABC. *Management Decision*, *26*(3), 20–24. doi:10.1108/eb001494

Frank, C., Garg, A., Raheja, A., & Sztandera, L. (2003). Forecasting women's apparel sales using mathematical modeling. *International Journal of Clothing Science and Technology*, *15*(2), 107–125. doi:10.1108/09556220310470097

Gunasekaran, A., Rai, B., & Griffin, M. (2011). Resilience and competitiveness of small and medium size enterprises: An empirical research. *International Journal of Production Research*, *49*(18), 5489–5509. doi:10.1080/00207543.2011.563831

Hanke, J., & Wichern, D. (2010). *Business forecasting* (9th ed.). Mexico: Pearson Education.

Haykin, S. (2009). *Neural Networks and Learning Machines*. Canada: Pearson Education.

Heizer, J., & Render, B. (2010). *Principles of operations management* (8th ed.). Mexico: Pearson Educación.

Hilera, J. R., & Martínez, V. J. (1995). *Redes neuronales artificiales: fundamentos, modelos y aplicaciones*. Madrid, España: RA-MA Editorial.

Hofer, P., Eisl, C., & Mayr, A. (2015). Forecasting in Austrian companies. *Journal of Applied Accounting Research*, *16*(3), 359–382. doi:10.1108/JAAR-10-2014-0113

Inter-American Development Bank [IDB]. (2010). *La logística de cargas en América Latina y el Caribe: una agenda para mejorar su desempeño*. Retrieved from: https://publications.iadb.org/bitstream/handle/11319/6269/La%20log%C3%ADstica%20de%20cargas%20en%20Am%C3%A9rica%20Latina%20y%20el%20Caribe%3A%20una%20agenda%20para%20mejorar%20su%20desempe%C3%B1o.pdf?sequence=1

Inter-American Development Bank [IDB]. (2011). *La logística como motor de la competitividad en América Latina y el Caribe. Banco Interamericano de Desarrollo*, Retrieved from: https://publications.iadb.org/bitstream/handle/11319/5091/La%20log%C3%ADstica%20como%20motor%20de%20la%20competitividad%20en%20Am%C3%A9rica%20Latina%20y%20el%20Caribe.pdf?sequence=1&isAllowed=y

Kourentzes, N. (2013). Intermittent demand forecasts with neural networks. *International Journal of Production Economics*, *143*(1), 198–206. doi:10.1016/j.ijpe.2013.01.009

Krajewski, L., Ritzman, L., & Malhotra, M. (2008). *Operation management process and values chain* (8th ed.). Mexico: Pearson Education.

Kumar, P., Herbert, M., & Rao, S. (2014). Demand forecasting Using Artificial Neural Network Based on Different Learning Methods: Comparative Analysis. *International Journal for Research in Applied Science and Engineering Technology*, 2(4), 364–374.

Lopez, E. Acosta, M. Espinoza, E. & Delfín, F. (2007). Globalization and its effects on the management of micro, small and medium enterprises and Latin America. *Equidad y desarrollo*, *8*, 113-125.

Mendez, G., & Lopez, E. (2014). Methodology to demand forecasting under multiproduct environments and high variability. *Tecnura*, *18*(40), 89–102.

Ministry of Economic. (2012). *Pequeñas empresas*. Retrieved from http://www.2006-2012.economia.gob.mx/mexico-emprende/empresas/mediana-empresa

Ministry of Economic. (2012). *Pequeñas empresas*. Retrieved from http://www.2006-2012.economia.gob.mx/mexico-emprende/empresas/pequena-empresa

Ministry of Economics. (2011). *Evolución del desempeño logístico de las cadenas de suministro en México*. Retrieved from http://www.elogistica.economia.gob.mx/swb/work/models/elogistica/Resource/3/1/images/evolucion2011.pdf

Morlidge, S., & Player, S. (2010). *Future ready: How to master business forecasting*. Chichester, England: John Wiley & Sons.

National Institute of Statistics and Geography [INEGI]. (2014). *Censos económicos 2014*. Retrieved from http://internet.contenidos.inegi.org.mx/contenidos/Productos/prod_serv/contenidos/espanol/bvinegi/productos/nueva_estruc/702825077952.pdf

National Institute of Statistics and Geography [INEGI]. (2018). *Sistema de Cuentas Nacionales de México*. Retrieved from http://www.inegi.org.mx/est/contenidos/Proyectos/scN/

Organization for Economic Co-operation and Development [OECD]. (2013). *Mexico: Key Issues and Polices*. Paris: OECD Publishing.

Organization for Economic Co-operation and Development [OECD]. (2017). *Enhancing the contributions of SMEs in a global and digitalised economy*. Retrieved from: https://www.oecd.org/mcm/documents/C-MIN-2017-8-EN.pdf

Raju, Y., Kang, P., Moroz, A., & Clement, R., Hopwell, A., & Duffy A. (2015). Investigating the Demand for Short-shelf Life Food Products for SME Wholesalers. *International Journal of Social, Behavioral, Educational, Economic, Business and Industrial Engineering*, *9*(6), 2051–2055.

Ravinder, H., & Misra, R. (2014). ABC Analysis For Inventory Management: Bridging The Gap Between Research And Classroom. *American Journal of Business Education*, *7*(3), 257–263. doi:10.19030/ajbe.v7i3.8635

Rubio, G. (2017). Multivariant perspective of forecasts in industrial SMEs of Ibague (Colombia). *Revista Facultad de Ciencias Económicas: Investigación y Reflexión*, *15*(2), 25–40.

Salazar, M., & Cabrera, M. (2007). Pronóstico de demanda por medio de redes neuronales artificiales. *Ingenierías. Universidad Autónoma de Nuevo León UANL*, *10*(35), 6–12.

Salles, M. (2006). Decision making in SMEs and information requirements for competitive intelligence. *Production Planning & Control: The Management of Operations*, *17*(3), 229–237. doi:10.1080/09537280500285367

Schreier, C., Udomkit, N., & Macchi, L. (2016). Cultural and Intuitive Intelligence of SME's Leaders. [GBR]. *GSTF Journal on Business Review*, *4*(3), 15–20.

Slimani, I., Farissi, I., & Achchab, S. (2015). Artificial Neural Networks/or Demand Forecasting: Application Using Moroccan Supermarket Data. *Proceedings of the 15th International Conference on Intelligent Systems Design and Applications (ISDA)* (pp. 266-271). Academic Press. 10.1109/ISDA.2015.7489236

Smith, H. III, Herbig, P., Milewicz, J., & Golden, J. (1996). Differences in forecasting behavior between large and small firms. *Journal of Marketing Practice: Applied Marketing Science*, *2*(1), 35–51. doi:10.1108/EUM0000000000006

Soto, P., & Dolan, S. (2004). *Las PYMES ante el desafío del siglo XXI: los nuevos mercados globales*. Mexico: Thomson.

Taiwo, A., Ayodeji, M., & Yusuf, A. (2012). Impact of Small and Medium Enterprises on Economic Growth and Development. *American Journal of Business and Management*, *1*(1), 18–22. doi:10.11634/21679606170644

Thomassey, S., Happiette, M., & Castelain, J. M. (2002). An automic textile sales forecast using fuzzy treatment of explanatory variables. *Journal of Textile and Apparel. Technology and Management*, *2*(4), 1–15.

van Kampen, T., Akkerman, R., & van Donk, D. (2012). SKU classification: A literature review and conceptual framework. *International Journal of Operations & Production Management*, *32*(7), 850–876. doi:10.1108/01443571211250112

Wei, S., Zhang, J., & Li, Z. (1997). A supplier-selecting system using a neural network. In *IEEE international conference on intelligent processing systems* (pp. 468–471). IEEE.

Wiratmadja, I., Rumanti, A., Reynaldo, R., & Kurniawati, A. (2016). Intuition Concept in Small and Medium Enterprises through Strategic Decision-Making. *International Review of Management and Marketing*, *6*(8), 322–326.

World Trade Organization [WTO]. (2016). *Informe sobre el comercio mundial 2016*. Retrieved from https://www.wto.org/spanish/res_s/booksp_s/world_trade_report16_s.pdf

Zeballos, E. (2003). Micro, Small and Medium-sized Enterprises in Latin America. *CEPAL Review*, *79*(79), 51–68. doi:10.18356/def1ede4-en

Zhang, G., Patuwo, B., & Hu, M. (1998). Forecasting with artificial neural networks: The state of art. *International Journal of Forecasting*, *14*(1), 35–62. doi:10.1016/S0169-2070(97)00044-7

Zuluaga, M., Molina, P., & Guisao, G. (2011). Supply chain management strategies based on demand planning in Colombia. *Revista Politecnica*, *12*, 1900–23510.

Chapter 2
Methodology Proposal for Logistics Management in an Automobile Company

Maidely Fernández-Rodríguez
UPAEP University, Mexico

Diana Sánchez-Partida
UPAEP University, Mexico

Patricia Cano-Olivos
UPAEP University, Mexico

José-Luis Martínez-Flores
UPAEP University, Mexico

ABSTRACT

Logistics impacts the costs and effectiveness of the client service system directly; therefore, it is of paramount importance to adequately design and operate warehouses in order to achieve lower costs and higher performance. The Pino Automotive México S.A de C.V for car supplies faces problems at the time of arranging goods as well as while measuring and controlling the logistics activity. This document presents a study case where the authors propose a methodology for logistics management that encloses the layout redesign and the proposal of indicators to improve the area and the company through the gathering of previous research and the use of good practices in warehouse management.

INTRODUCTION

Pino Automotive is an international enterprise specialized in passive safety components, soundproofing and thermal insulating for cars with 40 years of experience. One of the objectives of the company is to provide decent and stable jobs, protecting the environmental setting and the social progress of the territories that host the investments. Its production and development centers are certified according to IATF

DOI: 10.4018/978-1-7998-1831-1.ch002

ISO / TS 16949 2016 edition and have seven production plants placed in Spain, the Czech Republic, the Republic of South Africa and the Republic of Mexico.

The Pino Automotive México S.A de C.V plant located in Puebla, Mexico was founded in 2016 is in charge of manufacturing and commercializing polyurethane (PU) and polypropylene (PP) products. In the facility where the production plant is located, there are also included the offices and the warehouse where the raw materials, finished products, intermediate products, and the waste obtained in the production with a different destination than garbage.

Volkswagen, Mexico, is the primary customer of the company. Due to the growth that the company has had since its foundation, it is desired to make its goods storage more efficient; this storage services are empirically located without defined areas. Sometimes the corridors are obstructed by stack boxes or other obstacles that prevent the free access in the facility. There are corners and areas of difficult access with obsolete or little-used merchandise, and large or repeated displacements are made from one place to another, representing operational costs, time losses, and fatigue of the workers. Also, they do not have enough indicators to evaluate the storage and control of inventories to develop corrective actions.

Warehouses play a paramount role when combining de demand for goods with the supply chain and its different levels. There is no adequate design and chain supply management if the location, design, and warehouse organization are not taken into account. Today, warehouses operate not only as storage centers but also as centers of added value (De Koster, Le-Duc, & Roodbergen, 2007). The layout should address several issues besides the assignment of items to storage locations in order for a warehouse to be able to support its operations. These considerations are supported by several reasons, among them the following. First, the number of design decisions is significant, and many of them are combinatorial problems that are difficult to solve optimally. Second, many operations (e.g., picking, dual command, cross-docking, and value-added services) and factors (e.g., demand, physical characteristics of items and unit loads, serving global markets, material handling, and just-in-time (JIT)) impact travel time, material handling cost, and throughput in a warehouse. Such operations and factors should be accounted for in a comprehensive design of warehouse layout in order that it can support them (without resorting to frequent modifications), which will complicate the design. Third, the operations mentioned above and factors interact, and such interaction should be accounted for in the design (Mohsen, 2002). Designing warehouses is a complex problem, not only due to the factors above but also because of the variety of alternatives for possible layouts. As a consequence, it is challenging, perhaps impossible, to analytically identify the "optimal" solution design (Baker & Canessa, 2009).

Readings must be kept that reflect their behavior and evolution to maintain control over the different operations or activities that make up the processes within the company. Theoretically, all processes can be measured, and those who cannot be eliminated. This rule holds great significance since what is not measured cannot be enhanced. Hence, every process must transmit its current stage, improvement possibilities, and deviation from ideal parameters. These data can only be known through indicators that show the behavior of the variables within every process (Marín, 2014).

Research proves that understanding the principles of design and warehouse management can play a fundamental role when raising operational efficiency; reducing staff´s fatigue and rotation and enhancing client services levels. (De Koster, Johnson, & Roy, 2017). Market competition demands ongoing improvement in the design and operation of distribution and production networks and at the same time, more exceptional warehouse performance (Gu, Goetschalckx, & McGinnis, 2007). Nowadays, the concept of profitability that can be reached through a competitive warehouse has been included in terms of space optimization. Currently so costly, reduction of inventory levels by purging the old ones, having efficient labor force executing designed processes that drop activities that do not add any value such as displacement, which in many warehouses is what the staff do 50% of the time (Marín, 2014).

The purpose of this research is to enhance the logistic performance of the company through a methodology that fits and meets vulnerable points within the logistics area centered on the redesign of the warehouse layout and the proposal of indicators for controlling and monitoring the processes. The results would be focused on achieving organization, availability of products and activities, reduction of time, routes and contacts while handling the products as well as the elimination of bottlenecks, cutting costs, higher warehouse performance, enhanced levels of service and a more competitive company.

BACKGROUND

Warehouses as deposits of gods have gone through several denominations along with history. During the Roman Empire domination, Pretorian time, they were called "silos" (from the Celtic term *silon,* seed) and they were underground rooms used to store oil, wine, cereal, and so forth. Later, during the Muslim presence in the peninsula, Arab domination, barns and the deposits to store cereal were built on the ground and were called "*al-malizen,*" term from which Spanish word "almacén" is derived (Escudero, 2014). Warehouses are facilities that provide adequate environmental conditions for storing specific products and/or materials that require protection from the elements and guarantee commercial and social (Gunasekaran, Marri, & Menci, 1999). They are also essential in the operations of many organizations since they are used for storage, distribution, consolidation, and transition of different types of cargo. In today´s world, it is nearly impossible to think of efficient mass production or commercial distribution without careful consideration of the role of warehousing (Henneberry, 1987).

Warehouses are seen as vital liaisons between the current chains of supply. They can drive competitive differentiation and increase profitable growth. Delivering a product to a client in the correct amount and time depends on the fact that the warehouse can collect and dispatch the goods in precision. (Richards, 2014). They are an essential part of logistics operations and a significant contributor to speed and cost in supply chains. Lack of general knowledge on warehouses technical efficiency and associated causative factors limits the capacity of the industry to identify the best choices to enhance warehouse performance. (Johnson & McGinnis, 2010).

It may have state-of-the-art technology in handling and transportation equipment, qualified personnel and a well-structured information system but if these elements are not supported on a physical organization principle of the area, it will be hard to achieve positive operational results. While assessing the physical organization, the area´s first impression is critical, the flux of merchandise throughout the process as well as order and general cleanliness. It does not matter the size of the warehouse but the geometrical forms of the stacks that enable keeping good aesthetics and thus good care of the products (Marín, 2014). The process of deciding the design of a warehouse is similar to that of assembling a puzzle: it is hard to complete until all pieces have been defined and assembled (Frazelle & Sojo, 2007). As such, warehouses must be designed to satisfy a combination of several technical and operational requisites (Hassan, 2002).

Another element that should not be overlooked are the measures for operational performance in the design phase since operational efficiency is closely tied by design decisions. An excellent performance or performance evaluation model can help the designer decide the alternatives and spaces during the initial stage (Gu, Goetschalckx, & McGinnis, 2010) It is essential to obtain information from all agents that play a role in the supply chain. The objectives, results, and forecasts are factors that must be evaluated before starting the reconditioning of the logistics platform. Without an accurate analysis of each of these elements, it will not be possible to know if the needs are really justified or if they are simply errors that must be corrected in the supply chain (Mecalux. S A, 2010). The performance of a warehouse can be improved if the correct operations and methods are defined (Aminoff, Kettunen, & Pajunen-Muhonen, 2002). These methods may include: simplifying the flow by eliminating processes, or eliminating or reducing movements of material, labor force, and unnecessary machines (Tompkins, White, Bozer, & Tanchoco, 2010). The use of examples of warehouse designs with an excellent level of performance, or reference models, can serve to improve performance by addressing the problems from a general perspective by studying the adaptation of a system for a particular combination of flows (Rouwenhorst et al., 2000). These reference models can be obtained from a functional benchmarking, where the best physical and operational design practices are identified (Gu, Goetschalckx, & McGinnis, 2010).

In the literature, there are several investigations on methodologies for the management of warehouses, where most collect the design of warehouses and / or the proposal of indicators. Rouwenhorst et al. (2000) analyze the problems that are faced when designing warehouses and establish a frame of reference for them. Goetschalckx et al. (2001) contribute to the development of a methodology where the hierarchy of decisions is analyzed through an iterative algorithm using commercial linear programming software. Mohsen (2002) presents a general framework for warehouse design, describing each step and its relation to the other steps. Also, Emmett (2005) in his book establishes a model for planning the design of warehouses, which shows a general framework, detailing each of the aspects and links to be taken into account, as well as a series of indicators for the control of warehouses. Frazelle and Sojo (2007) propose a methodology for the design of warehouses that requires the information on the profile of activities of the warehouse, the objectives of performance of the operation, the definition and the configuration of the processes and systems of material handling of the storage. Baker and Canessa (2009) analyze and synthesize the works proposed by different authors to arrive at a methodology where their main contribution is the identification of necessary steps with a common denominator among them, accompanied by the main tools used. Then Gu et al. (2010) present a scheme of the available methodologies and tools to improve warehouse design. Also, Tompkins et al. (2010) suggest better physical and operational design practices for storage. Johnson et al. (2010) describe a methodology to evaluate the technical efficiency of the warehouse based on empirical data that integrate several statistical approaches and the results derived from the application of the method to a large sample of warehouses. Ghiani et al. (2013) establish different decisions that contemplate storage management; these aspects are mentioned previously by other authors. In Marín (2014), a technique is established for measuring the positioning of the warehouse known as World-class warehouse. It based on the evaluation of the critical operations of the warehouse by mean of the solution to the questionnaire that gives way to the diagnosis, valuation, and construction of an action plan. Richards (2014) considers a series of factors that must be taken into account for the operation of the warehouses, within which the design or redesign of the warehouse is concerned, and the proposal of some Key Performance Indicator (KPI) through studies of cases and the experience of several experts. Staudt et al. (2015) conduct a review of literature related to the performance of the stores and propose direct and indirect indicators for each dimension linked to the framework in which they can be developed. In De Koster et al. (2017) a compilation of works of design of warehouses and elaboration of orders are made to pay attention to the new technologies and methods with the impact in the design and the management of the warehouses. Chen et al. (2017) carried out case studies to develop KPIs for companies identifying the critical functions and processes in the warehouse management system. Finally, in Karakis et al. (2018) develop a systematic and hierarchical warehouse design approach, especially for distribution centers. They establish a methodology through a thorough systematic review of the literature, and detailed interviews with logistics service providers, companies in the retail industry in Turkey.

The decisions that are made about the general distribution must satisfy the needs of a storage system that allows: efficiently take advantage of the available space, minimize the manipulation of materials, achieve the maximum turnover rate of the merchandise and facilitate the control of the stored amounts (Escudero, 2014).The evaluation of the performance of the warehouse has also been explored in different ways by the researchers. Some of them focus on a specific area, while others try to cover all the activities of the warehouse. Performance measurement is commonly evaluated by the use of indicators, present in most jobs. However, in the literature, there is no common understanding of the definition of these indicators and on how to measure them. Many studies are developed using indicators that are classified and measured differently in each research (Staudt, Mascolo, Alpan, & Rodriguez, 2014).

METHODOLOGY

Methodologies and Methods Found

For the proposal of the methodology, there are many works related to storage logistics management in the literature. In Table 1, the investigations related to the design of the warehouse and the establishment of indicators as fundamental points for the work were selected.

Proposed Methodology

Based on the different researches consulted and to meet the needs of Pino Automotive México S.A de C.V, the following methodology is proposed for the logistics area.

1. Diagnosis of the logistics area of the company.
    a. Conduct the logistics audit.

In this step, the state of the area is diagnosed, and the main problems to be solved are determined.

2. Determination of layout redesign.
    a. Specification of the warehouse-type and purpose.
    b. Characteristics of the products and management groups.
    c. Analysis of the warehouse activity data: Establishment of cargo units and demand forecasts.
        i. Quantity of maximum products (SKU) received and sent by frequency (daily, weekly, monthly).
        ii. A maximum number of products (SKU) stored, considering the stock (daily, weekly, monthly).
        iii. Collection of orders by customers, by product line, a mix of orders or bylines and by-product volume (daily, weekly, monthly).
    d. Establishment of warehouse policies.
        i. Determination of operations, storage system, and collection of orders.
        ii. Material flows.

Table 1. Research on the design of warehouses and /or logistics indicators

Author	Year	Contribution or Methodology	Indicators
Mohsen	2002	They propose the following steps for the design of the warehouse: • Specify the type and purpose of the warehouse. • Forecasting and analysis of the expected demand. • Establishment of operational policies. • Determination of inventory levels. • Class formation. • Departmentalization and general design. • Storage partition. • Design of material handling, storage, and sortation systems. • Design of aisles. • Determination of space requirements. • Determination of the number and location of the entrances and exits. • Determination of the number and location of docks. • Arrangement of storage. • Zone formation.	None.
Emmett	2005	Model for warehouse design planning • Analyze the historical data of the demand: ○ SKU volumes by numbers received per day in minimum, average, maximum volumes. ○ SKUs volumes by numbers collected/sent per day in minimum - average - maximum volumes. ○ SKU by daily, weekly, monthly stock levels. ○ SKU range by product-handling groups. ○ Picking /issues bylines, by units, by order for daily, weekly, monthly periods. ○ Receipts for SKU by frequency and volume for daily, weekly, monthly periods. ○ Receipts by SKU by the supplier for daily, weekly, monthly periods. • Are there any future developments and/or changes that will alter this historical pattern? • Calculate the necessary internal space for Receiving, Storing, Picking, Despatching, Any special area necessary for quarantine, dangerous products, waste, and so forth. • Choose options for storage equipment and material handling, including aisle widths/heights. • If it is using an existing building, check the construction restrictions, floor loading, doors, columns, ceiling lighting, and all other obstructions. • Draw a sample disposition (s), not forgetting the fire exits and regulations related to the emergency evacuation of the personnel. • For each sample design, calculate people, equipment requirements, and then calculate costs. Check each design to detect bottlenecks, flows, and ergonomics. Especially examine the fast flows. • Decide to use the iteration for each of the above processes, if necessary. • Remember that the warehouse exists in an external environment and consider: Access for employees, customers, suppliers. Parking space for cars and vehicles. Rest/waiting area for visitors. • External perimeter / security.	They describe a series of indicators to take into account the categories: quality, costs, deliveries, security, motivation. No calculation method is proposed for them.
Frazelle y Sojo	2007	Methodology for the design of warehouses • Planning the requirements of the space: Calculation of the spaces for each process, take into account the storage mode and method of material handling. • Specification of material flow. • Planning of adjacent areas: Functions or upcoming processes. • Location of processes: processes with a high volume of storage and labor-intensive. • Planning expansion or contraction strategies.	They propose key indicators for the performance of the warehouse for each activity; only the general ones are considered: • Financial: Total cost per order, line, and item. • Productivity: Total lines dispatched by staff hours. • Utilization: percentage of utilization of storage capacity and total production. • Quality: Percentage of perfect orders in the warehouse. • Cycle time: Springtime to the shelf + cycle time of the warehouse order
Baker y Canessa,	2009	They propose a methodology for the design of warehouses • Define the system requirements • Define and obtain data • Analyze data • Set loading units • Determine operating procedures and methods • Consider possible equipment and features • Calculate equipment capacities and quantities • Define auxiliary services and processes • Prepare the Layout • Evaluate • Identify the design solution to be implemented	None.

continued on following page

Table 1. Continued

Author	Year	Contribution or Methodology	Indicators
Ghiani, Laporte, & Musmanno	2013	Proposal for warehouse design • Choice of the warehouse system • Choice of warehouse layout • Sizing of the storage area. • Sizing of the reception area • Sizing of the shipping area.	Performance indicators • Use of the surface: the ratio of the surface used to store concerning the total surface area of the warehouse. • Use of the volume: relation of the volume occupied by the load and the total volume of the warehouse. • Potential for receptivity: maximum number of storable cargo units; Static measurement of warehouse capacity. • Performance: maximum number of load units in transit in the warehouse within a given unit of time. • Potential receptivity saturation coefficient: represents a percentage of theoretical receptivity: it is defined as the ratio of the number of load units present on average in the warehouse within the period considered to the potential receptivity. • Selectivity index: the relationship between the number of load units directly accessible in the selection phase and the potential receptivity. • Access index: the relationship between the number of material handling operations within the reference period considered (day, month, or year) and the possible receptivity of the warehouse. • Index of turnover of the inventory: the relationship between the value of the outgoing goods in a specific period and the value of the average level of inventory in the warehouse at the same time horizon. In step 2 of the method collects control indicators: • Percentage of fulfillment in orders • Percentage of error in dispatches • Percentage of references delivered after the deadline. • Percentage of exhausted • Customer satisfaction (survey) • Percentage of returns (causes of returns). • Reliability of inventory records • The efficiency of the workforce (overtime) • Compliance with the budget of expenses • Labor climate (survey) • Percentage of errors in critical processes. They also suggest steps to follow to build indicators that add value to the processes, only when it is necessary to build them.
Marín	2014	World-class warehouse method • Recognition of crucial warehouse operations. •Recognition of warehouse support and control systems. • Evaluation of processes, procedures, and warehouse resources. • The solution to the questionnaire, diagnosis, evaluation, and construction of the action plan.	Second step control indicators: • Percentage of fulfillment in orders • Percentage of error in dispatches • Percentage of references delivered after the deadline. • Percentage of exhausted • Customer satisfaction (survey) • Percentage of returns (causes of returns). • Reliability of inventory records • Labor efficiency (overtime) • The expense of budget compliance • Work climate (survey) • Percentage of errors in critical processes. It also suggests steps to follow to build indicators that add value to the processes, only when it is necessary to build them.
Richards	2014	Tasks to facilitate the design or redesign of the layout. • Collect as much data as possible and analyze them. • Try to imagine the business in 5 to 10 years and develop flexibility. • Concentrate on the cubic capacity of the building. • Ensure that the design is compatible with existing equipment and storage media • Put the health and safety of personnel at the forefront of design. • Reduce the amount of movement required within the warehouse. • Standardize the packaging for both movement and storage. • Understand local building regulations and floor loading requirements • Do not forget the exterior of the building.	Metrics • Shipments on time (orders delivered according to customer requests); • Internal order cycle time; • Cycle time of coupling to stock in hours (time from arrival to stock available to pick up). • Total time of the order cycle; • Accuracy of order preparation (correct number of lines and units delivered); • Lines collected and sent per hour; • Lines received and saved per hour; • Percentage of orders from suppliers received without damage; • Average warehouse capacity used; • Order filling rate (orders completed entirely in the first shipment. • Productive hours of work concerning totals

continued on following page

Table 1. Continued

Author	Year	Contribution or Methodology	Indicators
Staudt et al	2015	They do not propose a methodology; they base their statements in previous research	Define • Direct indicators: simple mathematical expressions (the basis for measuring the performance of the warehouse) o Specific: it is defined for an activity o Transversal: these are defined for a process or group of activities o Related resources: they are divided into two categories (labor and equipment/building) • Indirect indicators: they consist of a measure of concept (they are not used for daily management, they need sophisticated tools of measurement: analysis of regression, diffuse logic) Dimensions of the indicators and examples **Time** Delivery order Reception time Order pick-up time Delivery term Wait time Time of departure Shipping time Coupling time to stock Equipment downtime **Cost** Cost of inventory Order processing cost Cost as% of sales. Labor cost Distribution cost Maintenance cost **Quality** Delivery time Customer satisfaction Order fill rate Accuracy of the physical inventory Depletion rate Accuracy of storage Picking accuracy Shipping precision Accuracy of delivery Perfect orders Waste rate Orders sent on time Load damage rate **Productivity.** Labor productivity Performance Shipping productivity Transportation utilization Warehouse utilization Picking productivity Use of inventory space Use of the exit space Reception productivity Turnover
Chen, Huang, Yu, & Hung,	2017	Develop KPI systems based on the following principles. • Identify the critical success factors for the industry. • Analyze the essential competencies of the company and the necessary capabilities related to these factors. • Evaluate the main competitors with the necessary skills (integrate the basic skills) • Determine the KPIs and their objectives to achieve the necessary capabilities and competitive advantage.	KPI based on: • Good quality • Precision of delivery • On-time delivery • Short-term delivery • Security of the goods delivered. • First quality service. • Acceptable price • Latent needs met Define those related to these for each store process.

continued on following page

Table 1. Continued

Author	Year	Contribution or Methodology	Indicators
Karakis, Tanya y Baskak	2018	The method that proposes the evaluation of the warehouse design according to three levels and assumes that the results of each level are used as inputs for the next level and the different steps or tasks are interrelated. **Strategic Level** Step 1 • Type of storage and purpose. • The industry that it serves. • Quantity of products or groups. • Several floors in the warehouse. • Objective (related to sustainability). • Number of ClientsPaso2 Step 2 Store location Step 3 Analysis of current data and for the future (Warehouse principles, movements, quantities) Step 4 Automated or conventional storage decision. Step 5 Determine the material flow, requirements of the high-level areas. Step 6 High-level design with an approximate estimate of the total storage area, although the estimate is specified with the decisions of the following levels. **Tactical level** Step 1 Determine storage policies, including the selection of a storage method and an order selection method. Step 2 Configuration of the corridors according to the storage policies. Step 3 Selection of shelving systems and equipment selections. Step 4 Decide the detailed design of the warehouse, with the dimensions of the warehouse. **Operative Level** Step 1 Detailed design of warehouse processes: • Reception (number of docks and warehouse doors). • Storage (Dispersion by department or rack). • Order preparation (appropriate collection methods). Step 2 Classification, storage, routes, sequence, and shipment. Step 3 Calculation of the number of personnel. **Evaluation and implementation** The different design alternatives are evaluated, and the implementation of the equipment management systems and the hiring of the required personnel are carried out.	Indicators that are proposed in the Evaluation stage. • Return on investment (ROI): Net profit / Investment cost. • Energy consumption per m2: Total energy consumption / m2. • Energy consumption ratio of renewable energy resources: Energy consumption of renewable energy resources / Total energy consumption. • Cost per storage unit: Total costs / Total number of storage units. • Relationship between operating. costs and total costs: (Operational Costs / Total Costs) × 100. • Cost per m2: Total Costs / Net m2. • Order picking time: Total time between start and finish.

   e. Sizing of the warehouse.
      i. Establish warehouse areas.
      ii. Calculation of the number of areas, locations, and corridors.
   f. Warehouse layout.
      i. Design of several layouts to identify the one that best suits the warehouse.
   g. Storage equipment and material handling.
3. Establishment of management indicators.
   a. Determine business goals.
   b. Know variables to control.
   c. Identification and selection of indicators.
4. Periodically perform the audit and calculation of the indicators to analyze the area or the emergence of some variation in the conditions.

APPLIED METHODOLOGY

The authors proceed to the application of the proposed methodology.

Diagnosis of the Logistics Area of the Company

The diagnosis of the logistics area is made with the audit taken from the book by Richards (2014). The audit checklist ensures complete coverage of issues relevant to logistics and creates the basis for identifying the most vulnerable aspects to focus on. The audit describes the current status of the area with 63% of overall logistics performance. Based on the problems detected in the performance of each of the groups covered by the audit, the results are expressed in Figure 1. It is perceived that the four most significant problems that represent 45% of the total must be prioritized by the company to achieve a significant improvement of the area according to the Pareto principle.

The main problems detected were:

- Performance management.

The company does not have a program for the management and control of storage; that is, they do not have metrics or performance indicators.

Figure 1. Pareto diagram of problems detected in the logistic audit

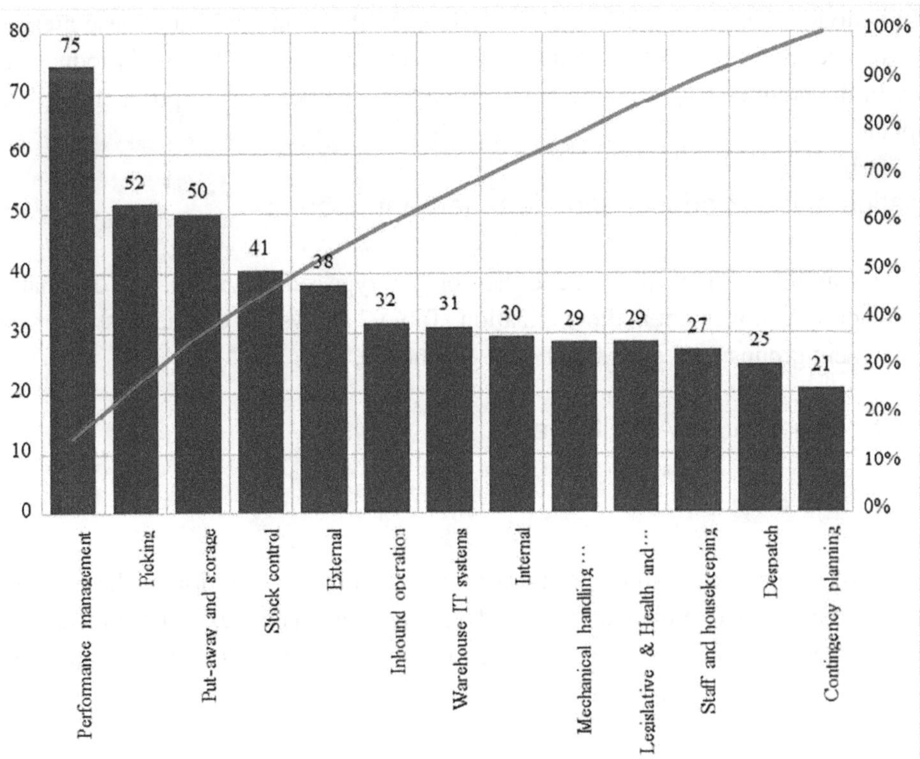

- Picking

In this process of the warehouse where the highest costs are involved, the areas are not delimited, the products are organized and ordered in the corridors.

- Put-away and storage

They do not have the prime locations to store, which causes irrigated products in different areas and also has no organization for the disposal of the products (ABC Classification or others). It is not possible to carry out the general monitoring of the products to know their traceability.

- Stock control.

They do not have inventory management indicators; they only perform physical and periodic counts.

The problems detected are related to the organization, layout of the warehouse layout, and the absence of control measures through indicators, which is developed in the following steps of the methodology for the logistics area. Taking into account basic principles and good practices focused on the redesign of the warehouse layout and indicators for the evaluation and control of the said area.

Determination of Layout Redesign

Specification of the Warehouse-Type and Purpose

The property is divided into warehouse and production. In the industrial warehouse area, mainly raw materials are located for the production plant, intermediate products, and finished products that are sent to customers. In the production part, there are two plants at each end, a PU production plant and a PP production plant.

Characteristics of the Products and Management Groups

There are several similar products that, due to their origin, type, employment, and destination, need to be grouped to facilitate movement and organization. Table 2 shows the characteristics of the products and the management groups corresponding to the warehouse.

Analysis of the Warehouse Activity Data

1. Demand forecasts.

A good forecast of the demand is fundamental for the design or organization of the warehouse. In the study, the data of the forecasts provided are considered adequate. The maximum forecast quantities for each SKU are worked according to the load unit, and possible increases or contractions of the demand are considered.

Table 2. Characteristics of the products and management groups

Products	Characteristics	Amount SKU	Management Groups
Raw materials	They are solid products and a few liquids; they are not very perishable. They weigh little, but of significant volume, They are conserved at room temperature.	17	Raw materials, packaging, obsolete and, production waste
Containers and packaging	They are generally metal and plastic containers. In addition to cardboard boxes and wooden pallets.	16	
Containers and unused packaging	They are containers or other means of packaging that are not used because they are defective or do not have enough for the consolidation of orders. (neither pallets nor cardboard boxes are taken into account)	11	
Obsolete	Solid products that are not occupied by project term or material change.	7	
Production waste	They are the rest of the production that is given a final destiny different from the rubbish, including the containers of the raw materials, the pallets, and the cardboard boxes were broken.	14	
Intermediate products	Solid products of polyurethane or expandable propylene. They are stored at room temperature. They are used in the PU plant	6	Intermediate or finished products
Finished products	Solid products of polyurethane or expandable propylene. They are stored at room temperature. From the PU and PP plants.	15	

a. Quantity of maximum products (SKU) received and sent by frequency (daily, weekly, monthly).
b. A maximum number of products (SKU) stored, considering the stock (daily, weekly, monthly).
c. Collection of orders by customers, by product line, a mix of orders or bylines and by-product volume (daily, weekly, monthly).

2. Establishment of the loading units.

Storage is identified per unit of cargo for each of the products depending on the containers, containers, packaging or packaging that are occupied either in the reception of the supplier, internal handling or in the orders of the customers in order to standardize the spaces and know the ideal storage media. Also, maximum load unit is handled in each of the areas based on historical and projections, for which the safety stock is included — all of the above based on the two management groups that were determined.

Considering the forecasts and the load units in Tables 3 and 4, the data of the warehouse activity for the previously determined management groups are detailed.

Establishment of Warehouse Policies

- Determination of operations, storage system, and collection of orders.

The warehouse, according to the management, characteristics, and movements of the products for each management group, the basic operations of:

- Reception
- Storage

Table 3. Data of the warehouse activity for raw materials, containers, packaging, packaging without use, obsolete, and production waste.

SKU	Place Where It Is Used	Unit of Time	Loading Unit	Frequency of Received Items by Supplier	Amount of Product Received per Unit of Charge (Maximum)	Amount of Product per Unit of Stored Load (Maximum)	# Orders Picking for Production (Maximum)
Raw Material							
AA	PU	weekly	Tank	1	2	5	2
CA	PU	weekly	Tank	1	1	4	1
BA	PU	weekly	Tank	1	1	1	1
DA	PU	quarterly	Tank	1	1	2	1
EA	PU	monthly	Tank	2	1	3	2
FA	PU	monthly	Tank	2	2	3	2
GA	PU	monthly	Pallet	2	8	15	6
GA	PU	bimonthly	Pallet	1	4	6	2
HA	PU	monthly	Box	1	4	8	1
IA	PP	bimonthly	Box	1	8	16	8
JA	PP	bimonthly	Box	1	30	40	20
KA	PP	monthly	Sacks	1	80	200	20
LA	PP	biweekly	Sacks	2	120	240	80
MA	PP	monthly	Sacks	2	2	10	2
NA	PP	monthly	Sacks	1		14	
ÑA	PU	monthly	Pallet	2	10	20	10
OA	PU	weekly	Idertia Box	1	20	50	16
Packaging or Container							
QA	PU	daily	container	1	40	70	10
RA	PU	daily	container	1	50	100	20
SA	PU	daily	container	1	20	50	10
TA	PP	daily	container	1	30	100	20
UA	PP	daily	container	1	15	50	10
VA	PP, PU	daily	container	1	60	100	20
WA	PP, PU	daily	container	1	80	100	30
XA	PU	daily	container	1	40	100	10
YA	PU	daily	container	1	20	100	10
AB	PP	daily	container	1	30	100	15
BB	PU	-	Box	-		269	70
CB	PU	-	Box	-		116	40
NB	PU	monthly	Pallet	1		100	
ÑB	PP	monthly	Pallet	1	50	400	20
QB	PU	-	Pallet	-		9	

Table 3. Continued

SKU	Place Where It Is Used	Unit of Time	Loading Unit	Frequency of Received Items by Supplier	Amount of Product Received per Unit of Charge (Maximum)	Amount of Product per Unit of Stored Load (Maximum)	# Orders Picking for Production (Maximum)
Unused Packaging or Container							
FC	Area for unused packaging or container	-	Pallet		2	50	-
ZA		-	Pallet		3	10	3
Obsolete							
PB	Obsolete area		Pallet			3	
UB	Obsolete area		Pallet			9	
EC	Obsolete area		Idertia Box			1	
OB	Obsolete area		Idertia Box			1	
SB	Obsolete area		Chattanooga Box			11	
TB	Obsolete area		Chattanooga Box			19	
VB	Obsolete area		Chattanooga Box			6	
Production Waste							
WB	Waste area	monthly	Tank			8	
XB	Waste area	monthly	Tank			4	
YB	Waste area	monthly	Tank			8	
ZB	Waste area	monthly	Tank			4	
AC	Waste area	monthly	Tank			4	
BC	Waste area	monthly	Tank			8	
CC	Waste area	monthly	Pallet			5	
DC	Waste area	monthly	Pallet			5	
FC	Waste area	monthly	Pallet			347	
GC	Waste area	monthly	Sacks			565	
HC	Waste area	monthly	Pallet			233	
IC	Waste area	monthly	Pallet			44	
LC	Waste area	monthly	Sacks			731	
MC	Waste area	monthly	Sacks			100	

Table 4. Warehouse activity data for finished or intermediate products

SKU	Classification	Place Where It Is Made	Unit of Time	Loading Unit	Quantity of Product Stored in Units (Maximum	The Quantity Stored in the load Unit (Maximum)	Dispatch of Products in Daily Load Unit (Maximum)	Frequency Delivered to Customers
OC	Finished	PU	weekly	Container 660055	6696	62	13	5
PC	Finished	PU	weekly	Container 660055	6696	62	13	5
QC	Finished	PU	weekly	Container 500375	5838	31	8	5
RC	Finished	PU	weekly	Container 500376	960	5	2	5
SC	Finished	PU	weekly	Container 500375	960	5	2	5
TC	Finished	PU	weekly	Container 500378	5838	31	8	5
DD	Finished	PP	weekly	Pallet	20800	13	4	5
UC	Finished	PP	weekly	Container 500553	12000	60	8	5
VC	Finished	PP	weekly	Container 500553	12000	60	8	5
WC	Finished	PP	weekly	Pallet	6840	46	10	5
XC	Finished	PP	weekly	Idertia Box	5200	65	6	5
YC	Finished	PP	weekly	Idertia Box	5200	82	6	5
ZC	Finished	PP	weekly	Idertia Box	5200	78	6	5
AD	Finished	PP	weekly	Pallet	4800	80	15	5
BD	Finished	PP	weekly	Pallet	7000	88	42	5
CD	Finished	PP	weekly	HBPO Container	8000	89	20	3
ED	Intermediates	PU	daily	Pallet	3503	4	-	-
HD	Intermediates	PU	daily	Pallet	576	2	-	-
ID	Intermediates	PU	daily	Pallet	576	2	-	-
JD	Intermediates	PU	daily	Pallet	3503	4	-	-
MD	Intermediates	PU	daily	Pallet	4017	2	-	-
ND	Intermediates	PU	daily	Pallet	4017	2	-	-

- Picking
- Shipping.

The storage system used is dedicated with a conventional method because, although they have an enterprise resource planning (ERP), they do not have it fully linked to all the warehouse functions because they do not handle a Warehouse Management System (WMS). Besides, they use pallets, or other means in stacked blocks for the placement of most of the products, except some liquid raw materials and little rotation that will be stored in two racks.

The orders are collected by customers — the principal requests all of the SKUs daily for a platform that is updated daily.

- Material flows.

The structure of the warehouse and the position of the loading and unloading doors will use a flow mainly in U that reveals the relationships between the functions and processes of the warehouse.

Sizing of the Warehouse

Establish the Areas and Areas of the Warehouse

Given the two large groups that work in the warehouse is divided into two areas, it is noteworthy that zone 2 is continuous to the production area, which determines the products assigned to it. Within each zone, the areas for each one are established. The determination of the zones, as well as the length and width of the warehouse and its areas, can be seen in Figure 2.

Figure 2. Warehouse areas and length and width considerations

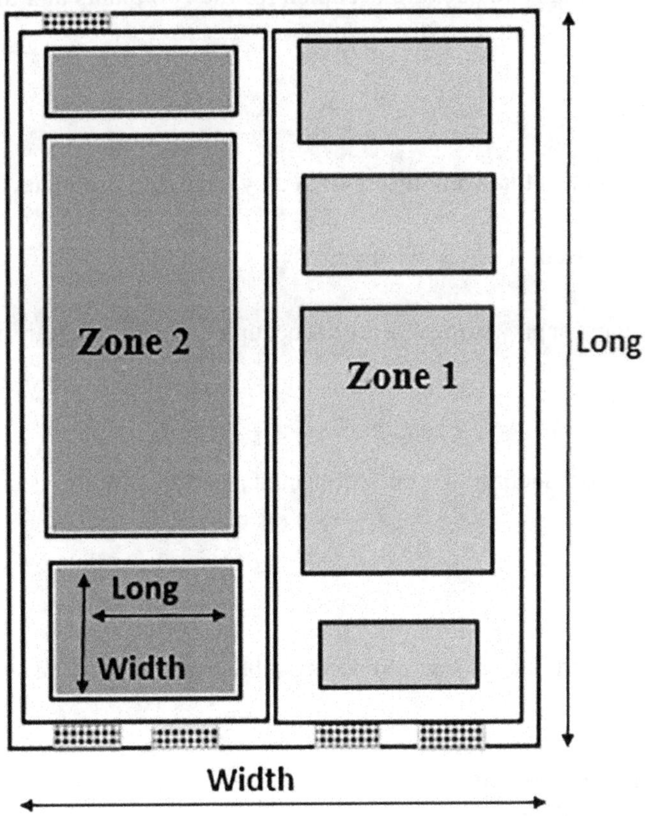

Each of the areas to be determined in each zone is shown below.

- Zone 1
  - Reception area for raw materials, packaging, and container.

The calculations are carried out, taking into account the products that are received in the quantity of load unit and frequency (Table 3).

- Raw materials área.

This area is divided into raw materials that supply the PU plant and raw materials that supply the PP plant. Due to the position of these plants, these areas are placed at the ends of zone 1, continuing to the reception area. The company has two racks, and they are used to store tanks with liquids (raw material) that supply the PU plant and another for raw materials of small volume and little rotation, remaining in the same 11 SKU of raw materials.

- Packaging area.

In this area, all the containers and packing occupied for the two plants and the shipment to the customers are placed.

- Obsolete area.

It is occupied by the products that were discontinued by changing products.

- Area of packaging not used.

It is occupied by containers or packaging that are not occupied because they are not enough or damaged.

- Waste production area.

There is all the waste of production that once collected a certain amount is collected by a customer.

- Zone 2
  - Dispatch area.

The calculations are carried out, taking into account the products that are sent in quantity per load unit and frequency (Table 4).

- Storage area for finished products,
- Storage area for intermediate products.

The intermediate products from the PU plant are stored momentarily to be sent to another process for their completion.

- Picking area

There are finished products that are consolidated for shipment when arriving at the finished product area, but others are prepared in the warehouse for which an area is destined for each product of this type.

- Reception area for some raw materials.

Some raw materials are received by the door opposite the loading docks, such as some packaging, ribbons, or other materials of low volume.

- Area of finished products in units for order completions (incomplete products).

Most of the time the orders are sent incomplete batches, but by not always send the products in the same container or packaging, the pieces are completed, so an area is stored where loose units of different products are stored in boxes.

Calculation of the Number of Areas, Locations, and Corridors

El The warehouse has a length of 108 m and a width of 27.5 m, its total area is 3078 square meters. A forecast of the need for storage space is calculated in the number of positions per load unit combining this information, taking into account product characteristics, dimensions of the load unit, maximum quantity to be stored, and routes.

Two racks are destined for the storage of some raw materials, and the dimensions and the space to occupy by them are calculated in table 5. Then the sizing of each area within the zones is done according to the management groups in Table 6 and seven respectively.

According to the previously calculated dimensions, the surface of each zone is calculated with its areas and corridors (Table 8). Also, the locations of the aisles are shown in Figure 3.

Table 5. Sizing for raw materials on racks

No	SKU	Dimensions of the Load Unit (m)		Dimensions of the Row (m)		Maximum Amount per Row at Floor Level	Maximum Space That Must Be Filled per Unit of Charge (Demand)	Assigned Rows	Stowage Level	Total, of Locations	Width by Located Products (m)	Total Width by Areas (m)
		Long	Width	Long	Width							
Raw Material in Blocks												
12	KA	1.3	1.3	12	1.75	10	200	10	2	200	17.5	49.5
13	LA	1.3	1.3	12	1.75	10	240	12	2	240	21	
14	MA	1.2	1.14	7	1.75	6	6	1	1	6	1.75	
15	NA	1.1	1.1	7	1.75	7	14	1	2	14	1.75	
16	ÑA	1.2	1.14	12	1.5	10	20	2	1	20	3	
17	OA	1	1.2	12	1.5	12	50	3	2	72	4.5	

continued on following page

Table 5. Continued

No	SKU	Dimensions of the Load Unit (m)		Dimensions of the Row (m)		Maximum Amount per Row at Floor Level	Maximum Space That Must Be Filled per Unit of Charge (Demand)	Assigned Rows	Stowage Level	Total, of Locations	Width by Located Products (m)	Total Width by Areas (m)
		Long	Width	Long	Width							
						Packaging or Container						
1	QA	1.485	1.139	12	1.5	9	75	1	10	90	1.5	21.25
2	RA	1.13	0.73	12	1.25	11	100	1	10	110	1.25	
3	SA	1.13	0.73	12	1.25	11	50	1	10	110	1.25	
4	TA	1.13	0.725	12	1.25	11	50	1	10	110	1.25	
5	UA	1.13	0.725	12	1.25	11	100	1	10	110	1.25	
6	VA	1.124	1.054	12	1.5	11	100	1	10	110	1.5	
7	WA	1.219	1.143	12	1.5	10	100	1	10	100	1.5	
8	XA	1.22	1.145	12	1.5	10	50	1	5	50	1.5	
9	YA	1.22	1.145	12	1.5	10	90	1	10	100	1.5	
11	AB	1.45	1.22	12	1.5	9	100	1	10	90	1.5	
12	BB	1.16	1.12	12	1.5	11	110	2	5	110	3	
13	CB	1.14	0.76	12	1.25	11	55	1	5	55	1.25	
14	NB	1.2	1.14	12	1.5	10	10	1	1	10	1.5	
15	ÑB	1.2	1.14	12	1.5	10	200	1	20	200	1.5	
16	QB	1.2	1.14	12	1.5					0	0	
						Unused Packaging						
1	FC	1.5	1.2	12	1.5	8	9	1	1	8	1.5	1.5
10	ZA	0.813	0.767	12	1.25	15	10	0.40	6	36	0.5	
						Obsolete						
3	PB	2.1	1.1	12	2	1	3	1	3	3	2	3.5
6	UB	1.2	1.14	12	2	9	9	1	1	9		
1	EC	1.16	1.12	12	2	1	1	1	1	1		
2	OB	1.16	1.12	12	1.5	1	1	1	1	1	1.5	
4	SB	1.14	0.76	12	1.5	4	11	1	3	12		
5	TB	1.14	0.76	12	1.5	7	19	1	3	21		
7	VB	1.14	0.76	12	1.5	2	6	1	3	6		
						Production Waste						
1	MC	1.3	1.3	12	1.7	10	27	3	1	30	5.1	20.7
2	LC	1.3	1.3	12	1.7	10	27	3	1	30	5.1	
3	WB	1.17	0.96	12	1.5	11	20	1	2	22	1.5	
4	AC	1.2	1.14	12	1.5	10	10	1	1	10	1.5	
5	CC	1.2	1.14	12	1.5	10	10	1	1	10	1.5	
6	FC	1.2	1.14	12	1.5	10	10	1	1	10	1.5	
7	GC	1.3	1.3	12	1.5	10	18	2	1	20	3	
8	IC	2.1	1.1	12	1.5	6	18	1	3	18	1.5	

Table 6. Sizing of raw materials, packaging, obsolete, and production waste.

No	Racks	SKU	Dimensions		Stowage Level	Total, Width by Area
			Long	Width		
1		AA				
2		CA				
3	Large rack	BA	9.17	1.26	2	
4		DA				
5		EA				
6		FA				2.48
7		GA				
8		GA				
9	Small rack	HA	4.67	1.22	3	
10		IA				
11		JA				

Table 7. Sizing for the finished product, intermediates, picking area, and incomplete products

No	SKU	Dimensions of the Load Unit (m)		Dimensions of the Row (m)		Maximum Amount per Row at Floor Level	Maximum Space That Must Be Filled per Unit of Charge	No	SKU	Dimensions of the Load Unit (m)	Dimensions of the Row (m)	Maximum Amount per Row at Floor Level
		Long	Width	Long	Width							
Finished Products PU												
1	OC	1.124	1.054	11.5	1.25	11	62	1	6	66	1.25	
2	PC	1.124	1.054	11.5	1.25	11	62	1	6	66	1.25	
3	QC	1.485	1.139	11.5	1.25	8	31	2	3	48	2.5	11
4	RC	1.485	1.139	11.5	1.5	8	5	1	3	24	1.5	
5	SC	1.485	1.139	11.5	1.5	8	5	1	3	24	1.5	
6	TC	1.485	1.139	11.5	1.5	8	31	2	3	48	3	
Finished Products PP												
1	DD	2.1	1.1	11.5	1.25	6	13	1	3	18	1.25	
2	UC	1.13	0.725	11.5	1.25	11	60	1	8	88	1.25	
3	VC	1.13	0.725	11.5	1.25	11	60	1	8	88	1.25	
4	WC	1.2	1.14	11.5	1.5	10	46	5	1	50	7.5	
5	XC	1.16	1.12	11.5	1.5	10	65	2	4	80	3	51.75
6	YC	1.16	1.12	11.5	1.5	10	82	3	4	120	4.5	
7	ZC	1.16	1.12	11.5	1.5	10	78	3	4	120	4.5	
8	AD	1.2	1.14	11.5	1.5	10	60	6	1	60	9	
9	BD	1.2	1.14	11.5	1.5	10	88	9	1	90	13.5	
10	CD	1.45	1.22	11.5	1.5	8	89	4	3	96	6	

continued on following page

Table 7. Continued

No	SKU	Dimensions of the Load Unit (m)		Dimensions of the Row (m)		Maximum Amount per Row at Floor Level	Maximum Space That Must Be Filled per Unit of Charge	No	SKU	Dimensions of the Load Unit (m)	Dimensions of the Row (m)	Maximum Amount per Row at Floor Level
		Long	Width	Long	Width							
Intermediate Products												
1	ED	1.2	1.14	11.5	1.5	10	4	0.5	1	5	0.75	4.5
2	HD	1.2	1.14	11.5	1.5	10	2	0.5	1	5	0.75	
3	ID	1.2	1.14	11.5	1.5	10	2	0.5	1	5	0.75	
4	JD	1.2	1.14	11.5	1.5	10	4	0.5	1	5	0.75	
5	MD	1.2	1.14	11.5	1.5	10	2	0.5	1	5	0.75	
6	ND	1.2	1.14	11.5	1.5	10	2	0.5	1	5	0.75	
Picking Area												
1	WC	1.2	1.14	11.5	3	10	10	1	1	10	3	21
2	XC	1.16	1.12	11.5	3	10	6	1	1	10	3	
3	YC	1.16	1.12	11.5	3	10	6	1	1	10	3	
4	ZC	1.16	1.12	11.5	3	10	6	1	1	10	3	
5	AD	1.2	1.14	11.5	3	10	15	1	1	10	3	
6	BD	1.2	1.14	11.5	3	10	42	2	1	20	6	
Incomplete Products												
1	WC	1.2	1.14	11.5	1.5	10	10	1	1	10	1.5	1.5
2	XC	1.16	1.12	11.5	1.5	10	6	1	1	10		
3	YC	1.16	1.12	11.5	1.5	10	6	1	1	10		
4	ZC	1.16	1.12	11.5	1.5	10	6	1	1	10		
5	AD	1.2	1.14	11.5	1.5	10	15	1	1	10		
6	BD	1.2	1.14	11.5	1.5	10	42	1	1	10		

Warehouse Layout

The arrangement inside the warehouse is made according to the flow, several movements, distances of routes, and necessary spaces of the areas in general. Within each of the areas, the disposition is made following the route of the product from its arrival to its destination, as well as the frequency of its movements and the facilities for its handling. For each of the products, the allocation of the locations is following a FIFO system mainly for raw materials, intermediate and finished products.

Design of Several Layouts to Identify the One That Best Suits the Warehouse

Since it has the data of the characteristics of the warehouse and in the subsequent steps, layout designs are made because each modified alternative generates a new design. In this step, it is decided which of the designs is the most convenient. It should be noted that the organization is based on the decrease in

Table 8a. Surface areas and corridors by zones

<table>
<tr><th>Zone 1
Storage</th><th>Width (m)</th><th>Long
(m)</th><th>Area
(m2)</th><th>Areas
Without
Storing
Zone 1</th><th>Width
(m)</th><th>Long
(m)</th><th>Area
(m2)</th></tr>
<tr><td>Reception area</td><td>3.57</td><td>12</td><td>42.84</td><td rowspan="2">Hall 1</td><td rowspan="2">5.5</td><td rowspan="2">12</td><td rowspan="2">66</td></tr>
<tr><td>Raw material racks</td><td>2.48</td><td>6.92</td><td>17.1616</td></tr>
<tr><td>Raw material blocks</td><td>49.5</td><td>12</td><td>594</td><td>Hall 2</td><td>1.22</td><td>7.33</td><td>8.9426</td></tr>
<tr><td>Obsolete</td><td>3.5</td><td>12</td><td>42</td><td>Other areas</td><td>1.26</td><td>2.83</td><td>3.5658</td></tr>
<tr><td>Packaging</td><td>21.25</td><td>12</td><td>255</td><td>Total</td><td>7.98</td><td>22.16</td><td>78.5084</td></tr>
<tr><td>Unused packaging</td><td>1.5</td><td>12</td><td>18</td><td rowspan="3">Main hallway</td><td rowspan="3">4.5</td><td rowspan="3">108</td><td rowspan="3">486</td></tr>
<tr><td>Production waste</td><td>20.7</td><td>12</td><td>248.4</td></tr>
<tr><td>Total</td><td>102.5</td><td></td><td>1217.4016</td></tr>
</table>

Table 8b. Surface areas and corridors by zones

<table>
<tr><th>Zone 2
Storage</th><th>Width (m)</th><th>Long
(m)</th><th>Area
(m2)</th><th>Areas
Without
Storing
Zone 2</th><th>Width
(m)</th><th>Long
(m)</th><th>Area
(m2)</th></tr>
<tr><td>Dispatch area</td><td>10.75</td><td>11.5</td><td>123.625</td><td rowspan="3">Hall 3</td><td rowspan="3">0.5</td><td rowspan="3">107.5</td><td rowspan="3">53.75</td></tr>
<tr><td>Finished products
PU</td><td>11</td><td>11.5</td><td>126.5</td></tr>
<tr><td>Finished products
PP</td><td>51.75</td><td>11.5</td><td>595.125</td></tr>
<tr><td>Intermediates products</td><td>4.5</td><td>11.5</td><td>51.75</td><td>Hall 4</td><td>0.5</td><td>11.5</td><td>5.75</td></tr>
<tr><td>Picking area</td><td>21</td><td>11.5</td><td>241.5</td><td>Hall 5</td><td>2.5</td><td>11.5</td><td>28.75</td></tr>
<tr><td>Incomplete products</td><td>1.5</td><td>11.5</td><td>17.25</td><td rowspan="2">Hall 6</td><td rowspan="2">2.5</td><td rowspan="2">11.5</td><td rowspan="2">28.75</td></tr>
<tr><td>Reception of small raw
materials.</td><td>2</td><td>11.5</td><td>23</td></tr>
<tr><td>Total</td><td>102.5</td><td></td><td>1178.75</td><td>Total</td><td>5.5</td><td>130.5</td><td>88.25</td></tr>
</table>

routes, manipulations, flow monitoring and greater use of space, no comparative basis of cost alternatives was made in each design because of the complexity of obtaining these costs with which they were not counted in the investigation.

The layout that is selected for implementation in the warehouse is shown in Figure 4.

Storage Equipment and Material Handling

Due to the type of warehouse and its storage system, material handling equipment is forklifts and pallet trucks. The forklift used assimilates the movement and loading of the products to be used today, and the pallet truck serves as an auxiliary for occasions of short distances or small quantities to be transported.

Figure 3. Halls by warehouse areas

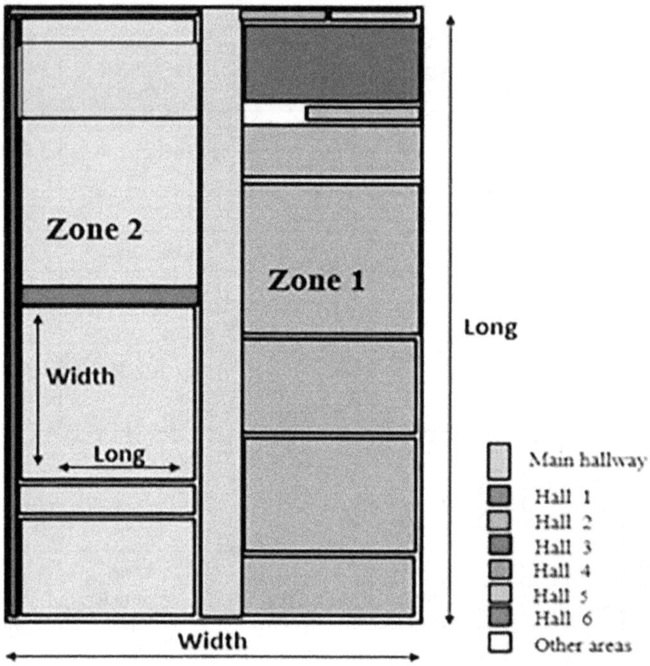

Figure 4. Selected layout

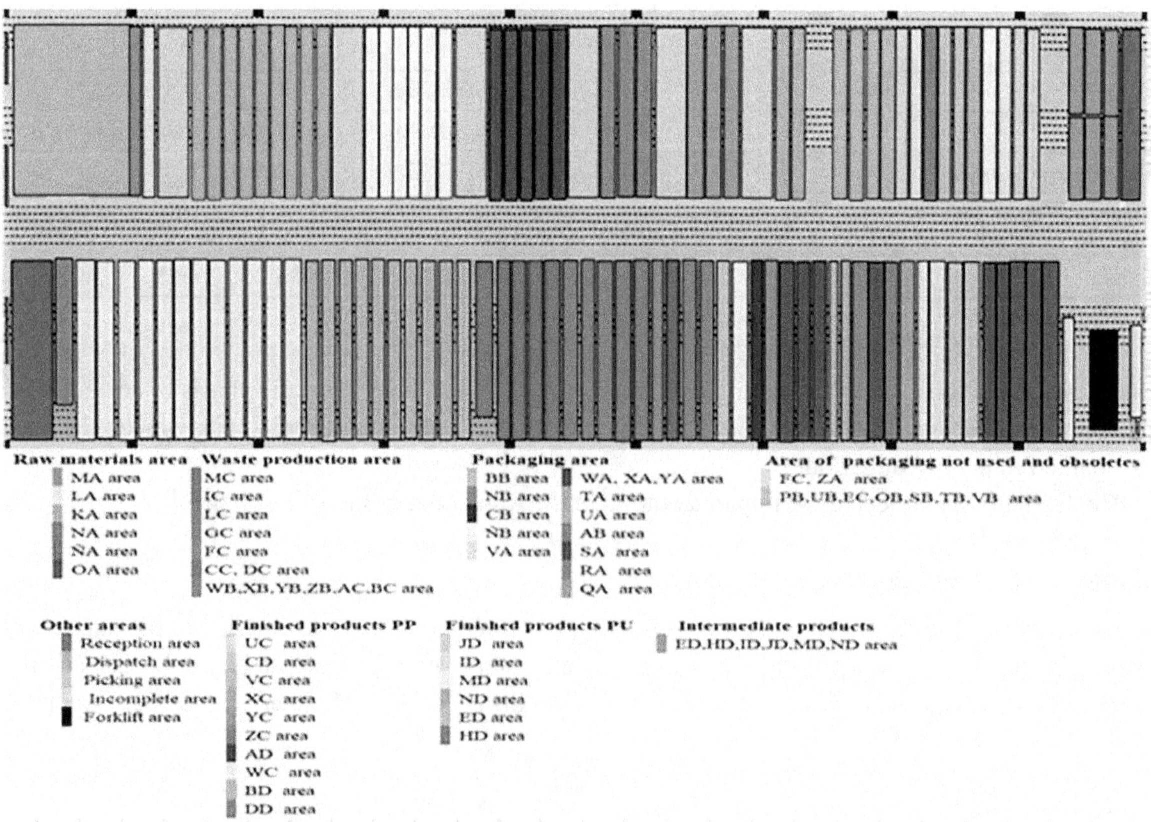

Establishment of Management Indicators

The logistic indicators are numerical data relationships that provide information about the resources we are using and the performance we are obtaining in each of the phases of the logistics chain. These have to be related to the mission, vision, corporate strategy, and competitiveness factors of the organization. With its interpretation, we will know if we arrive or are on the way to the objectives set. Also, they must be meaningful, coherent, and comparable to allow evaluations among diverse organizations (benchmarking).

For the establishment of indicators for the logistics area, it should be based on:

1. Determine business goals.

The business objectives are focused on compliance with current environmental management requirements for both processes and products focused on the quality, productivity, sustainability and responsibility of workers for the competitiveness of the company, customer satisfaction and the development of new products.

2. Know variables to control.

According to the objectives, the variables to be controlled for the logistics area would be Quality, Productivity, Time, Costs.

3. Identification and selection of indicators.

A group of indicators is identified for each variable. Also, clarity and continuity are needed in their calculations for the comparison analysis, evolution, and trend of the values.

- **Time**
    - Supply period (for raw materials).
        - Date of issue of the order - date of receipt of the order.
    - Cycle time of order delivery (by finished product).
        - Date of issue of the order - date of delivery of the order
    - Stock shortage time (by finished products).
        - Period without fulfilling the stocks.
    - Time of non-fulfillment of orders (for finished products)
        - Period without fulfilling orders.
- **Quality**
    - Reliability of the inventory (by type of products or in general).
        - 1 – (Difference in units / Total units in inventory).
    - Percentage of error in orders dispatched.
        - Amount of orders wrongly dispatched / Quantity of orders dispatched * 100.
    - Number or percentage of perfect orders.
        - Several orders shipped complete and on time / Total number of orders shipped.
    - Percentage of returns.
        - Several units or orders returned / Number of units or orders dispatched * 100.

    - Percentage of damaged inventory (raw material, containers or packaging, intermediate or finished products).
        - Amount of damaged inventory units / Total amount * 100.
- **Productivity**
    - Use of space
        - The area used in storage / Total warehouse area * 100.
    - Demand forecast (for finished products X).
        - Demand forecast of X + Real demand of X / Actual Demand of X * 100.
    - Rotation of the inventory (raw material, packaging or packaging, intermediate or finished products, obsolete or waste).
        - Cost of the product sold (material, labor, inventory cost) / average inventory (raw material, packaging, intermediate or finished products, obsolete or waste).
    - The oldness of the inventory (finished product or raw material).
        - Damaged units + obsolete + expired / available units of products * 100.
    - Unplanned shipments (for transportation of raw materials or finished product).
        - Amount of total shipments/quantity of planned shipments * 100.
- **Costs**
    - Inventory costs
        - The total cost of storage / Stored units.
    - Relationship between the cost of raw materials and sales.
        - The total cost of raw materials / Sales * 100.
    - Relationship between operating costs and sales.
        - Operating costs / Sales * 100.

Periodically Perform the Audit and the Calculation of the Indicators for Their Follow-Up

Se It is proposed to carry out the audit and the comparison of the indicators at the end of the month for the evaluation and correction of the area.

DISCUSSION

By gradually implementing the methodology, more organization and coordination is observed between supply, production, distribution and essential storage functions, adopting the principles of integral logistics because the warehouse is not an isolated entity from the rest of the company. The storage space was increased from 65% to 78%. The flow of goods improved, aisle clogging was eliminated, and costs caused by inefficient storage were reduced, which increases the performance of the area and the company. About logistics indicators, the company needs to know how it is going or what are the difficulties it faces in order to correct and reach the objectives set. When using the indicators, issues that were treated were seen, although not through any control measures and others that were not controlled and interfered in the activity and performance of the company. However, the most important thing was that they learned to calculate and that they understand the relevance and interpretation of each of these to compare and evaluate the area.

CONCLUSION

In the development of the work, a logistics management methodology was designed for Pino Automotive México S.A de C.V based on previous research and to cover the needs of the company. The complex combination of variables in this type of problems makes it difficult to ensure that the decision made is the optimal one. However, based on the general principles of good practices and the experience of the researchers with the redesign of the warehouse layout they were adapted and they improved the manipulations, distances, spaces, locations and risks that have repercussions in the reduction of operational costs and times. On the other hand, the correct implementation of the indicators will help analyze and control fundamental aspects for the organization, performance of the logistics area, and increase the profitability of the company.

ACKNOWLEDGMENT

This research was developed in the Master's program "Logistics and Supply Chain Management" awarded with the standard of excellence and supported with resources granted by the National Science and Technology Council (Consejo Nacional de Ciencia y Tecnología) – CONACyT. Scholarship holder number: 627285.

REFERENCES

Aminoff, A., Kettunen, O., & Pajunen-Muhonen, H. (2002). Research on factors affecting warehousing efficiency. *International Journal of Logistics, 5*(1), 45–57.

Baker, P., & Canessa, M. (2009). Warehouse design: A structured approach. *European Journal of Operational Research, 193*(2), 425–436. doi:10.1016/j.ejor.2007.11.045

Ballou, R. H. (2004). *Logística administración de la cadena de suministro*. Mexico: Pearson Educación.

Chen, P.-S., Huang, C.-Y., Yu, C.-C., & Hung, C.-C. (2017). The examination of key performance indicators of warehouse operation systems based on detailed case studies. *Journal of Information and Optimization Sciences, 38*(2), 367–389. doi:10.1080/02522667.2016.1224465

De Koster, R., Le-Duc, T., & Roodbergen, K. (2007). Design and control of warehouse order picking: A literature review. *European Journal of Operational Research, 182*(2), 481–501. doi:10.1016/j.ejor.2006.07.009

De Koster, R. B., Johnson, A. L., & Roy, D. (2017). Warehouse design and management. *International Journal of Production Research, 55*(21), 6327–6330. doi:10.1080/00207543.2017.1371856

Emmett, S. (2005). *Excellence in Warehouse*. England: John Wiley & Sons, Ltd.

Escudero, J. (2014). *Logística de almacenamiento. Madrid*, Spain: Ediciones Paraninfo.

Frazelle, E., & Sojo, R. (2007). *Logística de almacenamiento y manejo de materiales de clase mundial.* Bogotá: Norma S.A.

Ghiani, G., Laporte, G., & Musmanno, R. (2013). *Introduction to Logistics Systems Management.* Chichester: John Wiley & Sons, Ltd. doi:10.1002/9781118492185

Goetschalckx, M., McGinnis, L., Sharp, G., Bodner, D., Govindaraj, T., & Huang, K. (2001). Development of a design methodology for warehousing systems: hierarchical framework. *Proceedings of the Industrial Engineering Research.* Academic Press.

Gu, J., Goetschalckx, M., & McGinnis, L. F. (2007). Research on warehouse operation: A comprehensive review. *European Journal of Operational Research, 177*(1), 1–21. doi:10.1016/j.ejor.2006.02.025

Gu, J., Goetschalckx, M., & McGinnis, L. F. (2010). Warehouse performance measurement: A literature. *European Journal of Operational Research, 203*, 539–549. doi:10.1016/j.ejor.2009.07.031

Gunasekaran, A., Marri, H., & Menci, F. (1999). Improving the effectiveness of warehousing operations: A case study. *Industrial Management & Data Systems, 99*(8), 328–329. doi:10.1108/02635579910291975

Hassan, M. (2002). A framework for the design of warehouse layout. *Facilities, 20*(13/14), 432–440. doi:10.1108/02632770210454377

Henneberry, J. M. (1987). Occupiers and their use of accommodation on science parks and hightechnology developments. *Journal of Property Research, 4*(2), 109–144.

Johnson, A., & McGinnis, L. (2010). Performance measurement in the warehousing industry. *IIE Transactions, 43*(3), 220–230. doi:10.1080/0740817X.2010.491497

Karakis, I., Tanyas, M., & Baskak, M. (2018). Hierarchical Warehouse Design Approach for Distribution Centres. *International Journal of Business and Management Invention, 7*(2), 69–82.

Marín, R. (2014). *Almacén de clase mundial.* Medellín: Esumer.

Mecalux, S. A. (2010). *Manual técnico de almacenaje. Conceptos básicos sobre sistemas, equipos de manutención y procesos logísticos.* Barcelona: Mecalux S A.

Mohsen, H. (2002). A framework for the design of warehouse layout. *Facilities, 20*(13/14), 432–440. doi:10.1108/02632770210454377

Richards, G. (2014). *Warehouse Management: A Complete Guide to Improving Efficiency and Minimizing Costs in the Modern Warehouse* (2nd ed.). Londres: Kogan Page.

Rouwenhorst, B., Reuter, B., Stockrahm, V., Van Houtum, G., Mantel, R., & Zijm, W. (2000). Warehouse design and control: Framework and literature review. *European Journal of Operational Research, 122*(122), 515–533. doi:10.1016/S0377-2217(99)00020-X

Staudt, F. H., Alpan, G., Mascolo, M. D., & Rodriguez, C. M. (2015). Warehouse performance measurement: A literature review. *International Journal of Production Research, 53*(18), 5524–5544. doi:10.1080/00207543.2015.1030466

Staudt, F. H., Mascolo, M. D., Alpan, G., & Rodriguez, C. M. (2014). Warehouse performance measurement: classification and mathematical expressions of indicators. *Proceedings of the 5th International Conference: Information Systems, Logistics and Supply Chain ILS 2014* (pp. 1-9). Breda: Dinalog BETA.

Tompkins, J. A., White, J. A., Bozer, Y. A., & Tanchoco, J. M. (2010). *Facilities Planning*. New York: John Wiley & Sons.

Chapter 3
Methodology for Storage Location Allocation Based on the Planning of Material Requirements

Patricia Melchor-Pereda
UPAEP University, Mexico

Diana Sánchez-Partida
UPAEP University, Mexico

José-Luis Martínez-Flores
UPAEP University, Mexico

Patricia Cano-Olivos
UPAEP University, Mexico

ABSTRACT

The general goal of warehouse management is to ensure the correct supply of materials to the production line. In this context, this document presents a supposed instance for a general problem of a raw materials warehouse with poor control of inventories and poor internal traceability. An internal material requirement planning (MRP) and storage location-allocation problem (SLAP) for sixty materials was used to supply the restocking store that provides for the production line, to see the materials with the highest turnover and to reassign the materials in the general warehouse to minimize the internal long-term logistics costs. As a result of the methodology, the company determines the products with the highest turnover in a variable demand environment. Also, the company reallocates these products in the warehouse of raw materials reducing logistics costs and supply in 100% on-time the stock store that, in turn, supplies the production line.

DOI: 10.4018/978-1-7998-1831-1.ch003

INTRODUCTION

Some authors divide the logistics according to their dimension in internal and external logistics or their functions such as supply logistics, operations, and distribution. The logistics of operations, also known as production logistics or internal logistics, plays an essential role in the performance of organizations. Identifying the logistics of operations with activities such as the management of warehouses and inventories, transport and material handling, and production management (Cardenas & Urquiaga, 2007).

In Western culture and Japanese culture, warehouses and inventories lead to severe complications when there is not a management system that allows them adequate planning and control (Bustos-Flores & Chacón, 2007).

The warehouse and it is material handling capacity represent a mini system within the general logistics process (Bowersox, Closs, & Cooper, 2007). Even though warehouses can serve quite different ends, most share the same general pattern of material flow. Essentially, they receive bulk shipments, stage them for quick retrieval; then, in response to customer requests, retrieve and sort SKUs, and ship them out to customers. The reorganization of product takes place through inbound processes that include receiving ad put-away and outbound processes that includes activities such as order-picking, checking packing and shipping. Most of the expense in warehouse is in labor; most of that is in order-picking, and most of that is in travel (Bartholdi & Hackman, 2014).

Warehouse processes, unlike what happens in factories or workshops, usually do not add any value to the product from the customer's point of view, so it has to minimize the cost of the same through a proper rationalization of the resources employed (Anaya, 2011).

Thus, this work will focus on the area of raw materials warehouse; this warehouse has as its primary function the timely supply of raw materials, parts, and components to the production departments. Usually, it requires having three sections, such as reception, storage, and delivery (García, 1999).

Among the problems frequently faced by the raw materials warehouse with the production line, is the deficiency of control of the locations and internal traceability. These cause more time for collection and handling of materials and therefore a longer delivery time to the line; the obsolescence, consequence of the ignorance of the locations by the employees, the reliability of the information, and the lack of a First-In, First-Out (FIFO) treatment. Excess inventory of materials in storage due to lack of communication with other departments regarding scheduled production; Wastes due to unnecessary movements when having to request an input repeatedly for the same batch of production or the same run. Moreover, the biggest problem, a line stoppage due to the shortage of materials for production (Rajadell & Sánchez, 2010).

Given this situation, for some companies, there is a need to place pre-stocking warehouses between the central warehouse and the production line. The pre-assortment warehouses are those in which the raw materials from the central warehouse are protected before being processed or processed (Gonzalez, 2013). Faced with the urgent need to minimize costs and eliminate waste, the activities that consume resources of the organization, but do not create any value, activities of time, money and physical effort (Casanovas & Cuatrecasas, 2011). Through a reduction of material handling costs and eliminate constant and frequent trips to the warehouse during each shift to feed the production lines (Richards, 2011).

In this way, the relationship of efficiency and productivity of the warehouses with the production has allowed the emergence of different management philosophies, some of the most representatives are Just in Time (JIT) and Material Requirements Planning (MRP) (Cardenas & Urquiaga, 2007). The JIT goal is to establish times for activities so that the materials and components purchased to reach the point of manufacture or assembly just at the time they are required for the transformation process. The key to

JIT operations is that the demand for components and materials depends on the complete production program (Bowersox et al., 2007). On the other hand, the MRP consists mostly in the calculation of the net needs of the subsets, components and raw materials per period, from the net requirements per periods of the finished products, that is, the master production plan (Velasco & Campins, 2013). The basic idea is to have the right materials in the right place and times (Anaya, 2011). Each of these philosophies has specific characteristics that differentiate them from each other on the logistics of operations and storage treatment.

The main objective of this work is to offer a methodology that helps to calculate an initial level of raw material in the pre-stocking warehouse. That is, the "correct" quantity, according to the variations of the demand of the production program, and that in turn with the data obtained subsequently perform a reassignment of locations in the central warehouse. It solved by means of mathematical programming so that the raw material is available for the production line and thereby reduce times and avoid line shutdowns.

BACKGROUND

The general objective of warehouse management is to ensure the correct supply of materials; that is why a large number of studies have focused on the MRP. It is a tool used in different fields, and its implementation continues to expand also in small and medium enterprises since it is an entirely feasible solution when the company wants to have better control of materials, inventory reduction and timely supply (Rivera, Ortega, & Pereyra, 2014). The MRP used in the inventory management, allows adequate planning and control of the stocks, considering the dependent demands, planning the net needs of the components that make up a determined article (Bustos-Flores & Chacón, 2007). As an example of its application in the automotive industry was found the case of a body assembly on chassis, where a mixed whole programming model was applied, solved with Lingo. It aimed at minimizing costs, quantity of raw materials, storage of materials, control of levels of rotation of raw materials and finished product, the result allowed to reduce its inventory in a 49.7% allowing knowing the exact quantities of inventory, reducing the probabilities of delays in production (Cáceres, Reyes, García, & Sanchez, 2015).

In this same field, we have also worked on closed systems models with uncertainty in the production components, such as the necessary manufacturing capacity of each product. The delivery time and the availability of inventory, the treatment of uncertainty was worked through Fuzzy logic, modeling a closed deterministic MRP, for a company in the Colombian electricity sector, evaluating the total cost of the production plan, inventory levels, service level, and computational complexity. It allows finding parameters different from the external demand of the products when it works with fuzzy logic when compared with the results with completely deterministic parameters (Arango, Cano, & Álvarez, 2012), (Cano, 2011). Diffuse logic has been widely applied in the production field, specifically in MRP systems in the deterministic field and handling uncertainty (Arango Serna, Urán Serna, & Pérez Ortega, 2010).

Since its inception, the MRP has been a widely applied topic, and although it has evolved to overcome the limitations of the first system by integrating other essential areas in the company with the aim of making it more complete and also coordinate manufacturing activities, its essence lies in effective planning responding to the JIT philosophy (Medina, Nogueira, & Sosa, 2002).

A large number of studies have focused on this topic, as an example of its progress and application is the case of the production of medicines. The MRP II model was applied with the computer assistance of WinQSB software and a general method of solving problems, considering the given demand, the date of acquisition of materials and release of each production batch was obtained according to the master plan. The use of the MRP, combined with other tools, helps to improve its efficiency. MRP II systems offer the ability to compare forecasts made against actual results in order to analyze their performance and perform the necessary improvements that maximize their efficiency (Miño-Cascante, Saumell-Fonseca, Toledo-Borrego, & Moreno, 2015).

Recent studies introduced one of the most original concepts in this topic, the Demand Driven Materials Requirement Planning (DDMRP). It revolves around the use of strategic positions in inventory called "Buffers" based on the protection and acceleration of the flow of materials. The focus that is given to this concept is like an inventory management technique that establishes buffers at different points in the chain and generates replacement orders on real consumption, overcoming considerable improvements in consideration with the MRP II (Miclo, Fontalini, Lauras, Lamothe, & Milian, 2016).

The MRP system has been widely implemented in industries of different types. It has been used in the planning of production as an inventory control system, which works backward, to ensure the availability of materials for production with optimal inventory levels, with regarding the replenishment of materials by suppliers. However, no evidence was found of their use to manage the exact quantity of materials that should be supplied to supply the production line, the reason for this study.

Otherwise, as part of warehouse management is the Problem of Storage Location Assignment (SLAP), another subject reviewed continuously and used in different industries such as automotive, food, pharmaceutical, hardware, among others where the methods most used for its solution are exact methods and heuristic methods (Rojas, Solano-Charris, & Montoya-Torres, 2019).

The application of this SLAP problem was found in a supplier company of level one of the automotive industry with a Just In Sequence production system. The objective function for the solution of the allocation model was concentrated in minimizing the operating costs to find the optimal location, considering the distances of the location of the components inside the warehouse to the buffer and later of the buffer to the production line, as well as the frequency of consumption of each part. The computation time to obtain the solution of the model was carried out with a minimum duration, and the results were reflected in considerable savings for the company (Salazar & Martínez, 2017).

Another example of solving the problem was proposed through a Taboo search heuristic in which the average distance of the trip and optimization of the storage area is minimized to find the optimal zoning scheme. It based on the combination of storage of products between zones, that is, a class C product can be stored in a corridor closest to the tank, in order to minimize average distances (Le-Duc & Koster, 2005).

A similar problem is addressed in a dynamic storage system, where there are periods of high activity. Therefore, the warehouse positions were relocated in order to place the products with higher turnover near the entry/exit points. For this problem, an entire linear programming formulation was proposed, and the solution was obtained through a heuristic method developed over two stages, which generated an initial solution whose result was subsequently improved, through a taboo search algorithm and was tested in instances of different sizes (Chen, Langevin, & Diane, 2011).

Similarly, the SLAP with grouping restrictions (SLAP-GC), was treated through a restricted neighborhood Taboo search algorithm in a store of school and sports garments. The results obtained were compared with a mathematical programming method, and a genetic programming method and these were overcome by the proposed method in terms of solution quality and solution speed and execution time (Xie, Mei, T.Ernst, Li, & Song, 2015).

A two-stage heuristic algorithm is proposed to solve Storage space allocation and handling operation problems in container terminals., where the rolling planning horizon and a new update strategy are introduced (Chang, Zhu, & Haghani, 2019).

In this same context, among the aspects that should be considered are the activity profile of the stored products, that is, their storage position and the number of times the product is visited in a period, as well as the activities or operations that are performed in it. The distribution of the warehouse is measured through performance indicators to determine if the warehouse has the characteristics to be an efficient warehouse and perform the necessary corrective actions, which may include a reorganization of the warehouse (Arrieta, 2011).

Taking into consideration making the use of warehouse space more efficient can help speed up production, Zhang, Nishi, Turner, Oga, and Li (2016). Addressed the SLAP of the hand of planning production in a company of the food industry was proposed an integrated strategy that combines the problem of the size of batch trained with the problem of storage. It searches of effective coordination, through a mathematical model that aims to minimize the total costs of production and operation in storage by finding the optimal planning and best locations of the products based on a known demand. The model was tested in five instances of different sizes. The results derived from the application of the model were, what to produce for each period, how many locations to assign to each element, where to locate and recover the items, and what quantity should be inventoried of each item through the horizon of one year, where the best solutions were those derived from small instances.

In the same field and combined with the design of warehouses, little has been found about the method of configuring warehouses from the plan of the requirement of materials. It was until 2018 when a model was proposed that addresses these two areas of knowledge, production planning, and warehouse configuration using a mathematical model. It is determined from planning materials requirement in a liquid polyurethane coating factory in the construction sector. As a result of the application of the model, the storage option chosen based on area, actual capacity, and percentage of real use in response to the needs of the company, is selected from the five options generated by the model, in addition to the four configurations of reception-shipping docks (Suárez & Hualpa, 2018).

The importance of the proposal presented lies in providing a useful tool for the general warehouse of raw materials adaptable to the needs of each company. It allows calculating the right size of the pre-stocking warehouse to supply the production line in each of the shifts labor in order to eliminate deadlines translated, mainly in downtime. Also, allows a better organization and control of staff time in the warehouse of raw materials, likewise with the information obtained over three months realizing a reassignment of goods in the locations in the warehouse that allow a better flow of materials.

MAIN FOCUS OF THE ARTICLE

This section presents an overview of a stay based on assumptions, in which the proposed model can be applied. Consider the case of a company that produces nine types of plastic items. The department of demand planning prepares a weekly production program, which is made known to the production department and the latter in turn requests the materials from the raw materials warehouse so that these are taken from the company's stock. Once a production batch is completed, the sales department is responsible for distributing the product to customers.

The departments that can directly affect the production level are the following:

- **Department of Demand Planning:** Sales forecast, the capacity of distribution facilities, the effectiveness of advertising campaigns, and the effect of competition.
- **Production Department:** Production capacity expressed in terms of available staff hours and machine hours, inventory in process and quality control standards.
- **Materials Department:** Available stocks of raw materials, delivery programs from external suppliers, and storage limitations (Hillier & Lieberman, 2015).

For this instance, only one production line is considered, and the company works six days by week, from Monday to Friday, with three shifts of work of 8 hours each one and Saturdays with two shifts, the working days are organized as follow:

- **Shift 1** - 00:00 pm - 8:00 am
- **Shift 2** - 08:00 am - 16:00 pm
- **Shift 3** - 16:00 pm - 24:00 pm

The main problems that can arise in the enterprises when there does not exist good coordination between these areas are failures that affect the production process, and delays due to the lack of material in the line, these types of problems are detailed below:

- Inefficiency in the supply of materials to the production line.
- Delays due to the requisitions and search for authorization from the people in charge of the raw materials warehouse.
- Waste of Waiting, downtime, for example, workers waiting to receive the material, machines stopped trucks with incomplete orders in the middle of the production plant, and so forth (Sangüesa, Mateo, & Ilzarbe, 2006).
- Waste for transportation. It is the result of unnecessary movement or manipulation because of a poorly designed layout. The machines and production lines should be as close as possible, and the materials should flow (Rajadell & Sánchez, 2010).
- Excess inventory. When the inventory exceeds the minimum necessary lot of products, this waste is the number one enemy of quality and productivity (Casanovas & Cuatrecasas, 2011).
- Wastes due to unnecessary movements. The unproductive movements that the staff performs during the work, such as the search for parts, tools, documents, requests for support, and so forth (Casanovas & Cuatrecasas, 2011). The movements of searching, collecting, placing, or walking have to be reduced to the maximum (Sangüesa et al., 2006), as well as frequent trips when requesting the same materials for the same production lot.

Under this scenario, the proposed solution is the implementation of a pre-stocking warehouse that makes available to the production line the materials that will be used in each process, as illustrated in Figure 1.

The previous figure shows the layout that includes the raw materials warehouse. In the upper part it indicates the code of the locations in the warehouse and the lower part of the layout is the production area; in which the space allocated for the pre-stocking warehouse that supplies the production line is proposed. The areas are connected by a door where the transfer of materials is made.

Figure 1. Layout raw material warehouse and production area
Source: "Own Elaboration"

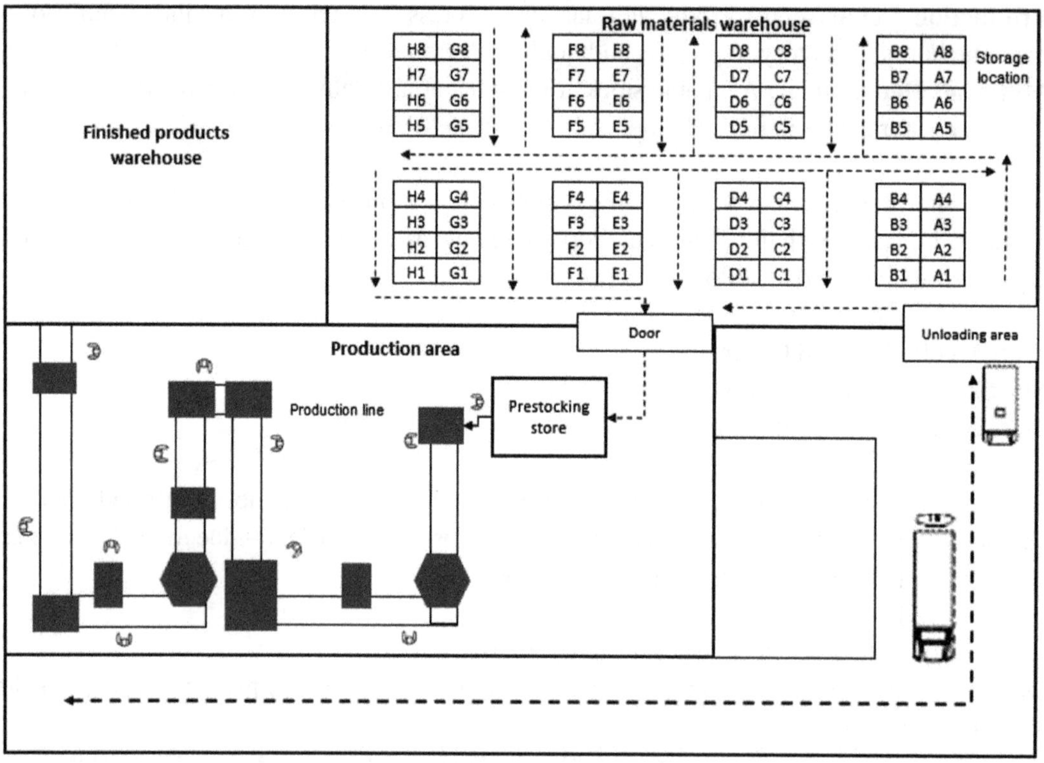

In this framework, the present work proposes a practical tool for the calculation of the size of the pre-stocking warehouse through the use of the MRP. It makes an analogy to the planning of materials concerning third parties, where the raw materials warehouse is the supplier. The production line is the customer, and the lead time of delivery is the lead time of pick up and/or production, respectively. Moreover, subsequently, the SLAP is used, bringing together two topics treated in an isolated manner such as the MRP and the Assignment Model in the same direction, and thus increase the efficiency of the warehouse of raw materials.

METHODS USED IN THE RESEARCH

Material Requirements Planning-MRP

In the first place, the research will use the MRP as a push system for the management of the flow of materials. These material flows are planned and controlled through a series of production programs that indicate when each batch of a PARTICULAR product must be finalized (Gaither & Frazier, 2000).

The MRP helps calculate the gross needs of a product. The gross needs are the result of the production program multiplied by the list of materials (Chiavenato, 1993).

The MRP is directed by a production program. Therefore, it works in a situation-dependent on demand, that is, "They are the orders that arise from different materials, components and semi-processed to be able to manufacture the products that the market requires, depending on the components that are necessary for the elaboration of the so-called final articles" (Anaya, 2011).

Although the elements that make up the MRP I are:

- PMP or Production Master Plan. It tells us when and how much of each final product has to be manufactured.
- BOM or list of materials, which indicates the parts or components and the quantities of each needed to manufacture an end product.
- Situation or state of the stock, available quantities of each item in the different time intervals and by difference, the quantities that must be bought or manufactured (Velasco & Campins, 2013).
- Delivery time of supplier and manufacturing (Lead time). The lead time is the total time taken from the decision to order until the moment when the article is available for issuance. There are different delivery times, such as Lead time of production, order processing, lead time of preparation (order is accepted to the date of manufacturing) and lead time of manufacture (wait time of machine start/operating time/inspection/ storage time) (Emmett, 2005).
- Minimum quantities to be manufactured from each reference (lot size).
- Safety stock of each reference at the end of the period.

Pochet & Wolsey (2006) consider the basic model of the MRP, as shown below:

$$\min \Sigma_i \Sigma_t \left(p_t^i x_t^i + q_t^i y_t^i + h_t^i s_t^i\right) \tag{1}$$

$$s_{t-1}^i + x_{t-\gamma^i}^i = \left[d_t^i + \sum_{j \in D(i)} r^{ij} x_t^j\right] + s_t^i \text{ for all } i, \text{t} \tag{2}$$

$$x_t^i \le M_t^i y_t^i \text{ for all } i, \text{t} \tag{3}$$

$$\sum_i \alpha^{ik} x_t^i + \sum_i \beta^{ik} y_t^i \le L_t^k \text{ for all } t, \text{k} \tag{4}$$

$$x \in R_+^{mn}, s \in R_+^{m(n+1)}, y \in \{0,1\}^{mn} \tag{5}$$

The objective function defined by (1) is the sum of the costs of unit production, fixed production, and unit inventory.

The structure of the bill of materials (BOM) is modeled in equation (2) for r^{ij} for all elements *i* and all $j \in D(i)$, that is, all the direct successors *j* of *i*, where r^{ij} is the number of units of *i* required per unit of *j*. The delivery time or constant production for the article *i* is represented by γ^i in equation (2). The dependent demand per article *i* in period *t* is represented by Σ *j* ∈D (i) r ^{ij} x_t^j in equation (2). Restriction (3) forces the configuration variable in period *t* to be 1 when there is production (i.e., $x_t > 0$) in period *t*.

In the formulation (1) - (5) the number of periods *n* is less than the total accumulated time from the order of raw materials to the completion of finished products, expressed in several periods. Now we can reformulate our condition of minimum length in the planned horizon. The number of periods *n* is at least as large as the sum of the values γ^i along any route in the bill of materials chart.

The unit production time of article *i* in resource *k* is denoted by α^{ik} in equation (4). In the same way, the time of preparation or preparation of the resource *k* to produce a batch of article *i* is denoted as β i k in Equation (4). The net capacity in resource k in period t is represented by L_t^k in equation (4) (Pochet & Wolsey, 2006).

In the case presented, the most critical elements are the production master plan, the list of materials, and the delivery terms of the supplier are homologated to the collection times of the materials in the warehouse locations.

Storage Location Assignment Problem-SLAP

Secondly, the SLAP Model, derived from the transport model, will be used for the reassignment of storage locations. One way to reduce the work of the workforce is to store the inconvenient product places to reduce travel (Bartholdi & Hackman, 2014). An allocation problem is a balanced transport problem in which products and locations are equal to 1. This model is used when it is necessary to assign *m* products to a location *n* (2).

$$X_{ij} = \begin{cases} 0 & product\, i\, is\, not\, assigned\, to\, location\; j \\ 1 & the\, product\, i\, is\, assigned\, to\, location\; j \end{cases} \tag{6}$$

Subject to the restrictions:

$$\text{Supply}\left\{\sum_{j=1}^{j=n} X_{i,j} = 1; i = 1,2,\ldots,m\right. \tag{7}$$

$$\text{Demand}\left\{\sum_{i=1}^{i=m} X_{i,j} = 1; j = 1,2,\ldots,n\right. \tag{8}$$

The objective function is:
Where:

Cij = Unit cost to assign from *i* to *j* (9)

Assignment model (Bartholdi & Hackman, 2014).

Suppose the following; there is a group with n of products that can be assigned to m locations, (8) in the same way you have m locations that can be assigned to n products (7), where Cij is the non-negative cost of assigning the resource i to activity j. The objective is to assign each product to each location, in such a way that the minimum possible cost is reached (9). Thus, we define the binary variables Xij with a value of 0 or 1 (2.1). When $Xij = 1$; this means that we should assign product i to location j (Martínez, Vértiz, López, Jiménez, & Moncayo, 2014).

Bartholdi and Hackman (2014) deal with the issue of movements in storage referring to the storage and collection of goods where the first restriction requires that each storage be paired with some recovery and the second restriction that each recovery is paired with some stowage. In this case, each product is assigned to a location, and each location is assigned to a product. If there are fewer products than locations, then add enough "fictitious" products to make them equal and find the balance.

RESEARCH METHODOLOGY

The methodology used is presented in two phases, the first phase is to calculate the planning of materials from the MRP and the second phase is focused on reallocating locations in the raw materials warehouse, the first phase is divided into five steps as detailed below:

Phase 1

1. Determine the list of materials.

Know the list of materials and quantity of raw materials necessary for each product to be prepared, (either per unit or lot, depending on the type of product being worked), based on the demand for the finished products programmed in the production master plan.

Due to the nature of the proposed pre-stocking warehouse, as it is adjacent to the production line. It is intended to be a kit type, with the number of raw materials necessary for the product to be manufactured. The materials considered are only those of level 1, "parents and children" of the final product, in the structure of the finished products, the article that is in the upper part is called father and the components that integrate it, those that are in the level lower are the "children."

In the presented case, the articles that are elaborated are PROD001, PROD002, PROD003, PROD004, PROD005, PROD006, PROD007, PROD008, and PROD009, as from 64 materials.

Table 1 shows the first 22 materials necessary for the elaboration of each of the nine products. The final products or parents are listed on the left side of the table, starting with product 1 (PROD001) up to Product 9 (PROD009) respectively. The parts used in the manufacture of each final product or children are located horizontally and are identified with a unique code previously assigned. It indicates the required quantity of each material per unit of the final product as can be seen for the manufacture of the PROD001 (Product 1), two pieces of CUE0562, six parts of VAS004, four pieces of MAN08, and so forth. Are necessary. (For the revision of the complete BOM review Appendix 1- Table 6).

2. Calculate the distance from the locations of the central warehouse to the pre-stocking warehouse. Consider the following layout of the raw materials warehouse (Figure 2) as an example of current locations.

Table 1. Bill of material

PROD/ COMP	CUE0562	TRI0143	VAS004	MAN089	TAP524	VAL145	SOP006	TUS098	RES014	EMB298	CAS017	ORO991	ARA146	ARO657	RUE398	TUC934	TOI471	REM355	SOT365	BOQ988	BOT641	ANI826
PROD001	2		6	4	2			8	12				8			3	9			5		
PROD002				5		14	4	6	3	13	1	5	3									
PRODU03	6		3	9		20	3	8	5		6	24	40	4	10				2			1
PROD004	3	1		2	6		14		10		25			30	6	8	14			6		3
PROD005		5	4		12					8			6	4		9	4		1		4	
PROD006							6	3			7				4				14		10	2
PROD007	6	10	3		5	7			2		1	11		6		8		2		2	6	
PROD008		7		2			5			4		11	10			5		13		9		1
PROD009	1		4			10		2	6					6			9		5	6	8	4

Figure 2. Layout raw materials warehouse
Source: "Own Elaboration"
Where:
A1, A2, A3 ... H8: are the locations in the warehouse.
Also, TOI471, RUE873, TAP524, TUC874, ARO870, REM355 ... and so forth: Represent the materials currently assigned to locations A1 ... .H8 in the warehouse.

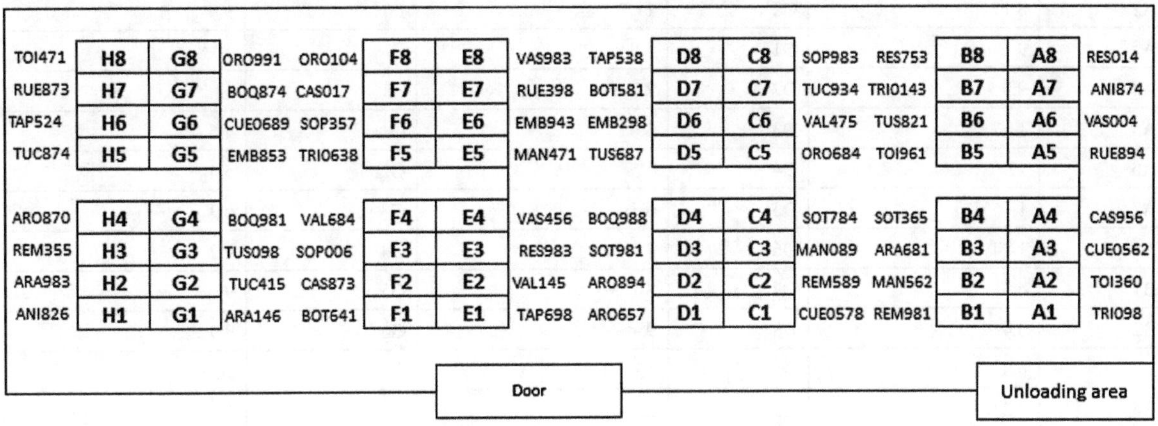

Based on the above, the distances of each location in the warehouse of raw materials to the pre-stocking warehouse are shown in Table 2, these distances are given in meters, for example the distance between the warehouse to the location A1, is 7.2 meters, the distance between the warehouse to the location A2 is 8.1 meters and so on as shown below:

3. Measure lead time of production to each final item and time of pickup of materials.

Lead time is measured in minutes, and the number of periods is the result of dividing the production lead time by 5 minutes, equivalent to one period. At this point, it is essential to remember that it is sought that the materials are placed at the disposal of the production line at the beginning of each shift.

Table 3 shows the lead time of production assigned to each product and its equivalence in periods.

The same procedure is carried out to establish the pick-up periods of each input. The first step is to measure the collection time of each one, the time it takes the warehouse personnel to collect the material from their current location until it is placed in the pressure store and later, convert that time in periods. For the presented instance the production corresponding to a working week that includes the days from Monday to Friday, and two Saturday shifts (as mentioned in the main focus) will be programmed, therefore a total of 1536 periods equivalent to 17 shifts with 8-hour each one will be considered.

For example, if the material's CUE0562 collection time is 20 minutes, the number of periods is obtained dividing 20 by 5, (the minutes corresponding to each period), in this way the collection periods or lead time is 4 periods for this material as detailed for every material in Table 4. (22 materials).

Table 2. Distances from the storage location to the pre-stocking store

Storage Location	Distances To The Pre-Stocking Store Meters	Storage Location	Distances To The Pre-Stocking Store Meters	Storage Location	Distances To The Pre-Stocking Store Meters
A1	7.2	C7	10.8	F5	9
A2	8.1	C8	11.7	F6	9.9
A3	9	D1	1.8	F7	10.8
A4	9.9	D2	2.7	F8	11.7
A5	11.7	D3	3.6	G1	4.5
A6	12.6	D4	4.5	G2	5.4
A7	13.5	D5	6.3	G3	6.3
A8	14.4	D6	7.2	G4	7.2
B1	4.5	D7	8.1	G5	9
B2	5.4	D8	9	G6	9.9
B3	6.3	E1	1.8	G7	10.8
B4	7.2	E2	2.7	G8	11.7
B5	9	E3	3.6	H1	7.2
B6	9.9	E4	4.5	H2	8.1
B7	10.8	E5	6.3	H3	9
B8	11.7	E6	7.2	H4	9.9
C1	4.5	E7	8.1	H5	10.8
C2	5.4	E8	9	H6	12.6
C3	6.3	F1	4.5	H7	13.5
C4	7.2	F2	5.4	H8	14.4
C5	9	F3	6.3		
C6	9.9	F4	7.2		

Table 3. Production lead time for each product

PRODUCT	LEAD TIME*MINUTES	PERIODS
PROD001	240	48
PROD002	300	60
PROD003	360	72
PROD004	300	60
PROD005	240	48
PROD006	210	42
PROD007	420	84
PROD008	480	96
PROD009	360	72

Table 4. Pick-up lead time for material

Material	Pick-Up Lead Time From Current Location	Periods	Material	Pick-Up Lead Time From Current Location	Periods	Material	Pick-Up Lead Time From Current Location	Periods
CUE0562	20	4.00	CUE0689	13	2.60	BOT581	8	1.60
TRI0143	10	2.00	TRI098	20	4.00	ANI874	5	1.00
VAS004	16	3.20	CUE0578	16	3.20	TOI961	10	2.00
MAN089	8	1.60	TRI0638	15	3.00	REM981	12	2.40
TAP524	9	1.80	VAS456	13	2.60	SOT981	14	2.80
VAL145	14	2.80	MAN562	14	2.80	BOQ981	15	3.00
SOP006	16	3.20	TAP538	5	1.00	VAS983	10	2.00
TUS098	18	3.60	VAL475	8	1.60	MAN471	8	1.60
RES014	20	4.00	SOP983	5	1.00	TAP698	10	2.00
EMB298	15	3.00	TUS821	12	2.40	VAL684	15	3.00
CAS017	15	3.00	RES753	11	2.20	SOP357	5	1.00
ORO991	14	2.80	EMB853	15	3.00	TUS687	10	2.00
ARA146	18	3.60	CAS956	20	4.00	RES983	13	2.60
ARO657	15	3.00	ORO684	24	4.80	EMB943	10	2.00
RUE398	13	2.60	ARA681	15	3.00	CAS873	15	3.00
TUC934	14	2.80	ARO894	13	2.60	ORO104	16	3.20
TOI471	10	2.00	RUE894	12	2.40	ARA983	14	2.80
REM355	8	1.60	TUC415	10	2.00	ARO870	8	1.60
SOT365	9	1.80	TOI360	16	3.00	RUE873	9	1.80
BOQ988	10	2.00	REM589	15	2.80	TUC874	10	2.00
BOT641	8	1.60	SOT784	14	3.20			
ANI826	12	2.40	BOQ874	16	1.60			

4. Calculate the Materials Requirement Planning of the weekly production area (size of the pre-stocking warehouse per shift).

The software used to solve the model was LINGO 10. Two files for its operation support this program, the first is a.LDT extension data file that contains all the information related to the demand, such as the BOM, the relationship product-input (father-children), the amount required in each relation, the lead time of production and / or collection of each part and the product-input matrix in relation to the periods available during the day (see Appendix 2). The second file is an executable document that contains all the programming of the model with which the calculation of the requirements planning is pèrformed. The programming can be consulted in Appendix 3.

Once is obtained all the information, the executable file program can be run to calculate the necessary materials according to the production plan, the program multiplies the number of final products according to the program given by the inputs to know the total requirement.

As a result of this calculation, a collection schedule is obtained just in time according to the demand, based on the production scheduled for each shift. This schedule is designed to work backward, that is, it indicates that and how many materials are needed and the appropriate time to start the collection in the raw materials warehouse as the purpose of placing them at the disposal of the line at the time required, that is to say, at the start of each shift.

For a better interpretation of the results, the data obtained is exported to an Excel file where the results can be better analyzed and processed for the next step:

For example, for the 2-day shift on Monday that starts at 8:00 am, whose scheduled planning is 10 pieces of Product 1, 15 pieces of Product 2, 14 pieces of Product 3, 20 pieces of Product 4, 36 pieces of Product 5, 50 pieces of the product 6, 30 pieces of the Product 7, 8 of the product 8 and 20 pieces of the Product 9. The personnel of the raw materials warehouse must collect 364 pieces of the material CUE0562 and the collection of that product in the warehouse of materials premiums should start 7:40 am, 556 components with part number TRI0143 whose collection should start 7:50 am, 416 pieces of VAS004 that should start collecting 7:50 am, 297 pieces of the MAN089 whose collection should start 7:55 am, in this way respectively for each the parts as shown in figure 3.

In this way, the collection time of all materials is the result of the sum of the backward periods of Teach part. That is five collection periods of CUE0562 plus three collection periods of TRI0143 plus four periods of collection of the part VAS004 and so on. So that the sum of all the periods, in this case, 206 periods for the collection of all the parts to be used for the second shift of the Monday must be mul-

Figure 3. Material planning for Monday's second shift

TIME	07:00	07:05	07:10	07:15	07:20	07:25	07:30	07:35	07:40	07:45	07:50	07:755	08:00
MINS	420	425	430	435	440	445	450	455	460	465	470	475	480
PERIOD	84	85	86	87	88	89	90	91	92	93	94	95	96
PROD001	0	0	0	0	0	0	0	0	0	0	0	0	10
PROD002	0	0	0	0	0	0	0	0	0	0	0	0	15
PROD003	0	0	0	0	0	0	0	0	0	0	0	0	14
PROD004	0	0	0	0	0	0	0	0	0	0	0	0	20
PROD005	0	0	0	0	0	0	0	0	0	0	0	0	36
PROD006	0	0	0	0	0	0	0	0	0	0	0	0	50
PROD007	0	0	0	0	0	0	0	0	0	0	0	0	30
PROD008	0	0	0	0	0	0	0	0	0	0	0	0	8
PROD009	0	0	0	0	0	0	0	0	0	0	0	0	20
CUE0562	0	0	0	0	0	0	0	0	364	0	0	0	0
TRI0143	0	0	0	0	0	0	0	0	0	0	556	0	0
VAS004	0	0	0	0	0	0	0	0	0	416	0	0	0
MAN089	0	0	0	0	0	0	0	0	0	0	0	297	0
TAP524	0	0	0	0	0	0	0	0	0	0	0	722	0
VAL145	0	0	0	0	0	0	0	0	0	0	900	0	0
SOP006	0	0	0	0	0	0	0	0	0	722	0	0	0
TUS098	0	0	0	0	0	0	0	0	0	472	0	0	0
RES014	0	0	0	0	0	0	0	0	615	0	0	0	0
EMB298	0	0	0	0	0	0	0	0	0	515	0	0	0
CAS017	0	0	0	0	0	0	0	0	0	979	0	0	0
ORO991	0	0	0	0	0	0	0	0	0	0	829	0	0
ARA146	0	0	0	0	0	0	0	0	0	981	0	0	0

tiplied by 5, which are the minutes corresponding to each period and then divided by 60 the equivalent in minutes to one hour. It results in a total of 17.2 hours; if four people are considered in the warehouse of raw materials per shift, in this case, the harvest would have to start 4 hours and 20 minutes ago, that is 3:40 am.

5. Process Lingo solution in Excel

Excel document is duly formulated in order to process the results obtained from lingo, consider the amount requested per input in each of the shifts and then sum them to obtain the total amount per material required at the end of the week, subsequently, sum them to know the requirement in the course of 3 months, as an example the status of ten components is shown below in Table 5.

This method basically proposes to execute the run of the weekly schedule foreseeing the variability of the demand in a lapse of 3 months as it is observed in the previous table in order to add it and obtain total requirement in the indicated time and in this way reallocate of locations on a quarterly basis according to the next phase.

Phase 2

1. Create a distance matrix

The first step of this phase after three months is to create a distance matrix in Excel that includes Location - Amount requested, where the locations must be placed on the horizontal axis and the products on the vertical axis. Each cell found at the intersection is the result of the product of the distance in meters from the location in the general raw materials warehouse to the pre-stocking warehouse by the quantity requested of each input in the period until completing the distance matrix (Location-Product).

Table 5. Total requirement per week

PROD	Week 1	Week 2	Week 3	Week 4	Week 5	Week 6	Week 7	Week 8	Week 9	Week 10	Week 11	Week 12	Week 13	Trimester
CUE0562	5353	3679	2703	2489	3174	3819	3358	4327	3359	3172	2555	3356	3655	44999
TRI0143	6502	4556	2941	3086	4632	4390	4502	5220	4579	3083	4335	2880	4002	54708
VAS004	6140	2833	3195	2953	3810	4332	3809	4334	3917	1930	3342	3755	4416	48766
MAN089	6274	4220	3893	4061	3624	3682	4015	5737	3919	2445	3412	4058	4634	53974
TAP524	8381	4052	1286	4063	4891	5701	4682	5138	5382	2323	4630	4603	4324	59456
VAL145	14007	10792	9584	7865	9165	9435	8696	11696	9226	7117	8266	10533	11416	127798
SOP006	9725	6376	3814	6445	5197	5691	6687	7467	6363	5856	5551	4586	4411	78169
TUS098	7886	4484	5645	4716	4671	5069	5130	6588	4951	2628	4495	5118	6151	67532
RES014	11680	5734	5773	6471	6418	8196	7967	8909	7359	4793	5946	7090	6722	93058
EMB298	7059	4623	3436	4687	4869	4371	4602	5635	5006	540	5582	4524	4488	59422

2. Perform locations reallocation

Once the three proposed months have elapsed, the next step is to carry out the reassignment of locations considering the products with the highest requirement, placing the most requested in the locations closest to the entrance to the warehouse. For this, the Lingo program is used again.

First, Lingo imports the data from the distance matrix, calculated in the previous step and then the program determines the most requested product during the period considered (3 months, 13 weeks) and assigns it the best location. The closer to the entrance of the warehouse and consequently, the product with less rotation will be placed in the location farthest from the entrance of the warehouse; the objective is to minimize the distances of the products with the most important requirement. The program used to reallocate goods can be found in Appendix 4.

RESULTS AND DISCUSSION

The results of the weekly run the reassignment of warehouse locations can be made adding the total requirement of each of the parts in the periods determined by the warehouse administration, in this case, the reassignment is proposed every 13 weeks to place the most requested products in the locations closest to the entrance.

Then, in Figure 4 the location of the materials is observed without considering their rotation, and in Figure 5 the reallocation of the materials according to their requirement is observed by the production line, placing the most requested in the most critical positions. Next to the door. Figure 4 shows in bold letters the materials RUE873, ORO991, BOQ874, ARA146, BOT581, ARO567, RES753 and TOI360, in their previous position away from the door through which goods are transferred from the warehouse, without considering its rotation, later in Figure 5 the reallocation is observed placing them in a strategic position that allows a quick collection.

In the same way, in Figure 4, the materials REM355, ARA983, CU30689, MAN471, TAP698, ANI874, CUE562, TRI098 are indicated in italics, using strategic locations that, due to their low turnover, were reassigned to farther locations as indicated in Figure 5.

Figure 4. Previous locations of the raw materials warehouse
Source: "Own Elaboration"

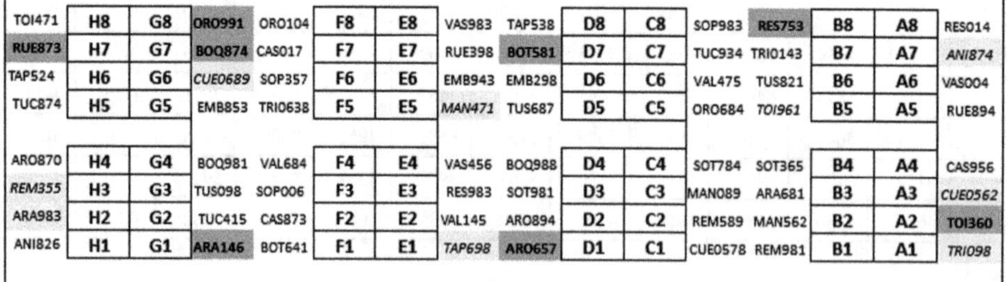

Figure 5. New locations of the raw materials warehouse
Source: "Own Elaboration"

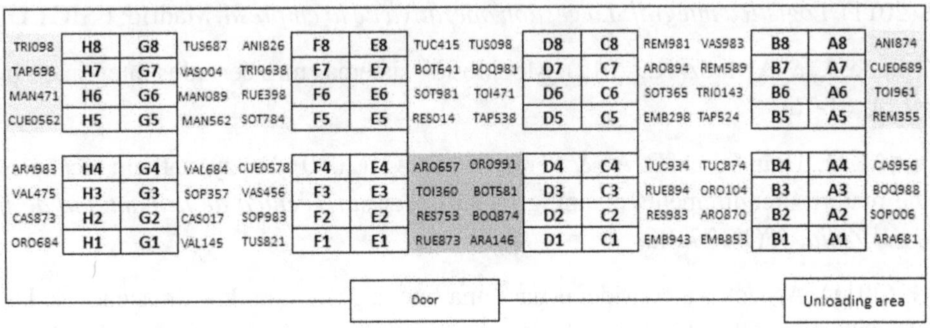

This same method was tested in two smaller instances more. The first includes three final products and 15 materials and the second with nine final products and 22 materials. Thus proving its effectiveness for small and medium enterprises in search of an efficient program that helps them have better control of their operations, leading to a substantial reduction of travel distances and preparation of internal orders. It is convenient for cases in which there is too much demand volatility since the proposed reassignment is a function of the products with the highest turnover in the previous period.

CONCLUSION

With the proposed method, the central warehouse is provided with a practical tool to know the movements that must be made, the material that must be collected and the time in which the activities must be carried out in order to avoid changes that do not create any value, such as unnecessary movements, downtime, and rework. It results from a lack of communication with the production department and on the other hand, reduces the risk of line stoppages and the tension in the line due to the lack of materials.

Likewise, having temporary dedicated spaces in the warehouse is intended to reduce the risk of loss and obsolescence of materials by placing them in a different place than the one that corresponds to them due to lack of space or lack of knowledge of the place that corresponds to them when integrating new elements to the organization. It sought to work in a standardized way and ordered in order to have an efficient warehouse.

ACKNOWLEDGMENT

This research was developed in the Master's program "Logistics and Supply Chain Management" awarded with the standard of excellence and supported with resources granted by the National Science and Technology Council (Consejo Nacional de Ciencia y Tecnología) – CONACyT. Scholarship holder number: 851922.

REFERENCES

Anaya, T. J. (2011). *Logística integral: La gestión operativa de la empresa.* Madrid: ESIC EDITORIAL.

Arango, M., Cano, J., & Álvarez, K. (2012). Modelos de sistemas mrp cerrados integrando incertidumbre. *Revista EIA*, 61-76.

Arango Serna, M. D., Urán Serna, C. A., & Pérez Ortega, G. (2010). Fuzzy mathematical programming applied to the material requirements planning (MRP). *Revista Técnica de la Facultad de Ingeniería. Universidad del Zulia, 33*(1), 77–86.

Arrieta, J. G. (2011). Aspectos a considerar para una buena gestión en los almacenes de las empresas (Centros de Distribución, CEDIS). *Journal of Economics, Finance and Administrative Science*, 84–86.

Bartholdi, J., & Hackman, S. (2014). *Warehouse & Distribution Science.* Atlanta, GA: The Supply Chain and Logistics Institute.

Bowersox, D. J., Closs, D. J., & Cooper, M. (2007). *Administración y Logística en la cadena de suministros.* México: McGraw-Hill Interamericana.

Bustos-Flores, C. E., & Chacón, P. G. (2007). *El MRP En la gestión de inventarios.* Visión Gerencial.

Cáceres, C. D., Reyes, V. J., García, C. M., & Sanchez, R. C. (2015). Modelo de Programación Lineal para Planeación de Requerimiento de Materiales. *Revista Tecnológica ESPOL*, 24-33.

Cano, A. J. (2011). *Modelo de un Sistema MRP Cerrado Integrando Incertidumbre en los Tiempos de Entrega, Disponibilidad de la Capacidad de Fabricación e Inventarios. Universidad Nacional de Colombia.* Medellín: Magister Tesis.

Cardenas, D. M., & Urquiaga, A. J. (2007). Logística de operaciones: integrando las decisiones estratégicas para la competitividad. *Ingeniería Industrial, 28*(1), 37–41.

Casanovas, A., & Cuatrecasas, L. (2011). *Logistica Empresarial. Gestión Integral de la información y material en la empresa.* Barcelona: Ediciones Gestión 2000.

Chang, Y., Zhu, X., & Haghani, A. (2019). Modeling and solution of joint storage space allocation and handling operation for outbound containers in rail-water intermodal container terminals. *IEEE Access: Practical Innovations, Open Solutions.*

Chen, L., Langevin, A., & Riopel, D. (2011). A tabu search algorithm for the relocation problem in a warehousing system. *International Journal of Production Economics, 129*(1), 147–156. doi:10.1016/j.ijpe.2010.09.012

Chiavenato, I. (1993). *Iniciación a la administración de materiales.* México: McGraw-Hill.

Chiavenato, I. (2006). *Introducción a la teoría general de la administración.* México: MCGraw-Hill Interamericana.

Demir, L., Tunali, S., & Tursel, E. D. (2014). The state of the art on buffer allocation problem: A comprehensive survey. *Journal of Intelligent Manufacturing, 25*(3), 371–392. doi:10.100710845-012-0687-9

Emmett, S. (2005). *Excellence in Warehouse Management: How to Minimise Costs and Maximise Value*. United Kingdom: John Wiley & Sons Ltd.

Gaither, N., & Frazier, G. (2000). *Administración de Producción y Operaciones*. México: International Thomson Editores.

García, C. A. (1999). *Almacenes: Planeación, organización y control*. México: TRILLAS.

Gonzalez, C. (2013). *Control de aprovisionamiento de materias primas*. Madrid, España: Paraninfo.

Hillier, S. F., & Lieberman, J. G. (2015). Introduction to Operations Research (10th ed.). New York: McGraw-Hill Education.

Le-Duc, T., & Koster, R. (2005). Travel distance estimation and storage zone optimization in a 2-block classbased storage strategy warehouse. *International Journal of Production Research*, *43*(17), 3561–3581. doi:10.1080/00207540500142894

Martínez, S. I., Vértiz, C. G., López, P. J., Jiménez, L. G., & Moncayo, M. L. (2014). *Investigación de Operaciones*. México: GRUPO EDITORIAL PATRIA.

Medina, L. A., Nogueira, R. D., & Sosa, N. E. (2002). *EL SISTEMA MRP*. Retrieved from https://www.academia.edu/1104284/EL_SISTEMA_MRP

Miclo, R., Fontalini, F., Lauras, M., Lamothe, J., & Milian, B. (2016). An empirical study of Demand-Driven MRP. *Proceedings of the 6th International Conference on Information Systems, Logistics and Supply Chain*, Bordeaux. Academic Press.

Miño-Cascante, G., Saumell-Fonseca, E., Toledo-Borrego, A., & Moreno, R. (2015). *Planeación de requerimientos de materiales por el sistema MRP. Caso Laboratorio Farmacéutico Oriente* (pp. 208–2019). Cuba: SCIELO.

Pochet, Y., & Wolsey, L. (2006). *Production Planning by Mixed Integer Programming*. New York: Springer.

Rajadell, M., & Sánchez, J. (2010). *Lean manufacturing- La evidencia de una necesidad*. Madrid: Ediciones Díaz de Santos.

Richards, G. (2011). *Warehouse Management: A Complete Guide to Improving Efficiency and Minimizing Costs in the Modern Warehouse*. London: Kogan Page.

Rivera, P. J., Ortega, P. E., & Pereyra, Q. J. (2014). Diseño e Implementacipon del Sistema MRP en PyMes. *Revista de la Faculta de Ingenieria Industrial*, 48-55.

Rojas, R. J., Solano-Charris, E., & Montoya-Torres, J. R. (2019). The storage location assignment problem: A literature review. *International Journal of Industrial Engineering Computations*, 199-224.

Salazar, E., & Martínez, J. (2017). Storage Location Allocation Model for a Just In Sequence (JIS) production System. *Proceedings of the International Conference on Industrial Engineering and Operations Management,* Bogota, Colombia (pp. 277-288). Academic Press.

Sangüesa, M., Mateo, R., & Ilzarbe, L. (2006). *Teoría y practica de la calidad*. Madrid: Paraninfo.

Suárez, C., & Hualpa, A. M. (2018). Warehouse Sizing based on Material Requirements Planning for a Polyurethane Coating Factory. *Ingeniería*, *3*(1), 48–69.

Velasco, S. J., & Campins, M. J. (2013). *Gestión de la producción en la empresa: Planificación, programación y control*. Madrid: Ediciones Pirámide.

Winston, W. (2005). *Investigación de operaciones. Aplicaciones y algoritmos*. México: Thomson.

Xie, J., Mei, Y. T., Ernst, A., Li, X., & Song, A. (2015). *A Restricted Neighbourhood Tabu Search for Storage Location Assignment Problem. Proceedings of the Congress on Evolutionary Computation (CEC),* Sendai, Japan *(pp. 2805-2812)*. IEEE.

Zhang, G., Nishi, T., Turner, S. D., Oga, K., & Li, X. (2016). An Integrated Strategy for a Production Planning and Warehouse Layout Problem: Modeling and Solution Approaches. *Omega*, *68*, 85–94. doi:10.1016/j.omega.2016.06.005

APPENDIX 1

Table 6. Complement- bill of material

PROD/ COMP	CUE0689	TRI098	CUE0578	TRI0638	VAS456	MAN562	TAP538	VAL475	SOP983	TUS821	RES753	EMB853	CAS956	ORO684	ARA681	ARO894	RUE894	TUC415	TOI360	REM589	SOT784
PROD001	2		4			8	12				8			3	9	6	4	2		6	4
PROD002			10	5		6	3	13	1	5	3						5		14		5
PRODU03	4		14		10	8	5		6	24	40	4	10			3	9		20	3	9
PROD004	3			5			10		25			30	6	8	14		2	6			5
PROD005			3		16			8			6	4		9	4	4		12		4	
PROD006	1			6		3			7				4				15		1		
PROD007		4			12		2		1	11		6		8		3		5	7	3	
PROD008		5	1	1				4		11	10			5			2				2
PROD009			2		2	2	6					6			9	4			10	4	

PROD/ COMP	BOQ874	BOT581	ANI874	TOI961	REM981	SOT981	BOQ981	VAS983	MAN471	TAP698	VAL684	SOP357	TUS687	RES983	EMB943	CAS873	ORO104	ARA983	ARO870	RUE873	TUC874
PROD001	2				2	8			2		4			6		6					6
PROD002		14									10	1			6		6		20		
PRODU03		20	3	1	4			1	4		14		15		1	5				60	12
PROD004	6			3	3	4		3	3			4		7	3		5		8	9	
PROD005	12					5	4				3			6	10	4		25		8	5
PROD006	20		2	2	1		10	2	1					13			10			2	
PROD007	14	7			2	1	6			4		20			6	9			14		10
PROD008				1		2		1		5	1			10	9	6	20			5	
PROD009		10	1	4		9	8	4			2	14			10	2			1		4

APPENDIX 2

For Material planning, Lingo's Material Requirements Planning (Model: MRP) has been used and modified; the original model can be reviewed at http://www.lindo.com

Data File Information (LTD) - Lingo

```
! Parts list;
AA,    !    PROD001;
BB,    !    PROD002;
CC,    !    PROD003;
DD,    !    PROD004;
EE,    !    PROD005;
FF,    !    PROD006;
GG,    !    PROD007;
HH,    !    PROD008;
II,    !    PROD009;
MA,    !    CUE0562;
MB,    !    TRI0143;
MC,    !    VAS004;
MD,    !    MAN089;
ME,    !    TAP524;
MF,    !    VAL145;
MG,    !    SOP006;
MH,    !    TUS098;
MI,    !    RES014;
MJ,    !    EMB298;
MK,    !    CAS017;
ML,    !    ORO991;
MM,    !    ARA146;
MN,    !    ARO657;
MO,    !    RUE398;
MP,    !    TUC934;
MQ,    !    TOI471;
MR,    !    REM355;
MS,    !    SOT365;
MT,    !    BOQ988;
MU,    !    BOT641;
MV,    !    ANI826;
MW,    !    CUE0689;
MX,    !    TRI098;
MY,    !    CUE0578;
MZ,    !    TRI0638;
```

```
NA,   !  VAS456;
NB,   !  MAN562;
NC,   !  TAP538;
ND,   !  VAL475;
NE,   !  SOP983;
NF,   !  TUS821;
NG,   !  RES753;
NH,   !  EMB853;
NI,   !  CAS956;
NJ,   !  ORO684;
NK,   !  ARA681;
NL,   !  ARO894;
NM,   !  RUE894;
NN,   !  TUC415;
NO,   !  TOI360;
NP,   !  REM589;
NQ,   !  SOT784;
NR,   !  BOQ874;
NS,   !  BOT581;
NT,   !  ANI874;
NU,   !  TOI961;
NV,   !  REM981;
NW,   !  SOT981;
NX,   !  BOQ981;
NY,   !  VAS983;
NZ,   !  MAN471;
OA,   !  TAP698;
OB,   !  VAL684;
OC,   !  SOP357;
OD,   !  TUS687;
OE,   !  RES983;
OF,   !  EMB943;
OG,   !  CAS873;
OH,   !  ORO104;
OI,   !  ARA983;
OJ,   !  ARO870;
OK,   !  RUE873;
OL~   !  TUC874;
!  The set of periods;
   1..1632 ~
! The parent-child use relationships and the number of child parts required in
each parent-child relationship, respectively;
! (i.e., PROD001 AA used CUE0562-MA- 2 pieces, etc.);
```

```
AA  MA 2, AA  MC 6, AA  MD 4,AA  ME 2, AA  MH 8, AA  MI 12,AA  MM 8, AA  MP 3,
AA  MQ 9, AA  MT 5, AA  MW 2, AA  MY 4, AA  NB 8,AA  NC 12, AA  NG 8, AA  NJ
3, AA  NK 9, AA  NL 6,AA  NM 4, AA  NN 2,AA NP 6, AA  NQ 4, AA  NR 2, AA  NV
2, AA  NW 8, AA  NZ 2, AA  OB 4, AA  OE 6, AA  OG 6, AA OL 6,
BB  MD 5, BB  MF 14, BB  MG 4, BB  MH 6, BB  MI 3, BB  MJ 13, BB  MK 1, BB  ML
5, BB  MM 3, BB  MY 10, BB  MZ 5, BB  NB 6, BB  NC 3, BB  ND 13, BB  NE 1, BB
NF 5, BB  NG 3, BB  NM 5, BB  NO 14, BB  NQ 5, BB  NS 14, BB  OB 10, BB  OC 1,
BB  OF 6, BB  OH 6, BB  OJ 20,
CC  MA 6, CC  MC 3, CC  MD 9, CC  MF 20,CC  MG 3, CC  MH 8, CC  MI 5,CC  MK
6, CC  ML 24, CC  MM 40, CC  MN 4, CC  MO 10, CC  MS 2, CC  MV 1, CC  MW 4,
CC  MY 14, CC  NA 10, CC  NB 8, CC  NC 5, CC NE 6,CC  NF 24, CC  NG 40, CC  NH
4, CC  NI 10, CC  NL 3, CC  NM 9, CC  NO 20, CC  NP 3, CC  NQ 9, CC  NS 20, CC
NT 3, CC  NU 1, CC  NV 4, CC  NY 1, CC  NZ 4, CC  OB 14, CC  OD 15, CC  OF 1,
CC  OG 5, CC  OK 60,CC  OL 12,
DD  MA 3, DD  MB 1, DD  MD 2, DD  ME 6, DD  MG 14, DD  MI 10, DD  MK 25, DD
MN 30, DD  MO 6, DD  MP 8, DD  MQ 14, DD  MT 6, DD  MV 3, DD  MW 3, DD  MZ
5,DD  NC 10, DD  NE 25, DD  NH 30, DD  NI 6, DD  NJ 8, DD  NK 14,DD  NM 2, DD
NN 6, DD  NQ 5, DD  NR 6, DD  NU 3, DD  NV 3, DD  NW 4, DD  NY 3, DD  NZ 3, DD
OC 4, DD  OE 7, DD  OF 7, DD  OH 5, DD  OJ 8, DD  OK 9,
EE  MB 5, EE  MC 4,EE  ME 12, EE  MJ 8, EE  MM 6, EE  MN 4, EE  MP 9, EE  MQ
4, EE  MS 1, EE  MU 4, EE  MY 3, EE  NA 16, EE  ND 8, EE  NG 6, EE  NH 4, EE
NJ 9, EE  NK 4, EE  NL 4, EE  NN 12, EE  NP 4, EE  NR 12, EE  NW 12, EE  NX 5,
EE  OB 4, EE  OE 6, EE  OF 10, EE  OG 4, EE  OI 25, EE  OK 8, EE  OL 5,
FF  MG 6, FF  MH 3, FF  MK 7, FF  MO 4, FF  MS 14, FF  MU 10, FF  MV 2, FF  MW
1, FF  MZ 6, FF  NB 3, FF  NE 7, FF  NI 4, FF  NM 15, FF  NO 1, FF  NR 20, FF
NT 2, FF  NU 2, FF  NV 1, FF  NX 10, FF  NY 2, FF  NZ 1, FF  OE 13, FF  OH 10,
FF  OK 2,
GG  MA 6, GG  MB 10, GG  MC 3, GG  ME 5, GG  MF 7, GG  MI 2, GG  MK 1, GG  ML
11, GG  MN 6, GG  MP 8, GG  MR 2, GG  MT 2, GG  MU 6, GG  MX 4, GG  NA 12,GG
NC 2, GG  NE 1, GG  NF 11, GG  NH 6,GG  NJ 8, GG  NL 3, GG  NN 5, GG  NO 7, GG
NP 3, GG  NR 14, GG  NS 7, GG  NV 2, GG  NW 1, GG  NX 6, GG  OA 4, GG  OC 20,
GG  OF 6, GG  OG 9, GG  OJ 14,GG  OL 10,
HH  MB 7, HH  MD 2, HH  MG 5, HH  MJ 4, HH  ML 11, HH  MM 10, HH  MP 5, HH  MR
13, HH  MT 9, HH  MV 1, HH  MX 5, HH  MY 1, HH  MZ 1, HH  ND 4, HH  NF 11, HH
NG 10, HH  NJ 5, HH  NM 2, HH  NQ 2, HH  NU 1, HH  NW 2,HH  NY 1, HH  OA 5, HH
OB 1, HH  OE 10, HH  OF 9, HH  OG 6, HH  OH 20, HH  OK 5,
II  MA 1, II  MC 4, II  MF 10,II  MH 2, II  MI 6, II  MN 6, II  MQ 9, II  MS
5, II  MT 6, II  MU 8, II  MV 4, II  MY 2, II  NA 2, II  NB 2, II  NC 6, II
NH 6, II  NK 9, II  NL 4, II  NO 10, II  NP 4, II  NS 10, II  NT 1, II  NU 4,
II  NW 9, II  NX 8, II  NY 4, II  OB 2, II  OC 14, II  OF 10, II  OG 2, II  OJ
1,II  OL 4 ~
! The lead times (pick up) for each part;
12,12,12,12,12,12,12,12,12,
```

```
4,2,3.2,1.6,1.8,2.8,3.2,3.6,4,3,3,2.8,3.6,3,2.6,2.8,2,1.6,1.8,2,1.6,2.4,2.6,4,
3.2,3,2.6,2.8,1,1.6,
1,2.4,2.2,3,4,4.8,3,2.6,2.4,2,3.2,3,2.8,3.2,1.6,1,2,2.4,2.8,3,2,1.6,2,3,1,2,2.
6,2,3,3.2,2.8,1.6,1.8,2~
```

APPENDIX 3

Executable Programming MRP in Lingo

```
MODEL:
! Data for this model is read from MRP.LDT;
SETS:
! The set of parts;
PART: LT;
! LT(i) = Lead time to produce part i;
! The set of periods;
TIME;
! A relationship called USES between pairs of parts;
USES(PART, PART): NEEDS;
! Parent part i needs NEEDS(i, j) units of
  child part j;
! For each part and period we are interested in;
PXT(PART, TIME): ED, TD;
! ED(i, j) = External demand for part i at time j;
! TD(i, j) = Total demand for part i at time j;
ENDSETS
DATA:
! Load the data from an external file;
! Parts list;
PART = @FILE('PERIODOS.LDT');
! Time periods;
TIME = @FILE('PERIODOS.LDT');
! Get the parent child relations and the
  number of parts required;
USES, NEEDS = @FILE('PERIODOS.LDT');
! Get the lead times from the file;
LT  = @FILE('PERIODOS.LDT');
! Get the external demands
 over time for each part;
!ED  = @FILE('MRP.LDT');
ED=@OLE('MRP.XLSX','matriz');
ENDDATA
```

```
! Set NP = no. of time periods in the problem;
NP = @SIZE(TIME);
! For each part P and period T, the total demand =
 external demand + demand generated by parents
 one lead time in the future;
@FOR(PXT(P, T) | T + LT(P) #LE# NP:
 TD(P, T) = ED(P, T + LT(P)) +
  @SUM(USES(P2, P): TD(P2, T + LT(P)) *
   NEEDS(P2, P));
);

DATA:
! Display a table showing the production schedule;
@TEXT() = ' The production schedule:';
@TEXT() = @TABLE(TD);
@OLE ('C:\Users\PATRICIA\Desktop\BUFER-ASIGNACION\INSTANCIA-1-SEMANAL\DATOS-
PRESURTIDO') = TD;
ENDDATA
END
```

APPENDIX 4

Programming in Lingo for Reassignment of Materials

```
MODEL:
SETS:
PRODUCTO;
UBICACION;
PAR(PRODUCTO,UBICACION): DISTANCIA,X;
ENDSETS
DATA:
!MIEMBROS DE LOS CONJUNTOS;
PRODUCTO=
MA   MB   MC   MD   ME   MF   MG   MH   MI   MJ   MK   ML   MM   MN MO   MP
MQ   MR   MS   MT   MU   MV   MW   MX   MY   MZ   NA   NB   NC   ND NE   NF   NG
NH   NI   NJ   NK   NL   NM   NN   NO   NP   NQ   NR  NS   NT   NU   NV   NW
NX   NY   NZ   OA   OB   OC   OD   OE   OF   OG  OH      OI   OJ   OK   OL
;
UBICACION=
A1        A2        A3        A4        A5        A6        A7
A8        B1        B2        B3        B4   B5 B6          B7        B
8        C1        C2        C3        C4   C5         C6        C7
```

```
C8          D1    D2 D3          D4          D5          D6          D7          D
8          E1          E2          E3          E4          E5          E6    E7 E8
F1          F2          F3          F4          F5          F6          F7          F
8          G1          G2
;
DISTANCIA=@OLE ('C:\Users\PATRICIA\Desktop\BUFER-ASIGNACION\INSTANCIA-1-SE-
MANAL\DATOS-PRESURTIDO-COMPLETO', DISTANCIAS);
ENDDATA
MIN=@SUM(PAR(I,J):DISTANCIA(I,J)*X(I,J));
@FOR(PRODUCTO(I):@SUM(UBICACION(J):X(I,J))=1);
@FOR(UBICACION(J):@SUM(PRODUCTO(I):X(I,J))=1);
END
```

Chapter 4
Radio Frequency Identification Technology and Its Security Concerns in the Manufacturing Industry

Kamalendu Pal
City, University of London, UK

ABSTRACT

The advent of Radio Frequency Identification (RFID) technique, low-cost wireless sensor devices, and web-based technologies are ushering new ways to take control of the plethora of data created by the Internet of Things (IoT) approach has gained popularity in the modern manufacturing industry. This mechanism provides an opportunity of remotely storing and retrieving data from RFID-tags. Products labelled with such tags can be scanned efficiently using RFID-readers that do not require line-of-sight. This type of identification and access to information on tags are used by manufacturing business operations, can lead to improving logistics, and better customer service. However, the widespread use of RFID technology also introduces serious security and privacy risks since the information stored in tags can easily be retrieved by hidden RFID-readers, eventually leading to the violation of genuine business operation privacy. This chapter discusses the open research challenges of RFID applications in the manufacturing supply chain operations.

INTRODUCTION

All businesses today understand the value and importance of building effective supply chains, as part of sustainable growth and profitability (Pal, 2019). In many global manufacturing businesses, such as automotive, pharmaceutical, apparel, consumer electronics, multitier supply chains, with different business-partners at different tiers of the chain, are common (Corbett & Karmarkar, 2001). Moreover, the operational structure of manufacturing supply chains can vary substantially. For example, there are handful of personal computer manufacturers, but only few microchips sellers dominate at their own tier.

DOI: 10.4018/978-1-7998-1831-1.ch004

The automotive industry has few final assemblers, but many manufacturers for most parts. However, the exact interactions between structural components, their pricing mechanisms, and entry and production decisions are very important for manufacturing companies. These production decisions are mostly based on day-to-day business operational information.

A manufacturing supply chain consists of interconnected activities, and their associated business processes together to provide value-added service to its customers. Customer-engaged manufacturing companies, from automobile to highly critical fighter aircraft makers, always need different stakeholders' information for their supply chains. An entire network of manufacturers and distributors, transportation and logistics agencies, financial institutions, warehouses and freight-forwarders work together to make sure that the right goods and services are available at the right price, where and when the customers want them. Having supplied value-added services (e.g. products and associated customer services), the supply chain does not terminate. The supply chain is comprised of several steps from the front end, through the customer request, supply chain order processing initiation, quality assurance assessment for products and services, relevant training processes for staffs, customer support facilities, to maintenance and replacement facilities. Manufacturers are investing in state-of-the-art operational practices to optimize both cost and efficiency to their supply chain.

In a typical supply chain, raw materials are purchased from suppliers and products are manufactured at one or more production plants (Pal, 2017) (Pal, 2018). Then, they are transported to intermediated storage facilities (e.g. warehouse, distribution centers) for packing and shipping to retailers or customers. The path from supplier to customer can include several intermediaries such as wholesalers, warehouse, and retailers, depending on the products and markets. In this way, supply chain management relates to business activities such as inbound and outbound transportation, warehousing, and inventory control. Importantly, it also embodies the information systems necessary to monitor these business activities. Figure 1 presents a simple diagrammatic representation of a manufacturing supply chain, which consists of two separate legal entities, a retailer and manufacturer. The retailer owns the first two nodes on the left-hand-side of the diagram, which consists of the retail store and the retail distribution center; and the manufacturer owns the last two nodes, the manufacturing distribution center and the plant.

Moreover, all supply chains share the following characteristics: (i) the supply chain comprises all business activities in order to supply a product or service to its end-customers; (ii) any number of supply chain partner organizations can be linked in the supply chain; (iii) a customer can be a supplier to another customer within the supply chain, which means that the total network of activities can consists of a number of suppliers / customer relationships; (iv) the path from supplier to customer, depending on the products and markets, can include a number of intermediaries (distributors) such as wholesalers, warehouses, and retailers. Products or service flows from supplier to customer are called downstream flows while demand information from customer to supplier is called upstream flows.

Figure 1. Diagrammatic representation of a supply chain

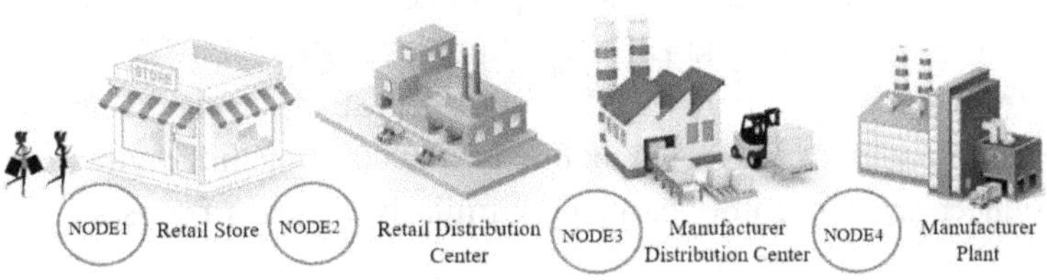

In this way, a supply chain creates a complex business network. Given process decentralization, the efficient performance of a supply chain requires a high degree of visibility – defined as the capability of sharing on time and accurate data throughout the entire supply chain, and coordination among supply chain partners. In today's global business environment, companies recognize the strategic importance of well-managed supply chains. For example, companies such as Dell, Nokia, Intel, Toyota, Zara, and Li & Fun have based their corporate strategy around achieving supply chain superiority over competitors (Copacino & Anderson, 2003). These global companies have gained competitive advantages by effectively managing the complex web of supply chain process interactions that extend across continents and across enterprises in product procurement, manufacturing and distribution. The requirements of modern supply chains are:

- **Connectivity**: With emphasis on the ability to make and maintain connection between business partners. It characterizes the ability to exchange information within supply chain partners in a way which provides inter-organizational collaboration.
- **Integration**: The ability to connect and coordinate business processes in a seamless way. It improves supply chain business processes performance by establishing collaborative connections among supply chain partners. For example, seamless integration increases information transparency among partners and allows pooling of inventories and sharing resources.
- **Visibility**: Visibility refers to the capability to access or view data or information related to logistics and the supply chains. For example, visibility is the ability of knowing where raw materials inventory for manufacturing, semi-finished products in the production line, and finished products are, at any time. But it is also actionable information that can help support customers at different interface points along the supply chain and improve business processes performance.
- **Responsiveness**: It is supply chain ability to react quickly to customer needs or specifications by delivering a product of the right quality, at the right time, at the right place at the lowest possible cost.
- **Lean and Agility**: As an aspect of lean production, lean supply chain depicts the state of business in which there is a dynamic competition and collaboration of equals in the supply chain, aimed at adding value at minimum total cost, while maximizing end customer service and product quality. Agility in contrast, refers to the ability to reconfigure supply chains with minimum effort.

The increasing interdependence of supply chain processes and the multiplicity of actors involved in them suggest that the full benefit of information and communication technology (ICT) adoption, for example in terms of end-to-end visibility, flexibility and global optimization, can only be captured if all supply chain stakeholders are aligned and coordinated in their efforts towards digitally transforming the system.

In the context of the digital transformation, supply chains are characterized by a high degree of cyber-physical interconnection, enabled by RFID tags (or sensors) that collect business functional data (known as Big Data) for large-scale, real-time decisions to optimize supply chain performance. The large-scale deployment of RFID tags (or sensors) and Big Data Analytics (BDA) enables preventive maintenance, avoiding disruptions from unexpected failures. Moreover, the use of sensors in conjunction with artificial intelligence provides automated inventory management, thus diminishing human error, input shortages and the high cost of unnecessary inventory carrying. Similarly, the use of sensors network, Big Data and artificial Intelligence in transport operations and infrastructure management allows for real-time route and asset optimization, hence improving reliability and efficiency in logistic operations.

Many researchers have proposed the existence of a positive correlation between supply chain business performances and corporate information systems integration and coordination of business partners activities (Pal & Karkostas, 2014; Pal, 2018a; Pal, 2018b). Information has been considered increasingly as a strategic asset for supply chains, in recent decades. As a result, many supply chain businesses are investing in new technology in order to harness the quicker information sharing provision in supply chain operations (Smith, 2005; Karkkainen, 2003). RFID technology is a promising supply chain information integration mechanism. It is a mainly a wireless technology that enable automatic identification and data capturing technology that gives the opportunity to monitor objects by using a tag that carries information.

In RFID systems, there are different software and hardware requirements for data gathering and management. One of the most important components of RFID systems is tag. A tag can be identified as a microchip that has an electronic circuit and antenna on it. For the purpose of tracking the movement of goods, tags can be placed anywhere, such as containers, pallets, materials handling equipment, cases or even on individual products. Tags can be classified as passive (no battery), active (with battery) or semi-passive according to their power supply. While active tags use an energy source that is integrated to a tag physically, passive tags obtain this energy from the readers in the communication field. The other component of RFID is reader which connects the tags to external world. Although readers can be classified as portable and mobile, all of them consist of same components. In every reader, there are some parts that read tags, gather data and handle communication. While the reader antenna receives / sends the radio waves, it builds the signal and decrypts the signal which is sent from tags. RFID tags are available with different memory sizes and encoding options. These tags can also incorporate sensors to record temperature, vibration, shock, or humidity, for example, providing the ability to track and report on an object's environmental characteristics dynamically.

RFID readers and tags are being increasingly used in various application areas in supply chain operations. The RFID-based business solutions are well associated with the powerful and dynamic data capturing ability (Singh, 2003). The real-time availability of physical product identity allows information related to the current state of the product and future actions required (Smaros & Holmstrom, 2000; Choi et al., 2018; Singh, 2003; Sabbaghi & Ganesh, 2008; Kärkkäinen, 2003); and this helps in accurate asset tracking and monitoring (Bansal, 2003; Johnson, 2002; Kinsella, 2003). In the context of supply chain operations, widespread introduction of such systems represents a major suitability to examine and enhance tracking and tracing systems, process control and inventory management. In this way, RFID-based business solutions can be an enriched business information source, which can bring to different level of organizational advantages. Thus RFID-based supply chain solutions should be viewed as an information facilitator that can directly enhanced the operational decision-making.

The motivation behind RFID-based supply chain applications is to eliminate the barriers by enabling the synchronization and sharing of valuable information among trading partners. Therefore, it has become an important objective in the development of RFID technology-based applications that how to ensure the safe, reliable and simultaneous recognition of multiple targets along supply chain operations. In RFID application systems, tag is mainly used to store the information encoded by the marked object and security encryption, the reader used to read, change and verify the label information.

In recent decades, RFID technology has been used in many supply chain coordination activities: manufacturing, transportation and logistics operations. Associated with the integration of RFID technology in business and in the day-to-day life of consumers are various security and privacy concerns. The potential to reduce purchase anonymity, embedded tags in products that can reveal sensitive information, violation of location privacy with no-contact, non-line-of-sight through non-conducting material (e.g. cardboard or paper), fast read rate at up to a few meters and invisible identification which can be done indiscriminately has raised privacy concerns.

Security and privacy aspects related to RFID applications are, however, gain significant importance as the absence of good security and privacy is partially responsible for holding back the large-scale implementations that are required for global SCM. Because RFID is a wireless system without any standard security controls, tag can be read, modified, manipulated, or disabled without physical contact. The privacy issues have been frequently hyped in the media by certain individuals and groups that are against the use of RFID technology, particularly in consumer products, as they expect it to violate their privacy. Their primary concern is that RFID tags are not disabled after purchase and can therefore still be scanned and read for unwanted purposes such as obtaining sensitive personal data or locating and tracking very valuable goods. Although the primary focus of RFID issues is on privacy, supply chain businesses that use RFID-based technological solutions must be aware of security as well. Unprotected RFID tags can be scanned to obtain business sensitive information and locate valuable products but can also be disabled to easily steal products or sabotage enterprise supply chains.

Researchers have been trying to find ways to prevent these RFID security and privacy threats. Most of the published academic papers on protection capabilities for RFID are independent studies, each presenting new security and privacy techniques with unique abilities. These independent studies are, however, not very useful for enterprises to determine which threats exist for their RFID implementation and how these threats need to be countered. This chapter therefore analyzed a large sample of academic research works from which it has selected the most common RFID threats.

Although there have been other classifications of RFID threats, these classifications do not include the possible security and privacy measures that can counter these threats. Additionally, there has been very limited research in how these threats relate to the risk management of organizations, which is required to determine that how risks are to be countered. This chapter provides an overview of the well-known privacy and security threats in wireless RFID communication. This chapter consists of seven sections. Section 2 presents the background information of RFID-based technological solution. Section 3 provides the overview of simple RFID-based system. It also includes the purpose of electronic product code (EPC), object name service (ONS), and physical markup language (PML). Section 4 explains the relevance of RFID standards, which consists of international standard organization (ISO) standards and EPC-based standards. Section 5 outlines the security and privacy related issues in RFID-based technological solutions for supply chain management. Section 6 introduces the future research directions. Section 7 concludes with concluding remarks.

BACKGROUND OF RFID-BASED TECHNOLOGY

In today's digital age, real-time data stream is the supply chain's survivable lifeblood. Whether contracting new fleets in London, or recruiting supply chain operation managers in Milan, or looking at adverse weather forecasting services for smooth transportation of goods through Dorah Pass in Hindu Kush area of Asia, data keeps everything running smoothly. In RFID-based applications, data are stored into the tags. A simple RFID tag (or transponder) and its reader (or integrator / scanner) are shown in Figure 2.

RFID tags are attached to objects (e.g. men, machines, and merchandise items) for identification purpose. Each tag typically consists of an antenna constructed from a small coil of wires; a microchip used to store information digitally about the companion object; and encapsulating material to enclose the chip and the coil. Like there are distinct types of barcode, RFID tags are available with different memory sizes and encoding options. These tags can also incorporate sensors to record temperature, vibration, shock, or humidity, for example, providing the ability to track and report on an object's environmental characteristics dynamically.

A RFID reader is the device generally used to communicate with the RFID tag(s). It discharges radio frequency (RF) signals to, and receives radio waves from, the tag via an antenna or antennas. The reader converts the received radio waves into digital information that is usually passed to a backend system. Reader, either as stationary or handheld devices, consists of a transmitter, receiver, antenna, microprocessor, controller, memory and power source.

The RFID tag can come in different shapes (e.g. Paper Sticker, Barcode Tags, Plastic Card, Chemical and Heat Resistance Plastic Capsule). The tag might be even powered by very small battery to support local functions such as storing temperature reading or enhancing the reach of the radio communication, and so on. In this way, RFID tags are at the heart of an RFID system, and can be categorized as active, semi-active / semi-passive and passive in relation to power, as well as read / write and read only in terms of their memory.

The major portion of RFID tags produced today are passive RFID tags, comprised basically of a micro-circuit and an antenna. They are referred to as passive tags because the only time at which they are actively communicating is when they are within relatively close-proximity of a passive RFID tag reader.

Another type of common RFID tag in the marketplace today is known as the active RFID tag, which usually contains a battery that directly powers radio frequency communication. This onboard power source allows an active RFID tag to transmit information about itself at great range, either by constantly beaconing this information to a RFID tag reader or by transmitting only when it is prompted to do so. Active tags are usually large in size and can contain substantially more information (because of higher amounts of memory) than do pure passive tag designs. Within these basic categories of RFID tags can be found subcategories such as semi-passive RFID tags.

Figure 2. A basic RFID tag and its interrogator

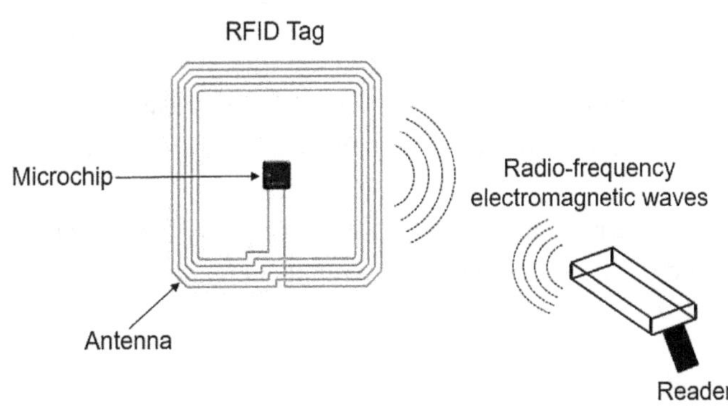

Read-Only vs. Read/Write or Smart Tags

Read-only tags have permanently encoded data set, usually in the form of a serial number. When it is in a reader field it starts to continuously broadcast this number. The flow is only one way, which means that the reader cannot access the transponder, it can only receive the information. It is important to make sure that there is only one transponder in the reader interrogation zone. Otherwise continuously transmit from the transponders creates a data collision and the reader cannot distinct one transponder from another. Read-only tags are much used as a substitute for barcodes in identification of products, containers or in other measures where very small amounts of data is needed. Read/Write-tags are wide diversity of different memory size.

The availability of low-cost RFID technology today allows for wider use beyond its traditional niche applications such as animal tagging and access control. In logistics, RFID competes with the omnipresent bar code in identifying arbitrary physical goods along the entire supply chain. Unlike bar codes, RFID allows for the unique identification of individual items and bulk reading with no line-of-sight required even under harsh environmental conditions. This gives rise to opportunities to collect the kind of fine-granular, real-time information about physical processes in the supply chain which often cannot be monitored using conventional approaches.

Besides its ever-broader diffusion across many industries, RFID has also become a fruitful research topic, not only in electrical engineering and computer science, but also in business management. Specifically, several success stories have recently been developed in the fields of information systems and operations management that not only support designers of RFID-based systems and processes but also explain how RFID can generate business value in organizations (Pal, 2019; Roussos et al, 2002; Jones et al, 2004a; Oghazi, et al., 2018; Alwadi, et al, 2017; Choi., et al., 2018). However, most of this work tends to consider RFID as a "next generation data capturing innovative technology" differing only slightly from its predecessor in its enhanced precision and the timeliness of collected data. This narrow view runs the risk of ignoring the various fundamental levels of data quality associated with the RFID-based technologies. However, the RFID-based system's data quality related issues are beyond the scope of this chapter. To initiate the exploration of the main theme of this chapter, the next section provides a basic description of an RFID-based system.

OVERVIEW OF A SIMPLE RFID SYSTEM

A simple RFID-based system solution consists of six main components as shown in Figure 3. These components are:

1. **Tag**: A tag is the data (or information) carrier part of RFID-based business solution. Generally, it contains a unique identification number, and specific *electronic product code* (EPC) programmed into the tag.
2. **Reader and Antenna**: A reader captures the data provided by the tag when tags come in the range of sensing area covered by the specific reader using its antenna (i.e. an electronic *signal receiving special circuit*).

3. **Middleware**: The middleware can be software as well as hardware dedicated to process data captured by the *tag reader*, then dispatch this information to backend servers.
4. **Backend Servers**: The backend servers hold the data collected from RFID-based application systems for processing purpose.
5. **Network Infrastructure**: This part of RFID-based solution infrastructure is basically providing data communication provision. It plays a very important role for RFID-based system solutions security and privacy related issues.
6. **Database for Tags Information**: This is the part of the enterprise information system.

Using an antenna, the information that is stored on the tag can be read or written from the tag. In a typical situation, the antenna is packaged with the transceiver into a larger structure called a reader that oversees the system's data communication and acquisition. The data that is obtained and analyzed by the reader can then be transported to a corporate data storage.

DIFFERENT TYPES OF RFID STANDARDS

Item management RFID tags play an important role in manufacturing and transport management systems in supply chain. ISO 18000 defines the air interface, collision detection mechanism and the communication protocol for item tags in different frequency bands.

SECURITY AND PRIVACY THREATS

In recent years, there have been several high-profile security breaches with global supply management. For example, retailer Target experienced a huge corporate data breach incident involving the theft of nearly 110 million customers' data and at least 40 million payment cards (Krebsonsecurity, 1014). In many of the most reported business cases, the attackers have targeted corporate strategic information,

Figure 3. A basic RFID-based technology solution architecture

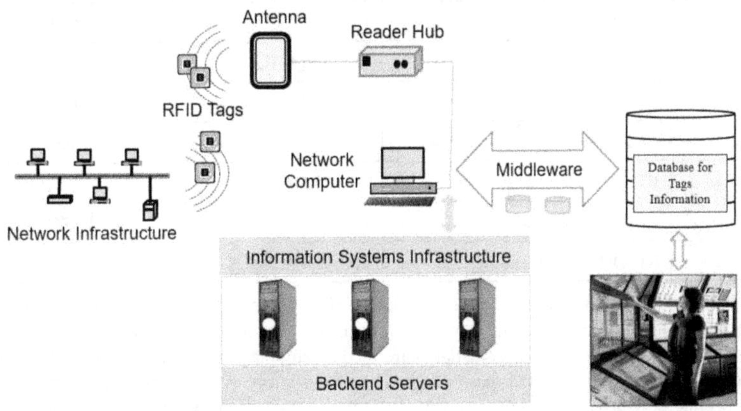

Figure 4. RFID technology standards and frequency bands

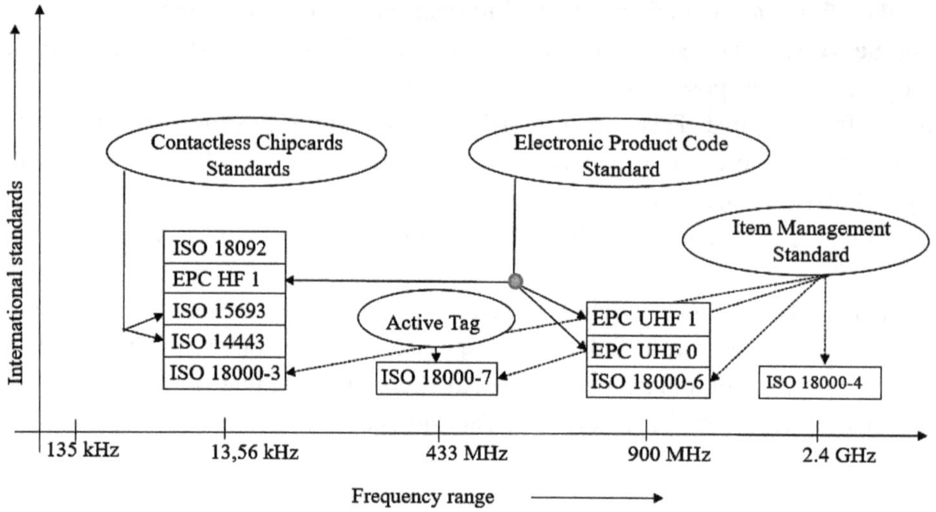

financial information, including debit and credit card details. Most of these security breaches initiated through the corporate ICT infrastructures and it includes RFID-based IS systems.

The radio communication between RFID tags and readers raises several security related issues. There are two main classes of RFID-based system's security attacks: privacy infringements and security infringements. In privacy infringements, the conspirator tries to find information from the objects by eavesdropping to the communications between the tagged-object and reader or by tracking them. In security infringements, an adversary counterfeits the behaviour of a tag or a reader for making undesirable communications. Such security attacks may target the physical tag, the communication channel between the tag and the reader, or the application or the system, which employs the RFID technology. This section classifies the existing security risks and threats according to their target into physical threats, channel

Table 1. RFID standards relevant for supply chain management

Specification	Description	Frequency	Sponsor
ePC HF Class 1	EPC tag class	13.56 MHz	EPC Global
ePC HF Class 2	EPC tag class	860-960 MHz	EPC Global
ePC HF Class 4	EPC tag class	860-960 MHz	EPC Global
ISO 14443/15693	Near Field	13.56 MHz	ISO
ISO 18000-3	RFID Air Interface	13.56 MHz	ISO
ISO 18000-4	RFID Air Interface	2.4 GHz	ISO
ISO 18000-6	RFID Air Interface	860-960 MHz	ISO

There are several existing ISO standards and proposed RFID standards (EPC – Electronic Product Code, Global) that deal with air interface protocol (how tag and readers communicate), data content (how data is organized and formatted), conformance (ways to test that products meet specifications) and applications (how standards are used on shipping labels, and so on). Table 1 shows a summary of RFID standards relevant for SCM applications. Each of these standards provides appropriate security and privacy related implementation frameworks. The following section provides a general overview of RFID-based application systems security and privacy challenges.

threats and system threats. However, threats, which supply chain RFID-based systems face today, are not limited to those listed below.

Physical Threats of RFID Systems

Physical threats are those threats that use physical means to attack the RFID-based information system's infrastructure to disable tags, modify their content data, or use these data for espionage purpose. Different types of physical threats for RFID-based system is shown in Table 2.

Channel Threats of RFID Systems

Channel threats refer to the attacks targeting the insecure channel between a reader and a tag. Since the RFID technology uses wireless means of communication between the reader and the tag, RFID systems may face eavesdropping, snooping, counterfeiting, playback, tracking threats, and other communication security issues that lead to privacy leaks. Different types of channel threats for RFID-based system is shown in Table 3.

System Threats of RFID Systems

System threats mainly refer to the attacks on the flaws existing in the authentication protocol and encryption algorithms. RFID tag data can be sensitive in terms of the security and privacy related issues. Operation managers need to make sure that the system has a proven security protocol to protect all pos-

Table 2. Types of physical threats for RFID systems

Threats Type	Description
Disabling RFID Tags	In these infringements, a predator takes advantage of the wireless nature of RFID systems in order to disable tags temporarily or permanently. To permanently disable a tag, the predator may remove the tag from one item with high price and switch it with a tag of an item with low price. The other way is sending a kill command to erase the memory of the tag. Removing the antenna or giving a high energy wave to a tag will destroy the tag permanently. To disable the tag temporarily, the predator can use a Faraday cage like an aluminum foil-lined bag in order to block electromagnetic waves from it. In other case, the predator may prevent tags from communicating with readers by generating a signal in the same range as the reader which is called active jamming.
Tag Modification	Since most RFID tags us writable memory, an adversary can take advantage of this feature to modify or delete valuable data from the memory of the tag. This information might be critical such as the data about a product's specification which any inconsistency between the data stored on the RFID tag and the corresponding tagged object may result in serious problems. In some cases, the reader may not even notice this inconsistency during the communication and thinks that the content of the tag is unaltered.
Cloning Tags	In these attacks, the attacker clones or imitates the tags after skimming the tag's information. Each RFID tag used for identification has a unique ID number. If the ID information is exposed by the attacker, the tag can easily be copied. Now that a lot of programmable read-write tags are put into use, cloning a tag is not challenging. This new tag can then act as the ordinary tag without being detected. Such cloned tags are used in counterfeiting and spoofing system-level attack.
Reverse Engineering and Physical Exploration	To maintain the tag cost low, most RFID tags are not equipped with a tamper-resistant mechanism for an estimated long period of time. An attacker with physical access to a tag can duplicate a tag with reverse engineering, and by means of physical probing, the attacker can get confidential information stored within tag. This is different from tag cloning which does not require physical exploration of the tag. However, they also are used in counterfeiting and spoofing system-level attack.

Table 3. Types of channel threats for RFID systems

Threats Type	Description
Eavesdropping	The wireless nature of RFID makes eavesdropping one of the most serious and widely deployment threats in supply chain management applications. In eavesdropping an unauthorized individual uses an antenna in order to record communications between legitimate RFID tags and readers. This type of attack can be performed in both directions tag-to-reader and reader-to-tag. In simple, the communication between reader and transponder via the air interface is monitored by intercepting and decoding the radio signals. The feasibility of this attack depends on many factors, such as the distance of the attacker from the legitimate RFID supply chain applications. This is one of the most important threats to RFID systems.
Snooping	Intruders can create a cloned tag by properly formatting an existing tag which can then be used for identity theft. For example, someone could clone an RFID transponder, using a sniffed and decrypted identifier. Then use that transponder to procure some commodity or service.
Sniffing	RFID tags are designed to be read by any generic reader device which might occur without the knowledge of the tag bearer and/or over a large distance. This could trigger the skimming of contactless identification card information theft.
Skimming	In this attack, the adversary observes the information exchanged between a legitimate tag and legitimate reader. Via the extracted data, the attacker attempts to make a cloned tag which imitates the original RFID tag. To perform this attack, the attacker does not need to have any physical access to the real tag. Skimming attack is precarious when documents like drivers' licenses are authenticated through RFID system. In these situations, the attackers observe the interactions between the RFID tag embedded in the document with the reader to make a fake document.
Replay Attack	One of the most serious threats which RFID systems face is the replay attack. The replay attack is when a malicious node or device replays that key information which is eavesdropped through the communication between reader and tag, in order to achieve deception. A typical application is when the illegal device playback the authentication between the reader and the tags, deceiving readers or tags to pass verification. Solutions to replay attacks include the use of stamp program, a one-time password and using the random number in authentication protocol or updating the ID information dynamically.
Relay Attacks	A relay attack, also known as man-in-the-middle attack, is when an attacker places an illegal device between the reader and the tag such that it can intercept the information between the two nodes and then modify it or forwarded directly to the other end. The information transmitted through illegal deices will encounter some delay, and hence, these attacks are called relay attacks.
Electromagnetic Interference	RFID channels can be the target of an adversary which aims at sabotaging the communication channel to prevent the tags from communicating with the reader. Such a communication channel threat can be either unintentional (passive interference) or intentional (active jamming). Passive Interference: Since RFID systems operate in an inherently unstable and noisy environment, their communication is rendered susceptible to possible interference and collisions from any source of radio interference such as noisy electronic generators and power switching supplies. This interference prevents accurate and efficient communication between the tags and the readers. Active Jamming: Although passive interference is usually unintentional, an attacker can take advantage of the fact that an RFID tag listens indiscriminately to all radio signals in its range. Thus, an adversary may cause electromagnetic jamming by creating a signal in the same range as the reader in order to prevent tags from communicating with readers.
Malware	The RFID tag can store small amount of data; but researchers (Rieback et al, 2006) proved that RFID malware, RFID worms and RFID viruses are a reality. The RFID malware can exploit RFID system components such as back-end databases and generic protocols.
Rogue Reader	Attacks based on malicious RFID readers are also a sincere concern which can manipulate, corrupt and hijack the connection between an authentic reader and a tag (Konidala, 2006). The rogue reader can obtain personal information about a user's performances.

sible security properties before one deploys RFID in a sensitive data exchange chain. Academics are working hard to establish a security standard for all types of RFID uses. To date, RFID industries have several standards applicable to some specific applications. One needs to be aware of these standards so that we can judge before using this technology. There are several well-established RFID security and privacy threats (Rieback, Crispo et al., 2006) such as: counterfeiting and spoofing attacks, tracing and tracking, password decoding, and denial of service (DOS) attacks.

RFID systems also may be subject to Denial of Service (DoS) attacks, which causes the system to not work properly. The attacker targets to block the reader from reading tags by using a blocker tag. Denial of service attacks are the threat to all modern communication systems. A set of mature anti-DoS solutions has developed for such threats. However, many of these solutions cannot be used in RFID systems due to the limited resources of RFID tags. For the RFID system to prevent denial of service attacks is still an area to be studied. Modern readers use anti-collision algorithms to support serving tags within their coverage areas. There are two main ant-collision algorithms; slotted ALOHA and binary search tree.

Security challenges raised by RFID can also be structured according to the traditional dimensions of information security: loss of availability, integrity and confidentiality. Availability, integrity and confidentiality are useful conceptual distinctions, but the separation is often artificial: a failure in one dimension can have consequences in another and therefore the concepts are interdependent and overlapping. Accountability and audit are often associated to these dimensions. They ensure allocation of responsibility for a security control and provide means for monitoring and measuring its effectiveness. Examples below help understand the nature of the general risk associated with each of these security dimensions and illustrate the potential consequences of attacks. Examples refer to logistic systems (e.g. tracking of containers, pallets and goods), and retail business (e.g. inventory management, check out, warranty services).

Risks Related to Tags and Readers

Many events can disrupt a RFID system. They can be categorized as interface or disruption of either the hardware, or the data components of the system. Interference or disruption to the physical components of readers and tags can be deliberate or accidental. Accidental or deliberate disruption of the radio components can be of the reader or tag or both. Deliberate attacks can be active or passive. Interference or disruption on the data components can be of the tag or reader data or both.

Availability

Availability is the assurance of timely and reliable access to data services for authorized users. It ensures that information or resources are available when required. For instance, Denial of Service (DoS) attacks target the availability of a system. Consequences of disruption of availability of typical RFID systems could be, for example, delays in processing identity document thus preventing individuals to access supply chain operation activity (e.g. warehouse access card) or work premises (e.g. access control cards), preventing automatic processing of international transportation document (e.g. loading control pass).

Threats to the availability of the physical components of a RFID system can be overt or covert. Overt attacks on tags include cutting the electrical circuit on the tag, detaching the tag from the tagged item, discharging the battery on an active tag, or masking the antenna (shielding) with a conductive material or paint. Such strategies could be pursued to evade anti-theft RFID-based systems in retail supply chain. They could also be used for privacy protection purpose: companies have developed RFID blocking wallets (for RFID cards) that is presented as privacy protective purpose.

Covert attacks can be conducted by overloading receiving components to stop them from functioning or to destroy them, for example by subjecting a passive tag to a high energy field in proximity. Hackers have demonstrated that a strong energy field generated by an inexpensive, modified, single-user camera flashlight can produce this result.

Integrity

Integrity is the underlying assurance that data has not been altered during a transmission from the point of origin to the point of reception. Consequences of loss of integrity in common RFID systems could include, for example, delays and misdirection in the supply chain or confusion in retail operations due to corrupt or erroneous information. In some cases, loss of integrity generates loss of availability. For example, corrupt access cards would not enable individuals to access the warehouse management system. If track ignition key information is altered, then it is likely that access to the track will be impossible.

A man-made unwanted signal can also be used to inject a false signal, compromising the system's integrity. Reader identity could be falsified to access, modify or kill a tag. For example, a "kill" command could be sent before the tag is read by legitimate reader and lead to fraud in a retail context or disrupt supply chain information. Tag cloning and emulation could be used to falsify the identity of goods, and, for example, replace them with cheaper item identifiers.

Confidentiality

Confidentiality is the assurance that information is accessible only to those who are authorized to have access. When the data relates to an individual, loss of confidentiality results in data protection violations. Consequences of loss of confidentiality in typical RFID systems could include, for example, stealing competitor information in the supply chain or in the retail environment, stealing a track by gaining access to electronic key information and cloning the chip. Unauthorized remote access to data is sometimes also called "skimming".

Any system based on radio technology is susceptible to eavesdropping of the radio signal between transmitter and receiver, thus raising confidentiality challenges (as well as integrity challenges if the data can be reinjected). RFID systems based on magnetic induction also generate radio waves that an attacker equipped with the appropriate radio equipment could, in theory, intercept. However, although theoretically possible, it is practically improbable because the energy levels would be relatively low and would be covered by noise, forcing the attacker to operate at short distance of the rags and reader.

RFID eavesdropping can be both passive and active. The attacker (or "interceptor") may actively send a signal to the tag to get a response, or simply passively listen to the response prompted by a reader activating the tag. Some tags can only reply with data (e.g. an identification number). More "intelligent" tags can send back a processed response akin to being actively interrogated with the objective of exploring a vulnerability.

Eavesdropping on the transmission can reveal the existence of a communication between RFID devices potentially leading to the disclosure of valuable information for the attacker (e.g. the number of goods arriving in a warehouse or the presence of an individual in each area). It enables tracking of the tag and the attached object. It can also reveal the content of the communications thus enabling unauthorized access to potentially valuable business, competitive or personal information and enabling other attacks such as unwanted modification of the data or "man in the middle attack" (where the interceptor modifies the data exchanged between the tag and the reader).

Moreover, a RFID system is vulnerable to Denial-of-Service (DoS) attack, privacy, and traceability. Applying a combination of these basic threats, offenders can try to clone RFID transponders, to prevent reading processes (e.g. in automated checkout systems), to obscure theft or manipulate data in production or business processes. In order to avoid misuse of RFID technologies, both the system reliability and system security must be optimized. Generally, the following security requirements must be considered in the context of RFID systems: functional reliability, authenticity, confidentiality, integrity, availability, liability, and data privacy.

Currently existing RFID transponders provide just very limited security functionalities. Further definitions of RFID tags as well as data and communication standards (e.g. EPC Class2) will address this issue. Therefore, new methods for authentication and authorization, encryption, integrity protection, pseudonymization, tag deactivation and prevention of unauthorized read/write operations must be addressed in ongoing research. In addition, the middleware must process the incoming RFID data intelligently to integrate into the business application. Major challenges of this layer are collision in singulation, data redundancy and data noise.

Since tag costs (e.g. tag complexity) must be kept in certain limits, new effective methods for mutual authentication of transponders and interrogators need to be developed. In this context the effectiveness of the communication in a limited timeframe must be considered. Classical methods like hased-based authentication demand a lot of computation effort on the reader side and even on the transponder side a comparatively large hardware effort is necessary. Symmetric encryption methods are not suitable, since the effort for a secure key management is comparatively high and not affordable at the required limited transponder complexities / costs. Furthermore, there is a certain risk that secret keys could be revealed by reverse engineering and this in effect would compromise the whole security concept. Therefore, new lightweight cryptography methods are required, including effective methods for the on-tag generation of random numbers and the computation of hash functions. Concepts as physically unclonable functions may show a way out of the security at high implementation cost dilemma.

RESEARCH CHALLENGES IN RFID SECURITY

Even though research on RFID security and privacy protection has advanced in recent years, but it is not still adequate. For practical deployment of RFID in wide scale, many issues need to be resolved. In addition, there are almost no works done for security in chip-less RFID. The followings are some of the research challenges in RFID security as noted by Duc et al. (2009):

Cryptography plays the core role in security protocol as described above. There are very strong and well-proven cryptographic techniques for wired networks, and in recent times, cryptographic algorithms for wireless networks have been significantly improved. However, RFID tags have very limited processing power and therefore, for practical deployment, one need cryptographic algorithms that are lightweight and fast, but at the same time possess strong cryptographic properties. The lack of the study on lightweight cryptographic primitives includes design of new primitives, analysis of the security of new primitives and their efficient implementation (Duc et al., 2009). Though few lightweight cryptographic algorithms are proposed in literature, much research needs to be done to evaluate them through cryptanalysis and investigate new algorithms for strong protection. As new types of attacks are reported, research need to focus to counter those attacks. Another important aspect that plays vital roles in cryptography is the regeneration of pseudorandom number so that it generation can be predicted. Further research needs to do to generate random number that satisfies the requirement of RFID.

FUTURE RESEARCH DIRECTIONS

In the physical world, human beings and animals are adept at sensing the environment. Their ears prick up strange sounds; their body hair gets spiked-up in the sense of danger; their skin tingles when something threatened their security of existence; they notice subtle body language in a stranger that makes them suspicious and venerable. Human being and intelligent animals are hardwired smartly to recognize the body-language of their neighbors; yet, most of RFID-based supply chain management systems operational environment's object identification occurs digitally, by voice, text, or using other biometric recognition technique. In addition, currently most of RFID-based supply chain operational community don't have full solutions when someone tries to extract potential business information by snooping from a distance or sending a group of smart armies of malware or block the whole network by jamming electromagnetically. In future, this research will review most of these challenges and will try bringing together potential solutions for naive operational managers.

CONCLUSION

This chapter focused on the issues and potential solutions for a range of RFID technology have generated significant interest in supply chain management. The use of RFID has grown across a variety of core industries, such as logistics, manufacturing, and retail. Although each application has its own specific requirements, security vulnerability will be always a major concern when deploying RFID applications. Like the Internet or mobile telephony, RFID is a wireless network technology. While the non-contact and non-line-of-sight properties of RFID systems increase the convenience and efficiency of their applications, these properties also increase the system's vulnerability.

With the increasing usages of RFID, such as manufacturing, retail transactions and consumer products, potential security threats and compliance risks in the future are enormous. It is necessary to pay attention to standardization of RFID systems.

Moreover, risk vary according to the type of technology used to perform a specific function. RFID is an umbrella concept for a variety of technologies with distinctive characteristics. The risks mentioned above are theoretically applicable to all RFID systems. However, on a continuum of risks, the likeliness of certain risks is higher in some technical configurations than in others. Parameters such as the use of passive or active tags, of electromagnetic induction or radio-wave communications, of read only or read-write capable memory influence the degree of likeliness that certain risks materialize.

A well-designed RFID policy can reduce the risk of attacks. When dealing with security and risk management, policy decisions also play a significant role in the security of an RFID system. An RFID security policy is a document that states how an organization plans to protect its physical RFID devices and information data assets. Since, eventually, new threats will appear, an RFID security policy should be considered a "living" document that needs to be continuously updated as the RFID technology and implementation requirements change. The policy also needs to consider how the users will be trained in the proper use of RFID and explain how security measures will be carried out and enforced.

REFERENCES

Alwadi, A., Gawanmeh, A., Parvin, S., & Al-Karaki, J. N. (2017). Smart Solutions for RFID Based Inventory Management Systems. *Survey, 18*(4), 347–360.

Alwadi, A., Kilby, J., & Gawanmeh, A. (2017). Tracking and Automating a Library System Using Radio Frequency Identification Technology. *International Journal on Smart Sensing and Intelligent Systems, 10*(2), 425–450. doi:10.21307/ijssis-2017-219

Baird, J. L. (1928). Improvements in or relating to apparatus for transmitting views or images to a distance. *Patent# GB292* 185.

Bansal, R. (2003). Coming Soon to a Wal-Mart Near You. *IEEE Antennas & Propagation Magazine, 45*(6), 105–106. doi:10.1109/MAP.2003.1282186

Choi, T., Wing-Kwan, Y., Cheng, T. C. E., & Yue, X. (2018). Optimal Scheduling, Coordination, and the Value of RFID Technology in Garment Manufacturing Supply Chains. *IEEE Transactions on Engineering Management, 65*(1), 72–84. doi:10.1109/TEM.2017.2739799

Choi, T., Wing-Kwan, Y., Cheng, T. C. E., & Yue, X. (2018). Optimal Scheduling, Coordination, and the Value of RFID Technology in Garment Manufacturing Supply Chains. *IEEE Transactions on Engineering Management, 65*(1), 72–84. doi:10.1109/TEM.2017.2739799

Corbett, C. J., & Karmarkar, U. S. (2001). Competition and Structure in Serial Supply Chains with Deterministic Demand. *Management Science, 47*(7), 966–978. doi:10.1287/mnsc.47.7.966.9799

Duc, D. N., Lee, H., Konidala, D. M., & Kim, K. (2009). Open issues in RFID security. *Proceedings of the International Conference on Internet Technology and Security Transactions (ICITST)*. Academic Press.

Harman, P. M. (1998). *The natural philosophy of James Clerk Maxwell*. Cambridge, England: Cambridge University Press.

Johnson, D. (2002). RFID tags improve tracking, quality on Ford line in Mexico. *Control Engineering, 49*(11), 16.

Kärkkäinen, M. (2003). Increasing efficiency in the supply chain for short shelf life goods using RFID tagging. *International Journal of Retail & Distribution Management, 31*(10), 529–536. doi:10.1108/09590550310497058

Kinsella, B. (2003). The Wal-Mart factors. *Industrial Engineering (American Institute of Industrial Engineers), 32–36*(November).

Knospe, H., & Pohl, H. (2004). RFID security. *Information Security Technical Report, 9*(4), 39–50. doi:10.1016/S1363-4127(05)70039-X

Krebsonsecurity. (2014). Target hackers broke in via HVAC company. Retrieved from http://krebsonsecurity.com/2014/02/target-hackers-broke-in-via-hvac-company

Luckett, D. (2004). The supply chain. *BT Technology Journal, 22*(3), 50–55. doi:10.1023/B:BTTJ.0000047119.22852.38

Oghazi, P., Rad, F. F., Karlsson, K., & Haftor, H. (2018). RFID and ERP Systems in Supply Chain Management. *European Journal of Management and Business Economics*, 171-182

Pal, K. (2017). Supply Chain Coordination Based on Web Services. In H. K. Chan, N. Subraanian, & M. D. Abdulrahman (Eds.), *Supply Chain Management in the Big Data Era* (pp. 137–171). Hershey, PA: IGI Global. doi:10.4018/978-1-5225-0956-1.ch009

Pal, K. (2018). A Big Data Framework for Decision Making in Supply Chain. In M.V. Kumar, G.D. Putnik, K. Jayakrishna et al. (Eds.), Emerging Applications in Supply Chains for Sustainable Business Development (Ch. 1, pp. 1-22). Hershey PA: IGI Global Publication.

Pal, K. (2019). Quality Assurance Issues for Big Data Applications in Supply Chain Management. In P.K. Gupta, T. Oren, & M. Singh (Eds.), Predictive Intelligence Using Big Data and the Internet of Things (Ch. 3, pp. 51-76). Hershey PA: IGI Global.

Rieback, M. R., Crispo, B., & Tanenbaum, A. S. (2006). Is your cat infected with a computer virus? Proceedings of PerCon (pp. 169-179). Academic Press. doi:10.1109/PERCOM.2006.32

Roberti, M. (2005). Financing for RFID Prison System. *RFID Journal*, December 31.

Sabbaghi, A., & Ganesh, V. (2008). Effectiveness and Efficiency of RFID Technology in Supply Chain Management: Strategic Values and Challenges. *Journal of Theoretical and Applied Electronic Commerce Research*, *3*(2), 71–81. doi:10.4067/S0718-18762008000100007

Sabbaghi, A., & Ganesh, V. (2008). Effectiveness and Efficiency of RFID Technology in Supply Chain Management: Strategic Values and Challenges. *Journal of Theoretical and Applied Electronic Commerce Research*, *3*(2), 71–81. doi:10.4067/S0718-18762008000100007

Singh, N. (2003). Emerging technologies to support supply chain management. *Communications of the ACM*, *46*(9), 243–247. doi:10.1145/903893.903943

Stockman, H. (1948). Communication by Means of Reflected Power. *Proceedings of the Institute of Radio Engineers*. Academic Press.

KEY TERMS AND DEFINITIONS

Active Tag: A tag with its own battery that can initiate communications.

Auto-ID: Automatic Identification (Auto-ID) systems automatically identify physical objects through optical, electromagnetic, or chemical means.

EAS: Electronic Article Surveillance. A radio frequency device that announces its presence but contains no unique identifying data. EAS tags are frequently attached to books or compact discs.

EPC: Electronic Product Code. A low-cost RFID tag designed for consumer products as a replacement for the UPC (Universal Product Code).

HF: High Frequency; 13.56 MHz.

LF: Low Frequency; 120-140 KHz.

Passive Tag: A tag with no on-board power source that harvests its energy from a reader-provided RF signal.

Reader: An RFID transceiver, providing real and possible write access to RFID tags.

RF: Radio Frequency.

RFID: Radio Frequency Identification. Describes a broad spectrum of devices and technologies and is used to refer both to individual tags and overall systems.

Skimming: An attack where an adversary wirelessly reads data from a RFID tag that enables forgery or cloning.

Supply Chain Management: A supply chain consists of a network of *key business processes* and facilities, involving end users and suppliers that provide products, services and information. In this chain management, improving the efficiency of the overall chain is an influential factor; and it needs at least four important strategic issues to be considered: supply chain network design, capacity planning, risk assessment and management, and performances monitoring and measurement. The coordination of these huge business processes and their performance improvement are the main objectives of a supply chain management system.

Tag: An RFID transponder, typically consisting of an RF coupling element and a microchip that carries identifying data. Tag functionality may range from simple identification to being able to form ad hoc networks.

Chapter 5
Ontology-Assisted Enterprise Information Systems Integration in Manufacturing Supply Chain

Kamalendu Pal
City, University of London, UK

ABSTRACT

Manufacturing communities around the globe are eagerly witnessing the recent developments in semantic web technology (SWT). This technology combines a set of new mechanisms with grounded knowledge representation techniques to address the needs of formal information modelling and reasoning for web-based services. This chapter provides a high-level summary of SWT to help better understand the impact that this technology will have on wider enterprise information architectures. In many cases it also reuses familiar concepts with a new twist. For example, "ontologies" for "data dictionaries" and "semantic models" for "data models." This chapter presents the usefulness of a proposed architecture by applying a theory to integrating data from multiple heterogeneous sources which entails dealing with semantic mapping between source schema and a resource description framework (RDF) ontology described declaratively using specific query language (i.e. SPARQL) queries. Finally, the semantic of query rewriting is further discussed and a query rewriting algorithm is presented.

INTRODUCTION

All manufacturing business today appreciates the value and consequence of building an effective supply chain as part of enterprise proliferation and profitability. There exist different types of the industry-specific supply chain (e.g. automotive, pharmaceutical, apparel, agriculture). In simple, the supply chain is a system with organization, people, technology, activity, information and resource involved in, to deliver a product or service from suppliers to customers. Supply chain activity transforms natural resources, raw materials and components into final products, and delivers them to customers. A network is composed of the enterprises and enterprise departments involved in this process. The most important requirements of supply chain operation are to minimize the inventory, create seamless material and information

DOI: 10.4018/978-1-7998-1831-1.ch005

flow, effective communication must exist among the market, sale, purchase, manufacturing plan and control, appropriate customer delivery service, after-sales service, and so on. Therefore, a supply chain is a network of facilities and distribution options that performs the functions of material procurement, the transformation of these materials into intermediate and finished products, and distribution of these finished products to customers (Pal, 2017). Supply Chain Management (SCM) aims at improving the allocation, management and control of logistical resources. In this way, manufacturing SCM is a set of synchronized activities for integrating suppliers, manufacturers, transporters, and efficient customer service so that the right product or service is delivered at the right quantities, at the right time, to the right places. The ultimate objective of SCM is to achieve sustainable competitive advantage (Pal, 2019).

The first signs of SCM were perceptible in Toyota Motor Manufacturing's Just-In-Time (JIT) procurement system. Particularly, JIT was used to control suppliers to the factory just in the right quantities, to the right location, and at the right time, in order to optimize system-wide costs and customer affordability. The main goal was to reduce inventory level drastically, and to regulate the suppliers' interaction with the production line more effectively. It consisted of two distinct flows through the supply chain organizations: material and information. The scope of the supply chain begins with the source of supply and ends at the point of consumption. It extends much further than simply a concern with the physical movement of materials. Equal emphasis is given to supplier management, purchasing, inventory-management, manufacturing management, facilities planning, customer service, information flow, transport and physical distribution. Some of the important business processes, along supply chain, are shown in Figure 1.

The ultimate objective with the implementation of SCM, suppliers and customers are viewed as partners and their relationship becomes a cooperative one as enterprises in the supply chain recognize that coordination among partners within the supply chain is a key factor of success. In order to operate a supply chain efficiently in a cooperative manner, all related functions across the supply chain must operate in an integrated way the various partners within the supply chain must be efficient with respect to service provisions. This balances Constraint Satisfaction Problem (CSP) with appropriate customer service, minimum inventory holding cost and optimal unit cost. In this complex CSP environment, the design and operation of an effective supply chain is of fundamental importance for global manufacturing business.

It is worth noting that purchasing process does not finish when the customer places an order using an existing sales channel. Customer queries, before or after order placement, are inevitable. At the same time, the seller might want to contact customers with purchase confirmation and shipping information. Customer service encompasses all points of contact between the seller and the customer and is an important output of SCM. It results from the accumulated value of all business processes along the supply chain. These business processes are responsible for offering an acceptable level of customer service. Moreover, these business processes are also interdependent, if one business function fails to provide the expected level of customer service the chain is disrupted, and the scheduled workload in other areas is destabilized thereby jeopardizing customer satisfaction (Pal, 2018) (Pal, 2019).

To provide better quality of customer service at no additional cost or workload, all business processes along the supply chain must be balanced. This requires trade-offs throughout the supply chain. It is essential to think in terms of a single interconnected chain rather than narrow functional business processes when considering effective trade-offs. Seamless integration along the supply chain is challenged when there is a conflict between a company's functional behaviours and objectives, as is often the case. For example, suppliers typically want manufacturers to purchasing in bulk quantities, in stable volumes, and with flexible delivery dates. However, although most manufacturers desire long production shifts, they

Figure 1. Diagrammatic representation of supply chain business process

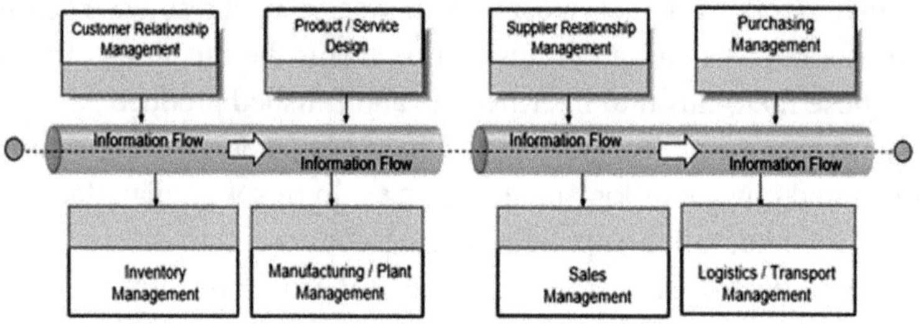

need to flexible to their customers' requirements and fluctuating market demands. Thus, the suppliers' objectives are in direct conflict with the manufacturers' wish for flexibility. Indeed, since manufacturing decisions are typically made without accurate information about customer demand, the ability of manufacturers to match supply and demand depends largely on their ability to change supply volume as information about demand arrives. In the same way, the manufacturers' goal of making bulk production batches typically conflicts with the objects of both distribution and warehouse facilities layout to reduce material inventory. To make the situation worse, this latter goal of reducing inventory typically implies an extra cost in transportation and distribution.

System fluctuations over time are also critical criteria that need to be considered. Even when the requisition is accurately known because of prior contractual agreements, say strategic decisions need to take demand and cost variations due to changes in market trends, market and sales logistics, competitive movement and the like. These time-varying demand and cost criteria make it more complex to figure out the most appropriate supply chain strategy – the one that optimizes system-wide management costs and complies with customer needs. Global optimization indicates that it is not only essential to optimize across supply chain resources, but also across business activities connected with the supply chain.

Managing a supply chain involves numerous decisions about the flow of information, product, funds, and coordination. Supply chain management (SCM) has been instrumental in connecting and smoothing business activities as well as forming various kinds of business relationships e.g. Customer Relationship Management (CRM), Supplier Relationship Management (SRM), among supply chain stakeholders. In this way, SCM is an integrative mechanism to manage the total flow of a distribution channel from supplier level to production, distribution and then ultimately to the end customer. The aim is to achieve goals related to total system performance rather than optimization of a single phase in a logistics chain. The aim of SCM is to enhance productivity by reducing total inventory level and cycle time for orders. It is essential for supply chain business-partners to create a network that is agile and able to respond rapidly to unpredictable changes in demand. To achieve cooperation among business partners is very essential.

SCM systems utilize modern Information and Communication Technologies (ICT) to acquire, interpret, retain, and distribute information. The software applications of SCM are ready-made packages, usually targeting a particular set of tasks, e.g., tracking product-related information during the transportation process. These ready-made package-software applications are mass-customized products that ignore the specific requirements of a certain business sector, and so they are quite problematic. The problem of the appropriate IT solutions for supporting collaboration between supply chain business-partners is not new and it has been approached with several standards and protocols implemented in numerous enterprise information systems. The application like ERP (Enterprise Resource Planning), CRM (Customer Re-

Figure 2. Heterogeneous data sources in a global supply chain

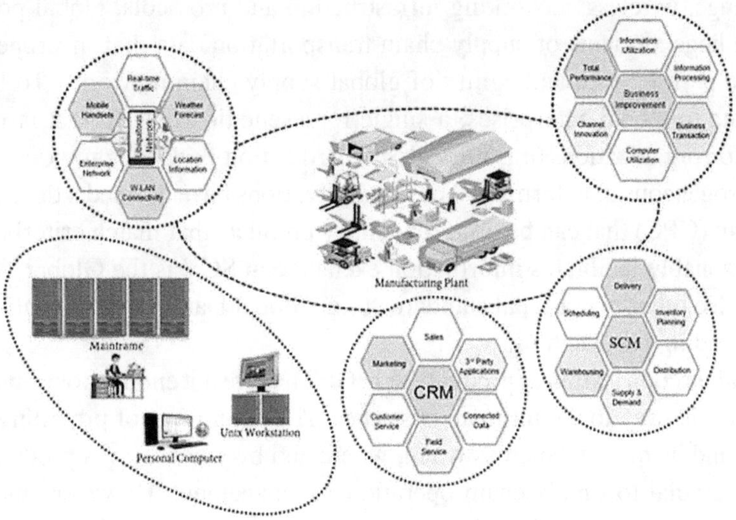

lationship Management), and WMS (Warehouse Management System) contains valuable data that can be utilized by the decision support systems (DSS). Moreover, the digital transformation of business and society presents enormous growth opportunities offered by technologies such as the Internet of Things (IoT), Big Data, advanced manufacturing, blockchain technologies, and artificial intelligence. This digital transformation is characterized by a fusion of advanced technologies and the integration of physical and digital systems, the predominance of innovative business models and new processes, and the creation of smart products and services (Baumgaertel, Ehm & Kasprzik, 2018). However, information integration problems often exist. Specifically, this problem is often attributed to heterogeneous hardware, different data formats, and pressure to improve a system's performance, as shown in Figure 2. The problem with information interchange is related to issues of data exchange between business-partners operating independently designed information systems. From a logistics point of view the issues in supply chain visibility (SCV) is the ability of parts, components or products in transit to be tracked from the manufacturer to their destination and traced from the customer to the manufacturer.

The aim of enhancing SCV is to improve the traceability of business process actions and reactions along the global supply chain. Innovative SCV technology promotes near real-time response to global supply chain business environments; and it improves operational performance of many high-tech industries (e.g. pharmaceutical, automotive, apparel). The objective of such data collection and analysis in the supply chain is to provide greater visibility of business operations. In other words, this greater visibility can be interpreted as "the ability of supply chain mangers to know exactly where things are at any point in time, or where they have been, and why". Collection of this data is not enough; the enterprises need new software tools, business process management practices, and properly trained personnel to interpret the hidden contextual meaning in the data.

In addition, the important challenges in supporting huge heterogeneous data integration in global supply networks are: (i) increasing number of business alliance partners due to globalization of business processes; (ii) different business practice and infrastructure facilities within participating business-partners; and (iii) differences in data exchange formats and standards among business-partners. Moreover, data capture and transmission mechanisms (e.g. bar coding, radio frequency identification technique, electronic data interchange, wireless networking infrastructure and protocols, global positioning system's capability) produce huge amounts of supply chain transportation data that, if properly controlled and shared, can enhance performance and agility of global supply chain networks. To harness this value-added service of data to global enterprises, a single representation data format is very essential. The global standardization organizations (e.g. Global Standardization 1 – GS1) provides a partial solution to the problem of heterogeneous data formats. These organizations have defined a detailed Global Product Classification system (GPC) that can be used to identify products that match criteria. The example of a standard, which reasonably facilitates information exchange in SCV is the Global Trade Item Number (GTIN). There are also other operational standards (e.g. Global Data Synchronization Network) available for improving trading partnerships.

However, this unique codification of products (e.g. GETIN) are not enough for transferring knowledge from the producer all the way to the ultimate customer. The main issue of providing detailed information on the product and its history (such as when, where and by whom it was produced, packaged, and transported) can be crucial to supply chain operational management. However, integrating data from multiple heterogeneous sources requires dealing with different data models, schemas, and query languages. The semantic web has provided several new methods for data integration. This chapter provides an integration of relational database and XML data with the help of ontology.

Section 2 of the chapter outlines the background of data integration problem. Section 3 describes the structure of a data integration framework. Section 4 presents different concepts of semantic mapping; it includes an RDF graph model, relational data model, and XML data model. In addition, some principles for semantic mapping between data source schemas and ontology have been introduced. After this, an algorithmic semantic query rewriting and optimization techniques are discussed. Section 6 concludes the chapter with summary remarks.

BACKGROUND OF DATA INTEGRATION PROBLEM

In today's digital age, data is the supply chain operation's lifeblood. Whether recruiting new staffs in London or New York, gaining new customers, or closing the accounting books at the end of every day, data keeps everything running smoothly. It enables business organizations to innovate, make smart and timely decisions, and maintain a quality level of supply chain operational service. To provide a certain level of operational service, business needs to capture data, clean them if it is appropriate, integrate different business activities data, and process them to make corporate decisions. Therefore, a database management system is an essential part of a supply chain's day-to-day operations.

Modern database management systems are heavily influenced by the relation database model (Codd, 1970). In this model, a database schema represents the logical configuration of all or part of a relational database. It can exist both as a visual representation and a set of operational conditions known as integrity constraints that govern a database. These constraints are expressed in a data definition language, such as SQL (Structured Query Language). As part of a central repository all the data and metadata, a database schema keeps track of the classes of objects (or entities) that make up the database, including tables or relations, different views, stored procedures, and more.

Typically, a database schema conveys – the logical constraints that apply to the stored data in the database. The mechanism of creating a database schema is called data modelling. When following the three-schema approach to database design, this step would follow the certain of a conceptual schema. Conceptual schemas focus on an organization's information. There are two main kinds of database schema:

- A logical database schema conveys the logical constraints that apply to the stored data. It may define integrity constraints, views, and tables.
- A physical database schema lays out how data is stored physically on a storage system in terms of files and indices.

At the most basic level, a database schema indicates which tables or relations make-up the database, as well as the attributes including on each table. Thus, database schema is an essential part of heterogenous information system.

Database technology was introduced in business communities since the late 1960s to support (initially rather simple using third generation procedural programming languages - e.g. COBOL (Common Business Oriented Language) and RPG (Report Program Generator)) business information providing applications. The software of the day provided a 'file-oriented record processing' model. Typical programs sequentially read several input files and produced new files as output. COBOL and several other programming languages were designed to make it easy to define these record-oriented sequential tasks. As the number of business applications and data repositories rapidly grew, the need for integrated data became apparent. Consequently, first integration approaches in the form of multi-database systems (Hurson& Bright, 1991) were designed around 1980s – e.g. MULTIBASE (Landers & Rosenberg, 1982). This was a first exemplar in a notable history of research in data integration. The development continued over mediators (e.g., Garlic (Carey et al., 1995)) and agent systems (e.g., InfoSleuth (Bayardo et al., 1997)), to ontology-based (e.g., OBSERVER (Mena et al., 1996)), peer-to-peer (P2P) (e.g., Piazza (Halevy et al., 2003)), and web service-based integration approaches (e.g., Active XML (Abiteboul et al., 2002); SWSDF (Pal, 2017)).

In many supply chain business applications, the number of data providers and amount of available data is increasing. Heterogenous data integration is now a very commonly used notion in supply chain operation management research. However, it appears that the problem of data integration resided initially in the lack of formalisms for representing data, and subsequently in the lack of standardization in the use of data formalism representing the same datasets. This latter case is known as logical heterogeneity (Hull, 1997) or semantic heterogeneity (Ceri & Widom, 1993) and stems from the fact that the same information resided in multiple overlapping data storages represented with different formats or different instance values.

Structured datasets from supply chain business customers can be used to explore some of the research issues in data heterogeneity. General concepts like name and surname may be represented by two attributes by some information system developers. Others may represent them in one unique attribute Name + Surname. The two cases highlight how even a simple and well understood concept can result in different representations. It is worth noting that the conceptual, logical and physical representation of data and related information is obtained after the requirements and specifications pass through users, business analysts, data specialists, systems developers and other software project workers, each having their level of knowledge and preference for data representation. In this way, different representations are possible for the same conceptual pieces of data; and all of them could be accurate and complete. This highlights the standardization problem in data representation.

In this way, there is a spectrum of approaches for resolving the problems arising from semantic heterogeneity. One key method is to standardize databases and data representations through the development of a global data model. Many practitioners and academics have provided detailed discussions on various aspects of semantic heterogeneity (Kim et al., 1993; Hammer & McLeod, 1993; Batini et al., 1986; Kim &Seo, 1991) for database systems. Most of these discussions highlight one crucial point that the types of heterogeneity between data sources to be integrated is a first step in developing an integration solution. Many other issues were identified e.g., (i) getting an integrated view of overlapping data sets; (ii) identification and specification of the relationship between replicated data; and (iii) keeping replicated data 'synchronized'.

Given that the schemas and instances of data sources to be integrated are usually different, any integration system must also handle instance inconsistency. This is generally achieved using two components: wrapper (Wiederhold, 1992) and mediator (Ullman, 1997). A wrapper is a tool that can access a local data source, and then extract and translate data from the local schema to an external global schema. An integration system is usually composed by two or more wrappers. They feed data to a mediator that reconciles and presents the local data to the final user, or to other systems as a unified view. The unified view is the only point of access for the final users who see the unified view independent of the integrated sources. This provides simplicity to the end user who can better focus on integration and manipulation of the repositories, rather than on the integration process.

The relation between the global schema and local schemas is characterized mainly by two types of approaches: Local As View (LAV) or Global As View (GAV) (Lenzerini, 2002). The first asserts that each local schema is a view over the global schema and can be used whenever there is a global reference model for the local repositories. The second considers the global schema as a view over the local schemas (Halevy, 2001). The global schema is obtained by analyzing, mapping and synthesizing the local schemas into a unique reference model that can simultaneously express the concepts of each independent data source, and offer users a transparent, unified overview.

A Global-Local As View (GLAV) approach (Friedman, Levy, & Millstein, 1999) has also been proposed to combine the advantages of the previous approaches. Without analyzing the approaches further, it must be said that the type of approach changes the semantics of the querying and integration processes.

One other important aspect of data integration is the architecture of the integration framework. Simplifying the numerous types of proposed architectures, one may conclude that the most frequently adopted solution for data integration is a central integration repository, federation approaches, and data exchange scenarios for distributed networks.

In the first case, a global point of reference is considered for integrating data from various sources (Beneventano et al., 2000; Chawathe et al., 1994; Carey et al., 1995; Roth et al., 1996), either through a materialized view, or through a mediator that accesses the individual data sources and extracts it from subsequent reconciliation and integration. A federation approach is like the first case but is focused on maintaining the physical division of the data sources. In distributed networks (think data sources in peer-to-peer networks) data is exchanged between single data sources (Arenas et al., 2013; Fagin et al., 2005b).

One of the main differences between building a central integration repository and data exchange is that, in the first case a single point of access is offered to the user who can seamlessly access the combined data sources through the interface (known as global schema architecture). In the second case, data is exchanged directly between two points of interest through a series of mappings and translation rules that convert data from the source schema to the target schema for critical discussion (Fagin et al., 2005a; Fagin et al., 2005b).

The different kinds of integration architecture radically change the semantics of the integration process. Without extending the discussion to the rigors of critical analysis, it must be emphasized that the main advantage of a mediator architecture is the unified global view through which the users have the possibility of accessing all data sources contemporaneously with a single query. The main disadvantage is the complexity required to analyze and synthesize a global scheme, as in real-life cases data integration systems need to combine a multitude of sources. Such systems rely on wrappers that can interface the mediator to each local source for rewriting the global query into a local query that is compatible to the schema of the local source and converting the result to the schema required by the mediator. The process introduces an overhead in accessing the data sources, both when rewriting the global query, as well as when translating and integrating data from all the data sources into a unified answer.

In this way, the reason for integration is twofold: First, given a set of existing information systems, an integrated view can be created to facilitate information access and reuse through a single information access point. Second, given a certain information need, data from different complementing information systems is combined to gain a more comprehensive basis to satisfy the need (Ziegler & Dittrich, 2007).

The other issue of heterogeneous data integration is data cleaning, also known as data cleansing or scrubbing, deals with detecting and removing errors and inconsistencies from data to improve the quality of data. These problems are present in single data collections, such as files and databases, and it can be due to misspellings during data entry, missing information or other invalid data. When multiple data sources need to be integrated, for example, in data warehouses, federated database systems or global web-based information systems, the need for data cleaning increases significantly. This is because the sources often contain redundant data in different representations. To provide access to accurate and consistent data, consolidation of different data representations and elimination of duplicate information become necessary. Data warehouses (Chaudhuri &Dayal, 1997) (Jarke et al., 2000) require and provide extensive support for data cleaning. They load and continuously refresh huge amounts of data from a variety of sources so the probability that some of the sources contain "dirty data" is high. Furthermore, data warehouses are used for decision making, so that the correctness of their data is vital to avoid wrong conclusions.

In the database design, semantics can be considered as end-user's interpretation of data and schema items according to their conceptualization of the worlds in a problem. In heterogeneous data integration, the type of semantics discussed is almost always real-world semantics that are involved with the mapping of entities in the model or computational world onto the real world; and it concern human interpretation of purpose and use of data and information. In this context, semantic integration is the activity of grouping, combing or completing data from various sources by considering explicit and precise data semantics to avoid that semantically incompatible data is structurally merged. That is, semantic integration must ensure that only data related to the same or very closely similar real-world object or concept is merged. A precondition for this is to resolve semantic ambiguity regarding integratable data by explicit metadata to find-out all relevant implications assumptions and underlying context information.

In one way to resolve semantic heterogeneity in the supply chain database research is to exhaustively specify the intended real-world semantics of all data and schema elements. However, it is impossible to completely define what a data or schema element denotes or means in the database world (Shethet al., 1993). Therefore, database schemas do typically not provide enough explicit semantics to interpret data always consistently and unambiguously (Sheth & Larson, 1990). These difficulties are further worsened by the fact that semantics may be embodied in data models, conceptual schemas, application programs, the data itself, and the minds of users. Moreover, there are no absolute semantics that are valid for all potential users; semantics are relatives ((Garcia-Solaco et al., 1996). These problems regarding semantics are triggering many heterogeneous data integration important research challenges.

An important step during database integration is to determine relations between schema elements from different local databases. Finding meaningful relations is in turn based on a sound understanding of the meaning of the schema elements, that is, their semantics. To that end, one can rely on formal ontologies (Gruber, 1993) available for local schemas. In simple, ontology can be defined as explicit, formal descriptions of concepts and their relationships that exist in a certain universe of discourse, together with a shared vocabulary to refer to these concepts – can contribute to solve the problems of semantic heterogeneity. Compared with other classification schemes, such as taxonomies, thesauri, or keywords, ontologies allow more complete and more precise domain models (Huhns& Singh, 1997). With respect to an ontology a user group commits to, the semantics of data provided by data sources for integration can be made explicit. Based on this shared understanding, the danger of semantic heterogeneity can be reduced. For instance, ontologies can be applied in the application of the Semantic Web to explicitly connected information from web documents to its definition and context in machine-processable form; therefore, semantic services, such as semantic document retrieval, can be provided.

In database research, single domain models and ontologies were first applied to overcome semantic heterogeneity. AS in SIM (Abiteboul & Polyzotis, 2007), a domain model is used as a single ontology to which the contents of data sources are mapped. Thus, queries expressed in terms of the global ontology can be asked. In general, single-ontology approaches are useful for integration problem where all information sources to be integrated provide nearly the same view on a domain (Wache et al., 2001). In case the domain views of the sources differ, finding a common view becomes difficult. To overcome this problem, multi-ontology approaches (e.g. OBSERVER (Mena et al., 1996)) describes each data source with its own ontology; then, these local ontologies must be mapped, either to a global ontology or between each other, to establish a collective understanding. This concept is changing the state of art of heterogenous database integration, in real-world applications.

DATA INTEGRATION FRAMEWORK

The proposed data integration framework is shown in Figure 3. The framework consists of four distinct components: application layer, mediating layer, wrapper layer, and heterogeneous data source layer. The application layer communicates with end-users; mediating layer contains a mediator which allows integration; wrapper layer contains software-based wrappers for each data resource; and source layer contains a set of heterogeneous data sources. The data integration system has a set of source descriptions that specify the semantic mapping between the mediated schema and the source schemas. It uses these source descriptions to reformulate a user query into a query over the source schemas.

- **Query Processor**: A database system needs to respond to requests for information from the end-user; and the dedicated software component which deals with the user-request to produce the required output is simply known as query processor. A database query is the vehicle for instructing a database management system (DBMS) to update or retrieve specific data to/from the physically stored medium. The actual updating, and retrieval of data is performed through various "low-level" operations. Examples of such operations for a relational DBMS can be relational algebra operations (e.g. select, project, join, and so on). How a DBMS processes queries, and the methods it uses to optimize their performance, are topics that will be covered in the later part of this chapter in the context of semantic data storage and querying facilities. There are three phases that a query passes through during the DBMS' processing of that query: parsing and translation, optimization, and evaluation.

Most queries submitted to a DBMS are in a high-level language such as SQL. During the parsing and translation stage, the human readable form of the query is translated into forms usable by the DBMS. These can be in the forms of a relational algebra expression, query tree, and query graph (Beg & Connelly, 2017).

The system in discussion parses, translates, rewrites and dispatches the user query to related data sources. When users pose their queries expressed in SPARQL (Ducharme, 2013) using terms from mediated schema, the parser analyzes the query, verifying if it is in accordance with SPARQL system. Rewrite implements the query rewriting work with reference to source descriptions.

- **Mediated Schema**: In the present framework, the mediated schema has two roles: (i) it facilitates the end-user access to the data with a uniform query interface to serve the formulation of a query on all sources; (ii) it provides a shared vocabulary set for wrappers (i.e. a software system working with a data source responsible for querying, processing the results and mapping them in the global schema according to the metadata from the data dictionary) to describe the content in every data sources. The mediated schema is expressed using RDFS (Resource Description Framework Schema). Moreover, ontology has been used in the mediated schema.
- **Semantic Module**: The queries are posed in terms of the mediated schema. To answer a query, the rewriter needs descriptions that relate the content of each data source to the classes, attributes and relations in the mediated schema. Each data source is described by one or more SPARQL queries. These semantically rich descriptions help the rewriter to form queries and direct the query dispatcher to distribute queries to specific data sources.
- **Wrapper Layer**: Initial work on data integration predates the recent efforts to standardize data exchanges. Thus, every data source might have its own format for presenting data (e.g. object database connectivity from relational databases, HTML from web servers, binary data from an object-oriented database). As a result, one of the major issues was the "wrapper creation problem".

In the present framework, the wrapper module provides an SPARQL view representing a data source, and a means to access and to query the data source. It translates the incoming queries into source-specific queries executable by the query processor of the corresponding sources.

- **Data Storage Layer**: The data storage layer, in this framework, is a conglomeration of pre-existing, heterogeneous, autonomous data sources.

SEMANTIC MAPPING

Everything on the web can be expressed in terms of object or resource: a webpage, a profile, person's name, a piece of content like raw material price, global trade item number (GTIN), and so on. In semantic web, each of these resources can be described using standard framework called resource description

Figure 3. Diagrammatic representation of the proposed framework

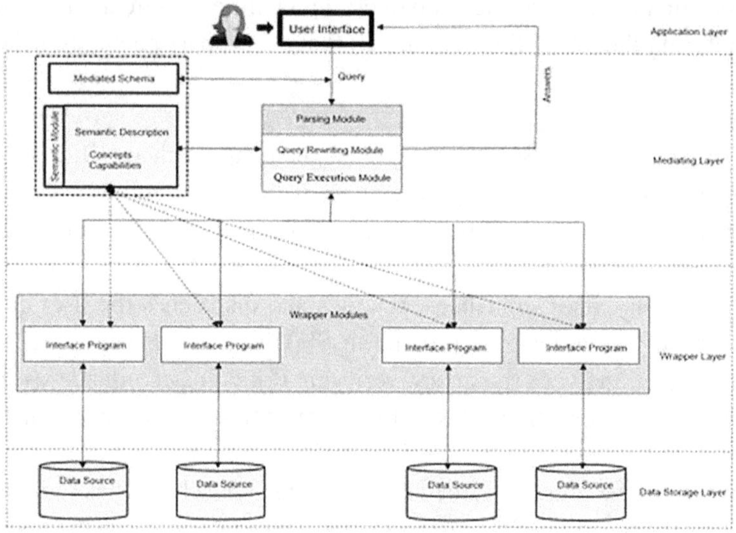

framework (RDF). One can think semantic web as a stack of different technologies, from the simplest to most powerful and expressive; RDF is in the medium level, just above XML and XML schemas (Lassila& Swick, 1998) as shown in Figure 4. RDF is a language for creating data model for expressing statements about objects and their relations. Statements are defined by triples that are composed of subject, predicate and values. Triples are used to store data and make it easier for machines to process and understand the data. Subject refers to a resource; predicate denotes the relationship between the subject and the object, where object is the value (Antoniou &Harmelen, 2004).

However, only describing resources is not enough; the relationship between resources and types of relations are important to make sense. Both syntax (structure) and semantics (meanings) need to be considered. RDF is represented in XML form. The RDF model consists of three things: resource, property and statements, as shown in Figure 5. Each resource has properties and resource represented by a class or a type. Like a website, URL represents a document of class (webpage) and has properties like created date, creator, language of the page, type of content, and so on. The values of the properties represent the current state of the resource. It is like an object (instance of a class) in object-oriented programming. Figure 7 shows the possible representation of resource. This is shown in Figure 6 in XML.

There are various names used in XML document for tags and attributes. Depending upon the domain, the names can have different meanings. For example, the tag 'title' may represent book name in an XML document that contains book details. The same tag may represent name of firm when information about the firms is stored in XML format. An XML spaces names to avoid element name conflicts. The primary use of such names in XML documents is to enable identification of logical structures in documents by software modules such as query processors, stylesheet-driven rendering engines (to convert into HTML), and schema-driven validators (to conform to structure). Multiple namespaces can be used in the same XML document.

RDF Schema is a vocabulary description language that extends RDF in order to include some basic features for defining application specific classes and properties. It enables definition of sub classes, sub properties and domain and range restrictions on properties (Antoniou & Harmelen, 2004).

Figure 4. Semantic web layer stack

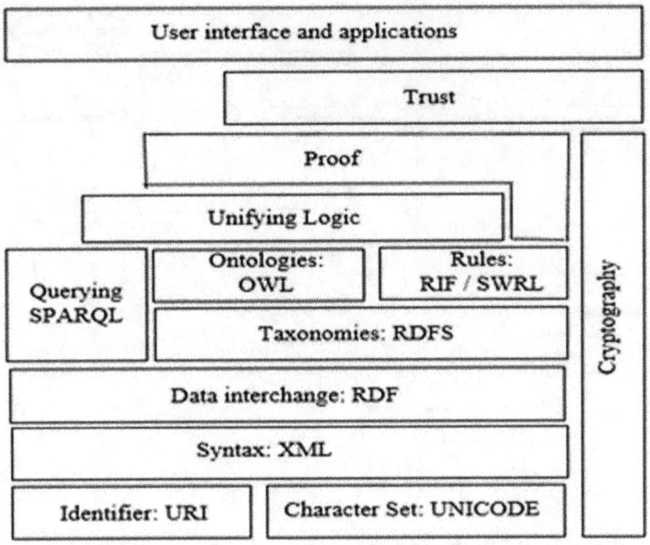

RDF statements show relationships between resources and properties. The RDF statement has a form called triple: subject-property-object (as shown in Figure 5c). For example, RDF statement 'English is the language of web page http://www.abc.co.uk/intro.htm'. In this statement, http://www.abc.co.uk/intro.htmis a resource (subject), language is the property (predicate) of the resource, whose value is English (object). The example in Figure 6a represents statements: URL http://www.abc.co.uk/intro.htm represents a web page; which is created by John Smith; it is created on 4 July 2018; the language of the webpage is English. As in natural language, composite sentences are possible. A statement itself can be the object for a composite statement. For example, the statement: web page http://www.abc.co.uk/intro.htm is created by John Smith who is a 34 year old teacher. In this example, [John Smith-age-34, John Smith-Occupation-Teacher] is object/value for property: creator of resource: http://www.abc.co.uk/intro.htm as shown in Figure 6c.

RDF statement in the form of subject-predicate-object is more general purpose or represents a meta-model. It just defines the structure in general. To model meaningful composite statements, RDF schemas are defined. The RDF schema provides information about the interpretation of the RDF statements and the relationships with other resources. It can be treated as a template to create similar kind of statements (Antoniou & Harmelen, 2004).

However, RDF lacks more advanced capabilities in defining the relationships. For example, it does not provide set cardinality, equality, disjointedness, and so on (Cure & Blin, 2015). These capabilities emerge in the semantic web world with the advance of Web Ontology Language by W3C, which is briefly explained below.

Web Ontology Language (OWL)

Due to the limitations of RDF, the semantic web technology needed a more expressive ontology language through the end of 1990s. There were several proposals for the new language such as Simple HTML Ontological Extensions (SHOE), the Ontology Inference Layer (OIL) and DML+OIL (Antoniou

Figure 5. Generic resource structure

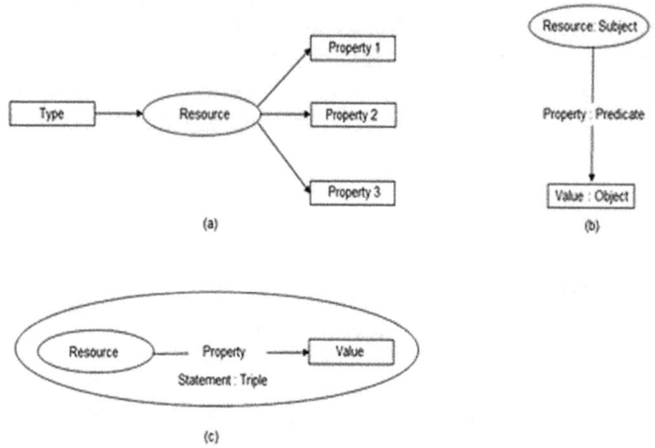

&Harmelen, 2004). Moreover, W3C launched the standard for a Web Ontology Language that is called OWL. W3C organization expanded the earlier work of OIL and improved the integration of it with RDF. OWL solves the deficiencies of RDFS by providing additional vocabulary like relations between classes (e.g. disjointedness), conjunction of classes, property characteristics (e.g. symmetry), cardinality (e.g. one or more, at most one), and so on (Cure & Blin, 2015).

RDF Graph Model

The RDF is a standard for representing knowledge on the web. It is particularly designed for building the Semantic web and has been widely adopted in database and data mining communities. The simple model of assertions leads to a network of information resources, interrelated by properties which establish relationships between resources and property values. In this way, RDF models a fact as a triple which consists of a subject (s), a predicate (p), and an object (o). The value of a statement is captured by URI references. Thus, one can intuitively understand a collection of information resources and RDF statements depicting them in a graph. To emphasize this characteristic, the term RDF Graph is defined as a set of RDF triples; hence, any collection of RDF data is an RDF Graph.

Definition 1 (RDF Terms, Triples, and Variables)

Formally, any RDF dataset is a set of RDF triples. One can consider an RDF triple <s, p, o> from the RDF triple-set <S, P, O>. Let us assume there are a pair of wise disjoint infinite sets I, B, and L (IRIs, Blank nodes, literals). A tuple (s, p, o) $\in \left(I \cup B\right) \times I \times \left(I \cup B \cup L\right)$ is called an RDF triple. In this tuple, s is the subject, p the predicate and o the object. We denote the union $I \cup B \cup L$ by T (RDF terms). Assume additionally the existence of an infinite set V of variables disjoint from the above sets.

Figure 6. Sample resource and schema

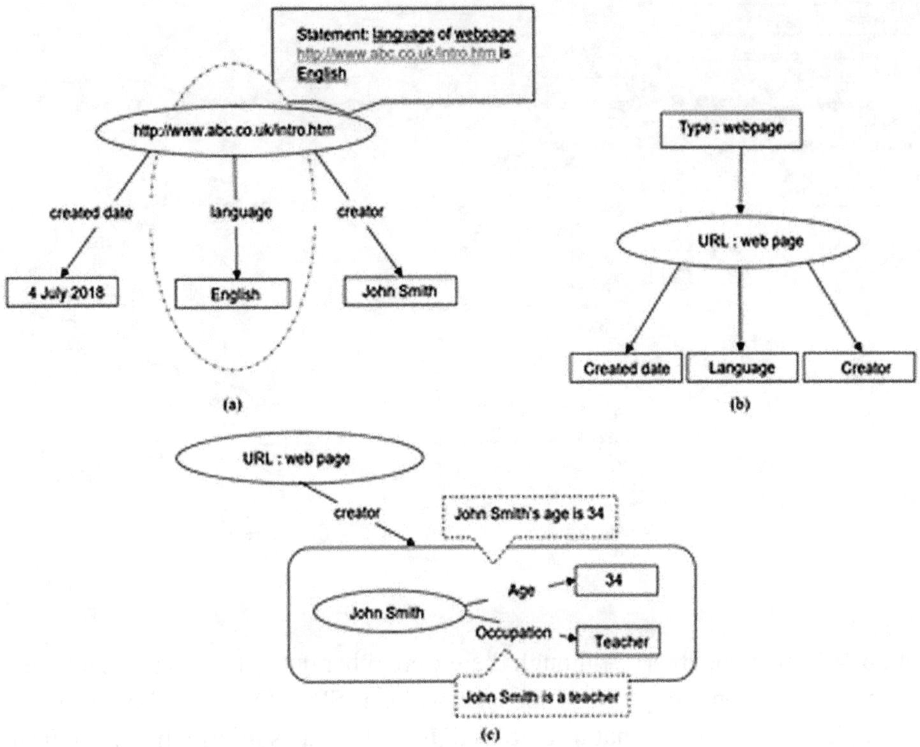

Definition 2 (RDF Graph)

An RDF graph is a set of RDF triples. Figure 7 shows a part of RDF ontology. In this diagram, there is a relation "write" between concepts "Manager" and "Publication", indicating the relationship between authors and their works. The relation "belong to" between "Manager" and "Organization" indicate what organizations manager belongs to. The relation "man-name" and "man-email" point a manager's name and email. The same applies to "pub-title", "pub-year" and "org-name".

As RDF databases increase in size to approach tens of thousands of triples, sophisticated graph matching queries expressible in language like SPARQL (Prud'hommeaux & Seabornr, 2013) become increasingly important. As more and more RDF database systems come 'online' and as RDF gets emphasized by both established companies (e.g. Oracle, Hewlett Packard), as well as from a slew of startups, the need to store and efficiently query massive RDF datasets is becoming increasingly important. Moreover, large parts of query languages like SPARQL increasingly require that queries (which may be viewed as graphs) be matched against databases (which may also be viewed as graphs) – the set of all possible "matches" is returned as the answer. The formal concepts of query graph and subgraph follow.

SPARQL Protocol of RDF Query Language (SPARQL)

After the improvements in Semantic Web during the last decade, there was a need to create a query language to process data that are stored in RDF format, as XML query language did not satisfy the

Figure 7. Part of RDF ontology

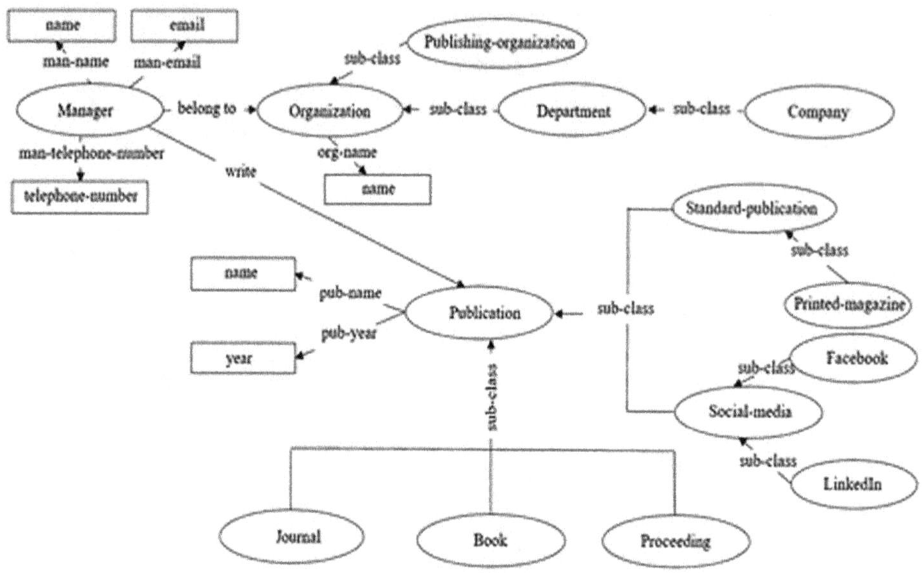

needs (Antoniou & Harmelen, 2004). Although there were other query languages before, in 2008 W3C announced the standard for ontology query language, which is SPARQL (Prud'hommeaux & Seabornr, 2013). It is a simple query language that resembles SQL and extracts information from RDF graphs. A query can consist of triple patterns that the RDF graph is composed of, and conjunctions and disjunctions (Prud'hommeaux & Seabornr, 2013).

Table 1 above is a small example of a SPARQL query, which is supposed to return all creators and their corresponding dates.

An SPARQL query usually contains a set of triple patterns, much like RDF triples, except that any of the subject, predicate and object may be a variable, whose bindings are to be found in the RDF data. This chapter addresses the SPARQL queries with 'SELECT/WHERE' option, where the predicate is always instigated as an URI (Uniform Resource Identifier). The SELECT clause identifies the variables to appear in the query results, while the WHERE clause provides triple patterns to match against the RDF data.

Overview of Query Graph and Subgraph

A network can be modeled as a graph $G = \{V, E, \sum, l\}$ where V is a set of vertices and $E \subseteq V \times V$ is a set of edges. $\sum$ is a vertex label set and $l: V \rightarrow \sum$denotes the vertex labelling function. For each notation, the vertex set of G is denoted as V(G) and its edge set is denoted as E(G). The *size* of G is defined

Table 1. SPARQL example

```
PREFIX archive: <http://www.abc.co.uk/archive >
SELECT ?date ?language ?creator
WHERE
{ ?creator archive:has-created ?date }
```

as $|V(G)|$, the size of its vertex set. Analogously, the graph queries posed upon the network can be modeled as a graph as well. This chapter focuses on the case of connected, undirected simple graphs with no weights assigned on edges.

A graph G' is a subgraph of G, denoted as $G' \subseteq G$, if $V(G') \subseteq V(G)$, $E(G') \subseteq E(G)$, and $\forall (u,v) = E(G'), u, v \in V(G')$. In other words, one can say that G is a supergraph of G' and G contains G'.

Definition 3 (Graph Query)

Given a network G and a query graph Q, the graph query problem is to find as output all distinct matching of Q in G.

Example: Figure 8(a) and Figure 8(b) show a network sample G and a query graph sample Q, respectively. Here the numeric identifiers have been used to distinguish different vertices in a graph. A subgraph G' of G with V(G') = {8,5,7,9} coloured in grey is isomorphic to Q and hence return as an answer to the graph query. It is worth noting that there exists multiple matching of Q in G. For example, given a triangle graph Q with A, B, C as the label for each vertex, respectively. All the matchings of Q in G, as shown in Figure 9(a), are {1,2,3}, (6,5,3}, {8,5,7} and {11,10,7}.

Relational Model

The relational model (Codd, 1970), built on a mathematical basis, provided the foundation for current relational database systems. A query language is a language in which a user requests information from the database. One can categorize query languages as being either procedural or non-procedural (Silberschatz, Korth, & Sudarshan, 1997). The difference lies in their approach to obtain the result. When a user wants to obtain a result using a procedural language, they need to instruct the system to perform a specific sequence of tasks on the database. In a nonprocedural language, the user only needs to describe the desired information without giving the system a specific procedure.

Relational algebra is a good example of procedural language, while relational calculus is representative of non-procedural languages (Silberschatz, Korth, & Sudarshan, 1997). However, most query languages used in current relational database systems combine elements of both the procedural and non-procedural approaches. The relations are stored in a database and the results from a database can be obtained by using database queries. In the field of relational databases, SQL became a standard that is now being used worldwide.

Figure 8. A Network G and a Query Graph Q

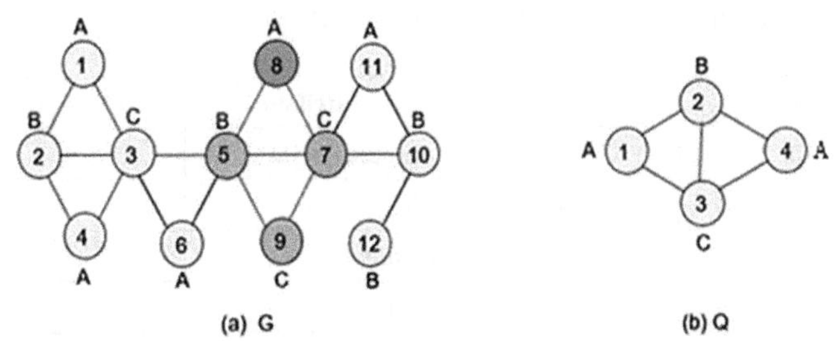

The term relation is used here in its accepted mathematical sense. Given sets D_1, D_2, …, D_n, R is a relation on these n sets if it is a subset of the cartesian product $D_1 \times D_2 \times \ldots \times D_n$. R is said to have degree n, often called n-ary. An n-ary relation R can be represented as a table with n column, which has the following properties: (i) each row represents an n-tuple of R; (ii) the ordering of rows is immaterial; and (iii) all rows are distinct. Normally, one column (or combination of columns) of a given relation has values which uniquely identify each element (n-tuple) of that relation. Such a column (combination) is called a primary key. A common requirement is for elements of a relation to cross-reference other elements of the same relation or elements of a different relation. We shall call a column of relation R a foreign key if it is not the primary key of R, but its elements are values of the primary key of some relation, S.

In the later part of this chapter, various concepts will be illustrated with reference to two example databases. Some of the relations in those databases, are managers and proceedings. These relations are shown in Table 2 (a) and 2 (b). Each manager and proceeding are unique; each manager can have *zero* or *more* papers in a proceeding. A manager can write and publish multiple papers.

The Relational Data-Base Management Systems (RDBMS) uses SQL (Standard Query Language) as the vehicle for instructing its database to update or retrieve specific data to/from the physically stored storage. The actual updating, and retrieval of data is performed through various 'low-level' operations. Examples of such operations for a relational DBMS can be relational algebra operations such as project, join, select, cartesian product, and so on (Silberschatz, Korth, & Sudarshan, 1997). With this back-ground information, it is worth considering the specific implementation issues for the proposed framework.

As shown in Figure 9(a), suppose there are two heterogeneous relational databases, each has several tables containing information about authors and papers.

XML Model

Extensible Markup Language (XML), a W3C recommendation, emerged as a standard for data representation and interchange among various web-applications, providing a simple means for more meaningful and understandable representation of web-contents. An XML document need only be well-formed, i.e., its tags be properly nested, but need not conform to a document type definition (DTD) or Schema. Hence, it is a variation of semi-structured data – data which may be varied and not restricted to any particular-schema. Management of semi-structured data by highly-structured modelling techniques, such as relational and object-oriented models, not only leads to a very complicated logical schema, but also demands much effort and frequent schema modifications, and thus obstructs the use of such approaches in modelling XML data. Consequently, development of an appropriate and efficient data model for XML documents has become an active research area with major current models based on directed, edge-labeled graphs, and Description Logic.

Table 2. (a) The manager relation; (b) The proceeding relation

Attribute	Length	Key	Attribute	Length	Key
ManagerID	10	YES	ProceedingID	15	YES
Name	30	NO	Year	8	NO
(a)			ISBN	13	NO
			(b)		

Figure 9. Two heterogeneous data sources

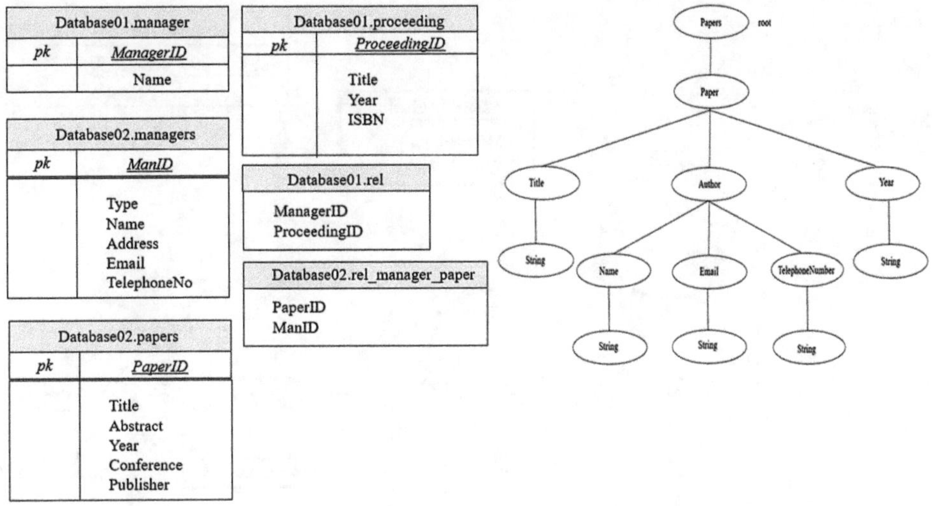

In addition, some systems offer XML views of non-XML data sources such as relational databases, allowing XML-based processing of data that are not physically stored as XML. The basic characteristic of the XML Schema is that it can define element attributes that can appear in a document, define which elements are child elements, the order of these elements, and their cardinality. XML Schema provides the user with ability to define an element or an attribute as a specific scope. The User can define complex or simple elements (depending on whether they have further structure on not) and cardinalities for them. Figure 9 (b) shows an XML tree, which describes the schema of an XML document about papers and their authors.

Mapping Relational Schemas to Ontology

To solve the heterogeneity problem, the meaning of the relational schema must be well described, which is called "source description". Before presenting how source descriptions are defined, it is worth considering how, in general terms, relational schema can be mapped to RDF ontology.

As mentioned above, RDF describes resources using a graph model. RDF Schema (RDFS) provides modelling primitives for defining classes and properties, range and domain constraints on properties, and subclass and sub-property relations.

The relational schema is based on entity-relationship diagram (ERD). Typically, each entity is represented as a database table, each attribute of the entity becomes a column in that table, and relationships between entities are indicated by foreign keys. Each table typically defines a class of entity, each column one of its attributes. Each row in the table describes an entity instance, uniquely identified by a primary key. The table rows collectively describe an entity set.

There are some similarities and differences between RDF and the ER model. Basically, ERD RDF Graphs have much in common. RDF can be viewed as a member of the Entity-Relationship model family. Class in RDFs corresponds to entity in ERD; property is kind of binary relationship; subclass and sub-property are subsumption relations. Therefore, some principles for semantic mapping between relational schema and RDF ontology are: (i) primary key of table is mapped to class, (ii) columns in

Figure 10. Semantic mapping between relational schemas and ontology

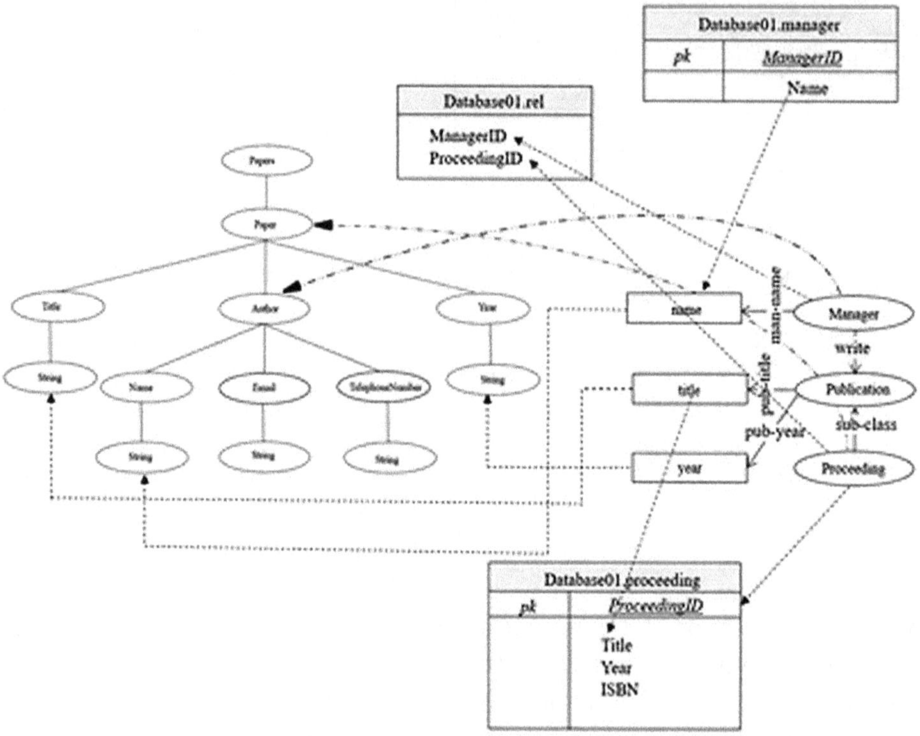

table are mapped to properties, (iii) column value is mapped to property value, (iv) each row key corresponds to an instance, (v) each row is represented in RDF by a collection of triples with a common subject. As shown in Figure 10, the semantic mapping between relational schemas and ontology are marked by arrows with broken line.

Mapping XML Schemas to Ontology

In the present framework, RDF metadata needs to be defined here because RDFS Class and RDF Property are enough for the specifications of classes and properties. Considering XML elements, attributes and their relationships, the proposed architecture proposes some principles below for semantic mapping between XML schema and RDF ontology: (i) attributes are mapped into properties; (ii) simple-type elements are mapped into properties; (iii) complex-type elements are mapped into classes.

As shown in Figure 10, the semantic mappings between XML schemas and ontology are marked by arrows with solid lines. In this example, attributes (e.g. title, name, year) are mapped into properties in ontology. And, complex-type elements (e.g. author, paper) are mapped into RDFS classes.

Source Description

After being mapped to ontology, source schemas must be described so that the query rewriting algorithm can use them to generate executable query plans efficiently. The functionality of SPARQL is used here.

Definition 4 (Source Description)

Given a data source P, its source description D_p is a tuple (Q_p, μ), where Q_p is a SPARQL query and μ is a mapping from variables appeared in Q_p to corresponding columns in P.

Each relational table is described by a SPARQL query over domain ontology. The semantic of the databases are thus explicitly defined, and it is easy to add and delete sources. Some of the relevant queries are presented in Table 3.

Semantic Query Rewriting and Optimization

The problem of query rewriting considers how to reformulate a query from the mediated schema to the underlying relational database. A survey and analysis of different algorithms to solve the problem is given in (Pottinger & Halevy, 2001). In this chapter, an algorithm is used to execute the query rewriting over the RDF Graph model.

The SPARQ query language is based on matching graph patterns. The Graph pattern contains triple patterns that are like RDF triples, but with the option of query variable in place of RDF terms in the subject, predicate or object positions.

Definition 5 (Minimal Connectable Units: MCU)

Given a SPARQLquery Q, a data source P and its source description $D_p = (Q_p, μ)$, a MCU m for P is a tuple of the form (Y, μ') where: Y is a subset of triple patterns in $Q_{p;}$ and μ' is a partial mapping from variables appeared in Q_p to corresponding component in P.

In the current system, as shown in ALOGRITHM01, an algorithm has been used to find minimal connectable units and joining them. This algorithm consists of two parts: *find* and *join*. It helps to reduce the number of rewritings in the heterogenous data integration, as shown in Figure 11.

Table 3. Different SPARQL queries used in the framework

database01.manager	**database01.proceeding**
(SELECT ?name WHERE {?X man_name ? name}, { ?name → Name}).	(SELECT ?title, ?year WHERE {?X pub_title ?title. ?X pub_year ?year}, {?title → Title, ?year → Year}).
database01.rel	**XML tree in Figure 9(b) described**
(SELECT ?manager, ?proceeding WHERE (?person Write ?proceeding}, {?manager → ManagerId, ?proceeding → ProceedingId}).	(SELECT ?title, ?name, ?year WHERE {?X man_name ?name. ?Y pub_title ?title. ?Y pub_year ?year. ?X Write ?Y}, {?name → /papers/paper/author/name, ?title → /papers/paper/title, ?year → /papers/paper/year}).

Algorithm 1.

```
// Find minimal connectable units  - FindMCUs //
// Q is a SPARQL query, D is the source description of a data
    source P has the form (Q_P, µ), where. µ is a subset. //
FindMCUs (Q, D_P)
Initialize M = Ø;
For each triple pattern t in Q, do
For each triple pattern t' in Q_P do
    If exist a mapping r that map t to t', then
    Find the minimal subset (denoted by Y) of triple pattern of Q_P that is con-
nectable.
    Find the subset (denoted by µ) of µ that relative to Y.
     M = M U <Y, µ'>;
End for
End for
Return M
// Join minimal connectable units part - JoinMCUs //
// Q is a SPARQL query, M is a set of MCUs, M = {m_1, …, m_n },
    where m = (Y_i, µ_i'), A is a set of rewritings. //
Initialize A = Ø;
For each minimal subset {m_1, …, m_k}of M such that Y_1 U Y_2 U Y_kcover all triple
patterns of Q.
 Create the conjunctive rewriting Q' contain all relative table to {m_1, …, m_k}
 Add Q' to A
End for
Return A
```

It is worth to note that the presented algorithm (i.e. ALGORITHM01) has got multiple parts that deal with the same triple partners (i.e. {1,4}, {2,5}, {3,6}). There are eight different ways one can join these partners as shown in left-hand side of the Figure 11. By using the proposed modification, the number of rewritings can be drastically reduced. In this example, rewritings are reduced from eight to only one.

Figure 11. Query rewriting optimization

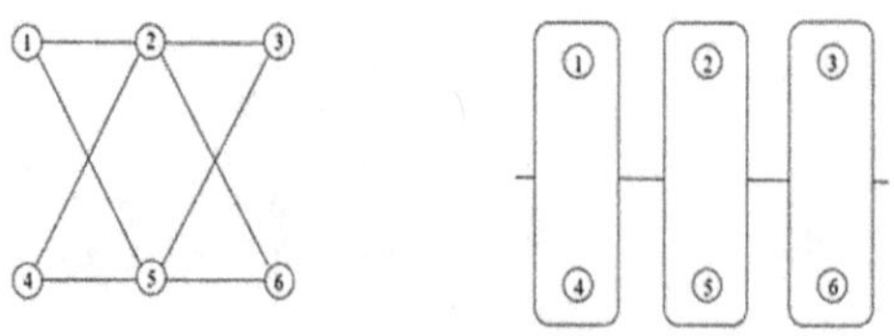

CONCLUSION

It is typical that global supply chain businesses run different but coexisting information systems. Using these information systems, supply chain organizations try to realize business opportunities in highly competitive markets. In this setting, the integration of existing information systems is becoming more and more indispensable to dynamically meet business and customer needs while leveraging long-term investments in existing information systems infrastructure. For example, in business intelligence (BI), integrated information can be used for querying and reporting on business activities, for statistical analysis, online analytical processing (OLAP), and data mining to enable forecasting, decision making, enterprise-wide planning, and, in the end, to gain sustainable competitive advantage. For CRM, integrated information on individual customers, business environment trends, and current sales can be used to improve customer services. Enterprise information portals preset integrated company information as personalized web sites and presents single information access points primarily for employees, but also for customers, business partners, and the public. Moreover, in e-commerce and e-business, integrated information enables and facilitates business transactions and services.

In addition, within this ambitious and broad aim, this chapter work focused on two related but independent aspects: semantic integration and query interoperability. Within the semantic integration problem, this chapter focused on XML-RDF integration, and the use of ontology. Within the query interoperability problem, the chapter queried restricted and heterogenous data interfaces with XML technologies.

The main problem highlighted is that integrating data from multiple heterogeneous sources needs to deal with different data models, database schema, query languages, and the chapter presents a framework. In this framework, ontology has been used as the mediated schema to represent data source semantics. It is also describing a method based on resource description framework graph patterns and query rewriting techniques.

The proposed solution first analyzes the similarities and differences among RDF schema, relational model and XML schema. It then discussed how to map relational schema and XML schema to ontology. A data source describing method based on SPARQL was defined. Heterogeneous data source schemas were described using queries defined by SPARQL. Based on an RDF Graph model, the semantic of query rewriting was defined, and a query rewriting algorithm is presented.

The essence of this research is to highlight semantic web computing application in industry. The semantic web helps to structure web-based information to make it more interoperable, machine-readable and thereafter to provide a means to relate various information concepts more easily and in a reusable way. The semantic web technology acts as an additional layer on the top of the web and is built around explicit representations of information concepts and their relationships such as ontologies and taxonomies. Furthermore, semantic web technologies are not only valuable on an open environment like the web, but also in closed systems such as in real-world application settings. Hence, these technologies can be efficiently deployed for domains including web services, enterprise application integration, knowledge management and electronic commerce, fulfilling existing gaps in current industrial applications.

REFERENCES

Abiteboul, S., Benjelloun, O., & Milo, T. (2002, December). Web services and data integration. *Proceedings of the Third International Conference on Web Information Systems Engineering WISE 2002*. (pp. 3-6). IEEE.

Abiteboul, S., & Polyzotis, N. (2007). The Data Ring: Community Content Sharing. Proceedings of the Third Biennial Conference on Innovative Data Systems Research (CIDR 2007), Asilomar, CA, January 7-10. Academic Press.

Antoniou, G., & Harmelen, F. V. (2004). *Semantic Web Primer*. MIT Press.

Batini, C., Lenzerini, M., & Navathe, S. B. (1986). A Comparative Analysis of Methodologies for Database Schema Integration. *ACM Computing Surveys*, *18*(4), 323–364. doi:10.1145/27633.27634

Baumgaertel, H., Ehm, H., & Kasprzik, A. (2018). Collaboration in Supply Chains for Development of CPS Enabled by Semantic Web Technologies. In M. Rabe, A. A. Juan, N. Mustafee et al. (Eds.), *Proceedings of the 2018 Winter Simulation Conference* (pp. 3627-3638). Academic Press.

Bayardo, R. J., Bohrer, B., Brice, R. S., Cichocki, A., Fowler, J., Helal, A., . . . Woelk, D. (1997). InfoSleuth: Agent-Based Semantic Integration of Information in Open and Dynamic Environments. *Proceeding of the 1997 ACM SIGMOD International Conference on Management of Data (SIGMOD 1997)*, Tucson, Arizona (pp. 195-206). ACM. 10.1145/253260.253294

Beneventtano, D., Bergamaschi, S., Castano, S., Corni, A., Guidetti, R., Malvezzi, G., . . . Vincini, M. (2000). Information Integration: the MOMIS Project Demonstration. *Proceeding of the 26th Very Large Database Conference*, Cairo, Egypt (pp. 611-614). Academic Pres.

Carey, M., Haas, L., Schwarz, P., Arya, M., Cody, W., Fagin, R., . . . Wimmers, E. (1995). Towards Heterogeneous Multimedia Information Systems: The Garlic Approach. *Proceeding of the 5th International Workshop on Research Issues in Data Engineering-Distributed Object Management (RIDE-DOM 1995)*, Taipei, Taiwan, March 6-7 (pp. 124-131). Academic Press. 10.1109/RIDE.1995.378736

Ceri, S., & Widom, J. (1993). Managing Semantic Heterogeneity with Production Rules and Persistent Queues. *Proceedings of the 19th Very Large Data Base Conference*, Dublin, Ireland (pp. 108-119). Academic Press.

Chaudhuri, S., & Dayal, U. (1997). An Overview of Data Warehousing and OLAP Technology. *SIGMOD Record*, *26*(1), 65–74. doi:10.1145/248603.248616

Chawathe, S. GarciaMolina, H., Hammer, J., Ireland, K., Papakonstantinou, Y, Ullman, J., &Widom, J. (1994). The TSIMMIS Project: Integration of Heterogeneous Information Sources. *Proceeding of the 1000th Anniversary Meeting of the Information Processing Society of Japan (IPSJ)*, Tokyo, Japan (pp. 7-18). Academic Press.

Codd, E. F. (1970). *The Relational Model for Database Management*. Addison Wesley.

Cure, O., & Blin, G. (2015). *RDF Database Systems Triples Storage and SPARQL Query Processing*. Waltham, MA: Morgan Kaufmann.

DuCharme, B. (2013). *Learning SPARQL: Querying and Updating with SPARQL 1.1*. Sebastopol, CA: O'Reilly Publishing Company.

Fagin, R., Kolaitis, P. G., Miller, R. J., & Popa, L. (2005a). Data Exchange: Semantics and Query Answering. *Theoretical Computer Science*, *336*(1), 89–124. doi:10.1016/j.tcs.2004.10.033

Fagin, R., Kolaitis, P. G., Popa, L., & Tan, W. (2005b). Composing schema mappings: Second-order dependencies to the rescue. *ACM Transactions on Database Systems*, *30*(4), 994–1055. doi:10.1145/1114244.1114249

Friedman, M., Levy, A., & Millstein, T. (1999). Navigational plans for data integration. *Proceedings of AAAI*, Menlo Park, CA (pp. 67-73). AAAI Press/ MIT Press.

Garcia-Solaco, M., Saltor, F., & Castellanos, M. (1996). Semantic Heterogeneity in Multidatabase Systems, In O.A. Bukhres & A.K. Elmagarmid (Eds.), Object-Oriented Multidatabase Systems, A Solution for Advanced Applications (pp. 129-202). Prentice-Hall.

Gruber, T. R. (1993). A translation approach to portable ontology specifications. *Knowledge Acquisition*, *5*(2), 199–220. doi:10.1006/knac.1993.1008

Halevy, A. Y. (2001). Answering queries using views: A survey. *The VLDB Journal*, *10*(4), 270–294. doi:10.1007007780100054

Halevy, A. Y. (2003). Data Integration: A Status Report. In *Datenbanksysteme in Business, Technology and Web (BTW 2003)*, Leipzig, Germany, February 26-28. Academic Press.

Hammer, J., & Mcleod, D. (1993). An Approach to Resolving Semantic Heterogeneity in a Federation of Autonomous, Heterogeneous Database Systems. *International Journal of Intelligent and Cooperative Information Systems*, *2*(1), 51–83. doi:10.1142/S0218215793000046

Huhns, M. N., & Singh, M. P. (1997). Agents on the Web: Ontologies for Agents. *IEEE Internet Computing*, *1*(6), 81–83. doi:10.1109/4236.643942

Hull, R. (1997). Managing semantic heterogeneity in databases: A theoretical perspective. *Proceedings of the sixteenth ACM SIGACT-SIGMOD-SIGART symposium on Principles of database systems*, Tucson, AZ (pp. 51-61). ACM. 10.1145/263661.263668

Hurson, A. R., & Bright, M. W. (1991). Multidatabase Systems: An Advanced Concept in Handling Distributed Data. *Advances in Computers*, *32*, 149–200. doi:10.1016/S0065-2458(08)60247-8

Jarke, M., Lenzerini, M., Vassiliou, Y., & Vassiliadis, P. (2000). *Fundamentals of Data Warehouses*. Springer. doi:10.1007/978-3-662-04138-3

Kim, W., Choi, I., Gala, S., & Scheevel, M. (1993). On Resolving Schematic Heterogeneity in Multidatabase Systems. *Distributed and Parallel Databases*, *1*(3), 252–279. doi:10.1007/BF01263333

Kim, W., & Seo, J. (1991). Classifying Schematic and Data Heterogeneities in Multidatabase Systems. *IEEE Computer*, *24*(12), 12–18. doi:10.1109/2.116884

Landers, T., & Rosenberg, R. L. (1982). An Overview of MULTIBASE. *Proceedings of the Second International Symposium on Distributed Data Bases (DDB 1982)*, Berlin, Germany, September 1-3 (pp. 153-184). Academic Press.

Lenzerini, M. (2002). Data integration: A theoretical preceptive. *Proceedings of Twenty-first ACM SIGACT-SIGMOD-SIGART Symposium on Principles of Database System*, Madison, WI (pp. 1-14). ACM.

Mena, E., Kashyap, V., Sheth, A. P., & Illarramendi, A. (1996). OBSERVER: An Approach for Query Processing in Global Information Systems based on Interoperation across Pre-existing Ontologies. *Proceedings of the First IFCIS International Conference on Cooperative Information Systems (CoopIS 1996)*, Brussels, Belgium, June 19-21 (pp. 14-25). IEE Computer Society. 10.1109/COOPIS.1996.554955

Pal, K. (2017). Supply Chain Coordination Based on Web Service. In H. K. Chan, N. Subramanian, & M. D. Abdulrahman (Eds.), *Supply Chain Management in the Big Data Era* (pp. 137–171). Hershey, PA: IGI Global. doi:10.4018/978-1-5225-0956-1.ch009

Pal, K. (2018). Building High Quality Big Data-Based Applications in Supply Chains. In A. Kumar & S. Saurav (Eds.), *Edited Book – 'Supply Chain Management Strategies and Risk Assessment in Retail Environments* (pp. 1–24). Hershey, PA: IGI Global. doi:10.4018/978-1-5225-3056-5.ch001

Pal, K. (2019). *Quality Assurance Issues for Big Data Applications in Supply Chain Management, Predictive Intelligence Using Big Data and The Internet of Things* (P. K. Gupta, T. Oren, & M. Singh, Eds.). Hershey, PA: IGI Global.

Pottinger, R., & Halevy, A. (2001). MiniCon: A Scalable Algorithm for Answering Queries Using Views. *The Very Large Database Journal*, *10*, 182–198.

Prud'hommeaux, E., & Seaborne, A. (2013). SPARQL Query Language for RDF. W3. Retrieved from http://www.w3.org/TR/rdf-sparql-query

Sheth, A. P., Gala, S. K., & Navathe, S. B. (1993). On Automatic Reasoning for Shema Integration. *International Journal of Intelligent and Cooperative Information Systems*, *2*(1), 23–50. doi:10.1142/S0218215793000034

Sheth, A. P., & Larson, J. A. (1990). Federated Database Systems for Managing Distributed, Heterogenous, and Autonomous Databases. *ACM Computing Surveys*, *22*(3), 183–236. doi:10.1145/96602.96604

Silberschatz, A., Korth, H. F., & Sudarshan, S. (1997). *Database System Concepts*. McGraw-Hill.

Ullman, J. D. (1997). Information integration using logical views. *Proceedings of International Conference on Database Theory*, Delphi, Greece (pp. 19-40). Academic Press.

Wache, H., Vogele, T., Visser, U., Struckenschmidt, H., Schuster, G., Neumann, H., & Hubner, S. (2001). Ontology-Based Integration of Information – A Survey of Existing Approaches. In *Workshop on Ontologies and Information Sharing IJCAI-2001*, Seattle, WA, April 4-5 (pp. 108-117). Academic Press.

Wiederhold, G. (1992). Mediators in the architecture of future information systems. *Computer*, *25*(3), 38–49.

Ziegler, P., & Dittrich, K. R. (2007). *Data Integration – Problems, Approaches, and Perspectives*. University of Zurich.

KEY TERMS AND DEFINITIONS

Linked Data: An approach taken to linking data such that it becomes more useful / accessible than it would be in isolation.

Ontology: Information sharing among supply chain business partners using information systems is an important enabler for supply chain management. There are diverse types of data to be shared across supply chain, namely – order, inventory, shipment, and customer service. Consequently, information about these issues needs to be shared in order to achieve efficiency and effectiveness in supply chain management. In this way, information-sharing activities require that human and / or machine agents agree on common and explicit business-related concepts (the shared conceptualization among hardware / software agents customers, and service providers) are known as explicit ontologies; and this help to exchange data and derived knowledge out of the data to achieve collaborative goals of business operations.

Relational Database: Relational database systems support processing of tuples of relations to generate a single result as a set of tuples. Relational algebra, relational calculus and structured query language (SQL) are used to specify queries on relational databases.

Resource Description Framework (RDF): The RDF is a standard for representing knowledge on the web. It is primarily designed for building the semantic web and has been widely adopted in database and datamining communities. RDF models a fact as a triple which consists of a subject (s), a predicate (p), and an object (o).

SPARQL: The SPARQL query language is a structured language for querying RDF data in a declarative fashion. Its core function is subgraph pattern matching, which corresponds to finding all graph homomorphism in the data graph for a query graph.

SPARQL Query: A SPARQL query usually contains a set of triple patterns, much like RDF triples, except that any of the subject, predicate and object may be a variable, whose bindings are to be found in the RDF data.

SQL: Structured Query Language (SQL) – a commonly-used language for querying relational database systems.

Structured Data: Data are stored in accordance with a strict schema for database management purpose.

Supply Chain Management: A supply chain consists of a network of key business processes and facilities, involving end users and suppliers that provide products, services and information. In this chain management, improving the efficiency of the overall chain is an influential factor; and it needs at least four important strategic issues to be considered: supply chain network design, capacity planning, risk assessment and management, and performances monitoring and measurement. Moreover, the details break down of these issues need to consider in the level of individual business processes and sub-processes; and the combined performance of this chain. The coordination of these huge business processes and their performance improvement are the main objectives of a supply chain management system.

XML: Extensible Markup Language (XML) is a simple, very flexible text format derived from SGML (Standard Generalized Markup Language). While XML was originally designed to meet the challenges of large-scale electronic publishing, it plays an increasingly significant role in the exchange of a wide variety of data on the web.

Chapter 6
Manufacturing, Control, and Automation

Shubhajit Das
https://orcid.org/0000-0002-4297-3324
National Institute of Technology, Arunachal Pradesh, India

Kakoli Roy
National Institute of Technology, Arunachal Pradesh, India

Tage Nampi
National Institute of Technology, Arunachal Pradesh, India

ABSTRACT

This chapter identifies the common needs for process controls and automation that include methodologies to enable in-situ-level process controls, optimization at the plant or industry level, open-architecture software tools, adaptive control systems, methods and diagnostic tools for condition-based maintenance of process equipment in a manufacturing industry.

INTRODUCTION

Manufacturing typically starts with raw materials, which are then subjected to a sequence of processes to make individual products of certain value. In more elaborate manner, we can say that manufacturing means the production of products for use or sale of the said product using engineered sequence of processes and machines with labor. The product is what we focus on for the profit of an industry. Manufacturing processes involved in the production of products are engineered sequences of operations through which the products are produced. In simple words, we can say manufacturing methods are the steps for converting the raw materials into a final product. Industries for years have gone through lots of evolution to what it is today and is further evolving as we read this sentence. We humans are creative creatures blessed with lots of curious logical problems and solutions. Every second new ideas pop out of our imagination. Industries today have to face a lot of competition in the market. For example, one of

DOI: 10.4018/978-1-7998-1831-1.ch006

the problems manufacturers and engineers face in this generation is the manufacturing of semiconductor. New generation phones have huge specification. The semiconductor chips in our phones can perform huge functions and with every new phone new specification pops out that, means more work done by the chips. However, the size does not increase with work and instead tinier chips are produced with the coming generation, i.e., minimum feature size increases with technology. This further means changes in the machines or processes of the manufacturing industry. The production and inspection of such small chips are difficult. We cannot risk any losses or waste whether it is labor or time etc. in this competitive market. Hence, to optimize the manufacturing processes we need to control these operations to automate it to some degrees. Generally, the aim of process control is to maintain the running processes at a preferred and steady operating condition (temperature, pressure, composition, etc.) in case of discrepancy or forces it to follow a desired route along the time of operation. From observations, we can say that control of manufacturing processes has direct impact on the product (Grover, 2002; Grady, 1986).

Over the years, rapid improvements in both computer software and hardware have resulted in a change leading towards automation in manufacturing industries. The word automatic comes from the Greek word "Autos" and "Matos" that mean self and moving respectively (Jouela, 2007). We can say that automation in manufacturing process can be said as the replacement of human interaction for decisions and operation by the use of logical programming instructions and computerized tools. Automation optimizes productivity by greatly reducing human intervention. In the present world of competitive technologies advancement, especially robots automation provides a compulsory tool for the survival of the company. They take a great role in enhancing the production flexibility, product quality and innovation in the international market and competition. Automation can and is being used to resolve the most fundamental control problems birthed by the rapid increase in product complexity and operational pace and decreasing product internal structure dimensions (Sadoyan, 2005). In modern big industries advanced automation tools used for the development of complex products, modeling, simulation, test and inspection and most importantly manufacturing in our case.

MANUFACTURING PROCESS CONTROLS

The conventional approach to process control involves primary equipment's calibration, crucial process parameters monitoring and final approval of the product. Those approaches mentioned are not capable of recognizing the interdependence of process parameters nor allows any modification or adjustment of the process to optimize yield. Traditional control methods are more like the control system and strategy for relatively simple processes. For modern machine tools, a more sophisticated control strategy was developed called feed forward that can compensate for the dynamic delay in the tools. Such control system can provide immense improvement in machining accuracy.

Computer and software usage have helped us to have an optimize control of the manufacturing in industries. Big industries have facilities and capital for high automation of manufacturing processes. The industry should be able to self-arrange the designing of process and tools, planning and controlling the production instructions, and assures product quality necessities. This ensures running the production of a product smoothly with minimum error and best quality. Manufacturing supports systems take a role in such area to accomplish these goals. Such support system consists of the people and actions, which help an industry to manage its manufacturing operations. However, most of these support systems have no direct connection with the product, but they help in planning and controlling of its progress (Sullivan, 2009). A cycle of information-processing activities or say information-processing cycle is usually involved. In the manufacturing support system (figure 1) the center of the figure consist of the overall production system while the outer rings consist of the four functions of the cycle that are: business functions, product design, manufacturing planning and manufacturing control.

The process control of the manufacturing system uses instantaneous control tactics and algorithms to communicate sensory response for making decision about the process. A very important factor we need to be careful is that manufacturing industries usually focuses on unit processes more rather than the bulk processes as a whole. Unit processes are the processes in a single line or say sector. The need of such focus on unit processes is necessary because it is easier to inspect and maintain the quality of each operation. Basic line operations formulated in the industry are

1. Processing operation
2. Assembly operations
3. Material handling
4. Testing and Inspection
5. Control and Coordination

Now let us briefly discuss the need for control of manufacturing process. The factors motivating the improvement of sensors and manufacturing process control technology are discussed below:

Need to Decrease Inconsistency and Improve Quality

The continuous change in market demand and supply has to deal with continuous variation within their process and operation. Control systems are able to monitor process and such variation to make high-

Figure 1. Typical processing cycle in an industry

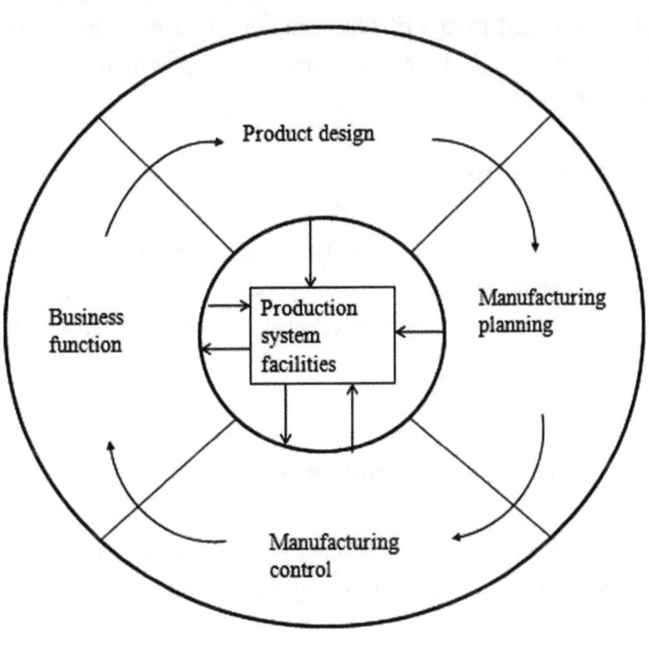

quality operations possible at a much lower expense. Software tools development for the design and advancement of smart process control software is necessary for both adaptability and reliability of the system. Such development of software includes instantaneous and implanted software. They are able to compensate for the variation occurring in processing conditions, machine performance, and materials properties. Other feature like self-configuration, self-learning, information transfer, reusability can be presented and offered by the advanced control system to reduce and adapt to the continuous changing conditions or say process variations.

Environmental Constraints

Improvement of energy efficiency and reduction of waste, lead-time and costs of the overall system can be achieved with advanced innovative control devices. Monitoring and information of the processes can be achieved directly, easily and accurately with lesser time. The durability and resistance of the devices are also increased with the ongoing advances in technology and control system making it able to work in extreme condition of the manufacturing operation.

Easier Servicing and Maintenance

Servicing and maintenance are important factors that are often neglected in theories but are a very crucial part of the manufacturing process and industry brand image. Better brand image leads to a better demand and hence productivity for the industry. Newer technologies provide more specifications compared to the conventional or existing technologies. Often time the manufacturers are eager to involve these new technologies and integrate them in manufacturing process control without even testing it. This leads to some defect and failure in the system. The error might not even be detected sometimes which make the process complex. Due to the difficulty in understanding the operation and some defective products might go to the market under the notch of the manufacturer. Thus, maintenance and servicing of product would help compensate the unsatisfied customer for such case. Faster the servicing, better will the trust of the company, better would be the brand image of the company. Process supervisors will have to provide positive maintenance capabilities, such as measurements of performance degradation, fault recovery, self-maintenance, and remote diagnostics.

Component of Control System

To understand the control processes of manufacturing system we first need to know the components or the backbone that guide and refer to the manufacturing process. There are three general crucial elements of process control configurations be it manual, automatic, or computer based. They are:

- Measurement
- A control strategy embedded in a controller
- A final element for implementing the control action

Every single systems and software that wield control over production processes falls under the components of manufacturing system control. A control system includes process sensors, data processing equipment, actuators, networks and algorithms to communicate process variables to product attributes. The components of a control system are shown in figure 2.

Process Sensors

Sensors play an important role in control manufacturing. Sensors as we know are a kind of measuring device. Together with the control technologies, they form an integral component of all modern processing industries. The sensor supplies the data for monitoring and controlling product or process varying parameters. This helps us to evaluate and monitor the control operation, its quality, material properties, security and energy efficiency. The process sensors usually address three operational issues, predictability, reducibility, and productivity (Hovhannes, 2005). The issues addressed need to be measured or are accounted from some measurement under a very harsh processing environment. A few of the required measurements of the process sensor for the above includes temperature profile measurement, chemical composition measurement, physical attributes at high line speeds and high temperatures measurement, combustion processes monitoring. The sensors used in manufacturing process control mostly used to maintain steady state operating conditions through the measurement and control of primary variables like temperature, pressure, flow, environmental control etc. Nowadays sensor are developed to focus more on the product condition, i.e. product quality, performance and optimization more than the operation condition control (Collins, 1980).

Data Processing Equipment

A data processing system is said to be a combination of machines, people and processes. The unique combination produces a defined set of outputs from a set of input. Depending on the interpreter, the input and output can be interpreted as data, facts, information, etc.

Actuators

An actuator is an element of a machine that is responsible for the movement and control of a mechanism or system. It has the ability to introduce a motion as well as prevent and eradicate it. Actuators covert electrical signals i.e. the control signal and convert it to mechanical motion that acts on the tools and

Figure 2. Components of the control system

product through an effectors. A control system acts upon a virtual environment through the instrument. Thus, it acts as the muscle of our body and is responsible for movement in the control system.

Networks

Networks are used to interconnect the various workstations and operation. They are a line of communication that is used to pass information of one station to the other in form of signal or direct verbal communication. They are a very important component of the control system as the situation of what is happening around and what needs to be done has to pass around continuously for smooth running of the system. Networks also help monitor the operation and thus take corrective measure in case it faces any hindrances.

Algorithm

Algorithm is mathematical and logical approach to solve complex computational problems within a specific time limit. It contains sets of self-controlled order of instructions and measures with finite space or sequence to give us an answer to any specific problems (Rajawat, 2017). Control algorithms are used to control the dynamic analog process with discrete digital logic. A common term we use in developing an algorithm for any processes is “data mining”. Data mining (Koyal, 2007) is the development and discovering of large data pattern to solve and control any processes say manufacturing process in our case concerning methods like the connection of machine learning, statistics, and systems. They further aspect and represent their collected knowledge in a form that is accessible to human operators. PID control algorithms (Proportional Integral Derivative) are the most common control algorithm used in manufacturing. However, they are relevant for a low order transfer function model and provide a low-level closed loop performance. This makes the PID controller more appropriate for low speed machining or say conventional machining operations. However, what do we do when the industry has high machining systems and operation? For eliminating, such we need to implement advanced control algorithms technology available with advanced computer workstations, real time operating systems and bus structure. Although manufacturing unit use latest electronics, CNC, and high-level algorithms the regular configurations existing cannot lodge non-machining devices such as sensors, fixtures and other holding devices etc. This arises due to the lack of hardware and operators in experience and inability to access such high-complicated data. Development of better computer devices, hardware and architecture buses serve as a decent future answer for the problems arising from the conventional machine tools and operation process. Innovative and latest architects are still under process and have long way to go (Halev, 2017).

Continuous Control Systems

Continuous control system is one of the basic types of control system in an industry. Realistic world consists of many separate feedback loops contradicting the theoretical continuous processes. The main objective of such system is to control, coordinate, and maintain the varying output to a constant value as such of a feedback control operation. For better understanding of this system let us consider few examples of continuous processes.

- Control of the input of a chemical reaction. The reaction of chemicals depends on temperature, pressure of the reactor; material where it is stored and experimented upon, flow rates of reactants etc.
- In a milling operation, a complex curved contoured surface of the tool motion is observed in control of the position of a work part relative to a cutting tool.

Continuous Control System Procedure

A continuous process control system can be used similarly to achieve a control objective in manufacturing process. Other approaches of a continuous system include the following control procedure.

Regulatory Control

The purpose of regulatory control is to retain process performance at a certain range of tolerance of the system. This control is useful for measuring the performance attribute of the product quality. The performance measures of the process are called the index of performance. However, the action compensating for the error faced in regulatory control is taken only after the variation the process output. An error is compulsory for any control action to be taken but such error indicates variations in the output from the preferred value. The following control mode, feed forward controls are used to addresses such issues.

Feed Forward Control

Feed forward control is a strategy applied to anticipate the consequences of bottlenecks in the process by sensing them and taking some corrective actions before they produce any side effects on the process. In the ideal case, the compensations are effective however; complete compensation is not likely to be possible due to the different imperfections in the feedback measurements, actuator operations, and control algorithms. To eliminate this, feed forward control is used with feedback control collectively. Among the systems, regulatory and feed forward controls are more closely related with the process industries than with discrete product manufacturing.

Steady State Optimization

Steady state optimization is a category of optimization techniques where the process operation exhibits some well-defined characteristics. They are index of performance, process variables and the index of performance relationship and optimizing values of the index of performance parameters. These characteristics help the control algorithm to drive the process towards an optimal position adapting various adjusting procedures. The control system observed in such technique is open loop. The problems can be solved through various mathematical technique and programming like differential calculus, calculus of variations, and a variety of other mathematical programs.

Adaptive Control

Steady-state optimizations as said above operate as an open-loop system. They operate successfully when there are no disturbances between process parameters and process performance. However, usually there is bound to be some variation existing on either the process or tolls. For such disturbances distracting the performance, a self-correcting optimal control can be used. This self-correcting system is called adaptive control. Adaptive control combines feedback and optimal control by measuring the appropriate variables during operation and uses a control algorithm to attempt or directly optimize some index of performance. Adaptive control has the potential to deal with a time-varying environment that affects the overall process unlike the above-discussed technique. The time varying environment that affects the performance of the structure includes variations in raw materials, tooling, atmospheric conditions etc. It is designed to compensate for such altering environment by self-monitoring its performance and shifting some feature of its control method to an optimal solution (Lozowski, 2016; Mahalik, 2014).

On-Line Searching Strategies

These are used for addressing a special class of adaptive control problem wherein the decision purpose cannot be satisfactorily defined, for example the relationship between the input parameters and the index of performance or even the adaptive control sometimes due to inadequate knowledge on the said topic. This makes it difficult to identify the variations in the internal parameters of the system for desired improvement in their performance. A number of experiments can solve such difficulties. Small methodical variations are made in the input parameters of the process and the effect of these changes are observed on the output variables. Based on the outcome of these experiments, the processes are improved by making larger change in the input parameters.

Discrete and Sequencing Control Systems

Discrete control system is yet another basic subsystem of the control system in industrial automation. In such control system, the parameters and variables of the system changes discretely at a time rather than in continuous matters. Discrete control system defined by way of program instructions in advance like the conventional binary command. There are two cases in this system through which the changes required are executed. These two cases can be illustrated as event driven change and time driven. A controller is used to execute an event-driven change. They drive the system to respond to some event that has causes the system to change. Such changes are used for both initiating and terminating an operation. Examples of event-driven changes are:

- A robot can be used for loading and fixing work parts loads. Altering of this system state is done by sensing of the parts present in the event by a limit switch. The event-driven change observed here is the commencement of the automatic machining cycle.
- Another example of event driven change is the system parts moving at the length of a conveyor past an optical sensor. The event driven changes here occurs due to thepart moving past the sensor.

A time-driven change consists of starting and stopping time as well as the changing process time. It is executed by the control system either at a specific point in time or after a certain time lapse has occurred. Two examples of time driven are:

- Factories use a shop clock to alarm the workers. The shop clock is set at a time to indicate the workers of their starting shift time, ending shift as well as break time. The clock rings a bell for alarming the worker about their specific routine time.

- Another example where we use a time driven system is heat treatment operations. Heating is allowed only for a certain amount of time in specific duration of the day. An automated heat-treating cycle consists of automatic loading and unloading of parts in the furnace (LaMarco, 2018).

Industrial Control System

Industrial control system deals with unit operations. The regulation of unit operation automatically and the relative equipment management in a manufacturing industry is called industrial control system. Computers are a vital part of such control system and are used indiscreetly to perform all the control functions of the unit operations. The industrial control system consist of two categories namely process industries and manufacturing industries. The unit operations for both the systems are different from each other. While the process industries work on the state of the material, the discrete industry works on the quantity and stage of the material.

INTEGRATED MANUFACTURING SYSTEM

Integrated Manufacturing System (IMS) can be said as method of integrating product and process design, execution and management of the approved design under one system. The prime aim of IMS is to combine or sometimes separate a different number of manufacturing processes for controlling them from a single source (Lewis, 2018). Early day's research and development on integration of manufacturing system started back with the remarkable development by U.K, U.S.A and Japan. IMS has always been a well-known experimentation and commercial research area for ages. A complete integrated system would require same degree of coordination and control required to be applied to other major operation areas, assembly and modification. By now almost all the companies trust their integration of manufacturing system with the advancement of computers. In other words, we can say integration of manufacturing processes today is done along with its management. Modern scenario of control along with business association, quality control, inspection and testing with the use of computers and logic comprises of the integration system. Computer integrated system is a manufacturing approach that uses computer of its overall production. Advance in computers allowed the use of complex algorithms to manage and control the manufacturing process, in so doing introduced a better planning, and control mechanism. Integration computers with manufacturing system i.e. Computer Integration Manufacturing (CIM) is a renowned IMS used in almost all the manufacturing industries. Manufacturing philosophy by CIM includes automatic production whether it is a small batch or a multi-batch production, fast and optimized design as well as fast product development, just-in-time (JIT) philosophy, and sturdy merchandising action. The CIM is related to the Strategic Information System (SIS) along with the human-centered flexible manufacturing process. Some of the benefits of integration system are:

- Reduce in lead time caused by interstate movement
- Component movements
- Improved machine tool utilization
- Reduced manpower requirement
- Work in progress reduction
- Superior flexibility of components, batching, loading

AUTOMATION AND MANUFACTURING

Every one of use has a basic knowledge about automation and manufacturing. So when we combine the words, then we can get a slight idea on what is automatic manufacturing. Something like self-manufacturing would be the common idea that pops out in our mind. Broadly or by definition Automation in manufacturing is defined as a technology consisting of a set of programmed instruction and control system through which the process is done without the assistance of human. They are a substitute of manual human assisting work with programmed machines and computer. In this fast economy and demand all over the world automation in manufacturing was a fast-growing trend and will be in trend in future as well. Automation creates job for skilled workers at the cost of unskilled and semi-skilled worker. They perform operations such as processing, assembly, inspection, or material handling. In some cases, they accomplish more than one operation in the same system. Compared to manual conventional processes they require reduced level of human interaction and participation. However, these days for some highly automated systems, where there is practically no human interaction. Examples includes processing of parts by automated machine system and devices, automatic transfer lines for series of manufacturing operations, automatic assembly system, processing or assembly operations by industrial robots, automatic material handling and storage operations, automatic quality control inspection (Soloman, 2011).

Automation Techniques

Automated manufacturing systems can be classified into three fundamental automation techniques. They are fixed automation, programmable automation, and flexible automation.

Fixed Automation

Fixed automation is a method by which the processing and assembling series operations are done by fixed configuration. The operations in the run are usually simple involving a plain linear or rotational motion or simple arrangement of both the motion. However, the integration and organization of many such operations into one single system makes it complicated. In fixed automation, the products are produced in large quantity with a high production rates but compared to other automation system it gives us the least flexibility. However, the overall unit cost of fixed automation is relatively lower compared to other automation processes. Machine transfer lines and automatic assembly system are some common examples of fixed automation.

Programmable Automation

In programmable automation, the production equipment has the skills to modify the series of operations to contain variations and variables accordingly. The operation series is controlled by a set of program. Programs are a set of implicit instructions that can be converted and interpreted by the system. Programmable automation systems are used for low to medium amount of production which are usually manufactured in batches. To manufacture each new separate batch of product, the manufacturing process system can be reprogrammed with the set of machine instructions corresponding to the new product. However, the substantial arrangement of the machine should also change accordingly like loading of tools, random fixture attachment, and the required machine setting. This changeover of such course of action takes time. As a result, the standard cycle in such system includes setup and reprogramming period followed by a production period of batches. Few examples of such automation are industrial robots, numerically controlled (NC) machine tools, and PLC.

Flexible Automation

Flexible automation is the expansion of programmable automation. Flexible automation is capable of producing a variety of parts with no wastage of time while changing. There is literally no waste of production time while reprogramming the system and changing the material and machine arrangements. The system is capable of producing the product directly without manufacturing it in batches. Hence, the operations changeover is optimized. Few of the features of flexible automation are custom manufacturing, high investment, variety product production, standard production rates and flexible production.

AUTOMATED MANUFACTURING SYSTEM

Production Planning and Control

Production planning and control (PPC) are responsible for solving the logistics problems encountered in manufacturing industry. It manages the information of what and how many products are to be produced and when it has to be produced. It is also responsible to obtain the raw materials, parts, and resources requirement details to produce those products. The computer is a very important factor for technically processing the huge amount of data and manufacturing resources involved in defining the products. PPC is the integrator in computer integrated manufacturing and should ensure them to be integrated functions. Both planning and control must be accomplished and coordinated with each other as well as with other functions in the manufacturing and processing firm. Now, having emphasized the integrated

Figure 3. Automation, product variety vs quantity produced

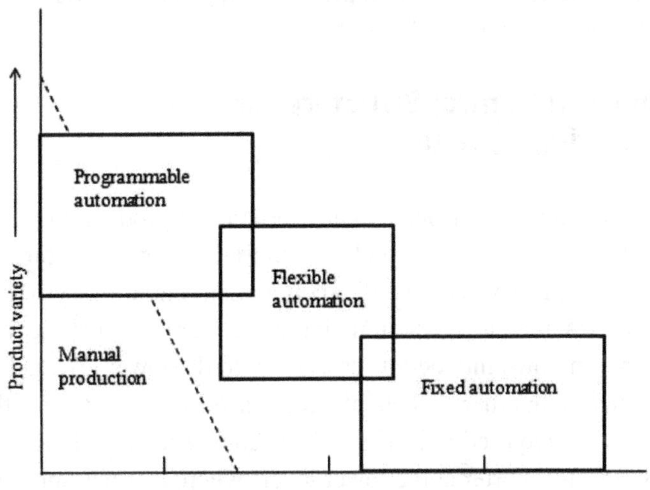

nature of PPC, let us nevertheless try to explain what is involved in each of the two function" production planning and production control. Production planning is concerned with: (1) Deciding products manufacturing and (2) Delivery scheduling and planning of manpower and material. Activities within the scope of production planning include:

1. Aggregate production planning involves planning of the production output for major product lines produced by the manufacturing unit. The plans must be synchronized with various functions in the unit, including product design, production, marketing and sales.
2. Master production planning is the next plan after the aggregate production plan, which is a precise plan of the quantities to be produced for individual product within each production line.
3. Material requirements planning (MRP) is a planning technique, usually implemented by computer to translate the MPS of products into a thorough plan for the raw materials and parts.
4. Capacity planning is used to determine the labor and tool resources needed to accomplish the master schedule. Production planning activities divide into two stages: (1) aggregate planning for MPS, and (2) detailed planning which includes MRP and capacity planning. Aggregate planning involves planning 6 months or more into the future whereas detailed planning is concerned with the short term planning.

Production control determines whether the required resources to implement the production plan are available, and if not, it takes corrective action against the problems. Production control includes the following systems and techniques:

- Shop floor control systems compare the progress and status of production orders in the factory to the production plans.
- Inventory control includes different techniques to manage the inventory of a company. Economic order quantity formula is considered one of the main tools.
- Manufacturing resource planning/MRP II combines MRP and capacity planning with shop floor control and other functions related to PPC.
- Just-in-time production systems refer to a scheduling where materials and parts are delivered to the next work cell or production line station prior to being used. It reduces inventory and different wastes in manufacturing (Xiaoyan, 2007).

Production Planning and Control Structure for Automated Manufacturing System

The previous section has shown few significant characteristics of production planning and control systems. The difficulty with production planning and control for batch manufacturing is very complicated. Various inter-relationships are present within a distinctive manufacturing system. This problem can be solved by means of hierarchies of control where major choices are taken at the highest levels and these decisions are gradually broken down into details as they go to the lower levels. The higher levels of the hierarchy are only concerned with other aspects of decision-making. It ensures that the manufacturing system is globally achieves the required objectives. The times tend to be longer at these higher levels, which are measured in years and shorter at the lower levels where it is measured in minutes. Therefore, the complexity of the production planning and control task for manufacturing systems can be approached via hierarchical planning and control. As we move down the hierarchy, the level of detail increases and the time decreases. The second characteristic is that production planning and control is significantly different in automated manufacturing systems. This difference is due to three fundamental factors: high capital investment, computer-oriented control and low manufacturing lead times. Since the capital investment is large, emphasis is made has to ensure that there is large usage of the system. Second, there is computer-oriented control; expensive existing software systems are integrated within the organization, to achieve efficient flow design, process planning and manufacture. Third, the reduced manufacturing lead times mean that decision cycle times will tend to be shorter. Another factor with shorter lead times lies in the provision of manufacturing resources: the short lead times of automated manufacturing systems mean that engineering details have to be ensured that these resources are available when necessary (Hitomi, 1991).

Other Types of Manufacturing System

There are four major types of manufacturing system namely:

Customer Manufacturing System

Company uses customer-manufacturing system for specified product lines. It is not appropriate for mass production businesses. In a custom manufacturing system, single operator with the help of machines makes each product. This gives highest-quality products but the low volume efficiency. Machines cannot produce more than one item at a time. The system has the maximum unit cost for manufacturing of the products. Thus, custom manufactured products are of highest quality but more expensive in the market. This system allows to customize individual products for the customer, but because of time-consuming, it is suitable for companies producing finely crafted, low volume, higher-priced item.

Intermittent Manufacturing System

The company produces several identical items at the same time. The products must be standardized and there is no customization of orders. Intermittent manufacturing system is also known as a “job shop”. The goods are produced in small quantities. Customization is done after purchasing of the products. This type of system is suitable for intermittent production. It is an effective alternative to continuous manufacturing. The intermittent manufacturing system allows companies to produce various goods in the same production line. Therefore, the manufacturing facility is designed to handle different product of different sizes and requirements. It is used for general-purpose machines and requires highly skilled labor.

Continuous Manufacturing Systems

Continuous manufacturing system is designed to facilitate the mass production of distinct products. This is the classic mass production system that people most generally think the word manufacturing. In a continuous manufacturing system, a product moves along an assembly line. The workers perform actions to mass product at stations along with this line. Continuous manufacturing system has higher output with a low unit cost. Nevertheless, such system requires a large capital investment due to the amount of labor and technology needed. They are used mostly for a very large operation with a large volume board of the same product line. This system was made famous by the ford company to produce TS model of cars in the 1920s during the famous industrial revolution.

Flexible Manufacturing Systems

Flexible manufacturing system is a recent manufacturing system that has become very popular due to its adaptability and flexibility in changing market. It has the ability and flexibility to produce a variety of high-volume goods, hence its name. The uniqueness of this flexible system includes the use of robots and other automated devices that replace or eliminate manual labor. It involves a considerable investment in equipment's and technology even though it reduces labor costs by using robots instead of paid human labor. The machines can be easily reprogrammed to manufacture different product with specific different quantity. The other advantage offered by flexible manufacturing system is its automated process quality control and low unit (Hitomi, 1991).

Cellular Manufacturing Systems

Cellular manufacturing is function of group technology wherein unrelated machines or processes are separated into different unique cells. Every single cell is dedicated to the manufacturing of a part and product or sometimes to a degree of group of family. The usual objectives of cellular manufacturing system are similar to that of a group technology. Some of the common objectives are:

- Curtailing manufacturing lead times: This is done by reducing setup, work part management, processing time, and operating system sizes.
- Decreasing work-in-process inventory: Smaller batch sizes and shorter lead times lead to reduced work progression.
- Improve in quality: Each cell is dedicated to produce only a specific number of different parts thereby improving the quality. This in turn reduces system variation.

Figure 4. Manufacturing system; flexibility vs. production

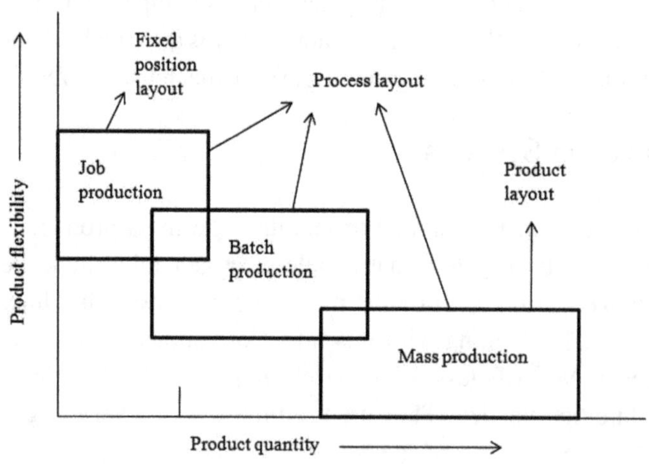

- Simplifying production schedule: The likeness among the parts in the cell reduces the complication of production schedule by simplifying scheduling through the cell instead of the overall machines.
- Reducing setup times: Using group tooling to process the part family, instead of the part tooling for a single part can help reduce the setup time effectively. This eliminates the number of individual tools required for operation as well as the time for changing the tools among parts.

Part family is said to be a grouping of related and similar parts and the grouping of machines and equipment's for process an individual part family is known as machine cell. Part families have similar design and/or manufacturing features. It is however not necessary for each element of a part family to be processed by each machine analogous to machine cell. The composite part for a known family includes all the design and manufacturing attribute of each part. Every single part of the family generally has some of the characteristics that distinguish the family but not necessarily all of them. Machine cell design is vital for all cellular manufacturing system that helps determine the performance of the overall cell. Group technology (GT) manufacturing cells can be classified according to the machines quantity and the measure of the material flow programmed between the machines. Here we categorize four general GT cell configurations (Soloman, 2011):

1. Single machine cell
2. Group machine cell with manual handling
3. Group machine cell with semi-integrated handling
4. Flexible manufacturing cell or flexible manufacturing system

Lean Production

Lean manufacturing is a term that embraces and follows many of the topics in manufacturing system, topics such as flexible manufacturing, minimizing work-in-process, "pull" systems of production control, and setup time reduction. MIT researchers to illustrate the collection of efficiency improvement that the Toyota Motors undertook to endure in the Japanese automobile business after World War II coined the term itself. Due to the collection of improvements has also been called the "Toyota production system'. Lean production can be defined as are vision of mass production that includes a better flexible and competent workers and work cells, by adopting methods to reduce waste in all forms. According to another author of "The Machine that Changed the World", lean production is based on four principles:

1. Minimize waste
2. Perfect first-time quality
3. Flexible production lines
4. Continuous improvement

Minimize Waste

All four principle of lean production are consequent of the first principle i.e., minimize waste. 'Taiichi Ohno's list of waste forms can be listed as follows:

1. Defective parts waste
2. Excess item waste
3. Inventories waste
4. Processing steps waste
5. People movement waste
6. Material transport waste
7. Workers waste

The various procedures, experiment, and theories were developed in the Toyota plants to minimize these forms of waste. The just-in-time production system we briefly discussed earlier was intended to produce no more than the least number of parts needed at the next workstation (waste 2). This reduced unnecessary inventories (waste 3).

Perfect First-Time Quality

The evaluation between mass production and lean production provides a sharp contrast in terms of quality and its control. For mass production, quality controls are defined in terms of an adequate quality level or AQT. This means that a definite point of fraction defects is sufficient and sometimes even satisfactory while lean production looks out for perfect quality of the product. This perfect quality is observed by the just-in-time delivery discipline for zero defect level part quality. There is minimum inventory in lean system to act as a buffer while in mass production, inventory buffers are used for the occurrence quality problems cases. The defective work parts are simply taken off the line and replaced with acceptable units. However, the problem is that such a policy tends to perpetuate the cause of the poor quality. Therefore, defective parts are continued to be produced. In lean production a lone defect draw attention to the quality problem forcing corrective action and a stable solution for minimizing the defects for the downstream production station (Lewis, 2018).

Flexible Production Systems

Mass production is predicated mainly on the principles of Frederick W. Taylor. As said by Taylor, workers had to be detailed about their working methods and were unqualified of planning their own tasks. By comparison, in lean production worker teams are able to organize their tasks and can participate to solve the technical problems. In mass production, the goal is to maximize efficiency, which is achieved by using a long production runs of like parts. This in turn tolerates long setup and speed changeover. Reduced setup times allow smaller batch sizes providing the production system with greater flexibility.

Continuous Improvement

Lean production supports the strategy of continuous improvement called kaizen by the Japanese. Continuous improvement is the constant search for better result and executing the resultant ways to reduce cost, improves quality, and increase productivity and even beyond that. Continuous improvement is carried out with the logic of one project at a time. The scheme are mostly concerned with the familiar problem areas such as cost reduction, quality improvement, productivity improvement, setup time reduction, cycle time reduction, manufacturing lead time and work-in-process, inventory reduction, and improvement of product design to increase performance and customer appeal.

APPLICATIONS

Inspection and Test Machines

We know that quality plays an important role in manufacturing. Precision and accuracy of the product is necessary for the quality of the product. Most of the time it is possible to inspect the product visually by humans or by testing it using x-ray detection, weight check, metal detection etc. however it is not possible to test every single product in very large production neither it is easy to inspect very tiny products like semiconductor diodes. In addition, inspection of every operation in units every time is impossible especially in fast and complex operation line. Such inspection may be tedious, difficult and might even take up lots of useful time. Modern technologies offer us a choice for automated inspection and testing system. We do not need to rely on humans for each inspection detail. Automatic inspection has high and better detection and inspection rates than manual human inspection. They provide high advantages over humans. Processes that are more complex can be detected. Better accuracy and precision are observed in such automatic system in faster time. This in turn increases the efficiency and the quality of the product for a better brand image.

Packaging

The packaging of products this day is an important factor both for keeping away the product from environmental and physical harm to the aesthetic properties of the product. Packaging of large and heavy products is very tedious. Not just that packaging of products in big industries had a very fast and mass production line. Manual and traditional approach to packaging is impossible to cope up with such production with precision. Automatic packaging and use of robots changed the packaging process by a par. The complex control system in a manufacturing unit makes it possible to load, unload and pack the products easily with high precision and accuracy at a much faster rate.

Assembly Machines

Manufacturing and assembly machines sustain a wide range of purposes and applications. Manufacturing machines such as stamping, forming and casting, packaging etc. are used to convert raw materials to a final product. Assembly machines bring together parts into final products or subcomponents. Automating manufacturing system allows automatic control assembly of machines. This provides better flexibility, faster lead-time and less repeatability. Automation also allows quick response to changes and increases the modification functionality of the system. This in turn increases the efficiency of the machines with faster response to the demand and market.

Chemical Processing

The manufacturing of chemicals has to be done very carefully and cautiously. The risk of coming in contact with the chemicals may be harmful to the workers and manufacturer as well as customers (in case of defective product). With the help of automation and control of the manufacturing processes the risk of such declines a lot. The entire life cycle of processes, including new installations, monitoring rebuilds and services can be automated to a certain extent. The automation and control of manufacturing provides better information system and tracking. It allows the storage of the data and process operation. The safety interlocking systems are also interfaced to the automation system. This provides a safe environment for the workers and manufacturers. Some of the benefits of automatic manufacturing of chemical processes are:

- High automation and production process availability
- Improved operational efficiency
- Easy repeatability of process operations
- Open system structure
- Reliability
- Scalable for further extensions
- Life cycle services close to the customer

Industrial Robotics

An industrial robot is an automatically controlled, reprogrammable, multipurpose manipulator, programmable machines with three or more axes. The most common applications of robots in industries are welding, painting, assembly, material handling, transportation, packaging and labeling, palletizing, product inspection, and testing. They are programmed to accomplish and assist their work with high endurance, accuracy, speed and precision. The most commonly used robot configurations are articulated robots, SCARA robots, delta robots and Cartesian coordinate robots. Generally, most types of robots in manufacturing fall into the category of robotic arms or manipulator with varying degrees of independence. The robots are programmed to carry out unambiguous actions several times repeatedly without variation with a high degree of precision. The following actions are determined by programmed routines that identify the direction, acceleration, deceleration, velocity and distance of a series of coordinated motions. Other robots are much more flexible to the course and orientations of the object on which they are operating. They can show their flexibility even the on task that has to be performed on the object itself that the robot might need to identify. For example, for accurate and better guidance, robots often contain machine vision sub-systems acting as their visual sensors connected to their controllers.

LATEST TRENDS IN AUTOMATIC CONTROL SYSTEM

Modern age or say computer era is highly competitive. IIOT or simply Industrial Internet Of Things is a well-known topic for the manufacturers in this era. Every second the customer demand changes in the market or they want something more, something new, and something better. This high ever-changing demand in the globalization together with the upcoming new companies or manufacturer in the market provides a high and tough competition. Sustainability is a common term for the researchers these days. All the new inventions and products have to follow certain sustainability criteria. But with the present processes and machines and skills creating such a sustainable product satisfying both the market and the supplier causes a hindrance. On top of that time is the most important aspect a customer looks while investing. To cope up and survive in the current competition company has to be fast, fast in production, fast in processes, fast in the market, fast in inspection etc. The faster the delivery option better would be the utilization of resources and man. This is where the development of automatic control in manufacturing comes in the area. Automatic control systems allow us to somewhat optimize and integrate all the processes from one end to the other. Such integration allows us to have a consistent, well-organized, efficient, accurate production. The development of AI and remote communication is the key factor in popularizing the term automatic manufacturing system. It has been and will always be a topic of interest for researchers, inventors, manufacturers and the market.

In this section let us discuss briefly various trends available in the automation and control of the manufacturing system.

Improved Accuracy

AI development and automatic control system allow us to analyze the changing demands of the market. Such analyzing can be used for predicting the yield rates of the overall production and even correct it. In this way we can have an accurate view point of the manufacturing unit which would eventually reduce the mistakes of human and reduce the time of inspection, design, analyze and overall production.

Optimization

Integration and optimization of all the processes and automization of all the machines, process and control system provide efficient and better use of space and resources, commission of work and machines, time, and design. Development of automatic control system in manufacturing industry would help to give an overall view of the industry and its related resources and to control the optimization. These provide us easy monitoring of the unit and finding the defaults as well as correct them. Thus, they overall reduce the use of prototypes and the time of manufacturing.

Artificial Intelligence

It is no surprise that computer is a must necessity of the modern industry. They are the core of the modern industry. Development of artificial intelligence has further increased the scope of automation and control system in manufacturing. Artificial Intelligence most commonly Robots provide a very flexible alternate to human labor and cost. Robots can be made to do works that would be impossible for man as well as work in an environment much harsher than a human can tolerate. For example, AGV development has helped an industry to transport heavy object to a large distance compartment of the industry without getting tired. Robots are a very trendy research area. Demand of flexible and cheap resource has led to the demand of better more advanced robots. Even though much developed countries uses robots for their overall manufacturing processes, India lacks behind and is still using conventional machines and processes.

Virtual and Artificial Reality

Prototype testing and manufacturing sometimes has a very complex and difficult procedures. Virtual Reality helps us to stimulate the product and the working environment where it is to be held digitally in computers. Augmented Reality is then used to project the digital information into the actual world. Thus, development of such software in automation would help visualize the complex and difficult product, analyze it under different condition and predict the processes and product accordingly.

Turnkey Development

This term is used for the product or says services designed to operate under a single step of turning the keys of the product. Even a manufacturer and its worker require an easy operation. Turnkey development provides easy starting of operation thereby reducing any accidents from complex machines or products. Turnkey automation would thus provide a simple yet very effective improvement in the control and manufacturing industry. However, countries all over the world especially developing countries like India and other south Asian countries lack such equipment. This is mainly due to the fact that there is less knowledge and engineers for the development of such turnkey deliveries.

CONCLUSION

Automation and control of manufacturing process is very popular topic for both the present and future age. Automation in terms of manufacturing means taking course tasks and designing a system that performs them in a total involuntary way without any human intervention and assistance. However, we should note that even if automation provides lesser human activities they would still need human assistance to deal with failures and recognize the problems. When integrated from one end to the other such automatic manufacturing produces reliable, efficient, transparent, and predictable output. Full automation of machining and process would provide better mechanical precision and safety of product and workers as well. It would provide better consistency in quality and less operating and material expenses. Development of technologies like wireless network artificial intelligence (AI), bus architects etc. are proving as a great asset for improvement of efficiency of the system. Development of such remote communication is breaking the barriers of information and communication in automation industry for the, analysis, modeling and design of the control system. Out of all the development of AI has far greater opportunity in the manufacturing industry together with the remote communication. AI provides synchronized equipped intelligence that supports product lifecycle management. It includes process integration, automation and collaboration insights, monitoring, and diagnostics. The control of manufacturing system and automation can be developed to respond automatically to the variation occurring in the market and processes. The digitalization of plants and machines ensures well-organized commissioned optimized machine design, smooth operations and short trading time. This reduces the dependence on expensive sample while reducing the time waste in the markets. The increasing intelligent industrial robotics introduction on the factory ground is a win for the fourth industrial revolution (Industry 4.0). Even though manufacturers have been using robots for decades, the constant development of robotic technologies every day has undoubtedly extended the budding applications for smart industrial robots. Thus, today robots are driven by revolutionary software and vision systems can be programmed to perform a variety of tasks, which fits in efficiently with the demand for flexible manufacturing.

REFERENCES

Collins, J. (1980). *Integrated Manufacturing- a state of art*. IET.

Groover, M. P. (2007). *Automation, production systems, and computer-integrated manufacturing*. Prentice Hall Press.

Gunasekaran, A., Goyal, S. K., Virtanen, I., & Yli-Olli, P. (1993). Production planning and control in automated manufacturing systems. *Production Planning and Control*, *4*(1), 2–16.

Halev, G. (2017). *Integrated manufacturing System. In Expectation and Disappointment of Industrial Innovation*.

Hitomi, K. (1991). Strategic integrated manufacturing systems: The concept and structures. *International Journal of Production Economics*, *25*(1-3), 5–12. doi:10.1016/0925-5273(91)90125-D

Jouela, S. L. J. (2007). *Future trends in automation*. Finland: Helsinki University of Technology.

Key, A. (2015, July 22). 5 Types of inspection equipment every manufacturer should know.

LaMarco, N. & Shoenberger, E. (2018). Types of Manufacturing System.

Lewis, A. (2018, February). A Breakdown of manufacturing. *System*, 26.

Li, P. (2019). Introductory Chapter: New Trends in Industrial Automation. In New Trends in Industrial Automation. IntechOpen.

Lozowski, D. (2016). Advancing with automation. *Chemical Engineering (Albany, N.Y.)*, *123*(3), 5.

Mahalik, N. P. (2014). Advances in packaging methods, processes and systems. *Challenges*, *5*(2), 374–389.

National Research Council. (1998). *Manufacturing Process Controls for the Industries of the Future*. National Academies Press.

O'Grady, P. J. (Ed.). (2012). *Controlling automated manufacturing systems*. Springer Science & Business Media.

O'Sullivan, D. (2009). *Industrial Automation*. National University of Ireland Galway.

Popovic, D. (n.d.). Automation and control in production processes. *Automation and Control, 3*, 2.

Rajawat, S.S. (2017). *What is an Algorithm and what are the types of algorithm with examples*.

Rockwell Automation. (n.d.). Manufacturing and assembly Machine Automation And Control.

Sadoyan, H., Zakarian, A., & Mohanty, P. (2006). Data mining algorithm for manufacturing process control. *International Journal of Advanced Manufacturing Technology*, *28*(3-4), 342–350.

Soloman, S. (2011). *Industrial sensors and control--The basics--Part I.* EE Times.

Xiaoyan, C., & Zhenliang, L. (2007, August). Integrated Computer-aided Control and Optimization of Béchamel Production Processes. *Proceedings of the 2007 IEEE International Conference on Automation and Logistics* (pp. 1646-1651). IEEE.

Chapter 7
Performance Evaluation of Cloud Systems by Switching the Virtual Machines Power Mode Between the Sleep Mode and Active Mode

Sudhansu Shekhar Patra
https://orcid.org/0000-0001-9996-7681
Kalinga Institute of Industrial Technology, India

Veena Goswami
https://orcid.org/0000-0002-4260-4721
Kalinga Institue of Industrial Technology, India

G. B. Mund
https://orcid.org/0000-0002-7845-3486
Kalinga Institue of Industrial Technology, India

ABSTRACT

Data centers are cost-effective infrastructures for storing large volumes of data and hosting large scale service applications. Cloud computing service providers are rapidly deploying data centers across the world with a huge number of servers and switches. These data centers consume significant amounts of energy, contributing to high operational costs. In this chapter, we study energy savings of data centers by consolidation and switching off of those virtual machines which are not in use. According to this policy, c virtual machines continue serving the customer until the number of idle servers attains the threshold level d; then d idle servers take synchronous vacation simultaneously, otherwise these servers begin serving the customers. Numerical results are provided to demonstrate the applicability of the proposed model for the data center management in particular, to quantify theoretically the tradeoff between the conflicting aims of energy efficiency and Quality of Service (QoS) requirements specified by cloud tenants.

DOI: 10.4018/978-1-7998-1831-1.ch007

1. INTRODUCTION

The advancement of virtualization technology made the new emerging field Cloud Computing (Durao et al., 2014; Buyya et al., 2009; Rimal et al., 2009) as an egressing computation epitome that offers unlimited number of testing and staging servers which can be dynamically provisioned on the basis of pay-per-use (Tsai et al., 2013). The emerging computing assures to provide on-demand, elastic and compromising IT services, which leaves the traditional programming models far behind and adopts the new ones. The team of agile development seamlessly combines the many developments, production and testing environments with the other services of the cloud using the Cloud computing and virtualization technology. This emerging service epitome help the users and eradicate the burden of creating and managing the composite infrastructure. Rather than using the traditional own-and-use pattern, the consumers are now diverting towards this new computing paradigm in which the computing paradigm is developed as a utility computing which offers a pool of computing resources in a pay-per-use manner pay-as-you-go basis (Sivathanu et al., 2010). Virtualization offers more infrastructure services rather than platform and application services. But Cloud computing providers offer supply service resources based on several fundamental models, including infrastructure as a service (IaaS), platform as a service (PaaS) and software as a service (SaaS) (Armbrust et al., 2009; Begnum, 2012). For example, Amazon Elastic compute cloud (EC2), Google's App Engine, Amazon S3, Salesforce, etc. are all existing business models to provide computing infrastructure, programming platforms, data storage and software applications as services respectively. Cloud computing has gained enormous popularity in both business and scientific communities because of its cost-effectiveness, reliability, and scalability (Buyya et al., 2009; Armbrust et al., 2009). For the procurement of any new infrastructure, the users do not require further investment. The users can get the required demanded services just by payment and without worrying about the complexity of the IT infrastructure. The potential of cloud computing can only be realized if the cloud service providers can flexible in their service delivery to meet the various customer's requirements by keeping the consumers isolated from the underlying infrastructure.

Recently, every service provider has given their attention towards high-performance computing in data centre deployments, but without paying much attention to energy consumption, these demands have been fulfilled. However, the energy consumption by a data centre in an average is as much as that of 25,000 households (Kaplan et al., 2008; Sofia et al., 2015). As energy costs are increasing day by day while availability dwindles, there is an urgent need for optimizing the energy efficiency of data centres without violating performance. There is a chance that the cloud service providers profit margin may dramatically reduce due to high energy costs. The risk of rising energy cost may increase the Total Cost of Ownership (TCO) and reduce the Return on Investment (ROI) of Cloud infrastructures. Therefore, the service provider needs to adopt measures to ensure their profit margins. The major power consumed by data centres are by the physical machines during their computation and air-conditioning for cooling the servers which are almost the same as that used by the server itself. The researchers and the industries are using the energy consumption index to measure the energy efficiency of Cloud data centres. This index is used to measure the percentage of energy consumed by the data centre computing devices out of all of the energy consumed by the data centre. Here the energy consumption by the data centres includes the energy consumption of computing devices, as well as the energy used for heating, temperature control, ventilation, lighting and much more. Most of the power consumption in data centres is caused by computational processing, disk storage, networking, and cooling systems, and it is given as

$$P_{total} = P_{AC} + P_{pm} + P_{additional}$$

where P_{total}, P_{pm}, P_{AC} and $P_{additional}$ refer to the data centre total energy consumption, the total physical servers energy consumption, the cooling system energy consumption, and the other energy consumption, respectively. Worldwide there is a rising pressure from the government agencies to reduce the carbon footprints which have a substantial effect on climate change. The Japan Data Centre Council established by the Japanese government addresses the gliding energy consumption of data centres (Ministry, 2010). The traditional data centers are competing to find ways to increase the data centre resource efficiency and reduce energy consumption. Recently, various service providers have formed, The Green Grid (green grid, 2011), a global consortium to minimize the environmental impact and to promote energy efficiency for data centres. Thus, providers necessitated minimizing the energy consumption of cloud infrastructures while ensuring service delivery.

Due to the quick growth of computing applications and data, there is a constant necessity to increase the larger servers and disks to process fast within the required period which makes it challenging and complex issue to lower the energy usage by the data centres. Green Cloud computing is visualized to achieve efficient processing and utilization of computing infrastructure as well as to minimize energy consumption (Beloglazov et al., 2010). It is crucial to maintain the sustainable growth of Cloud computing shortly. Otherwise, Cloud computing with increasingly permanent front-end client devices interacting with back-end data centres will cause a tremendous escalation of energy usage. To address this problem and drive Green Cloud computing, data centre resources need to be managed in an energy-efficient manner. In particular, Cloud resources need to be allocated not only to satisfy Quality of Service (QoS) requirements specified by users via Service Level Agreements (SLAs) but also to reduce energy usage. Energy efficiency may be more desirable on quite different levels.

Worldwide the energy consumption in data centres is expected to double in every five years, which is going to add a huge cost to the environment as well as to the business. Unfortunately, most of the energy is wasted only to maintain extra servers and excess service capacity during periods of low load. A statistical study shows that during the low utilization of resources the servers operate at 10 to 15 percent of their full capacity, which leads to wastage of energy. Therefore, it is a great challenge for cloud providers to manage the capacity of data centres to use power as per need. So, the data centre operators want to manage data centres in a cost operative manner by reducing the energy consumption as well as by meeting the SLAs. Virtualization is the most promising technology to implement energy-efficient right-sizing of data centres.

As shown in Fig. 1, at the lowest layer, there is a pool of physical servers organized as the data centre infrastructure consisting of a large amount of computing and storage resources. Above this layer is the virtualization layer, where users access the resources in the form of virtual machines (VMs) and submit their tasks with different Service-Level Agreements (SLAs). From this architecture, it can be seen that with the benefits of VMs, the goal of right-sizing the data centres can be accomplished through the consolidation of computation load into a smaller number of servers while powering off the idle ones. Driven by this idea, the basic problem underlying the right-sizing can be seen as adapting the number of active servers in the data centres to match the current workload. This chapter addresses energy efficiency management and analyses the performance of the system by putting the VMs in an idle state or in vacation when no customer is waiting in the system.

The rest of the chapter is organized as follows. Section 2 briefly reviews the related works. Section 3 presents Cloud power management architecture. Model description and its analysis are carried out in Section 4. Various performance measures and the power consumption by VMs are evaluated in Section 5. Section 6 contains computational experiences with a variety of numerical results in the form of graphs to show the effectiveness of the model parameters. Section 7 concludes the chapter.

2. RELATED WORK

Improvements in power consumption and the relationship between resource usage and energy consumption has been widely studied (Fan et al., 2007; Wang et al., 2010; Lien et al., 2006; Nathuji et al., 2007; Bekelman et al., 2008; Song et al., 2009; Bekelman et al., 2008; Torres et al., 2008). Elnozahy et al. (2003); Sasikala (2013); Kharat et al. (2016) have investigated the problem of power-efficient resource management in a single web-application environment with fixed SLAs (response time) and load balancing handled by the application. As in Chase et al. (2001), two power-saving techniques were applied: switching power of computing nodes on/off and Dynamic Voltage and Frequency Scaling (DVFS). The main idea of the policy is to estimate the total CPU frequency required to provide the necessary response time, determine the optimal number of physical nodes and set the proportional frequency to all the nodes.

Several excellent surveys on single server vacation models have been done by Kalloniatis et al. (2014), Chebiyyam et al. (1986), Teghem (1986), and Takagi (1993). Tian and Zhang (2006) investigated the multi-server vacation models of both synchronous (servers taking the same vacation together) and asynchronous types and provided some algorithms for computing the stationary probability distributions and expected performance measures. Recently, Tian et al. (2006) have established the conditional stochastic decomposition properties and have provided the stationary distributions for the queue length and the waiting time in the Markovian multi-server synchronous vacation queueing system. They have proved that given that all servers are busy, the queue length or the waiting time can be decomposed into the sum of two independent random variables. Jain et al. (1989) presented the modelling and analysis of unreliable Markovian multi-server finite buffer queue with discouragement and synchronous working vacation policy. Power and performance analysis for web server cluster have been carried out in Wang et al. (2012). In Khazaei et al. (2012), the authors have described a novel approximate analytical model for performance evaluation of cloud server farms and solved it to obtain an estimation of the complete probability distribution of request-response time and other important performance indicators. They also pointed out that accommodated heterogeneous services in a cloud centre might impose longer waiting time for its clients than a homogeneous equivalent.

In the cloud-based IoT applications Energy consumption is a challenging problem in due to massive data transfer between several geographically distributed cloud servers. Baker et al. (2017) proposed an algorithm to reduce the energy consumption during massive data exchange between servers. Further, Baker et al. (2015) proposed a network-based routing algorithm which aims to search the lowest energy consumption path for processing and to store the big data in cloud data centre. Energy consumption is increasing day by day due to high computational demand of extra servers, switches, routers, etc. To solve this issue Baker et al. (2013) have proposed an autonomic meta-director framework that searches the optimal (energy efficient) path between users and data centre to reduce the energy consumption by using the linear programming approach. Aldawsari et al. (2015) proposed a new framework which aims to find a suitable data centre to allocate the application for processing and optimize the energy consumption as well as other QoS parameters. Patra (2018) proposed an algorithm that optimizes energy consumption and resource utilization parameters. Shen et al. (2018) have proposed certificate less authentication protocol for wireless body area network that reduces the execution time. Li et al. (2015) have proposed an algorithm to protect the confidentiality of the sensitive data using convergent encryption technique. However, no power saving methodology was discussed in their work. In this chapter, we use the asynchronous vacation queueing model to analyse the server cluster. We evaluated the number of VMs that may go to sleep mode without affecting the performance of the cloud system, maintaining the QoS parameters and following the SLA.

Wang et al. (2018) proposed virtualization based unified optimization framework to improve the energy efficiency of data centre networks by using the concept of multi path routing protocol and hierarchical feature of the topology. Wang et al. (2014) proposed an efficient energy-saving technique for data centre networks by scheduling and routing "deadline constrained flows" where the transmission of every flow has to be accomplished before a rigorous deadline, being the most critical requirement in production data centre networks. Shu et al. (2016) proposed architecture of cloud-integrated cyber-physical systems to execute complex industrial application in a controlled manner. Wang et al. (2013) proposed an energy-efficient framework (GreenDCN) by assigning virtual machines to servers to reduce the amount of traffic and to generate favourable conditions for traffic engineering. Further, GreenDCN reduces the number of active switches and balances traffic flows which improve energy-efficient of the data centre network. Gill et al. (2017) proposed a resource management algorithm that optimizes the parameters time, execution cost and SLA violation. Mishra et al. (2018) have proposed an algorithm for efficient energy consumption using the technique of virtual machine migration in cloud computing, but none of the algorithms has proposed the method of switching off the VMs when the machines are not in use. To improve the performance of the cloud system and also to reduce energy consumption of the data centre, we propose a technique to switch the VM's between the sleep mode and the active mode.

3. CLOUD POWER MANAGEMENT ARCHITECTURE

In recent years, significant research has been done to reduce energy cost in cloud data centres. The two high-level approaches to reduce power consumption in data centres are by dynamic server provisioning and dynamic VM consolidation (Mathew et al., 2012).

Dynamic server provisioning proposes to power on the required minimum VMs according to the current workload to save power. Additional VMs are brought online if the amount of resource requests increases (Farahnakian et. al., 2014). If request demand decreases, unused VMs should be switched off or put into the energy preserving mode like sleep state. The characteristics of the dynamic server provisioning in data centres are as follows:

- If a server becomes idle, it will be set in a sleep mode with low power consumption from which it can be reactivated quickly without booting.
- If a server becomes idle, it will be completely switched off and booted again if server activation is required.
- The cloud system should automatically oscillate between activations and deactivations of server resources to serve the volatile load.
- Lowering of new server activations upon short load burst during the scalable upper threshold.

- The task requests should be served as per the service level agreement (SLA) with maximum service rate by the activated servers to keep delay time as low as possible.
- The SLA should be guaranteed except the overload situations of the cloud system. When all the servers are activated, and the new requests could not be served under the overload situations, the new requests can be handled by the multi-cloud system.
- Server deactivation takes place when the waiting queue requests fall below a scalable lower threshold.
- The server deactivation can be considered in two different modes:
- Sleeping or deactivated servers are activated again when the FIFO queue waiting length increases to a certain threshold.

The other way of energy-saving is by dynamic VM consolidation. The main idea in the living VM consolidation approach (Vogels, 2008; Feller et al., 2012; Murtazaev et al., 2011; Chebiyyam et al., 2011) is to use live migration to consolidate VMs periodically. Virtual machine consolidation has proven to be

Figure 1. The architecture of cloud data centre using virtualization technology

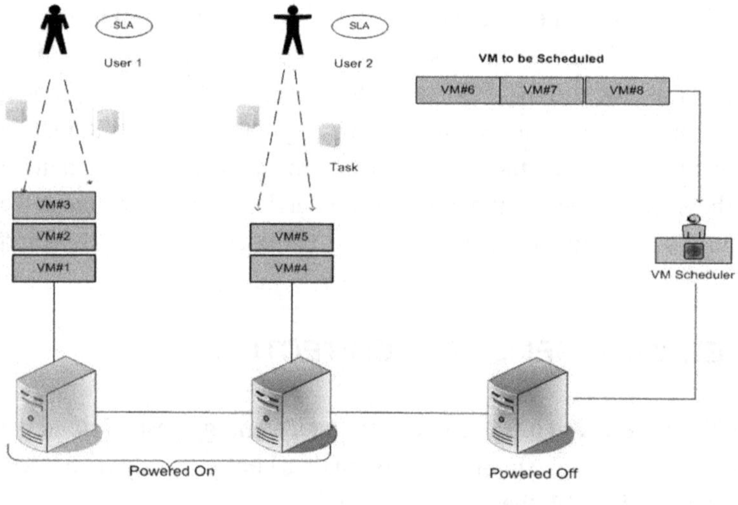

Figure 2. Server consolidation operation

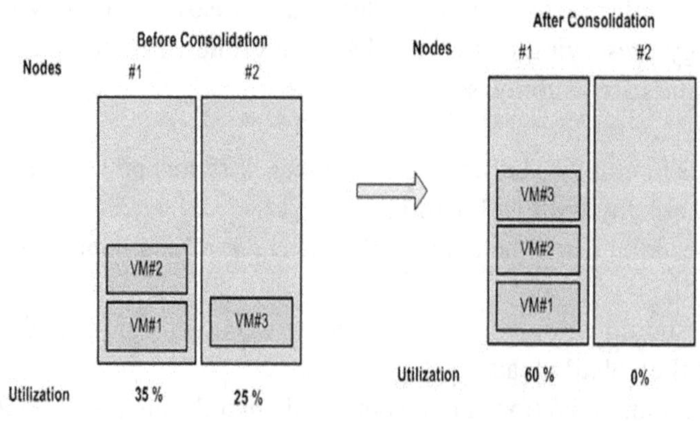

an effective technique for reducing the power consumption by switching the idle or under-utilized physical servers off or to a low power modes (Pedram, 2012). It leverages virtualization technology such as Xen (Barham et al., 2003) and VMware to provide significant benefits in data centres to reduce energy consumption. This technology allows a Physical Machine (PM) to be shared between multiple VMs, where each VM can run multiple application tasks. A dynamic VM consolidation method packs the existing VMs into fewer PMs as possible and switches the idle PMs into a power-saving mode. However, achieving the desired level of Quality of Service (QoS) between users and a data centre is critical. The QoS requirements are formalized via a Service Level Agreement that describes such characteristics as minimal throughput, maximal response time or latency delivered by the deployed system.

The traditional approach for server consolidation involves manually mapping VMs from source hosts to destination hosts (Chebiyyam et al., 2009). However, this is not feasible where periodic server consolidation operations are performed. Server consolidation operation starts from some intermediate state and performs several VM migrations to reach such a state where minimal nodes are used as illustrated in Figure 1.

Numerous research work has been done on server consolidation, which not only minimizes the number of nodes but also minimizes the number of migrations. Server consolidation operation starts from some intermediate state and performs several VM migrations to reach such state where minimal nodes are used as shown in Figure 2. It is better to turn off some VMs so that only the minimum number of VMs is kept active to support the workload, as depicted in Figure 3. Nodes are brought active after the completion of sleep state.

4. DESCRIPTION OF THE MODEL

Given the dynamic nature of cloud data centres, we first introduce a queueing model, which captures the main characteristics of the cloud data centre environment. We then derive the steady-state system-length distributions of the model. Figure 4 shows a generic queueing model with dynamic activation/deactivation for Markovian task requests.

Figure 3. Cloud power management model

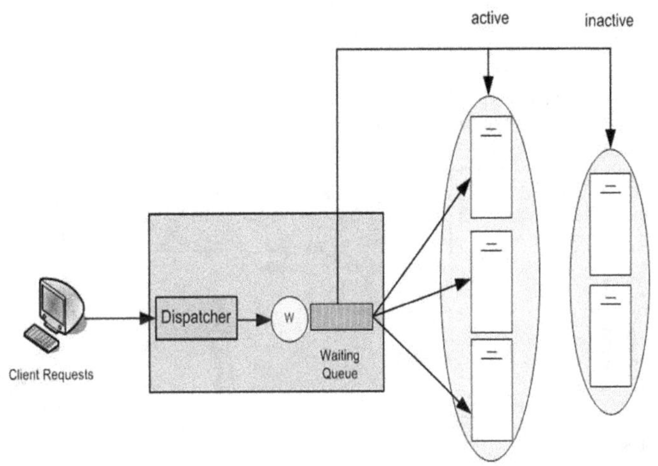

We model the system as a multi-server queueing system with finite waiting spaces in which the arrival of customers is a Poisson process with rate λ and service times for each c server are exponentially distributed with rate μ. The sleep state specifies that any d ($1 <= d < c$) of the c servers may take synchronous sleep when these d servers become idle at a service completion instant and c-d servers are available. During the synchronous sleep of d servers, the rest c-d servers are available. If the number of customers is not more than c-d at a sleep completion instant, the d servers take another sleep simultaneously. Otherwise, theses servers start serving the customers. The sleep times are assumed to be exponentially distributed with rate ϕ. We assume that the arrival process, the service and sleep times are mutually independent. The state of the system at time t is described by following random variables:

- $Ns(t)$ = *number of customers present in the system,*
- $\delta(t) = \begin{cases} 0, \textit{if the d servers are on sleep(inactive)} \\ 1, \textit{if the servers are not on sleep(active)} \end{cases}$

The process $\{(Ns(t); \delta(t)): t \geq 0\}$ is a continuous-time Markov chain (CTMC) with state

space $\Omega = \{(n,0): 0 \leq n \leq c\text{-}d\} \cup \{(n, i): c\text{-}d + 1 \leq n \leq N, i = 0,1\}$.

We define the stationary probabilities of the system as

Figure 4. General queueing model for the data center

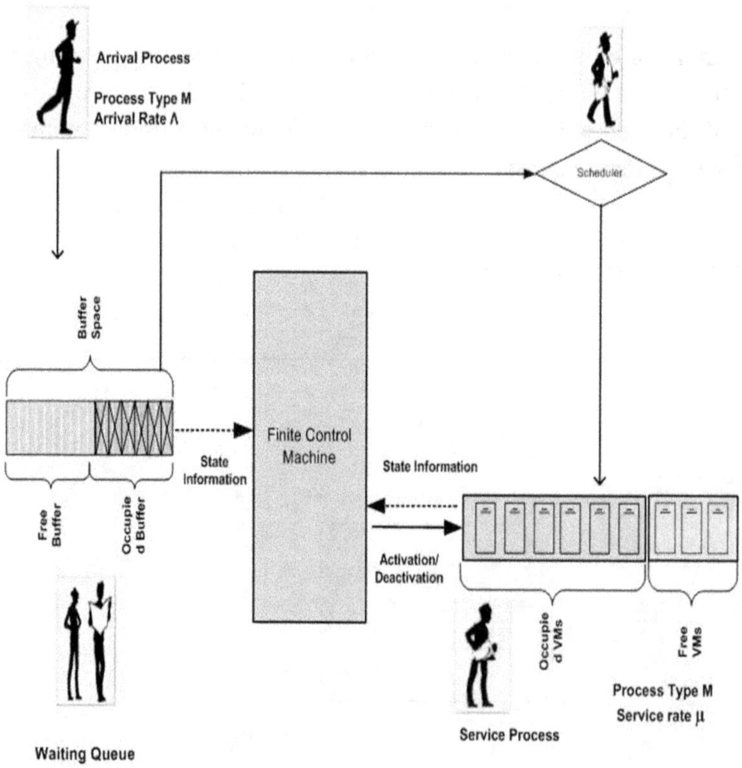

$$\Pi_{n,0} = \lim_{t->\infty} P\{N_s(t) = n, \delta(t) = 0\}, 0 \le n \le N,$$

$$\Pi_{n,1} = \lim_{t->\infty} P\{N_s(t) = n, \delta(t) = 1\}, c-d+1 \le n \le N$$

The steady-state balance equations for the model are given by

$$\lambda \Pi_{0,0} = \mu \Pi_{1,0} \tag{1}$$

$$(\lambda + n\mu)\Pi_{n,0} = \lambda\Pi_{n-1,0} + (n+1)\mu\pi_{n+1,0,} 1 \le n \le c-d-1 \tag{2}$$

$$\left(\lambda + (c-d)\mu\right)\Pi_{c-d,0} = \lambda\Pi_{c-d-1,0} + (c-d)\mu\pi_{c-d+1,0} + (c-d+1)\mu\pi_{c-d+1,1} \tag{3}$$

$$\left(\lambda + \phi + (c-d)\mu\right)\Pi_{n,0} = \lambda\Pi_{n-1,0} + (c-d)\mu\pi_{n+1,0}, c-d+1 \le n \le N-1 \tag{4}$$

$$(\Phi + (c-d)\mu)\Pi_{N,0} = \lambda\Pi_{N-1,0} \tag{5}$$

$$(\lambda + (c-d+1)\mu)\Pi_{c-d+1,1} = \phi\Pi_{c-d+1,0} + (c-d+2)\mu\pi_{c-d+2,1} \tag{6}$$

$$(\lambda + n\mu)\pi_{n,1} = \phi\pi_{n,0} + \lambda\pi_{n-1,1} + (n+1)\mu\pi_{n+1,1}, c-d+2 \le n \le c-1 \tag{7}$$

$$(\lambda + c\mu)\pi_{n,1} = \varphi\pi_{n,0} + \lambda\pi_{n-1,1} + c\mu\pi_{n+1,1}, c \le n \le N-1 \tag{8}$$

$$c\mu\pi_{N,1} = \varphi\pi_{N,0} + \lambda\pi_{N-1,1} \tag{9}$$

Using equations (4) and (5) recursively, we get

$$\pi_{N,0} = \psi_n \pi_{N,0}, c-d \le n \le N$$

where the unknowns $\psi_n \left(c-d \le n \le N\right)$ are given by

$$\Psi_n = \begin{cases} 1 & n = N \\ \dfrac{\varphi + (c-d)\mu}{\lambda} & n = N-1 \\ \dfrac{\lambda + \varphi + (c-d)\mu}{\lambda}\psi_{n+1} - \dfrac{(c-d)\mu}{\lambda}\psi_{n+2}, & n = N-2, \ldots\ldots, c-d \end{cases} \tag{10}$$

Using the backward substitution method in the equations (7)-(9), we get

$$\pi_{n,1} = u_n \pi_{N,0} + v_n \pi_{N,1}, c-d+1 \le n \le N,$$

where u_n and v_n, $c-d+1 \le n \le N$ are

$$u_n = \begin{cases} 0 & n = N \\ -\left(\dfrac{\varphi}{\lambda}\right), & n = N-1 \\ \dfrac{\lambda + (n+1)\mu}{\lambda} u_{n+1} - \dfrac{(n+2)\mu}{\lambda} u_{n+2} - \dfrac{\varphi}{\lambda}\psi_{n+1}, & n = N-2, \ldots . c-d+1 \end{cases} \tag{11}$$

$$v_n = \begin{cases} 1 & n = N \\ \dfrac{c\mu}{\lambda}, & n = N-1 \\ \dfrac{\lambda + (n+1)\mu}{\lambda} v_{n+1} - \dfrac{(n+2)\mu}{\lambda} v_{n+2}, & n = N-2, \ldots . c-d+1 \end{cases} \tag{12}$$

Now, using equation (6), we get

$$\pi_{N,1} = TN_0 \tag{13}$$

where

$$T = \left(\frac{(\lambda + (c-d+1)\mu)u_{c-d+1} - (c-d+2)\mu u_{c-d+2} - \varphi\psi_{c-d+1}}{(c-d+2)\mu v_{c-d+2} - (\lambda + (c-d+1)\mu)v_{c-d+1}} \right)$$

From equations (3) and (2), we yield

$$\psi_n = \begin{cases} \frac{(\lambda + (c-d)\mu)}{\lambda}\psi_{c-d} - \frac{(c-d)\mu}{\lambda}\psi_{c-d+1} - \frac{(c-d+1)\mu}{\lambda}\left(u_{c-d+2} + Tv_{c-d+1}\right), n = c-d-1 \end{cases} \quad (14)$$

Using the normalization condition, we get $\pi_{N,0}$ as

$$\pi_{N,0} = \left(\sum_{n=0}^{N}\psi_n + \sum_{n=c-d+1}^{N}\left(u_n + Tv_n\right)\right)^{-1} \quad (15)$$

5. PERFORMANCE ANALYSIS

In order to derive the optimal energy consumption $(F(d^*))$ by the system, several performance measures are essential including the mean number of client requests in the system (L_s), the mean queue length of the buffered request (L_q), the expected number of busy VMs in the sleep mode E(V) and the expected number of idle VMs in the system E(I). Based on the stationary state probabilities given in section 4, the expressions for L_s, L_q, E(B), E(V) and E(I) are obtained as follows:

- The mean number of client requests in the system is given by

$$L_s = \sum_{n=0}^{N} n\pi_{n,0} + \sum_{n=c-d+1}^{N} n\pi_{n,1}$$

- Mean queue length of the buffered request is given by

$$L_q = \sum_{n=c-d+1}^{N}(n-c+d)\pi_{n,0} + \sum_{n=c+1}^{N}(n-c)\pi_{n,1}$$

- The average number of busy VMs in the system is given by

$E(B) = L_s - L_q$

- The average number of VMs on sleep mode is given by

$$E(V) = \sum_{n=0}^{N} d\pi_{n,0}$$

- The average number of idle VMs in the system is given by

$$E(I) = c - E(V) - E(B)$$

POWER CONSUMPTION BY VIRTUAL MACHINES

The total energy consumption F(d) by the system can be defined as

$$F(d) = C_s E(B) + C_s E(I) + C_h E(V) / \lambda + C_\mu \mu$$

Where

C_s = Constant energy cost when one VM is in the setup state.
C_μ = the energy cost of serving a VM at service rate μ.
C_h = the cost of holding each VM per unit time

Our main objective is to find the optimal value of d so that F(d) should be optimum, and the procedure to find the optimal solution is described in Algorithm 1.

6. EXPERIMENTAL RESULTS

We evaluate the proposed model through simulation run with synthetic data sets. The experiments were carried out using MATLAB R2012a version 7.14.0.739 on an Intel Core 2 Duo processor, 2.20 GHz CPU and 4 GB RAM running on Microsoft Windows 7 platform. Numerical results for various system performance measures are presented. Table 1 shows the expected waiting time in the system *Ws* for varying number of VMs (*c*) and idle VMs (*d*) in the cloud system. It is observed that the expected waiting time increases when the idle VMs in the system increases and decreases when the number of VMs increases. Various system performance measures under the optimal operating conditions for various values of $\lambda, \mu\, and\, \varphi$ are given in Table 2. Table 3 shows the various performance measures such as the average length of the system (*Ls*), the expected number of VMs in sleep mode (*E*[*V*]), the expected number of the idle VMs (*E*[*I*]) and the expected number of busy VMs (*E*[*B*]) in the system.

Figure 5 depicts the average number of idle VMs E(I) in the system versus the number of VMs c by varying the number of VMs undergone sleep mode d. It is seen that the average number of idle VMs increases when c increases for a fixed d. As d increases, E(I) decreases for a fixed c. Figure 6 shows the variation of arrival rate λ with the average length of the system L_s for various sleep rate of the VMs ϕ. It is observed that L_s increases when λ increases for a fixed ϕ. Figure 7 shows the relationship among λ, d and Ploss. For a fixed d, the P_{loss} increases when λ increases. Similarly, for a fixed λ, the Ploss increases when d increases. The expected number of VMs on sleep mode versus λ for various d is shown in Figure 8. The graph shows that E(V) increases as d increases. When a fixed number of VMs are in sleep mode, the energy consumption by the system is optimal. Figure 9 and Figure 10 depicts energy consumption versus the number of VMs in sleep mode. As the graphs show when the value of d= 6 (the

Algorithm 1. Optimization Algorithm F(d)*

```
PROCEDURE MEASURE
INPUT: F(d), c, λ, μ, φ and N
OUTPUT: Approximate Solution d*
F* =∞
  for (d=1; d<c; d++) do
Calculate F(d*)
   if (F(d*) < F*) then
      F* = F(d*)
S= d*
end if
end for
return S
```

Figure 5. Impact of c on E[I] for various d.

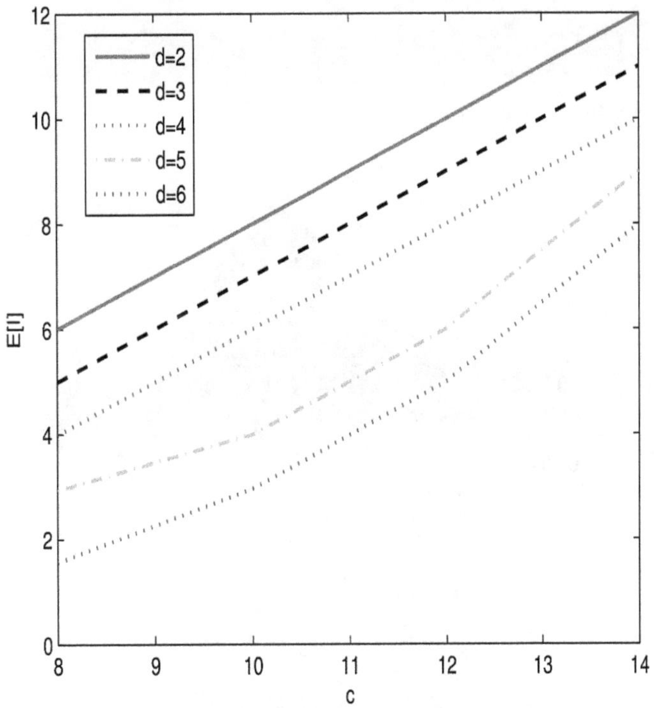

Figure 6. Impact of λ on Ls for various ϕ.

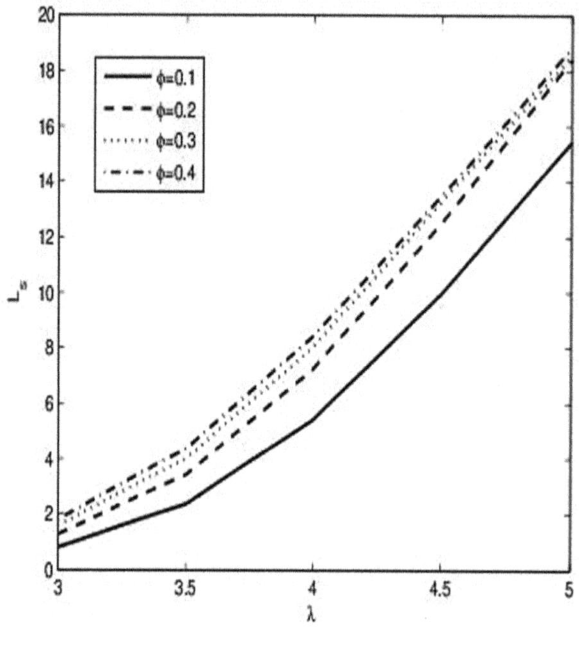

Figure 7. P_{loss} versus λ vs. d.

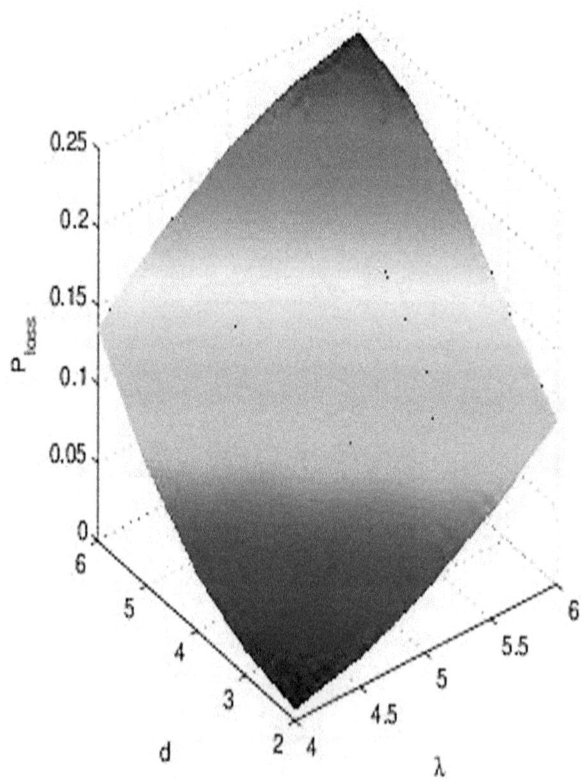

Figure 8. Expected number of VMs on sleep mode

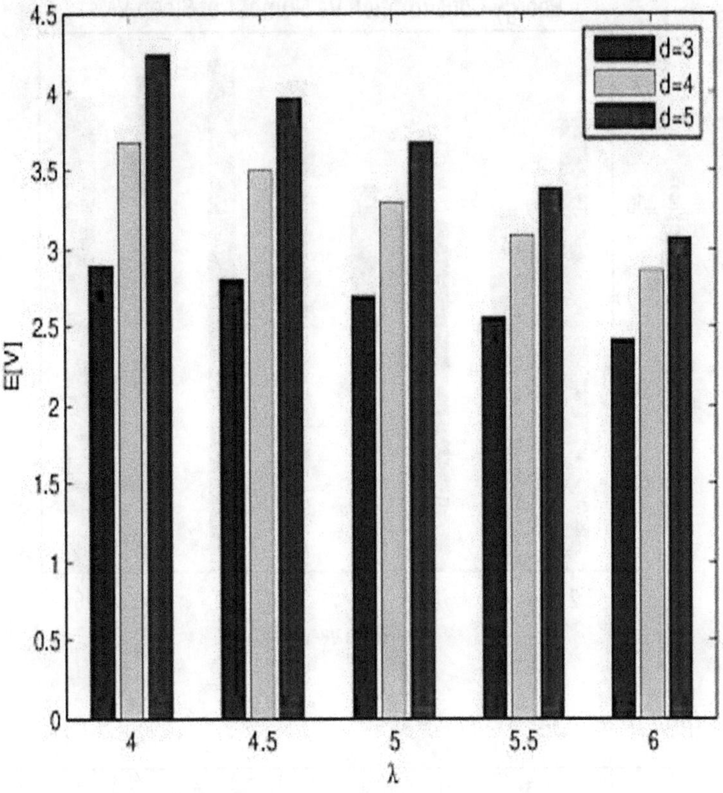

Figure 9. Energy consumption by the system Vs Number of VMs on vacation

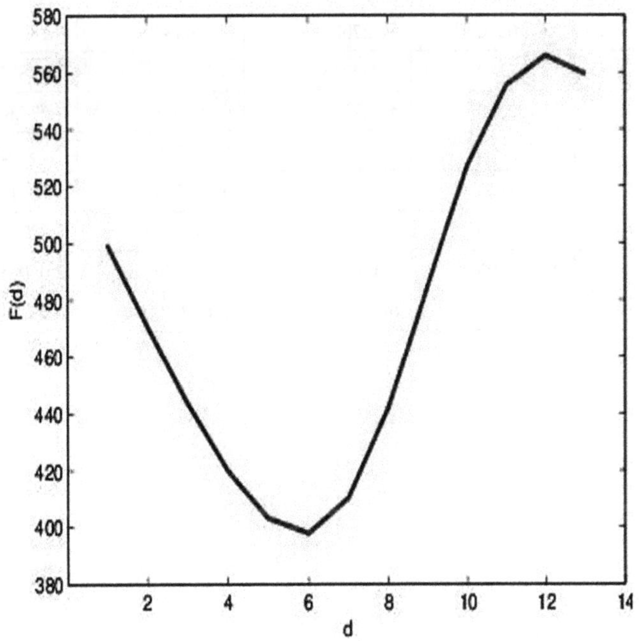

Figure 10. Energy Consumption Vs Number of VMs in vacation

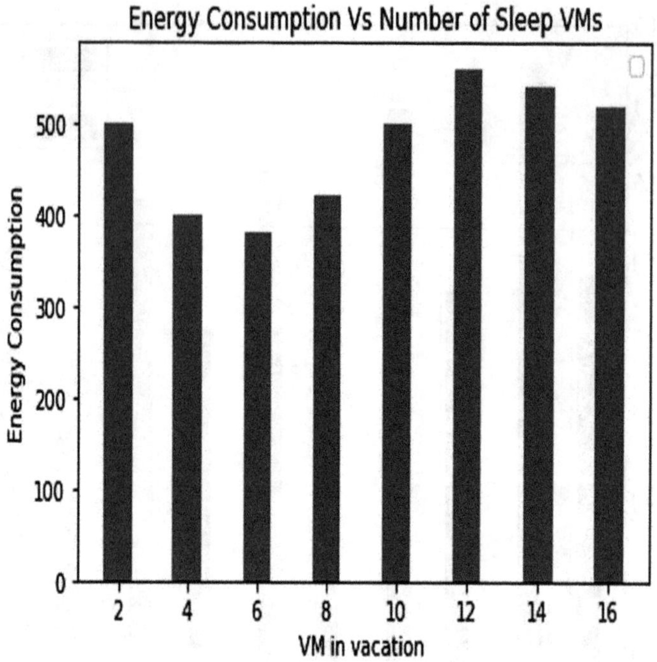

Table 1. Expected waiting time in the system Ws

c/d	2	3	4	5	6	7
3	2.7383738	4.8393409	7.8749383	8.9373488	9.9283838	9.8937383
4	2.4583837	4.2877383	7.1837484	8.3773830	9.0389783	9.4748748
5	2.2452828	3.8191738	6.2827383	7.6253728	8.4527287	8.9638388
6	1.7618628	3.6730380	5.9273633	7.1244339	8.1383839	8.5263788
7	1.6784582	3.1927270	5.3522738	6.8937373	7.8289370	8.3536738
8	1.6123568	2.9182638	4.9273738	6.4646738	7.5373689	8.0928373
9	1.4160700	2.7455800	4.5455200	6.2601500	7.2390570	7.6743100
10	0.5395200	1.1835100	2.3803200	4.0793400	5.7941300	6.8872500
11	0.2014790	0.4719220	1.0630400	2.1867400	3.8223600	5.5262000
12	0.0740314	0.1833190	0.4387650	1.0037700	2.0892100	3.6882000
13	0.0260473	0.0695078	0.1751920	0.4240460	0.9771190	2.0440500

Table 2. System performance measures for various values of λ, μ and ϕ

ϕ	μ	λ	c	d^*	$F(d^*)$	$E(I)$	$E(B)$	L_q	P_{loss}
0.1	0.7	5	12	2	377.750176	2.892137	7.13407	0.471775	0.0016697
			15	5	407.194264	2.918037	7.13516	0.449175	0.0015872
			17	7	427.608276	2.937325	7.13539	0.446355	0.0015797
			20	10	458.488184	2.967940	7.13550	0.445469	0.0014757
0.1	0.7	7	12	1	472.700676	1.313574	9.76868	1.965525	0.0324269
			15	3	472.730913	2.240552	9.83697	0.976960	0.0163374
			17	5	492.319524	2.273584	9.84222	0.957433	0.0156852
			20	7	523.737815	5.053436	9.95720	0.128575	0.0044373
0.1	1.2	5	12	6	290.390613	1.839392	4.22418	0.619120	0.0006359
			15	9	322.433236	1.870267	4.22421	0.600966	0.0007359
			17	11	343.908136	1.891460	4.22421	0.605948	0.0008369
			20	13	375.400167	2.812266	4.22987	0.214401	0.0009046
0.1	1.2	7	12	4	343.668418	2.203591	5.84018	0.676912	0.0016468
			15	7	374.059400	2.231646	5.84037	0.677181	0.0018452
			17	9	394.601889	2.251727	5.84041	0.675093	0.0018445
			20	12	425.500742	2.281763	5.84044	0.677072	0.0014646
0.3	0.7	5	12	3	368.287183	2.056485	7.12750	0.738406	0.0028667
			15	6	397.768325	2.187982	7.13110	0.659848	0.0023767
			17	8	420.213016	2.287198	7.13184	0.640935	0.0022565
			20	11	454.744658	2.442360	7.13223	0.645719	0.0022506
0.3	0.7	7	12	1	464.211829	1.256983	9.91576	1.807315	0.0288755
			15	3	464.194351	2.307196	9.87089	0.798270	0.0129554
			17	5	483.376302	2.386059	9.88256	0.739570	0.0117586
			20	10	492.129822	1.861477	9.75472	1.705152	0.0245768
0.3	1.2	5	12	6	290.390613	1.925040	4.22626	0.469859	0.0001886
			15	9	322.433234	1.998460	4.22629	0.469592	0.0001984
			17	11	343.908133	2.048150	4.22630	0.464565	0.0001387
			20	13	369.400166	2.879258	4.23025	0.188086	0.0000536
0.3	1.2	7	12	4	337.333948	2.261560	5.84372	0.543857	0.0009626
			15	7	368.430909	2.328972	5.84408	0.528073	0.0008764
			17	9	389.772550	2.377060	5.84417	0.526352	0.0008754
			20	12	421.951476	2.450367	5.84422	0.525973	0.0008447

Table 3. Various performance measures of the system by varying λ and d

c	d	λ	L_s	E[V]	E[I]	E[B]
10	1	2	0.0008383010	0.999973	8.99919	0.0008382720
		2.5	0.0067951400	0.999828	8.99338	0.0067937700
		3	0.0358056000	0.999268	8.96495	0.0357775000
		3.5	0.1399840000	0.997620	8.86272	0.1396640000
		4	0.4368560000	0.993710	8.57175	0.4345360000
		4.5	1.1340800000	0.985934	7.89167	1.1224000000
	2	2	0.0031175500	1.999780	7.99710	0.0031173500
		2.5	0.0210497000	1.998850	7.98011	0.0210414000
		3	0.0973467000	1.995750	7.90705	0.0971983000
		3.5	0.3453800000	1.987690	7.66843	0.3438810000
		4	0.9940890000	1.970480	7.04509	0.9844330000
		4.5	2.3835400000	1.939770	5.71934	2.3408900000
	3	2	0.0117528000	2.998630	6.98962	0.0117513000
		2.5	0.0675772000	2.994020	6.93845	0.0675257000
		3	0.2791030000	2.980700	6.74101	0.2782890000
		3.5	0.9049340000	2.950130	6.15217	0.8977040000
		4	2.3790400000	2.893010	4.76785	2.3391500000
		4.5	5.1341700000	2.805090	2.20744	4.9874700000
12	1	2	0.0000387119	0.9999990	11.0000	0.0000387118
		2.5	0.0004388090	0.999990	10.9996	0.0004388050
		3	0.0029257100	0.999947	10.9971	0.0029255700
		3.5	0.0134960000	0.999793	10.9867	0.0134939000
		4	0.0474250000	0.999369	10.9532	0.0474042000
		4.5	0.1349800000	0.998421	10.8667	0.1348460000
	2	2	0.000169995	1.999990	9.99984	0.0001699940
		2.5	0.001592230	1.999920	9.99848	0.0015922100
		3	0.009223280	1.999640	9.99113	0.0092225300
		3.5	0.038293900	1.998750	9.96296	0.0382831000
		4	0.124147000	1.996530	9.87942	0.1240510000
		4.5	0.331324000	1.991910	9.67733	0.3307580000
	3	2	0.000732537	2.999930	8.99934	0.0007325340
		2.5	0.005713220	2.999560	8.99472	0.0057130600
		3	0.028938700	2.998200	8.97287	0.0289347000
		3.5	0.108744000	2.994360	8.89695	0.1086910000
		4	0.326480000	2.985620	8.68833	0.3260590000
		4.5	0.817304000	2.968940	8.21603	0.8150260000

number of VM in sleep mode) the energy consumption by the system is minimal and is approximately F(d) = 380 and then as d increases the energy consumption by the system increases. So, the service provider should maintain the value of d such that F(d) is minimum.

7. CONCLUSION

In cloud computing, data centre's resources consume a large amount of energy. To reduce the power consumption and heat generation for green computing VM consolidation in data center plays a crucial role for resource usage optimization. This chapter addresses the problem of VM activation and de-activation to achieve energy efficiency by taking the consideration of various factors such as VM in vacation and the overhead introduced by virtualization. A stochastic model of data centres has been proposed using Queueing theory to understand the dynamic nature of the data centres. As there exists a trade-off between the performance of the system and the energy consumption we proposed a heuristic algorithm to find the optimal parameter values. A variety of numerical results are presented to display the effect of the system parameters on the performance measures.

REFERENCES

Armbrust, M., Fox, A., Griffith, R., Joseph, A. D., Katz, R. H., Konwinski, A., & Zaharia, M. (2009). Above the clouds: a Berkeley view of cloud computing. UC Berkley. Retrieved from http://www.eecs.berkeley.edu/Pubs/TechRpts/2009/EECS-2009-28.html

Baker, T., Al-Dawsari, B., Tawfik, H., Reid, D., & Ngoko, Y. (2015). GreeDi: An energy efficient routing algorithm for big data on cloud. *Ad Hoc Networks*, *35*, 83–96. doi:10.1016/j.adhoc.2015.06.008

Baker, T., Aldawsari, B., & England, D. (2015). Trusted energy-efficient cloud-based services brokerage platform. *International Journal of Intelligent Computing Research*, *6*(4), 630–639. doi:10.20533/ijicr.2042.4655.2015.0078

Baker, T., Asim, M., Tawfik, H., Aldawsari, B., & Buyya, R. (2017). An energy-aware service composition algorithm for multiple cloud-based IoT applications. *Journal of Network and Computer Applications*, *89*, 96–108. doi:10.1016/j.jnca.2017.03.008

Baker, T., Ngoko, Y., Tolosana-Calasanz, R., Rana, O. F., & Randles, M. (2013, December). Energy efficient cloud computing environment via autonomic meta-director framework. *Proceedings of the 2013 Sixth International Conference on Developments in eSystems Engineering* (pp. 198-203). IEEE. 10.1109/DeSE.2013.43

Barham, P., Dragovic, B., Fraser, K., Ho, A., Neugebauer, R., Pratt, I., & Warfield, A. (2003). Xen and the art of virtualization. In Proceedings of the nineteenth acm Symposium on Operating Systems Principles(SOSP) (No.5, pp. 164-177 10.1145/945445.945462

Begnum, K. (2012). Simplified cloud-oriented virtual machine management with MLN. *The Journal of Supercomputing*, *61*(2), 251–266. doi:10.100711227-010-0424-0

Buyya, R., Beloglazov, A., & Abawajy, J. (2010). Energy-efficient management of data center resources for cloud computing: a vision, architectural elements, and open challenges. *Proceedings of the 2010 International Conference on Parallel and Distributed Processing Techniques and Applications, PDPTA 2010*. Academic Press.

Buyya, R., Yeo, C. S., Venugopal, S., Broberg, J., & Brandic, I. (2009). Cloud computing and emerging IT platforms: Vision, hype and reality for delivering computing as the 5th utility. *Future Generation Computer Systems*, *25*(6), 599–616. doi:10.1016/j.future.2008.12.001

Chao, X., & Yiqiang, Z. (1998). Analysis of multi-server queues with station and server vacations. *European Journal of Operational Research*, *110*(2), 392–406. doi:10.1016/S0377-2217(97)00253-1

Chase, J. S., Anderson, D. C., Thakar, P. N., Vahdat, A. M., & Doyle, R. P. (2001). Managing energy and server resources in hosting centers. *Proceedings of the 18th ACM Symposium on Operating Systems Principles* (pp. 103-116). New York, NY: ACM. 10.1145/502034.502045

Chebiyyam, M., Malviya, R., Bose, S. K., & Sundarrajan, S. (2009). Server consolidation: Leveraging the benefits of virtualization. *Infosys Research*, *7*(1), 65–75.

Doshi, B. (1986). Queueing systems with vacations-a survey. *Queueing Systems*, *1*(1), 29–66. doi:10.1007/BF01149327

Durao, F., Carvalho, J. F. S., Fonseka, A., & Garcia, V. C. (2014). A systematic review on cloud computing. *The Journal of Supercomputing*, *68*(3), 1321–1346. doi:10.100711227-014-1089-x

Elnozahy, E., Kistler, M., & Rajamony, R. (2003). Energy-efficient server clusters. *Proceedings of the International Workshop on Power-Aware Computer Systems* (pp. 179-197). Springer. 10.1007/3-540-36612-1_12

Fan, X., Weber, W. D., & Barroso, L. A. (2007). Power provisioning for a warehouse sized computer. *Proceedings of the 34th Annual International Symposium on Computer Architecture* (pp. 13-23). 10.1145/1250662.1250665

Farahnakian, F., Pahikkala, T., Liljeberg, P., Plosila, J., & Tenhunen, H. (2014). Multiagent based architecture for dynamic vm consolidation in cloud data centers. *Proceedings of the 40th IEEE Euromicro Conference on Software Engineering and Advanced Applications Software Engineering and Advanced Applications* (pp. 111-118). IEEE Press.

Farahnakian, F., Pahikkala, T., Liljeberg, P., Plosila, J., & Tenhunen, H. (2014). Multi agent based architecture for dynamic vm consolidation in cloud data centers. *Proceedings of the 40th IEEE Euromicro Conference on Software Engineering and Advanced Applications* (pp. 111–119). IEEE Press. 10.1109/SEAA.2014.56

Feller, E., Morin, C., & Esnault, A. (2012). A case for fully decentralized dynamic VM consolidation in clouds. *Proceedings of the IEEE International Conference on Cloud Computing Technology and Science* (pp. 26-33). IEEE Press. 10.1109/CloudCom.2012.6427585

Gill, S. S., Chana, I., Singh, M., & Buyya, R. (2018). CHOPPER: An intelligent QoS-aware autonomic resource management approach for cloud computing. *Cluster Computing*, *21*(2), 1203–1241. doi:10.100710586-017-1040-z

Kalloniatis, C., Kavakli, E., Gritzalis, S., Forrest, W., & Kindler, N. (2014). Addressing privacy in Traditional and Cloud-Based Systems [IJAIE]. *International Journal of Applied Industrial Engineering*, *2*(1), 14–40. doi:10.4018/ijaie.2014010102

Kaplan, J., Forrest, W., & Kindler, N. (2008). *Revolutionizing Data Center Energy Efficiency (Tech. Report)*. McKinsey and Company.

Kharat, M. G., Khadke, S. S., Raut, R. D., Kamble, S. S., Kamble, S. J., & Kharat, M. G. (2016). Application of hybrid firefly algorithm-tabu search technique to minimize the makespan in job shop scheduling problem. *International Journal of Applied Industrial Engineering*, *3*(2), 1–21. doi:10.4018/IJAIE.2016070101

Khazaei, H., Misic, J., & Misic, V. B. (2012). Performance Analysis of Cloud Computing Centers Using M/G/m/m+r Queuing Systems. *IEEE Transactions on Parallel and Distributed Systems*, *23*(5), 936–943. doi:10.1109/TPDS.2011.199

Lee, Y. C., & Zomaya, A. Y. (2010). Energy efficient utilization of resources in cloud computing systems. *The Journal of Supercomputing*, *60*(2), 268–280. doi:10.100711227-010-0421-3

Li, J., Li, Y. K., Chen, X., Lee, P. P., & Lou, W. (2014). A hybrid cloud approach for secure authorized deduplication. *IEEE Transactions on Parallel and Distributed Systems*, *26*(5), 1206–1216. doi:10.1109/TPDS.2014.2318320

Lien, C. H., Bai, Y. W., Lin, M. B., Chang, C. Y., & Tsai, M. Y. (2006). Web server power estimation, modeling and management. *Proceedings of 14th IEEE International Conference on Networks* (Vol. 2, pp. 1-6). IEEE Press. 10.1109/ICON.2006.302688

Liu, X. (2013). A Programming Model for the Cloud Platform. *International Journal of Advanced Science and Technology*, *57*, 75–82.

Marzolla, M., Babaoglu, O., & Panzieri, F. (2011). Server consolidation in clouds through gossiping. *Proceedings of the IEEE International Symposium on World of Wireless, Mobile and Multimedia Networks (WoWMoM)* (pp. 1-6). IEEE Press. 10.1109/WoWMoM.2011.5986483

Mathew, V., Sitaraman, R. K., & Shenoy, P. (2012). Energy-aware load balancing in content delivery networks. *Proceedings of the 31st Annual IEEE International Conference on Computer Communications (IEEE INFOCOM)* (pp. 954-962). IEEE Press. 10.1109/INFCOM.2012.6195846

Ministry of Economy. (2010). Trade and Industry, Establishment of the Japan data center council [Press Release].

Mishra, S. K., Puthal, D., Sahoo, B., Jayaraman, P. P., Jun, S., Zomaya, A. Y., & Ranjan, R. (2018). Energy-efficient VM-placement in cloud data center. *Sustainable computing: informatics and systems*, *20*, 48-55.

Murtazaev, A., & Oh, S. (2011). Sercon: Server consolidation algorithm using live migration of virtual machines for green computing. *IETE Technical Review*, *28*(3), 212–231. doi:10.4103/0256-4602.81230

Nathuji, R., & Schwan, K. (2007). Virtual Power: coordinated power management in virtualized enterprise systems. *Proceedings of Twenty-First ACM SIGOPS Symposium on Operating Systems Principles* (pp. 265- 278). ACM. 10.1145/1294261.1294287

Patra, S. S. (2018). Energy-efficient task consolidation for cloud data center. *International Journal of Cloud Applications and Computing*, *8*(1), 117–142. doi:10.4018/IJCAC.2018010106

Pedram, M. (2012). Energy-efficient datacenters. *IEEE Transactions on Computer-Aided Design of Integrated Circuits and Systems*, *31*(10), 164–177. doi:10.1109/TCAD.2012.2212898

RightScale. (2012). Designing Private and Hybrid Clouds: Architectural Best Practices [white paper]. Retrieved from http://www.rightscale.com/lp/private-hybrid-cloud-white-paper.php

Rimal, B. P., Choi, E., & Lumb, I. (2009). A taxonomy and survey of cloud computing systems. *Proceedings of the Fifth International Joint Conference on INC, IMS and IDC* (pp. 44-51). Academic Press. 10.1109/NCM.2009.218

Rivoire, S., Ranganathan, P., & Kozyrakis, C. (2008). A comparison of high-level full system power models. *Proceedings of the 2008 Conference on Power Aware Computing and Systems* (*Vol. 8*, p. 3). Academic Press.

Sasikala, P. (2013). Energy efficiency in cloud computing: Way towards green computing. *International Journal of Cloud Computing*, *2*(4), 305–324. doi:10.1504/IJCC.2013.058095

Shen, J., Gui, Z., Ji, S., Shen, J., Tan, H., & Tang, Y. (2018). Cloud-aided lightweight certificateless authentication protocol with anonymity for wireless body area networks. *Journal of Network and Computer Applications*, *106*, 117–123. doi:10.1016/j.jnca.2018.01.003

Shu, Z., Wan, J., Zhang, D., & Li, D. (2016). Cloud-integrated cyber-physical systems for complex industrial applications. *Mobile Networks and Applications*, *21*(5), 865–878. doi:10.100711036-015-0664-6

Sivathanu, S., Liu, L., Yiduo, M., & Pu, X. (2010). Storage management in virtualized cloud environment. *Proceedings of the IEEE 3rd International Conference on Cloud Computing (CLOUD)* (pp. 204-211). 10.1109/CLOUD.2010.27

Sofia, A. S., & Kumar, P. G. (2015). Implementation of energy efficient green computing in cloud computing. *International Journal of Enterprise Network Management*, *6*(3), 222–237. doi:10.1504/IJENM.2015.071135

Song, Y., Zhang, Y., Sun, Y., & Shi, W. (2009). Utility analysis for internet-oriented server consolidation in VM-based data centers. *Proceedings of IEEE International Conference on Cluster Computing* (pp. 1-10). IEEE Press. 10.1109/CLUSTR.2009.5289190

Srikantaiah, S., Kansal, A., & Zhao, F. (2008). Energy Aware consolidation for cloud computing. *Proceedings of the 2008 Conference on Power Aware Computing and Systems* (*Vol. 10*). Academic Press.

Takagi, H. (1993). Queueing Analysis—a foundation of performance evaluation. *Discrete time Systems*, *3*(1), 141-158.

Teghem, J. Jr. (1986). Control of the service process in a queueing system. *European Journal of Operational Research*, *23*(2), 141–158.

The green grid consortium. (2011). Trade and Industry, Establishment of the Japan data center council. Retrieved from http://www.thegreengrid.org

Tian, N., & Zhang, Z. G. (2006). *Vacation queueing models: theory and applications*. Springer. doi:10.1007/978-0-387-33723-4

Tian, N., & Zhang, Z. G. (2006). Synchronous working vacation policy for finite-buffer multiserver queueing system. *Applied Mathematics and Computation*, *217*(24), 9916–9932.

Torres, J., Carrera, D., Hogan, K., Gavaldà, R., Beltran, V., & Poggi, N. (2008). Reducing wasted resources to help achieve green data centers. *Proceedings of IEEE International Symposium on Parallel and Distributed Processing* (pp. 1-8). IEEE Press. 10.1109/IPDPS.2008.4536219

Tsai, J., Fang, J., & Chou, J. (2013). Optimized task scheduling and resource allocation on cloud computing environment using improved differential evolution algorithm. *Computer Operation Research Elsevier*, *40*(12), 3045–3055. doi:10.1016/j.cor.2013.06.012

Varshney, K., Jain, M., & Sharma, G. C. (1989). A multi-server queueing model with balking and reneging via diffusion approximation. *Journal of Physical and Natural Science*, *10*, 10–15.

Velte, A., Velte, T., & Elsenpeter, R. (2010). Your organization and cloud computing. In *Cloud Computing-A practical Approach*. McGraw-Hill.

Vogels, W. (2008). Beyond server consolidation. *Queue*, *6*(1), 20–26. doi:10.1145/1348583.1348590

Wang, K. & Li, Z. (2012). Power and performance analysis for web server cluster. *Advances in information Sciences and Service Sciences*, *4*(12), 285-293.

Wang, L., Zhang, F., Aroca, J. A., Vasilakos, A. V., Zheng, K., Hou, C., & Liu, Z. (2013). GreenDCN: A general framework for achieving energy efficiency in data center networks. *IEEE Journal on Selected Areas in Communications*, *32*(1), 4–15. doi:10.1109/JSAC.2014.140102

Wang, L., Zhang, F., Zheng, K., Vasilakos, A. V., Ren, S., & Liu, Z. (2014). Energy-efficient flow scheduling and routing with hard deadlines in data center networks. *Proceedings of the 2014 IEEE 34th International Conference on Distributed Computing Systems* (pp. 248-257). IEEE. 10.1109/ICDCS.2014.33

Wang, S., Zhou, A., Bao, R., Chou, W., & Yau, S. S. (2018). Towards green service composition approach in the cloud. *IEEE Transactions on Services Computing*, 1. doi:10.1109/TSC.2018.2868356

Wikipedia. (n.d.). Wiki. Retrieved from http://en.wikipedia.org/wiki/

Yao, D. Q. (2012). Application of the Theory of Constraints (TOC) to batch scheduling in process industry. *International Journal of Applied Industrial Engineering*, *1*(1), 10–22. doi:10.4018/ijaie.2012010102

Zhang, R., & Liu, L. (2010). Security models and requirements for healthcare application clouds. *Proceedings of the IEEE 3rd International Conference on Cloud Computing (CLOUD)* (pp. 268-275). IEEE Press. 10.1109/CLOUD.2010.62

Chapter 8
QFD-Based TOPSIS Methodology for Material Selection

K. G. Durga Prasad
https://orcid.org/0000-0002-1450-159X
Gayatri Vidya Parishad College for Degree and P. G. Courses (Autonomous), India

B. L. Manasa
Jawaharlal Nehru Technological University, Kakinada, India

P. Krishna Murthy
Gayatri Vidya Parishad College for Degree and P. G. Courses (Autonomous), India

K. D. S. Sravani
Andhra University College of Engineering for Women, India

ABSTRACT

Material selection is one of the interdisciplinary activities involved in product design and development. It is an integral part of concept phase and configuration phase of mechanical design process. Therefore, a successful product design should take into consideration of material properties to meet the design requirements which reflect the expectations of the end users of a product. This chapter presents a user-centered material selection methodology which paves a path for reflecting user preferences into design requirements and then into material selection decision. In the proposed approach, weights of the material selection criteria are obtained through Quality Function Deployment (QFD) and then the TOPSIS (Technique for Order Preference by Similarity to Ideal Solution) methodology is employed to achieve the optimum selection of material under a multi-attribute decision-making environment. An illustrative example is taken from the literature and presented in this chapter to demonstrate the proposed methodology.

DOI: 10.4018/978-1-7998-1831-1.ch008

INTRODUCTION

The successful design and development of an engineering component is highly dependent on appropriate selection of material to manufacture the component. It is essential to focus on material selection during the conceptual stage of mechanical design process instead of embodiment stage leads to achieve more optimal design solutions. The initial stage of product development process is known as conceptual design stage in which design solution concepts are generated to meet the design requirements of a product (Prasad et al., 2014).The conceptual design stage is more important and critical stage compared to the other design stages in product development process as it involves complex decision making activities such as the selection of design concepts, materials selection and manufacturing process selection. The conceptual design stage plays a vital role in the overall success of the product as once the conceptual design process has been completed, the majority of product cost and quality has been fixed by selecting particular concepts (Rehman and Yan, 2003). It is reported in the literature that 70-80% of a product cost is committed during the conceptual phase of product development process (Rush and Roy, 2000). Material cost is one of the main cost components while determining the total cost of a product. In the conceptual stage, the designers are usually planned for selecting a material to satisfy design requirements of the product and available at minimum cost. If the material selection issues are not properly addressed in conceptual stage, there is no scope for product designers in detailed design stage to make any modifications for consciously considered material issues. Therefore, design team has to take astute decision on selection of materials which leads to a lot of time saving in product development and minimizing the probability of late detection of materials related critical characteristics.

The objective of material selection process is the transformation of product design requirements into a list of required material properties so as to find an appropriate material for manufacturing a product. Usually product design team considers various parameters such as functional requirements, manufacturability, environmental and economic considerations, mechanical properties, physical and chemical properties etc. In addition to these parameters, product designers need to consider user requirements during material selection so as to achieve competitive advantage in the present market scenario. The improper material selection may lead not only to failure in achieving manufacturer requirements but also users' expectations. Therefore, it is necessary to capture users' needs and then deploy them into material selection process. The users of a product are generally expressing their requirements in the form of unclear and non-measurable statements as they may not have sufficient technical knowledge. Therefore, in a conceptual design stage, it is required to translate their need statements into appropriate design requirements and then it is possible to identify necessary criteria for material selection in accordance with the design requirements. To deal with this issue Quality Function Deployment (QFD) technique is adopted in the proposed methodology.

Material selection is a process of choosing the most appropriate material for a product from a finite set of available materials on the basis of different attributes and hence it is a multi-attribute decision making (MADM) problem. To tackle the materials selection issue under MADM situation various mathematical approaches have been reported in the literature with a view to derive the best decisions on selecting materials suitable to different engineering applications. It is observed from the literature that most of the researchers adopted TOPSIS (Technique for Order Preference by Similarity to Ideal Solution) methodology for material selection as it is quite capable and computationally easy to evaluate and select the proper material under MADM environment. During the implementation of TOPSIS methodology, the weightages assigned to various criteria through different methods such as AHP, entropy method, compromised method etc. Though these methods provide the weightages to criteria for solving material selection problem, the methods are not going to give scope to deploy users' requirements into the selection process of material for developing product as desired by the users. In this context QFD technique is employed in the proposed methodology to obtain priority structure of material selection criteria and then the same is used in the TOPSIS methodology for selecting a suitable material to attain the satisfaction of users and manufacturers. The subsequent section provides the brief overview of techniques used to develop the proposed hybrid methodology. To demonstrate the methodology, an illustrative example is also presented in this chapter.

BACKGROUND

The proposed hybrid methodology has been developed with the support of techniques namely QFD, AHP and TOPSIS. These techniques are briefly discussed in the following sections.

Quality Function Deployment

Quality Function Deployment (QFD), a customer focused quality engineering technique and it was developed in Japan in the late 1960's by Prof. Yoji Akao and Prof. Shigeru Mizuno. It is a customer-driven product development tool employed for efficiently translating customer needs into design requirements and parts deployment, as well as manufacturing plans and controls in order to achieve higher customer satisfaction (Yang et al., 2012). It has gained global acceptance as one of the most important tools in the product design and development process (Raharjo et al., 2006). It enables an organization to instill quality into a product and control the development process from conception to the commencement of manufacturing operations (Chen, 2009). It has been widely used in various industries with a view to add quality, value and customer satisfaction during the design and development of new products and services. Prasad (1998), Chan and Wu (2002) discussed the definitions, contents and applications of QFD along with historical developments from its inception. Sharma et al. (2008) reviewed the literature of more than 400 QFD publications (Subbaiah et al., 2011). In fact, QFD is a structured methodology consists of four inter-linked matrices namely product planning matrix, parts planning matrix, process planning matrix and production planning matrix. These are the correspondence matrices which are sequenced to reflect a typical path in the product development process such that outputs of a matrix are inputs in a succeeding matrix (Fogliatto et al., 2003). The product planning matrix of QFD system is popularly known as House of Quality (HOQ) which is a conceptual map that assists for inter-functional planning and communications (Hauser and Clausing, 1988). The HOQ is a basic design tool of QFD (Durga Prasad et al., 2012) appears like an ordinary house with foundation, walls and roof. The major components involved in HOQ matrix are customer needs (WHATs), product design requirements (HOWs), prioritized customer needs, inter-relationship between customer needs and product design requirements (WHATs vs HOWs), relationship among product design requirements (HOWs vs HOWs) and prioritized design requirements. The inter-relationships among customer needs and design requirements form the main body of the house. Product design requirements occupy the ceiling portion of the house. The correlations among the design requirements and the priority structure of design requirements form the roof and foundation of the house respectively (Islam et al., 2007).The exact translation between customer needs and design requirements takes place in inter-relationship matrix of HOQ. The inter-relationship values are further reflected in the computation of relative importance values of the design requirements. The prioritized design requirements are the outcome of HOQ and the customer perceptions are reflected in the priorities of the design requirements (Prasad et al., 2019). The second matrix of QFD approach is the parts planning matrix which maps the prioritized design requirements to critical parts characteristics. The priority ratings of the parts characteristics are further used to develop the subsequent matrices of QFD methodology.

Earlier QFD approach has been widely applied to design and development of new products or improve existing one. Over the years QFD has been successfully implemented with its customized versions to assist decision makers for solving numerous issues in various engineering fields, different service sectors, supply chain management, e-commerce, construction and other sectors. Many researchers developed QFD-based hybrid methodologies to address complex decision-making issues. The review of literature revealed that a limited number of papers reported on the application of QFD for solving material selection problem. Jalham (2006) developed computer-aided QFD method for quantitative evaluation of the optimum material under post-selection phase of the material selection process. He claimed that the proposed method can substitute the existing weighted properties method and found to be a good tool for selecting the optimum material especially when there is huge number of candidate materials under consideration. Mayyas et al. (2011) proposed a procedure developed by using QFD and AHP techniques for determining the suitable material for automotive Body-In-White (BiW) panels at the conceptual design stage without scarifying customer needs as well as engineering requirements. Prasad and Chakraborty (2013) made a note-worthy attempt for solving material selection problems using QFD-based approach with a view to integrate the voice of customers for a product with its technical requirements. They developed a software prototype with the support of QFD concept to automate the material selection decision making process. Cavallini et al. (2013) introduced a material selection decision model through Integrated Aided Material Selection (IAMS) approach based on integrating HOQ and the comprehensive VIKOR algorithm. To justify their proposed model, they presented a case study concerns the identification of the best coating for the protection of an aluminum alloy substrate from the effects of abrasive wear against an alternating counterpart made by a high-strength cast iron. Kasaei et al. (2014) proposed a method which combines the Ashby's materials selection concepts with the support of QFD. In their work, they suggested material indices which assures the generality of the results and finally concluded that their proposed QFD method addresses all the requirements for effective materials selection for mechanical design. Prasad et al., (2016) designed and developed a real time expert system on the basis of QFD methodology for selecting the most suitable materials for varying engineering components. Their expert system maps the customer requirements with technical characteristics of materials and assists the designers to have any exhaustive knowledge on different physical properties of the considered materials. Dursun and Arslan (2018) developed a fuzzy MCDM method based on QFD which allows dealing with the relationships between the customer needs and material selection attributes, and also the relationships between material selection attributes for assuring higher customer satisfaction. It is observed from the literature that most of the organizations and researchers were confined only to apply their customized version of the first matrix of QFD. In the present work the methodology has been developed by considering first and second matrices of QFD with slight modifications for solving material selection problem with a view to develop products as desired by the users.

Analytic Hierarchy Process

Analytic hierarchy process (AHP) is one of the most commonly used multi-criteria decision making (MCDM) techniques and it was developed by Prof. T.L. Saaty in 1971 (Wind and Saaty, 1980). The AHP method consists of four major operations namely hierarchy construction, performing pair-wise comparison of all the elements, priority analysis, and consistency verification. In the first operation decision makers need to focus on break down the complex multiple criteria decision problem into its component parts of which every possible attribute are arranged into multiple hierarchical levels. Afterwards they have to compare each cluster in the same level in a pair-wise fashion based on their own experience and knowledge. To perform pair-wise comparison between the elements, Saaty proposed a fundamental scale of absolute numbers. This comparison scale allows the decision-makers to incorporate experience and knowledge intuitively and indicate how many times an element dominates another with respect to the criterion. After all pair-wise comparisons are performed the vector of weights is to be computed using Satty's eigen vector procedure (Saaty, 2008). The final operation in AHP procedure is the consistency verification which helps to check the judgments made during pair-wise comparisons. The consistency verification is regarded as one of the main advantages of AHP as there is a possibility of some degree of inconsistency while subjective judgments carried in establishing pair-wise comparisons. In order to measure the degree of consistency, it is required to find the consistency ratio. If the consistency ratio exceeds the limit, the decision makers should review and revise the pair-wise comparisons. Once all pair-wise comparisons are examined to be consistent, and then the judgments can be synthesized to find out the priority ranking of each criterion and its attributes (Ho, 2008).

Extensive studies have been carried on AHP and used in almost all the applications related to MCDM situations for the last 30 years. In the recent years it is observed that the AHP is integrated with other techniques with a view to strengthen the concerned decision-making procedure. As the prioritization of customer needs can be viewed as complex MCDM problem, AHP gives the best solution to determine the priorities of customer needs in HOQ (De Felice and Petrillo, 2010; Mastura et al., 2018). A number of QFD-AHP integrated approaches are reported in the literature to address various decision-making situations. A few attempts are reviewed here. Hanumaiah et al., (2006) proposed QFD-AHP approach for solving rapid tooling process selection. They employed AHP to prioritize the tooling requirements against a set of die development attributes such as product material, geometry, die material and production order. They used these priority ratings for selecting the most appropriate tooling process with the help of QFD in the subsequent phase. Chen et al., (2007) developed a key model for knowledge management system using AHP-QFD for a semiconductor industry in Taiwan. The importance weights of models related to knowledge management are obtained through AHP with the consideration of the opinions from the experts. Then QFD was integrated for the system models of a knowledge management system and correlation weights of key objectives are to be improved. Das and Mukherjee (2008) made an attempt to develop an AHP-QFD framework for designing a tourism product, which takes care of the touristic needs of tourists. They made pair-wise relationships among the operating strategies using AHP procedure to find out whether the strategies support each other or in conflict with each other. The AHP supports in prioritizing the needs of tourists with reference to a tourist destination and then QFD is employed to formulate operating strategies by eliciting views from tourism professionals. Ho et al., (2011) developed an integrated analytical approach for selecting suppliers using a combined QFD and AHP. They adopted AHP to prioritize the evaluating criteria derived from the requirements of company stakeholders using HOQ. On the basis of ranked criteria, the potential suppliers are evaluated using AHP again to make an optimal selection. Scott et al. (2014) proposed a decision support system using the integration of AHP and QFD for solving supplier selection problem and order allocation in stochastic, multi-stakeholder and multi-criteria environments. Chadawada et al. (2015) implemented AHP-QFD model to assist organizations through observing various factors and selecting the best location among different alternatives. This approach uses AHP method to match the preferences required by decision-makers and these preferences are applied to the characteristics of QFD. Ucler (2017) employed an iterative AHP-QFD-AHP approach for brainstorming is implemented into a concurrent engineering environment in the early phases of layout conception. In the case study, the proposed model delivered the product definition of the cryoplane concept successfully, which promises clean operation. He believed that his proposed approach can be used in front-end product conception, where high levels of innovation are required. Abdel Basset et al. (2018) presented an integrated neutrosophic AHP-QFD approach for selecting best supplier from different alternatives. They used neutrosophic QFD to obtain the requirements of the stakeholders and neutrosophic AHP for calculating the weights of stakeholder's requirements and the weights of available suppliers. In the present work AHP is employed to obtain the priority ratings of the user requirements and are incorporated in HOQ to obtain the weightages of the design requirements so as to reflect the user desires in material selection for developing a product. The AHP procedure consists of four main steps which are discussed below.

Step 1: Establishment of pair-wise comparison matrix

Establish the pair-wise comparison matrix of order $n \times n$ consists of n elements (requirements) in the rows and columns whose priorities are to be computed.

Step 2: Perform pair-wise comparisons of all the elements

A pair-wise comparison is the process of comparing the relative importance, preference, or likelihood of two elements (objectives) with respect to another element (the goal). To perform pair-wise comparisons of the elements, the decision maker has to express opinion on each paired comparison of the elements. Saaty proposed a scale which enables the decision-maker to express his preference between each pair of elements verbally as equally preferred, moderately preferred, strongly preferred, very strongly preferred and extremely preferred. These descriptive preferences would then be translated into numerical values 1, 3, 5, 7, 9 respectively. The Table 1 shows the fundamental scale proposed by Saaty.

The intermediate values for comparisons between two successive judgments are indicated by 2, 4, 6 and 8. Reciprocals of these values are used for the corresponding transposed judgments (Prasad et al., 2017).

Step 3: Estimation of eigen values of the matrix

The eigen values can be estimated by averaging over normalized columns method proposed by Saaty. In eigen value method, first sum the values in each column of the pair-wise comparison matrix and then divide each element in a column by the sum of its respective column. The resultant matrix is termed as the normalized pair-wise comparison matrix. Finally sum the elements in each row of the normalized pair-wise comparison matrix and divide the sum with the number of elements. The result of this computation is referred to as the priority matrix and is an estimation of the eigen values of the matrix.

Table 1. Saaty's fundamental scale of absolute numbers (Tsinidou et al., 2010)

Intensity of Preference	Verbal Judgments of Preferences	Explanation
1	Equally preferred	Two activities contribute equally to the objective
3	Moderately preferred	Experience and judgment slightly favor one activity over another
5	Strongly preferred	Experience and judgment strongly favor one activity over another
7	Very strongly preferred	Experience and judgment very strongly favor one activity over another
9	Extremely preferred	Experience and judgment are of the highest possible order of affirmation in terms of favor one activity over another
2,4,6,8	Intermediate preferences between the two adjacent judgments	When compromise in preference is needed

Step 4: Check the consistency of pair-wise judgments

In order to verify the consistency of the pair-wise comparison matrix, Saaty proposed consistency index (CI) and consistency ratio (CR). The expressions for CI and CR are given as follows.

$$CI = \frac{\lambda_{max} - n}{n - 1} \quad (1)$$

$$CR = \frac{CI}{RCI} \quad (2)$$

Where

λ_{max} = maximum principal eigen value of the comparison matrix
n = number of elements (order of the pair-wise comparison matrix)

The value of λ_{max} is obtained by first multiplying the pair-wise comparison matrix with the priority matrix. Then divide the first element of the resulting matrix by the first element of the priority matrix, the second element of the resulting matrix by the second element in the priority matrix, and so on. A single column matrix is obtained and the average of the elements of the matrix gives the value of λ_{max}. The Random consistency index (RCI) in the equation (2) represents consistency index of a randomly generated pair-wise comparison matrix. The value of RCI depends on the order of matrix. The Table 2 shows the RCI values for the corresponding order of a matrix.

In actual practice, perfect consistency rarely occurs in judgments. In AHP the pair-wise comparisons in a judgment matrix are considered to be adequately consistent if the corresponding consistency ratio (CR) is less than 10% (Saaty, 1980). Hence, if CR ≤ 0.1, then the estimate is accepted; otherwise, a new comparison matrix is solicited until CR ≤ 0.1 (Chang et al., 2007).

Table 2. Average value of RCI for corresponding matrix order

n	1	2	3	4	5	6	7	8	9
RCI	0	0	0.58	0.90	1.12	1.24	1.32	1.41	1.45

(Saaty, 1980)

TOPSIS

Multi-attribute decision making (MADM) refers to making decision preferences, such as evaluation, prioritization and selection of available alternatives. A number of techniques reported in the literature to address MADM problems. The technique for order preference by similarity to ideal solution (TOPSIS) is one of the traditional MADM techniques introduced by Hwang and Yoon in 1981. It is used to solve MADM problems through identifying the solutions from a finite set of alternatives. TOPSIS is a simple logic-based computation method which provides sensible decision-making, preferences and an index that accounts for the most excellent alternatives as well as poor alternatives (Mittal et al., 2016). It is a practical and useful technique for ranking and selection of a number of possible alternatives through measuring Euclidean distances. This technique is primarily based on the principle that the chosen alternative should have the shortest distance from the positive ideal solution and the longest distance from the negative ideal solution (Opricovic and Tzeng, 2004). The positive ideal solution is identified with a 'hypothetical alternative' that has the best values for all considered criteria whereas the negative ideal solution is identified with a 'hypothetical alternative' that has the worst criterion values. TOPSIS has been successfully applied in practice to solve various selection/evaluation problems with a finite number of alternatives because it is intuitive and easy to understand and implement. A sound logic involved in TOPSIS methodology that represents the rationale of human choice and hence TOPSIS has been proved to be one of the best methods in addressing the issue of rank reversal (Roghanian et al., 2010).To apply TOPSIS technique, attribute values must be numeric, monotonically increasing or decreasing, and have commensurable units. TOPSIS procedure involves the following simple steps (Behzadian et al., 2012).

Step 1: Formulation of decision matrix

When *n* alternatives (A) and *m* criteria (C) are involved in a decision-making process, then the decision matrix has to be formed as shown in Table 3.

Step 2: Establishment of normalized decision matrix

The normalized decision matrix can be developed by determining normalized ratings using the following formula.

$$r_{ij}(x) = \frac{x_{ij}}{\sqrt{\sum_{i=1}^{n} x_{ij}^2}} \quad (3)$$

Table 3. Decision matrix format

		Criteria (C)					
		C_1	C_2	C_3	.	.	C_m
Alternatives (A)	A_1	x_{11}	x_{12}	x_{13}	.	.	x_{1m}
	A_2	x_{21}	x_{22}	x_{23}	.	.	x_{2m}
	A_3	x_{31}	x_{32}	x_{33}	.	.	x_{3m}
	.	.	.	.	.	.	.
	.	.	.	.	.	.	.
	A_n	x_{n1}	x_{n2}	x_{n3}	.	.	x_{nm}

Where A_i = the ith alternative (i = 1, 2, …, *n*)
C_j = the jth criterion (j = 1, 2, …, *m*) and x_{ij} = individual performance rating.

Step 3: Construction of weighted normalized decision matrix

The weighted normalized matrix is developed by determining the weighted normalized ratings using the following expression.

$$v_{ij}(x) = w_j\left[r_{ij}(x)\right] \tag{4}$$

Where w_j = weightage of jth criterion.

Step 4: Deriving positive ideal solution and negative ideal solution

The positive ideal solution V^+ and the negative ideal solution V^- are determined as follows:

$$V^+ = \max\left\{v_1^+(x), v_2^+(x),,..,v_j^+(x)\right\}, j = 1,2,...m \tag{5}$$

$$V^- = \min\left\{v_1^-(x), v_2^-(x),,..,v_j^-(x)\right\}, j = 1,2,...m \tag{6}$$

Step 5: Calculation of separation measures for each alternative

The separation measures from the positive and the negative ideal solution between the alternatives are calculated. The separation (distances) of each alternative from positive ideal solution can be calculated by using the following equation.

$$D_i^+ = \sqrt{\sum_{j=1}^{m} \left\{ \left[v_{ij}(x) - v_j^+(x) \right] \right\}^2} \tag{7}$$

Similarly, the separation (distances) of each alternative from negative ideal solution can be calculated by using the following equation.

$$D_i^- = \sqrt{\sum_{j=1}^{m} \left\{ \left[v_{ij}(x) - v_j^-(x) \right] \right\}^2} \tag{8}$$

Step 6: Computation of closeness coefficient

The closeness coefficients for all the alternatives are calculated by using the following expression.

$$C_i^+ = \frac{D_i^-}{D_i^+ + D_i^-} \tag{9}$$

Where $i = 1, 2, 3, \ldots, n$

The value of closeness coefficient $\left(C_i^+\right)$ lies in between 0 and 1.

Step 7: Rank the alternatives

The alternatives have to be ranked on the basis of closeness coefficient values.

Step 8: Select the best alternative

The preferred alternative is the one which has the highest value of closeness coefficient (closest to 1).

TOPSIS procedure has been successfully applied to the areas of human resources management, transportation, construction, product design, manufacturing, water management, supply chain management, soft computing, environmental research, operation research, quality control and location analysis (Shih et al., 2007; Zavadskas et al., 2016). In recent years, material science researchers are widely using TOPSIS for solving material selection problem as the implementation of TOPSIS is simple and involves only a limited amount of subjective input from decision-makers and the method has an ability to identify the best alternative quickly. TOPSIS is usually integrated with other techniques to strengthen the methodology suitable to select the desired materials. During the implementation of TOPSIS, the techniques such as AHP, entropy method, ANP, grey relation analysis, preference selection index technique etc. are adopted to determine the weightages to criteria for material selection. Kumar and Roy (2014) evaluated optimum material for engineering design using integrated approach developed by using entropy method and TOPSIS. The criteria weights are computed by using entropy whereas ranking of the materials is computed by using TOPSIS. Arep et al. (2016) presented the integration of AHP and TOPSIS in material selection process of an automotive fender design. The main objective of their integrated AHP-TOPSIS approach is to identify the important material selection criteria and to determine the best material for automotive fender. The purpose of AHP employed in their work is to obtain the weightages of the criteria followed by TOPSIS method to performing ranking of the materials chosen during material selection process. Zhang et al. (2017) developed a hybrid decision-making approach through combining the techniques DEMATEL (Decision making and evaluation laboratory), ANP (Analytical network process), GRA (Grey relational analysis) and TOPSIS for selecting the optimum green material for sustainability based on the needs of the product. The DEMATEL is employed to analyze the inter-relationship and influence of each criterion, whereas ANP is employed to calculate the final weights of the material selection criteria. Finally, GRA with TOPSIS is proposed to obtain the final rank of each alternative and select the optimal green material for a product. Tian et al (2018) formulated a hierarchical structure with interior environmental characteristics and presented a grey-correlation-based hybrid decision-making approach that integrates AHP, Grey correlation and TOPSIS to overcome the deficiencies of single method. As the weights vector of each decision criterion has a great influence on decision-making of green decoration materials. They applied AHP approach to obtain the weights vector of each criterion of environmental characteristics. Sobhanallahi et al. (2019) suggested a hybrid QFD-TOPSIS solution in MCDM methods for solving supplier selection problem in the IT department of financial organizations. They adopted their approach successfully in a case study conducted in a private bank in Iran. Yadav et al., (2019) proposed a hybrid methodology established by using TOPSIS with the aid of Preference selection index (PSI) method for solving material selection problem. They used PSI for determining the weights of criteria as it provides less numerical calculations. They demonstrated well in applying their hybrid TOPSIS - PSI approach for effective material selection in marine applications. It is noticed that successful attempts made on solving material selection issues using integrated approaches developed by using TOPSIS and other techniques. However, the intervention of user conception about a particular product has not been taken place during material selection. In view of incorporating the user needs and expectations, QFD technique is employed to obtain the weightages of the material selection criteria. The weightages are further utilized in TOPSIS methodology for selection of a material to make a product. The sequential steps involved in the proposed methodology are discussed in the subsequent section.

QFD-BASED TOPSIS METHODOLOGY FOR MATERIAL SELECTION

The hybrid methodology proposed in this chapter provides a scope to reduce the gulf between users' perception and designers' conception during material selection process for manufacturing a product as desired by the users. The Figure 1 shows the outline of the proposed hybrid methodology. The step by step methodology to handle the material selection process is discussed below:

Step1: Identification of users' requirements

The identification of users' requirements is the initial step in user-centered product development. The requirements of the users of a product are usually identified by interacting directly with the users by asking what they are expecting in the final product.

Step 2: Prioritization of users' requirements using AHP

The priority structure of users' requirements indicates the weightages of the requirements and also helps to understand the relative importance that users' attitude towards each requirement. AHP is the best method to determine the degree of importance of users' requirements (Bhimaraya, 2003) and the AHP discussed in the earlier section can be carried to obtain the priorities of the users' requirements. The priority structure gives a scope on understanding about the intensity of users' desires about the product which is required to know the design team while establishing design requirements.

Step 3: Identification of design requirements

The vertical portion of HOQ chart begins with the establishment of design requirements and completed with the determination of priority structure of the design requirements. The users are generally expressing their expectations in their own language. As the users may not have sufficient knowledge of the individual functions of the product, they are unable to describe the features clearly in terms of technical specifications that are to be incorporated in the final product. But the core philosophy of QFD is the exact translation of users' requirements into design requirements for developing products to meet those requirements (Prasad et al., 2019). Therefore, each customer's needs may be translated into one or more design requirements. Therefore, it is essential for a design team to establish necessary design requirements to incorporate users' views in the final product. The design requirements are explored through brainstorming sessions conducted among the focus group members. The outcome of the focus group discussions provides the list of design requirements.

Figure 1. Outline of the QFD-based TOPSIS methodology for material selection

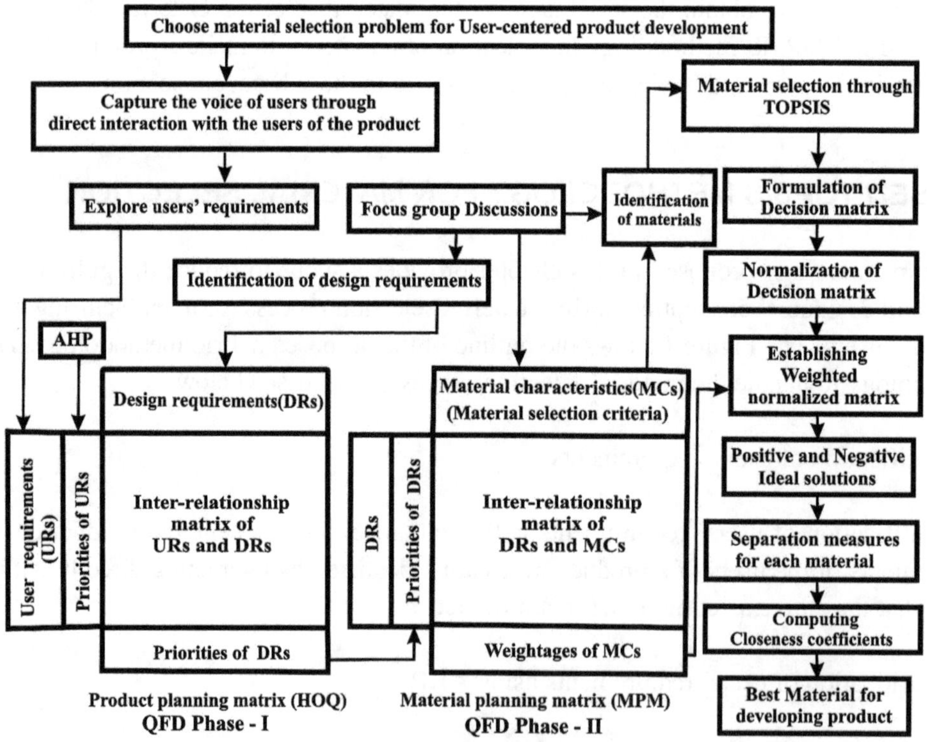

Step 4: Calculation of priority ratings of the design requirements

After identifying the design requirements, the next step is to establishment of inter-relationship matrix of HOQ. The exact translation of users' requirements into product design requirements takes place in this matrix. The cells of the inter-relationship matrix are filled with the strength of the relationships among users' requirements and design requirements, which are usually expressed by means of a rating scale 0-1-3-9 to indicate no, weak, medium and strong relationships respectively (Lin et al., 2006). The absolute importance values of each design requirement are computed by summing the products of the weightages of the users' requirements and the corresponding inter-relationship values in the respective design requirement column. The normalization of the absolute importance values of the design requirement gives the relative importance values of the design requirements. These are the priority ratings of the design requirements which are the outcome of HOQ.

Step 5: Identification of material characteristics (selection criteria)

The design team has to discover the necessary material characteristics which have the most impact on meeting the product design requirements. The material characteristics indicate the criteria which provide a basis for material selection. The materials selection criteria are specific properties of material derived from the product design requirements identified by the design team.

Step 6: Establishment of material planning matrix

In the proposed methodology the second phase of QFD is considered as material planning matrix (MPM) instead of parts planning matrix. The MPM appears like HOQ, but the design requirements will occupy in the place of users' requirements column and the design requirements are replaced with material characteristics. The inter-relationship matrix of MPM influences more on priorities of the material characteristics of the product. The design team has to fill the cell values of the inter-relationship matrix of MPM same as the procedure adopted for establishing inter-relationship matrix of HOQ.

Step 7: Determine the weightages of material selection criteria

The absolute importance values of each material characteristic are computed by summing the products of the relative importance values of the design requirements and the corresponding inter-relationship values in the respective material characteristic column. The weightages of the material characteristics are obtained through normalization of the absolute importance values of material characteristics. These weightages are further utilized in the selection of material through TOPSIS.

Step 8: Identification of materials

Materials identification is a subtle task as different materials are available and new materials are also emerging regularly. Once the required material characteristics are established, the design team and materials research experts need to involve in brainstorming sessions to explore different materials possess the required characteristics. After identification of the materials, it is necessary to analyze and test them to determine whether they do meet the selection criteria or not.

Step 9: Selection of best material using TOPSIS

As the selection of optimum material under different materials against the required attributes, the application of multi-attribute decision-making tool is necessary for solving material selection problem. As TOPSIS is one of the most popular MADM tools and it serves as decision making tool for selection of best material. The detailed procedure of TOPSIS discussed in the background section of the chapter helps to identify the best material to meet the product design requirements.

ILLUSTRATIVE EXAMPLE

A decision-making problem on material selection for making an engineering component has been taken from the literature (Çalışkan et al., 2013) with a view to demonstrate the proposed hybrid methodology. The authors (Çalışkan et al., 2013) have defined clearly the decision-making problem on selection of material for the tool holder working under hard milling conditions. Hard milling is an advanced material machining technology, which has been attracted the manufacturers around the world for the last few years as it offers potential benefits such as higher material removal rate with more flexibility, shortest route with minimum capital investment, less environmental hazards etc. As hard milling requires a large range of rotational speeds and high amount of forces are involved, tool holders play an important role in hard milling operation. The users (operators) may have some expectations on tool holder as they are directly associated with machining activity. In this context, the users' opinions and requirements are captured through direct interaction with them. The Table 4 shows the users' requirements on tool holder under hard milling conditions.

To obtain the priorities of the URs, pair-wise comparisons have been carried with the involvement of users. The subjective judgments made over their requirements are quantified using Saaty scale for preparing pair-wise comparison matrix (PWCM), which is shown in Table 5.

Then the normalized pair-wise comparison matrix (NPWCM) is prepared by using eigen value method as discussed in the step 3 of AHP procedure presented earlier in the background section of the chapter. The Table 6 shows the NPWCM.

The normalized principal eigen vector is also called priority vector. It is computed by averaging the elements in each row of the normalized pair-wise comparison matrix. The priority vector indicates the relative weights of the users' requirements considered under comparison. The priority ratings of all the URs are computed and are presented in Table 7.

The consistency index (CI) and consistence ratio (CR) are computed as discussed in the earlier section. The computation of the values of CI and CR are presented below.

Table 4. User requirements for a tool holder suitable to hard milling

Sl. No	Users' Requirements (URs)
1	UR-1: able to withstand under vibrations
2	UR-2: will not be deflected during machining
3	UR-3: capacity to withstand loads
4	UR-4: no premature failure
5	UR-5: available at affordable price

Table 5. Pair-wise comparison matrix (PWCM) of users' requirements

URs	Able to Withstand Under Vibrations	Will Not Be Deflected During Machining	Capacity to Withstand Loads	No Premature Failure	Available at Affordable Price
able to withstand under vibrations	1	3	1/3	5	2
will not be deflected during machining	1/3	1	1/3	5	4
capacity to withstand loads	3	3	1	6	5
no premature failure	1/5	1/5	1/6	1	1/3
available at affordable price	1/2	1/4	1/5	3	1

Table 6. Normalized Pair-wise comparison matrix (NPWCM) of users' requirements

URs	Able to Withstand Under Vibrations	Will Not Be Deflected During Machining	Capacity to Withstand Loads	No Premature Failure	Available at Affordable Price
able to withstand under vibrations	0.198	0.403	0.164	0.250	0.162
will not be deflected during machining	0.066	0.134	0.164	0.250	0.324
capacity to withstand loads	0.596	0.403	0.492	0.300	0.405
no premature failure	0.039	0.027	0.082	0.050	0.027
available at affordable price	0.099	0.034	0.098	0.150	0.081

Table 7. Priority ratings of the users' requirements (URs)

URs	Able to Withstand Under Vibrations (UR-1)	Will Not Be Deflected During Machining (UR-2)	Capacity to Withstand Loads (UR-3)	No Premature Failure (UR-4)	Available at Affordable Price (UR-5)
Priority ratings	0.24	0.19	0.44	0.04	0.09

$$\begin{bmatrix} 1 & 3 & \frac{1}{3} & 5 & 2 \\ \frac{1}{3} & 1 & \frac{1}{3} & 5 & 4 \\ 3 & 3 & 1 & 6 & 5 \\ \frac{1}{5} & \frac{1}{5} & \frac{1}{6} & 1 & \frac{1}{3} \\ \frac{1}{2} & \frac{1}{4} & \frac{1}{5} & 3 & 1 \end{bmatrix} \times \begin{bmatrix} 0.24 \\ 0.19 \\ 0.44 \\ 0.04 \\ 0.09 \end{bmatrix} = \begin{bmatrix} 1.356 \\ 1.008 \\ 2.442 \\ 0.234 \\ 0.480 \end{bmatrix}$$

$$\begin{bmatrix} 1.356 / 0.24 \\ 1.008 / 0.19 \\ 2.442 / 0.44 \\ 0.234 / 0.04 \\ 0.480 / 0.09 \end{bmatrix} = \begin{bmatrix} 5.757 \\ 5.369 \\ 5.560 \\ 5.182 \\ 5.195 \end{bmatrix}$$

$$\text{Maximum principal eigen value } \left(\lambda_{\max}\right) = \frac{5.757+5.369+5.560+5.182+5.195}{5} = 5.413$$

$$\mathrm{CI} = \frac{5.413-5}{5-1} = 0.103$$

$$\mathrm{CR} = \frac{0.103}{1.12} = 0.092$$

The value of CR is 0.092, which is less than 0.10 and proved that the judgments made on pair-wise comparisons and AHP results were consistent. Afterwards, focus group discussions made with six experts in which three of them have expertise in the field of manufacturing and two have sound knowledge on product design and one has expertise in Industrial engineering for the purpose of investigating design requirements to meet the requirements of the users. The outcome of the discussions are summarized and presented in the form of design requirements shown in Table 8 to design and development of tool holder suitable to hard milling conditions.

Once the design requirements are established, the next step is to establish inter-relationship matrix of HOQ. To this end the strength of the relationships among the URs and DRs are obtained on the basis of rating scale1-3-9 which indicates weak, medium and strong relationships respectively. If no relationship exists, zero is assigned. Then the values of absolute importance and relative importance of all the DRs are computed through procedure as discussed in the step 4 of the proposed methodology. The Table 9 shows the customized version of the HOQ comprising URs and their priority ratings along with DRs and their absolute, relative importance values. The relative importance values of the DRs indicate the priorities of DRs.

Table 8. List of design requirements for tool holder suitable to hard milling

Sl. No	Design Requirements (DRs)
1	DR-1: High rigidity (HR)
2	DR-2: Resistance to bending (RBE)
3	DR-3: Resistance to breaking (RBR)
4	DR-4: High energy dissipation rate (HEDR)
5	DR-5: Good specific damping capacity (GSDC)
6	DR-6: Less cost (LC)

Table 9. House of quality

		Priorities of Users' Requirements	Design Requirements					
			HR	RBE	RBR	HEDR	GSDC	LC
Users' requirements	able to withstand under vibrations	0.24	9	1	1	3	9	1
	will not be deflected during machining	0.19	3	9	1	1	1	1
	capacity to withstand loads	0.44	3	3	3	0	0	3
	no premature failure	0.04	3	9	9	3	3	3
	available at affordable price	0.09	3	3	3	3	0	9
Absolute importance			4.17	3.63	2.11	1.03	2.47	1.87
Relative importance			0.27	0.24	0.14	0.07	0.16	0.12

In order to identify the necessary material requirements a focus group has constituted that contains experts who have sound knowledge and experience in material research and product design. The several discussions in focus group meetings explored six requirements of material, i.e., Young's modulus (YM), Compressive strength (CS), Fracture toughness (FT), Mechanical loss coefficient (MLC), Hardness (H) and Cost (C)with a view to meet the design requirements established earlier in this work. Later material planning matrix is developed as discussed in the step 6 of the proposed methodology. The Table 10 shows the material planning matrix of QFD.

In fact, one of the important aspects in the methodology is the quest for materials which possess required material characteristics. The trial and error method or simply on rely the basis what has been used earlier may not give fruitful result in identifying the materials all the time. The identification of materials for a specific purpose is a complex, time consuming and expensive process that require efforts from metallurgists and experts in the field of material science and engineering. In the present work, the list of materials for manufacturing of tool holders suitable to work under hard milling conditions is directly taken from the literature. The authors (Çalışkan et al., 2013) provided nine materials required to meet the demands from the tool holder design. The quantitative data of material characteristics of all the materials (alternatives) are presented in Table 11.

Table 10. Material planning matrix of QFD

		Priorities of Design Requirements	Material Characteristics (Material Selection Criteria)					
			YM	CS	FT	MLC	H	C
Design requirements	HR	0.27	9	3	3	0	3	3
	RBE	0.24	9	3	3	0	1	3
	RBR	0.14	3	9	9	1	9	3
	HEDR	0.07	1	1	3	9	0	0
	GSDC	0.16	0	0	3	9	0	0
	LC	0.12	3	3	3	0	3	9
Absolute importance			5.44	3.21	3.83	2.20	2.67	3.05
Weightages (priorities)			0.267	0.157	0.188	0.108	0.131	0.149

Table 11. Quantitative data on required properties of different materials demanded from the tool holder

Sl.No	Material	Quantitative Data on Properties of the Materials					
		YM (GPa)	CS (MPa)	FT (MPa m)$^{1/2}$	MLC	H (HV)	C ($/kg)
1	AISI 1020	210	330	54.5	0.00111	150	0.673
2	AISI 1040	212	632.5	46	0.00117	355	0.7045
3	AISI 4140	212	655	87.5	0.000515	305	0.864
4	AISI 6150	206.5	1575	38	0.00026	483	1.175
5	AISI 8620	206.5	360	111.5	0.00089	190	0.8665
6	Maraging steel	187.5	1825	80	0.00071	532.5	6.97
7	AISI 55	210	1930	21	0.00002055	771	7.99
8	Tungsten carbide-cobalt	593	4405	14.05	0.00135	1250	79.6
9	Fe-5Cr-Mo-V	212.5	1655	120	0.00113	448.5	1.73

Source: (Çalışkan et al., 2013)

The decision matrix has been established by considering nine materials as alternatives and six mechanical characteristics as criteria for solving material selection problem through TOPSIS. The normalized decision matrix is developed as discussed in step 2 of the TOPSIS procedure in background section and is presented below.

$$\begin{bmatrix} 0.251839 & 0.057667 & 0.248609 & 0.409992 & 0.084481 & 0.008376 \\ 0.254237 & 0.110529 & 0.209835 & 0.432154 & 0.199938 & 0.008768 \\ 0.254237 & 0.114461 & 0.399143 & 0.190222 & 0.171777 & 0.010754 \\ 0.247641 & 0.275230 & 0.173342 & 0.096034 & 0.272028 & 0.014624 \\ 0.247641 & 0.062910 & 0.508623 & 0.328732 & 0.107009 & 0.010785 \\ 0.224856 & 0.318918 & 0.364931 & 0.262247 & 0.299907 & 0.086750 \\ 0.251839 & 0.337266 & 0.095794 & 0.007590 & 0.434231 & 0.099446 \\ 0.711145 & 0.769771 & 0.064091 & 0.498639 & 0.704006 & 0.990721 \\ 0.254837 & 0.289210 & 0.547397 & 0.417379 & 0.252597 & 0.021532 \end{bmatrix}$$

The weighted normalized matrix is developed by multiplying the normalized decision matrix with the weightages obtained for material characteristics (material selection criteria) through material planning matrix of QFD which were given in Table 9. The weighted normalized decision matrix is given below.

$$\begin{bmatrix} 0.067217 & 0.009074 & 0.046673 & 0.044221 & 0.011045 & 0.001252 \\ 0.067857 & 0.017391 & 0.039393 & 0.046612 & 0.026140 & 0.001310 \\ 0.067857 & 0.018010 & 0.074933 & 0.020517 & 0.022458 & 0.001607 \\ 0.066097 & 0.043306 & 0.032542 & 0.010358 & 0.035565 & 0.002185 \\ 0.066097 & 0.009898 & 0.095486 & 0.035457 & 0.013990 & 0.001611 \\ 0.060015 & 0.050180 & 0.068510 & 0.028286 & 0.039210 & 0.012962 \\ 0.067217 & 0.053067 & 0.017984 & 0.000819 & 0.056772 & 0.014859 \\ 0.189807 & 0.121119 & 0.012032 & 0.053783 & 0.092042 & 0.148031 \\ 0.068017 & 0.045506 & 0.102765 & 0.045018 & 0.330250 & 0.003217 \end{bmatrix}$$

The positive ideal solution (PIS) and negative ideal solution (NIS) are determined by considering the two approaches i.e., larger is better and smaller is better. The PIS (V^+) and NIS (V^-) are expressed as follows.

$$V^+ = \{0.189807, 0.121119, 0.102765, 0.053783, 0.092042, 0.001252\}$$

$$V^- = \{0.060015, 0.009074, 0.012032, 0.000819, 0.011045, 0.148031\}$$

The separation measures i.e., the distance of each alternative (material) from positive ideal solution (D_i^+) as well as negative ideal solution (D_i^-) and closeness coefficients for all the nine materials are calculated.

For instance, the computation of the separation measures and closeness coefficient for AISI 1020 is given below.

$$D_i^+ = \sqrt{\begin{matrix}(0.067217-0.189807)^2+\left(0.009074-0.121119\right)^2+(0.046673-0.102765)^2\\+(0.044221-0.053783)^2+(0.011045-0.092042)^2+(0.001252-0.001252)^2\end{matrix}}$$
$$= 0.19334$$

$$D_i^- = \sqrt{\begin{matrix}(0.067217-0.060015)^2+\left(0.009074-0.009074\right)^2+(0.046673-0.012032)^2\\+(0.044221-0.000819)^2+(0.011045-0.011045)^2+(0.001252-0.148031)^2\end{matrix}}$$
$$= 0.15709$$

$$C_i^+ = \frac{0.15709}{0.19334+0.15709} = 0.4483$$

In the similar manner the separation measures and closeness coefficients for all the remaining eight materials are calculated and the priority ranks of all the materials are also summarized in Table 12.

It is observed from the Table 12, the closeness coefficient of the material Fe-5Cr-Mo-V is higher than that of the remaining materials. According to TOPSIS, Fe-5Cr-Mo-V is the best material to attain the desired mechanical properties.

RESULTS AND DISCUSSION

In the present numerical illustration, the priorities of the users' requirements are obtained using AHP. From the Table 7, it is noticed that the most preferred requirement is UR-3, i.e., the tool holder should withstand loads. Orderly the next priority requirements are UR-1 (able to withstand tool holder under vibrations), UR-2 (will not be deflected during machining), UR-5 (available at affordable price) and UR-4

Table 12. Separation measures and closeness coefficients and ranks of all the materials

Sl. No	Material	Separation Measures		Closeness Coefficient	Rank Order
		D_i^+	D_i^-	C_i^+	
1	AISI 1020	0.19334	0.15709	0.4483	9
2	AISI 1040	0.18450	0.15726	0.4601	7
3	AISI 4140	0.17952	0.16142	0.4735	5
4	AISI 6150	0.17711	0.15359	0.4645	6
5	AISI 8620	0.18481	0.17219	0.4823	4
6	Maraging steel	0.16319	0.15707	0.4905	3
7	AISI 55	0.17630	0.14781	0.4561	8
8	Tungsten carbide-cobalt	0.17256	0.19689	0.5329	2
9	Fe-5Cr-Mo-V	0.15529	0.18175	0.5393	1

(no premature failure).The priorities of the URs are incorporated in HOQ and are involved in computing the relative importance values (priorities) of design requirements. The appropriate mapping of URs and DRs in the inter-relationship matrix of HOQ provide the priorities of DRs. The outcome of HOQ indicates that the most important design requirements in developing tool holder are HR (high rigidity), RBE (resistance to bending), GSDC (good specific damping capacity) and RBR (resistance to breaking).

The priorities of the DRs forms a basis for deciding the material characteristics and then they are used in determining the weightages of the material characteristics (material selection criteria). The material planning matrix revealed that the most important material characteristics in an order are Young' modulus, fracture toughness, compressive strength, cost, hardness and mechanical loss coefficient. The weightages of the material selection criteria are used to identify the suitable material among the available nine materials for manufacturing tool holder. The Table 12 shows that the best material identified through QFD-based TOPSIS approach is Fe-5Cr-Mo-V steel as it possesses good amount of fatigue resistance due to its high strength and also its cost is affordable to users. Moreover, the outcome of the proposed approach matches well with the results obtained by Caliskan et al. (2013). The proposed QFD-based TOPSIS methodology also reveals that the Fe-5Cr-Mo-V steel is the best material for making hard milling tool holders. This material will satisfy both users' as well as manufacturers.

CONCLUSION

The proposed QFD-based TOPSIS methodology is an appropriate approach to address material selection problem during user-centered product development. The methodology helps the decision makers for identifying a material to meet designers' conception as well as users' perception. The users' requirements are translated into appropriate design requirements and then the design requirements are influenced in determining the weightages of the material selection criteria. In this way users' preferences are reflected in decision making on material selection. The methodology integrates product designers, manufacturing engineers and materials engineers and enables them to make appropriate decision on selection of best material to satisfy the users of a product and to meet the design requirements. The illustrative example presented in this chapter helps to understand the application of the proposed structured methodology for solving material selection problem under user-centered product development of any product. Further, the work may be extended through considering fuzzy environment as ambiguity is inevitable during subjective judgments made in the work.

REFERENCES

Abdel-Basset, M., Manogaran, G., Mohamed, M., & Chilamkurti, N. (2018). Three-way decisions based on neutrosophic sets and AHP-QFD framework for supplier selection problem. *Future Generation Computer Systems*, *89*, 19–30. doi:10.1016/j.future.2018.06.024

Arep, H. B., Min, G. C., Tar, A. F. B. M., & Jaafar, R. B. (2016). Material selection for automotive fender design using integrated AHP-TOPSIS. *Journal of Advanced Manufacturing Technology*, *10*(2), 29–44.

Behzadian, M., Otaghsara, S. K., Yazdani, M., & Ignatius, J. (2012). A state-of the-art survey of TOPSIS applications. *Expert Systems with Applications*, *39*(17), 13051–13069. doi:10.1016/j.eswa.2012.05.056

Bhimaraya, A. M. (2003). Impact of house of quality on buyer satisfaction. *Productivity*, *43*(4), 603–610.

Çalışkan, H., Kurşuncu, B., Kurbanoğlu, C., & Güven, Ş. Y. (2013). Material selection for the tool holder working under hard milling conditions using different multi criteria decision making methods. *Materials & Design*, *45*, 473–479. doi:10.1016/j.matdes.2012.09.042

Cavallini, C., Giorgetti, A., Citti, P., & Nicolaie, F. (2013). Integral aided method for material selection based on quality function deployment and comprehensive VIKOR algorithm. *Materials & Design*, *47*, 27–34. doi:10.1016/j.matdes.2012.12.009

Chadawada, R., Sarfaraz, A., Jenab, K., & Pourmohammadi, H. (2015). Integration of AHP-QFD for selecting facility location. *Benchmarking: An International Journal*, *22*(3), 411–425. doi:10.1108/BIJ-06-2013-0064

Chan, L. K., & Wu, M. L. (2002). Quality function deployment: A comprehensive review of its concepts and methods. *Quality Engineering*, *15*(1), 23–35. doi:10.1081/QEN-120006708

Chang, C. W., Wu, C. R., Lin, C. T., & Lin, H. L. (2007). Evaluating digital video recorder systems using analytic hierarchy and analytic network processes. *Information Sciences*, *177*(16), 3383–3396. doi:10.1016/j.ins.2007.02.010

Chen, C. C. (2009). Integration of quality function deployment and process management in the semiconductor industry. *International Journal of Production Research*, *47*(6), 1469–1484. doi:10.1080/00207540701549616

Chen, S. C., Yang, C. C., Lin, W. T., Yeh, T. M., & Lin, Y. S. (2007). Construction of key model for knowledge management system using AHP-QFD for semiconductor industry in Taiwan. *Journal of Manufacturing Technology Management*, *18*(5), 576–597. doi:10.1108/17410380710752671

Das, D., & Mukherjee, K. (2008). Development of an AHP-QFD framework for designing a tourism product. *International Journal of Services and Operations Management*, *4*(3), 321–344. doi:10.1504/IJSOM.2008.017297

De Felice, F., & Petrillo, A. (2010). A multiple choice decision analysis: An integrated QFD–AHP model for the assessment of customer needs. *International Journal of Engineering Science and Technology*, *2*(9), 25–38.

Durga Prasad, K. G., Venkata Subbaiah, K., & Narayana Rao, K. (2012). Aligning the competitive strategy with supply chain strategy through QFD. *Journal of Advances in Management Research*, *9*(2), 189–198. doi:10.1108/09727981211271931

Dursun, M., & Arslan, Ö. (2018). An integrated decision framework for material selection procedure: A case study in a detergent manufacturer. *Symmetry*, *10*(11), 657. doi:10.3390ym10110657

Fogliatto, F. S., Da Silveira, G. J., & Royer, R. (2003). Flexibility-driven index for measuring mass customization feasibility on industrialized products. *International Journal of Production Research*, *41*(8), 1811–1829. doi:10.1080/1352816031000074991

Hanumaiah, N., Ravi, B., & Mukherjee, N. P. (2006). Rapid hard tooling process selection using QFD-AHP methodology. *Journal of Manufacturing Technology Management*, *17*(3), 332–350. doi:10.1108/17410380610648290

Hauser, J. R., & Clausing, D. (1988). The House of Quality. *Harvard Business Review*, *66*(3), 63–73.

Ho, W. (2008). Integrated analytic hierarchy process and its applications–A literature review. *European Journal of Operational Research*, *186*(1), 211–228. doi:10.1016/j.ejor.2007.01.004

Ho, W., & Prasanta, K. (2011). Strategic sourcing: A combined QFD and AHP approach in manufacturing. *Supply Chain Management*, *16*(6), 446–461. doi:10.1108/13598541111171093

Islam, R., Ahmed, M., & Alias, M. H. (2007). Application of Quality Function Deployment in redesigning website: A case study on TV3. *International Journal of Business Information Systems*, *2*(2), 195–216. doi:10.1504/IJBIS.2007.011619

Jalham, I. S. (2006). Computer-aided quality function deployment method for material selection. *International Journal of Computer Applications in Technology*, *26*(4), 190–196. doi:10.1504/IJCAT.2006.010764

Kasaei, A., Abedian, A., & Milani, A. S. (2014). An application of Quality Function Deployment method in engineering materials selection. *Materials & Design*, *55*, 912–920. doi:10.1016/j.matdes.2013.10.061

Kumar, R., & Ray, A. (2014). Selection of material for optimal design using multi-criteria decision making. *Procedia materials science, 6*, 590-596.

Lin, M. C., Wang, C. C., & Chen, T. C. (2006). A strategy for managing customer-oriented product design. *Concurrent Engineering*, *14*(3), 231–244. doi:10.1177/1063293X06068390

Mastura, M. T., Sapuan, S. M., Mansor, M. R., & Nuraini, A. A. (2018). Comparative Study on Weighting Customer Requirements using Fuzzy Analysis Hierarchy Process with Extent Analysis and Analytic Hierarchy Process. *Journal of Advanced Manufacturing Technology*, *12*(1), 477–490.

Mayyas, A., Shen, Q., Mayyas, A., Shan, D., Qattawi, A., & Omar, M. (2011). Using quality function deployment and analytical hierarchy process for material selection of body-in-white. *Materials & Design*, *32*(5), 2771–2782. doi:10.1016/j.matdes.2011.01.001

Mittal, K., Chandra Tewari, P., Khanduja, D., & Kaushik, P. (2016). Application of Fuzzy TOPSIS MADM approach in ranking & underlining the problems of plywood industry in India. *Cogent Engineering*, *3*(1), 1155839. doi:10.1080/23311916.2016.1155839

Opricovic, S., & Tzeng, G. H. (2004). Compromise solution by MCDM methods: A comparative analysis of VIKOR and TOPSIS. *European Journal of Operational Research*, *156*(2), 445–455. doi:10.1016/S0377-2217(03)00020-1

Prasad, B. (1998). Review of QFD and related deployment techniques. *Journal of Manufacturing Systems*, *17*(3), 221–234. doi:10.1016/S0278-6125(98)80063-0

Prasad, K., & Chakraborty, S. (2013). A quality function deployment-based model for materials selection. *Materials & Design*, *49*, 525–535. doi:10.1016/j.matdes.2013.01.035

Prasad, K., Maity, S. R., & Chakraborty, S. (2016). Quality function deployment-based expert system for materials selection. *International Journal of Materials Science and Engineering*, *4*(2), 80–93.

Prasad, K. D., Sravani, K. D. S., & Manasa, B. L. (2019). Kano-HOQ-GRA Hybrid Methodology for Customer-Driven Product Development. In Optimizing Current Strategies and Applications in Industrial Engineering (pp. 120–139). Hershey, PA: IGI Global. doi:10.4018/978-1-5225-8223-6.ch005

Prasad, K. D., Subbaiah, K. V., & Rao, K. N. (2014). Multi-objective optimization approach for cost management during product design at the conceptual phase. *Journal of Industrial Engineering International*, *10*(1), 48. doi:10.100740092-014-0048-8

Prasad, K. G. D., Subbaiah, K., & Prasad, M. (2017). Supplier evaluation and selection through DEA-AHP-GRA integrated approach-A case study. *Uncertain Supply Chain Management*, *5*(4), 369–382. doi:10.5267/j.uscm.2017.4.001

Raharjo, H., Xie, M., & Brombacher, A. C. (2006). Prioritizing quality characteristics in dynamic quality function deployment. *International Journal of Production Research*, *44*(23), 5005–5018. doi:10.1080/00207540600547414

Rehman, F., & Yan, X. T. (2003). Product design elements as means to realise functions in mechanical conceptual design. *Proceedings of ICED 03, the 14th International Conference on Engineering Design DS 31*, Stockholm (pp. 213-214). Academic Press.

Roghanian, E., Rahimi, J., & Ansari, A. (2010). Comparison of first aggregation and last aggregation in fuzzy group TOPSIS. *Applied Mathematical Modelling*, *34*(12), 3754–3766. doi:10.1016/j.apm.2010.02.039

Rush, C., & Roy, R. (2000). Analysis of cost estimating processes used within a concurrent engineering environment throughout a product life cycle. *Proceedings of the 7th ISPE International Conference on Concurrent Engineering: Research and Applications*, Lyon, France (pp. 58-67). Technomic Inc.

Saaty, T. L. (1980). *The Analytic Hierarchy Process*. New York: McGraw-Hill Company.

Saaty, T. L. (2008). Decision making with the analytic hierarchy process. *International Journal of Services Sciences*, *1*(1), 83–98. doi:10.1504/IJSSCI.2008.017590

Scott, J., Ho, W., Dey, P. K., & Talluri, S. (2015). A decision support system for supplier selection and order allocation in stochastic, multi-stakeholder and multi-criteria environments. *International Journal of Production Economics*, *166*, 226–237. doi:10.1016/j.ijpe.2014.11.008

Sharma, J. R., Rawani, A. M., & Barahate, M. (2008). Quality function deployment: A comprehensive literature review. *International Journal of Data Analysis Techniques and Strategies*, *1*(1), 78–103. doi:10.1504/IJDATS.2008.020024

Shih, H. S., Shyur, H. J., & Lee, E. S. (2007). An extension of TOPSIS for group decision making. *Mathematical and Computer Modelling*, *45*(7-8), 801–813. doi:10.1016/j.mcm.2006.03.023

Sobhanallahi, M. A., Zendehdel Nobari, N., & Pasandideh, S. H. R. (2019). An Aggregated Supplier Selection Method Based on QFD and TOPSIS (Case Study: A Financial Institution). *Journal of Optimization in Industrial Engineering*, *12*(1), 31–40.

Subbaiah, K. V., Durga Prasad, K. G., & Narayana Rao, K. (2011). Customer-driven product planning using conjoint analysis and QFD-ANP methodology. *International Journal of Productivity and Quality Management*, *7*(3), 374–394. doi:10.1504/IJPQM.2011.039353

Tian, G., Zhang, H., Feng, Y., Wang, D., Peng, Y., & Jia, H. (2018). Green decoration materials selection under interior environment characteristics: A grey-correlation based hybrid MCDM method. *Renewable & Sustainable Energy Reviews*, *81*, 682–692. doi:10.1016/j.rser.2017.08.050

Tsinidou, M., Gerogiannis, V., & Fitsilis, P. (2010). Evaluation of the factors that determine quality in higher education: An empirical study. *Quality Assurance in Education*, *18*(3), 227–244. doi:10.1108/09684881011058669

Ucler, C. (2017). Brainstorming the cryoplane layout by using the iterative AHP-QFD-AHP approach. *Aviation*, *21*(2), 55–63. doi:10.3846/16487788.2017.1344138

Wind, Y., & Saaty, T. L. (1980). Marketing applications of the analytic hierarchy process. *Management Science*, *26*(7), 641–658. doi:10.1287/mnsc.26.7.641

Yadav, S., Pathak, V. K., & Gangwar, S. (2019). A novel hybrid TOPSIS-PSI approach for material selection in marine applications. *Sadhana*, *44*(3), 58. doi:10.100712046-018-1020-x

Yang, C. L., Huang, R. H., & Ke, W. C. (2012). Applying QFD to build green manufacturing system. *Production Planning and Control*, *23*(2-3), 145–159. doi:10.1080/09537287.2011.591632

Zavadskas, E. K., Mardani, A., Turskis, Z., Jusoh, A., & Nor, K. M. (2016). Development of TOPSIS method to solve complicated decision-making problems-An overview on developments from 2000 to 2015. *International Journal of Information Technology & Decision Making*, *15*(3), 645–682. doi:10.1142/S0219622016300019

Zhang, H., Peng, Y., Tian, G., Wang, D., & Xie, P. (2017). Green material selection for sustainability: A hybrid MCDM approach. *PLoS One*, *12*(5), 1–26. doi:10.1371/journal.pone.0177578 PMID:28498864

Chapter 9
Task Allocation and Path Planning of a Multi-Robot System Using Heuristic Coupled Particle Swarm Optimization Algorithm

Arindam Majumder
National Institute of Technology, Agartala, India

Rajib Ghosh
National Institute of Technology, Agartala, India

ABSTRACT

This study deals with a plant layout where there were ninety predefined locations which have to be inspected by using three multiple robots in such a way that there would not be any collisions between the robots. A heuristic integrated multiobjective particle swarm optimization algorithm (HPSO) is developed for allocating tasks to each robot and planning of path while moving from one task location to another. For optimal path planning of each robot the research utilized A algorithm. The task allocation for each robot is carried out using a modified multiobjective particle swarm optimization algorithm where the earliest completion time (ECT) inspired technique is used to make the algorithm applicable in multi robot task allocation problems. At the later stage of this study, in order to test the capability of HPSO an instance is solved by the algorithm and is compared with the existing solutions of a genetic algorithm with the A* algorithm. The computational results showed the superiority of the proposed algorithm over existing algorithms.*

DOI: 10.4018/978-1-7998-1831-1.ch009

INTRODUCTION

Multi robot system is a group of robots working together and which can coordinate and communicate to each other to perform certain task through co-operation. With the use of multi robot system we can perform several tasks continuously that couldn't be possible by using a single manipulator. For example, with an appropriate design of control and coordination multiple robot manipulators could move a long or heavy object, no matter how capable or advance a single mono robot system is it has always limitation on many aspects. Some task requires the use of tool that can handle two or more robots in coordination in these cases the use of these tools may be possible because of the rigidity that provides by the multi robot system. Multi robot system can be either centralized one or de centralized one. In centralized multi robot system control is done by using central computer every agent maintains a connection to the central agent. So, there is a supervisory control by the computer over the system whereas in de centralized multi robot system there is no central connection between the agents to the computer. For the following benefits of the MRS this topic has become more popular among the researchers. Some task which are quite complex for a single to do can be easily be done by multi robot system, some complex task can be of distributed nature or the diversity in terms different require requirements. The time taken to complete the task can be reduced if multiple robots cooperates together par alley. It increases the system reliability because using single robot may work as a bottleneck during critical conditions. But when we implement multiple robots doing a similar task if one gets fails the task can be easily completed by using others. Having a single powerful robot will be complex in design and costlier too use than small simple multiple robots. Thus, multi robot systems have been used in various missions like Distributed controllers for the deployment of mobile sensor networks, agile coordinated mobile robot control, multi robot control with adversaries and environment hazards, persistent monitoring persistent environmental sampling with robots, information based active sensing estimation, multi robot manipulation, path planning of multiple AUV's under marginal acoustic communication condition, detection and tracking any object in different unwanted environment conditions, surveillance, reconnaissance, and continuous area sweeping. A numerous number of problems can be solved by deploying and developing multi robot systems (MRS) in real world applications. Davidson and Murry (1998) proposed a fully automated mobile robot system which can detect, store and track suitable landmark features during goal directed navigation. Liu and Bucknlal (2018) introduced multi unmanned surface vehicle for carrying out dangerous missions such as the maritime portal, protection in hazardous waste environments and rescue in post-disaster scenarios. Caloud et al. (1990) implemented multi robot systems for transportation of objects, performing operations, detection of hazard, maintenance and cleaning in indoor environments. Parker (1994) showed the applicability of multi robot system for cleaning up of hazardous waste situated in different locations. Ju and Son (2018) introduced multi aerial robot system for farming application. The other problems to be solved through MRS include, task allocation, group formation, cooperative object detection and tracking, communication relaying and self-organization.

Multi robot task allocation problem is one of the most distributed and challenging problems of multi robot system where the tasks should be allocated in an optimal way so that the cost can be reduced. Multi robot task allocation is an optimal assignment problem where the main objective is how optimally a set of tasks can be assigned to a set of robots so that the overall system performance can be increased. A proper allocation of tasks to a set of robots not only increases the productivity but also reduces the resource consumption of the corresponding industry. One of the challenges of multi robot system is to find the best way of taking advantage from the variety of available resources. Agents in the environment must communicate and coordinate to determine what job has to be done and who is the best suited to perform them. Multi robot planning can be divided into two categories coupled and decoupled. A coupled algorithm seeks to find a path in the full configuration space of a system. While the full configuration space contains all possible paths, it increases the complexity exponentially with the number of robots in the system. As a result, couple planners may require more computation (CPU time) for finding the optimal path. On the other hand, the decoupled algorithm searches a portion of the full configuration space. The searching in a portion of the full configuration space reduces the probability of capturing all the possible ways of path planning and task allocation while solving problems related to multi robot systems. As a result, decoupled algorithm generally produces inferior results with quicker convergence. Therefore, it becomes very much necessary to develop an algorithm with better capability to solve multi robot task allocation (MRTA) problem at comparatively low computational cost.

Previously a number of efforts have been given by the researchers in order to solve the multi robot task allocation problem efficiently. In most cases the researches implemented heuristics or meta-heuristics for solving multi robot task allocation problems. The reason behind using heuristics and meta heuristics for solving multi robot system is its NP-hard nature. Liu and Bucknlal (2018) proposed a novel algorithm based on a self-organizing map (SOM) for the assignment of multiple tasks and a route planning algorithm based on the first gearbox is used to calculate trajectories and avoid collisions. Yoon and Jinwhan (2017) proposed a two phase task allocation algorithm for the optimal route planning multi robot system. In first phase the first graph search method is used (Hungarian Algorithm) and in the second phase the optimal feasible solution is resolved by edge swapping. Lee (2018) introduced a resource-based task allocation technique for multi robot system. The technique enables the robots to consume their resources continuously during their operations. Robot calculates the task performance estimation considering all the different paths of refill stations based on their refill stations. Wagner and Choset (2015) implemented a sub dimensional expansion for multi robot path planning which creates one dimensional search space for one particular robot. In their study they used M* algorithm for an implementation of sub dimensional expansion. Binetti, Naso and Turchiano (2013) presented an extension of decentralized task assignment algorithm for heterogeneous agents with the limited capacity of critical task. The algorithm is able to find solutions fulfilling all the constraints at the expense of increase in computational and communication cost. However, it preserves the polynomial nature of its predecessor. Suemitsu et al. (2016) implemented cell manufacturing systems of several robots for determining the positions of the manufacturing components and the programming of tasks simultaneously. During this study task assignments for each robot were considered discrete design variables. Qian and Cheng (2018) implemented two metaheuristics namely genetic algorithm and ant-colony algorithm for solving problems of multi robot system with the objective to form coalition. The results of their study showed canonical GA as better performer with respect to ant-colony algorithm while solving this problem. Park and Wang (2009) combined TABU search algorithm with genetic algorithm for allocating and scheduling of jobs in collision free multiple robot system. In their study they implemented genetic algorithm for global search while local search is carried out using TABU search algorithm. Wang, Gu and Li (2012) applied ant colony algorithm for task allocation in multi-robot system. Li et al. (2009) developed an improved version of quantum evolutionary algorithm to form robot coalition effectively in multi-robot system. The results of their study indicate the superiority of the proposed algorithm over existing genetic algorithm and classical quantum evolutionary algorithm. Zhu, Tang and Yuan (2017) proposed a modified version of particle swarm optimization algorithm to allocate task in multi robot system efficiently. In their modified PSO a feasibility-based rule is implemented for handling constraints of multi robot task allocation problem. Muhuri and Rauniyar (2017) developed two variants of immigrants based genetic algorithms namely elitism-based immigrants genetic algorithm and random immigrants genetic algorithm for solving problems related to multi-robot system with the objective of coalition formation. Jianping, Yimin, and Yunbiao (2009) applied genetic algorithm for allocating tasks to the robots of multi robot system. In their study they considered maximization of utility as objective. Hussein et al. (2013) proposed a generic market-based technique for solving problems related to multi-robot task allocation problem. Huang et al. (2018) suggested a niching immune inspired technique to solve multi robot task allocation problems efficiently. Zhou et al. (2019) introduced an exact flow-based integer program for multi robot task allocation problem. Later on, they proposed a tree inspired genetic algorithm for obtaining approximate solution of the considered problem. However, in the recent days these existing approaches are not capable enough to solve task allocation problems of multi robot system (MRS). This is due to the fact that the recent real-life task allocation problems are mostly multi objective problems in nature and the approaches available in the literatures are incapable to handle multi objectives at a time. This motivates the authors to introduce a modified variant of multi objective particle swarm optimization algorithm which can handle multiple objectives of MRTA problems efficiently. To the best of author's knowledge, no such effort has been given by the previous researchers for implementing multi objective PSO in multi robot task allocation problem.

Therefore, the present study proposes a heuristic coupled multi objective particle swarm optimization algorithm for solving multi robot task allocation problem. The algorithm applied A* algorithm for path planning and a modified multi objective particle swarm optimization algorithm to allocate tasks. An earliest completion time (ECT) based heuristic is integrated with multi objective particle swarm optimization algorithm for making the algorithm capable to solve multi robot task allocation problem. The algorithm is tested by solving a three-robot task allocation problem having ninety locations to be inspected. The solutions obtained by the proposed algorithm finds better while comparing with Genetic algorithm with A* algorithm (Liu et al., 2012) and Genetic algorithm with A* algorithm (Jose et al., 2016).

PROBLEM DESCRIPTION

Multi robot task allocation (MRTA) is an optimal assignment problem where the main objective is how optimally a set of multiple robots can be assigned to a set of multiple tasks so that it can reduce the task completion time and increase the efficiency of system performance. Multi robot task allocation problem is concerned about the question of how optimally or sequentially the task should be allocated to the robots to achieve the system goal. This MRTA problem can be divided into two sub problems. First one is how to find the optimal path or shortest path for the robots and the second one is how the task should be assigned to the set of robots. Thus, the robot team must be coordinate in such a way that it can reduce the overall time. The task allocation problem can vary with time and environment. Thus, it is a very complex problem to tackle. However, the present study considers a problem related to tank farm layout applied in petroleum refinery, where there is a least chance of change in task location with respect to time and farm environment (Liu et al, 2012). Therefore, in this study a static environment has been considered where the location of task and the types of task does not change with time and the environment condition remains constant.

The other constrains considered for the MRTA problem is given below:

1. One robot is capable of executing one task at a time.
2. Each task required exactly one robot to achieve it.
3. The assignment is instantaneous allocation with no planning for future.
4. All tasks can be executed by using all robots.
5. At the same time each of the robots starts from the starting nodes.

Let a group of m robots R={R_1,R_2,R_3...R_i...R_m} are to be assigned to a group of n tasks T={T_1,T_2,T_3...T_j...T_n} with relative weights w={w_1,w_2,w_3......w_n}, and C_{ij} is the cost of the robot R_i while executing task T_j. The goal of MRTA is to find an ordered set of tasks for minimizing the total cost of the mission. According to the application requirements the cost of a problem can be determine from the fuel consumption and the operating time of the robots. The robots have to choose the tasks, its sequence and planning of path in such a way that both the task completion time and fuel consumption can be reduced. The task completion time which is the time span between the first robot starting its task and the last robot finishing its task is defined as:

$$TCT = \max\left\{\sum_{j=1}^{n} C_{1j,} \sum_{j=1}^{n} C_{2j}, \ldots, \sum_{j=1}^{n} C_{ij}, \ldots, \sum_{j=1}^{n} C_{mj}\right\} \quad (1)$$

Where, C_{ij}> 0, if robot R_i is allocated task T_j else C_{ij} =0.

Whereas, the fuel consumption can mathematically be defined as:

$$FC = sum\left\{\sum_{j=1}^{n} \sigma_1 C_{1j,} \sum_{j=1}^{n} \sigma_2 C_{2j}, \ldots, \sum_{j=1}^{n} \sigma_i C_{ij}, \ldots, \sum_{j=1}^{n} \sigma_m C_{mj}\right\} \quad (2)$$

Where, σ_j represents the fuel consumption coefficient of jth task, which is assumed as same for every robot.

The task allocation and path planning should also satisfy two constraints, each task is exactly executed once and the robot needs to move inspection positions without colliding with each other or with obstacles.

The inspections positions of the inspected objects are determined from the object size, the sensor characteristics and measurement method (e.g. the sensor range). For example, in a gas leakage problem, the goal is not only finding an ordered set of inspected objects for each robot providing for minimum total cost, but also finding a shortest feasible path between the inspection positions. Therefore, the cost C_{ij} of ith robot during execution of jth task is composed of the travelling time C_{ijv} from its previous position to the current gas inspection positions, and the inspection time C_{ijs} which is the time to complete the required gas measurements at the inspection position. The mathematical representation of C_{ij} is:

$$C_{ij} = C_{ijv} + C_{ijs} \quad (3)$$

Where, travelling time, $C_{ijv} = \alpha \cdot \left(d_{ij} / v_i\right)$ and inspection time, $C_{ijs} = \beta \cdot M_{ij}$, d_{ij} is the distance traveled by i[th] robot to reach at j[th] task, v_i is the velocity of ith robot, $\alpha \in \left[1:10\right]$ is the ground condition coefficient which is related is the robot velocity, $\beta_j \in \left[1:10\right]$ is the sensor performance coefficient related to the capabilities of sensor for inspecting different types of objects and M_{ij} is the measurement range related to the sensor range, the size of the measurement range related to the sensor range.

In this study there are ninety predefined locations where tasks are allocated and these tasks are executed by using three multiple robots which are kept in three different positions. The plant layout for this study has shown in this figure 1. The plant layout has been divided into 40×45 equally square grids, where each square grid has given an equal value of one unit. In this figure the dark shadow grids represents the inadmissible areas through which robots cannot move and the remaining white grids are admissible areas from where the robots moves to execute the tasks. If only one single robot would have been utilized for executing all the tasks in 90 different locations then there would have 90 ways through which it could inspected. Therefore, it would take more time to complete all the tasks. Thus, by providing three robots in different locations the task completion time can be dramatically reduced.

A* ALGORITHM

A* (pronounced "A star") algorithm was developed by Nil Nilsson (1998) in the year 1968. It is one of the most popular methods for finding the closest path between the two locations which are called as starting node and goal node respectively. A* algorithm is a combination of formal approach like Dijsktra's algorithm (Dijsktra et al., 1959) and heuristic approaches like best First Search technique (BFS). At the beginning of the A* algorithm it maintains a priority queue by finding the most promising path to visit by surveying the nodes. While visiting a node to search the optimal path it surveys the neighboring nodes around it and adding them to the priority queue. This process is repeated until it finds the goal node. The priority queue consists of two types of records, one is open list or fringe list and another is closed list. The algorithm starts from the starting node and then starts discovering the other nodes. The nodes which have been discovered near to the evaluated nodes are stored into the open or fringe list and the nodes which are already evaluated after exploration, are stored into the closed list. The closed list is basically a record of all the locations that have been discovered and evaluated by the algorithm. Initially fringe list contains only one element i.e. the starting node. The closed list is considered as empty at the initial stage. The function used to evaluate the nodes which are included in open or fringe list is expressed as:

$$f\left(a\right) = g\left(a\right) + h\left(a\right) \tag{4}$$

Where, g(a) is the distance from the current node to the 'a' node and h(a) is the heuristic value that is the estimated distance from the node n to the goal node. The value of h (a) can be found by using the Euclidean distance or straight line distance and can mathematically be represented as:

$$h\left(a\right) = \sqrt{\left(x_2 - x_1\right)^2 + \left(y_2 - y_1\right)^2} \tag{5}$$

Where, x_1 and y_1 are the coordinate of the current location, and x_1 and y_1 are the coordinates of the goal location. The heuristic function guides the search process through the most profitable or optimized path among all the available paths, and helps to minimize the alternatives.

Figure 1. Plant layout

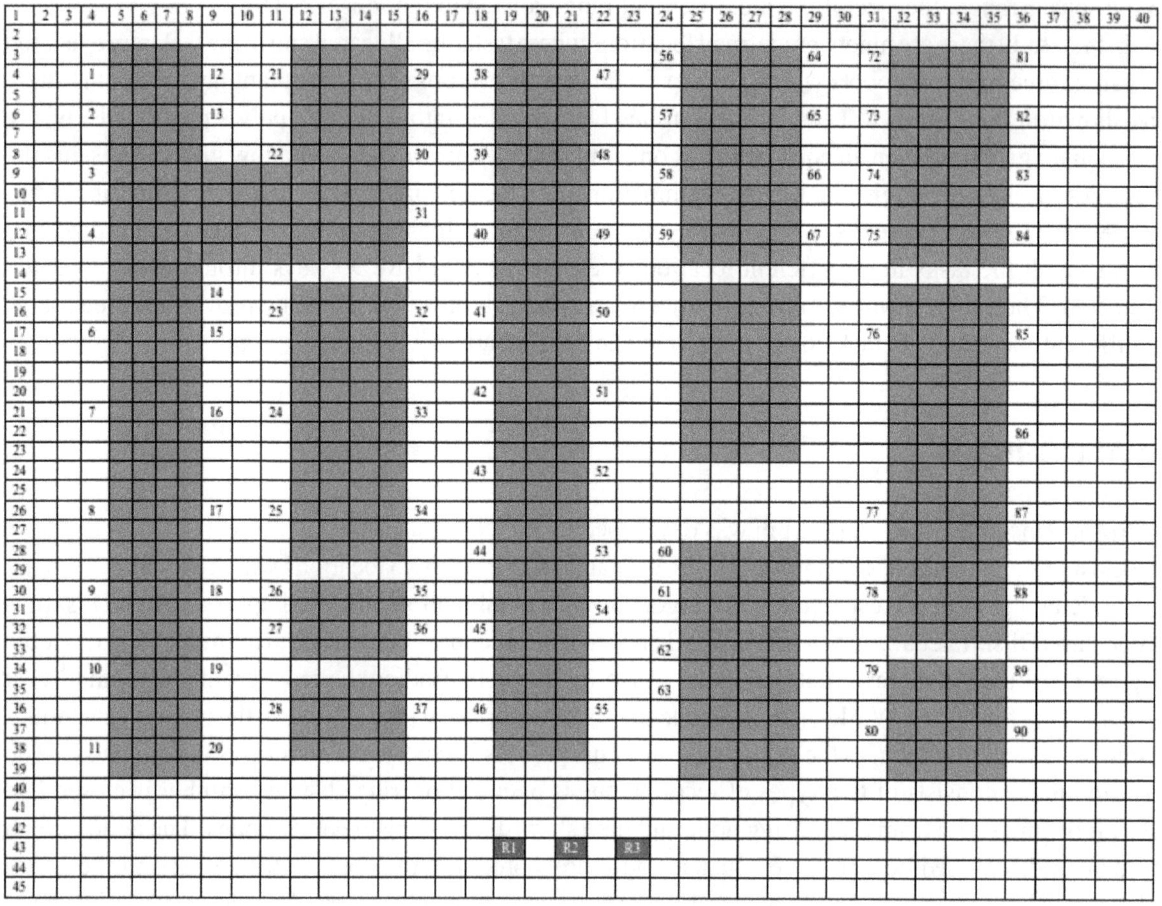

PSO ALGORITHM

Particle swarm optimization algorithm is a nature inspired population based evolutionary and stochastic optimization technique to solve computationally hard optimization problems. It has inspired from the social behavior of nature such as swarms of bird flocking, fish schooling etc. This algorithm has a population of particles where each particle provides a solution and the solution is associated with a random velocity which is used in the problem space. In the particle swarm optimization algorithm, all the particles communicate with each other either directly or indirectly by using search directions. In this algorithm all the individuals are moving through the search space to find the global optima. This algorithm initialized with a group of random individual particles and then searches for optima by updating generations. In each iteration the particles are updated by two best values. The first one, known as P-best, is the best position achieved so far by an individual particle. On the other hand, the best position named as G-best is the position where the best one among all the particles is situated. Thus, the movement of the particle are guided by their individual best known position (P-best) in the search space as well as the entire swarm best known position (G-best). In each time the velocity of particle, flying towards P-best and G-best value, changes. In each step the position and velocities of the particles are changes and the function is evaluated with new coordinates accordingly. The difference between the P-best and particular current position is added to the current velocity causing the trajectory to oscillate around the point. The weighted difference between the best position found by the swarm (G-best) and individual current position is also added to this velocity before implementing it to change the position of the individual iteratively. Thus, the mathematical equation used to update velocity for a particle in each iteration is given below:

$$V_{id}^{(t+1)} = W \times V_{id}^{(t)} + C_1 rand_1 \left(P_{i,best}^{(t)} - X_i^{(t)}\right) + C_2 rand_2 \left(G_{best}^{(t)} - X_i^{(t)}\right) \quad (6)$$

Where, V_{id}^{t} and X_{id}^{t} are the velocity and the position of the ith particle in the dth dimension respectively. $P_{i'best}^{t}$ is the best position of ith individual in d dimension t[th] generation and g_{best}^{t} is the best position of the group of swarm in d dimension in t[th] generations. W is the inertia weight factor which controls the dynamics of flying. C_1 is the cognitive parameter and C_2 is the social parameter, $rand_1$ and $rand_2$ are random variables in range (0, 1).

Later on the following equation is used to update the particle position in (t+1)[th] iteration:

$$X_{id}^{t+1} = X_{id}^{t} + V_{id}^{t+1} \quad (7)$$

In the algorithm, the inertia effect dynamically adjusts the velocity and increasing or decreasing it from V_{min} to V_{max} as per requirement. It also controls the exploration of the search space. The inertia weight is used to control the velocity and by controlling the velocity it compromises between the local exploitation and global exploration ability of the groups of individuals of the swarm. The small inertia weight facilitates the local exploitation whereas the large inertia weight provides the global exploration. The perfect inertia weight can maintain a proper balance between the local exploitation and global exploration of the group of individual or swarm which can give a better solution. Initially the weight must be set a large value so that it can make better global exploration of the search space and later it should decrease gradually for best solution. The continuously decreasing inertia effect moves the search space from global exploration to local exploitation. The following equation can be used to find the proper inertia weight (W) for the next iteration.

$$W = W_{\max} - \frac{W_{\max} - W_{\min}}{iter_{\max}} \times iter$$

Where, W_{max} is the maximum value of inertia weight and W_{min} is the minimum value of inertia weight; *iter* represents the current iteration value and $iter_{max}$ is the maximum iteration numbers.

PROPOSED ALGORITHM

In this present study an integrated approach is developed to determine the path planning and task allocation for centralized multi robots system. The integrated technique comprised of two algorithms, namely A* algorithm and modified multi objective PSO algorithm for path planning and allocating tasks respectively. The detailed description of these algorithms is given below:

A* Algorithm

The planning of robot path during task inspection is carried out by A*algorithm, described in section 2. The approach basically determines the approximated shortest path between the robot and the next task to be inspected by the robot.

Modified Multi-Objective PSO

In this study the Multi-objective PSO (MOPSO), proposed by Coello, Pulido and Lechuga (2004) is used for task allocation to each robot. However, the algorithm suffers from a major drawback that it is generally used to solve problems having continuous search domain due to its equations used for updating particle velocity and position. Since Multi Robot Task Allocation problems belong to those problems related to discrete search space, therefore the MOPSO is unable to solve such problems directly. Thus, the present study proposes a heuristic coupled modified MOPSO for efficient allocation of tasks to the robots by converting continuous particle position into discrete position. The detailed description of major steps used by the proposed algorithm are given below:

Population Initialization

In this stage the initial population is generated by assigning weight value to each task. The assigned weight value indicates the priority of the task to be inspected. A mapped approach named as smallest possible variable (SPV) is implemented in order to sequence the task based on the inspection importance. According to this rule more inspection importance is given to the task having less weight value. For example, if the assigned weight matrix for five tasks is [0.2, 0.34, 0.13, 0.4, 0.02] then the sequence of tasks to be inspected is [5, 3, 1, 2, 4].

Heuristic for Assigning Tasks to the Robots

The section describes the approach used by the algorithm for allocating tasks to the machines. An earliest completion time (ECT) based heuristic is applied for this purpose. The step wise description of this heuristic is given below:

Step 1: Arrange the tasks based on the increasing order of weight values.
Step 2: Find the total number of task.
Step 3: Initialize the number of tasks not yet complete (U) as the total number of task.
Step 4: While $U \geq 1$
Step 5: If $U \leq$ number of machines (N) then assign i^{th} task to i^{th} Robot. Where, i=1, 2… N. Else find the robot which completes its previous task at the earliest and assign it the nearest incomplete task. During this step the task completion time for each robot is calculated based on the path chosen by the robot using A* algorithm.
Step 6: End while

External Repository

The stage deals with the storing of non-dominated particles in repository during the search process. The considered repository comprised of two components namely: achieve controller and adaptive grid technique. The achieve controller function enables the repository for taking decision of adding or discarding of certain solution. The addition or elimination of any solution is based on following conditions:

1. If the repository is empty then the current solution is added to the repository.
2. If the newly obtained solution is dominating over any individual solution stored in the repository then the individual solution is replaced by the new solution.
3. If the repository achieves its maximum storing capacity then adaptive grid technique is applied.

The adaptive grid technique helps the algorithm to distribute the points uniformly in the pareto front. It is basically a space made with hypercube which indicates the concentration level of points in the pareto front. If a newly obtained individual is belonging to a less concentrated zone then the individual is added to external repository by replacing another individual lying in the most concentrated zone of pareto front.

Generation of new Particle Position

The step deals with the generation of new particle positions in each iteration by updating the previously obtained particle velocities and positions. The equations used to update the particle velocities and positions are given below:

$$V^{t+1}\left(i\right) = w \times V^{t}\left(i\right) + r_1\left(Pbest\left(i\right) - P^{t}\left(i\right)\right) + r_2\left(R\left(h\right) - P^{t}\left(i\right)\right)$$

$$P^{t+1}\left(i\right) = P^{t}\left(i\right) + V^{t+1}\left(i\right)$$

Where, w represents the inertia weight, $V^{t+1}(i)$ denotes the velocity of ith particle at t^{th} iteration, r_1 and r_2 are the random number between 0 and 1, Pbest(i) represents the personal best of ith particle, $P^{t}(i)$ denotes the position of ith particle in t^{th} iteration, and R(h) denotes the value picked from the repository.

Thus, based on the modifications mentioned above the various steps of the proposed algorithm are given below:

Step 1: Initialize the process parameters namely, Population size (n), repository capacity, inertia weight value (w),

Step 2: Generate initial population of particles. Let $P \in p_{ij}$ where i = [1, 2, …, n] j = [1,2, …, d]

Step 3: Initialize the speed of each particle as equal to 0. Let $V \in v_{ij}$ where i = [1, 2, …, n] j = [1, 2, …, d]

Step 4: Assign the tasks to the robots for each particle using earliest completion time (ECT) based heuristic mentioned in section ….

Step 5: Compute the objective function of each particle. Let $F \in f\left(l\right)$ where l = [1, 2, …, n]

Step 6: Store the particles having non-dominated solutions in a repository R
Step 7: Initialize the initial particle position as Pbest (i, j) position of that particle. Let Pbest(i, j) = p_{ij}
Step 8: While (t <N) do
Step 9: For i=1 to n

$$v_{ij} = w \times v_{ij} + r_1 \times \left(Pbest\left(i,j\right) - p_{ij}\right) + r_2 \times \left(R\left(h,j\right) - p_{ij}\right) \forall j \in d$$

Where, the index h is determined by implementing following strategy: (a) Choose a hypercube using roulette-wheel selection procedure. (b) Select a particle from the chosen hypercube randomly. r_1 and r_2 are the random values within range of [0-1].

$$p_{ij} = p_{ij} + v_{ij} \forall j \in d$$

Step 10: End of for loop
Step 11: If the particle positions are situated beyond the corresponding boundary then: (a) Set the corresponding boundary as the position of the particle, and (b) Multiply the corresponding velocity by (-1). Otherwise go to Step 11.
Step 12: Assign the tasks to the robots for each particle using earliest completion time (ECT) based heuristic mentioned in section: Heuristic for assigning tasks to the robots
Step 13: Compute the objective functions of each particle.
Step 14: Update the repository (R) content by inserting all the current particle positions having non dominated solution. During insertion if the number of particle positions excides the capacity of the repository (R) then the particles placed in the least concentrated region are given more priority to retain with respect to the particles placed in highly dense area.
Step 15: For all particles if the current particle position i^{th} particle is better than the previous particle position then $Pbest\left(i,j\right) = p_{ij} \ \forall j \in d$.
Step 16: t = t + 1
Step 17: End while loop

COMPUTATIONAL RESULTS

In this section, an instance of tank farm layout, used in petroleum refinery [2], was solved by the proposed algorithm in order to find its capability. The plant layout consists of ninety inspection positions and three homogeneous robots for inspection. The inspection time required to inspect each position was assumed as negligible as compare to the robot travel time. In other words the inspection time was considered as zero for this study.

The proposed algorithm was coded in MATLAB 20017a programming environment and was executed in a PC equipped with Intel i5-2450M CPU with 4 GB in RAM running at 2.50 GHz. The following input parameters are utilized by the proposed algorithm during this study: Population size=100, Maximum number of iterations=100, Repository size= 50, C_1=C_2=1.5, w_{max}=0.9, w_{min}=0.4.

Figure 1 and 2 show the path planning and task allotment considered in the two non-dominated solutions obtained by the algorithm. The route planning and task allotment shown in figure 1 and 2 give shorter completion time with higher fuel consumption and longer completion time with lesser fuel consumption respectively. It is observed from the figure 2 that there is an over lapping of robot-1 (blue path) and robot-2 path (red path) in between the coordinates (16, 11) and (16, 8). While the collision will not happen as because of the visiting time of robot-1 and robot-2 lie between 54.7989 s ~ 56.7989 s and 43.0413 s ~ 46.0413 s respectively. From figure 3 it has been seen that there are overlapping of robot-1 path (blue path) and robot-2 path (red path) in several coordinates. However, there will be no collision at the overlapping coordinates: (15, 39), (17, 4), (16, 11) and (18, 40) as the robots will visit these coordinates in different times (blue path: coordinate (15, 39) at 4.2426 s, coordinate (17, 4) at 61.2131 s, coordinate (16, 11) at 55.7989 s, coordinate (18, 40) at 126.5978. red path: coordinate (15, 39) at 130.012 s, coordinate (17, 4) at 56.8699 s, coordinate (16, 11) at 44.6273 s, coordinate (18, 40) at 2.8284 s and 138.2546 s.

For finding the efficiency of this proposed algorithm a comparison between the solutions obtained by the proposed and existing algorithms are compared at the later stage of this study. The following two existing algorithms are used for this comparison: Genetic algorithm with A* algorithm (GA1-A*) proposed by Liu and Kroll, 2012 and Genetic algorithm with A* algorithm (GA2-A*) proposed by Jose and Pratihar, 2016. Table 1 presents the results obtained by the proposed and the considered existing algorithms. From the results it can be observed that the proposed algorithm can improve the completion time and total fuel consumption up to 9.84% and 9.24% respectively while comparing with the existing techniques.

Figure 2. Task allotment and route of three robots for the Completion time=141.2 s and Total Fuel Consumption= 413.54 units

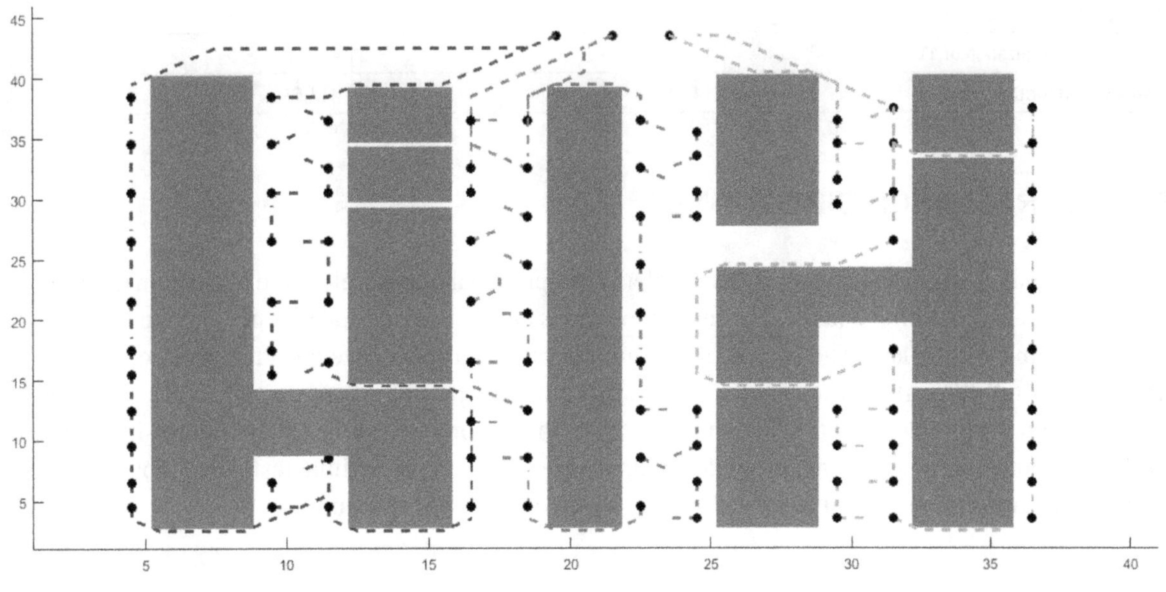

Figure 3. Task allotment and route of three robots for the Completion time=143.24 s and Total Fuel Consumption= 412.05 units

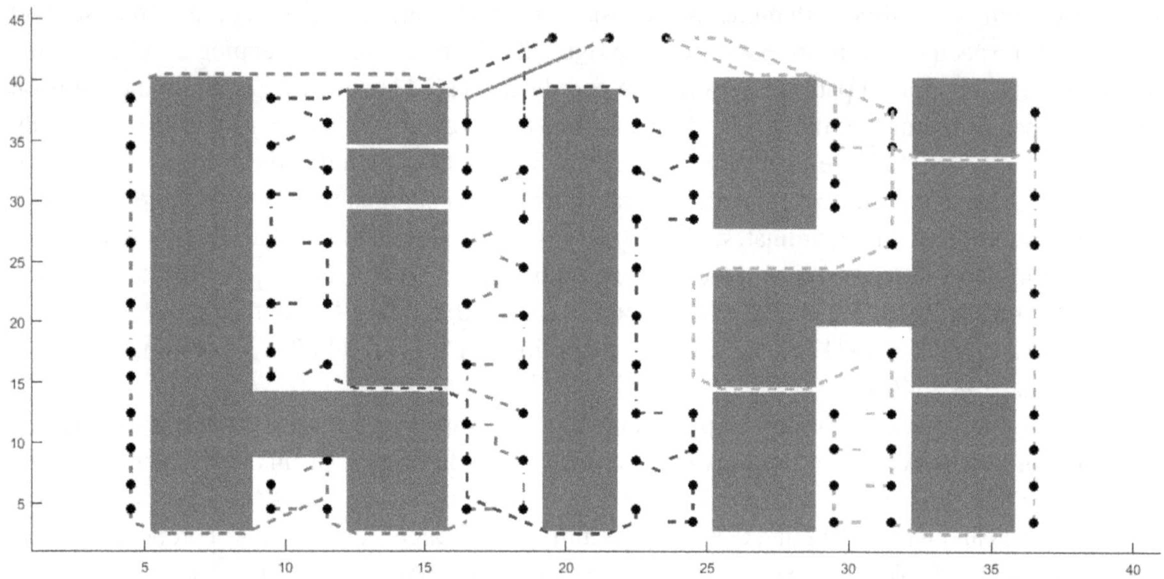

Table 1. Results of completion time and total fuel consumption of the existing algorithms and the proposed method

Algorithm	Completion Time (s)	Total Fuel Consumption
GA1-A* (Minimization of CT)	164.4696	490.0381
GA1-A* (Minimization of TFC)	176.6416	484.9670
GA2-A* (Minimization of CT)	156	462
GA2-A* (Minimization of TFC)	172	454
Proposed Algorithm	**141.2**	**413.54**

FUTURE RESEARCH DIRECTIONS

Since this study is based on a hypothetical problem not a practical one therefore the inspection times have been taken as zero for all the tasks. The inspection times can be consider later depending on the type of tasks in the industrial plant. The speeds of the three inspecting robots have been taken as constant (one unit per second) throughout the problem. However, in actual problem they can have different speeds depending on the ground condition and the capacity of the motor, etc., which can be studied in future. In the present study particle swarm optimization algorithm was used for optimal task allocation, however, there is numerous other evolutionary techniques which can be used in future.

CONCLUSION

This study deals with the path planning and task allocation for the multi robot system where three robots have been used for inspection of ninety predefined locations. A* algorithm has been used for optimal path planning and a modified version of multi objective particle swarm optimization algorithm is used for task allocation. The modified version of multi objective particle swarm optimization algorithm is constructed by integrating earliest completion time (ECT) based technique with multi objective particle swarm optimization algorithm. The implementation of ECT based technique enables the HPSO to solve multi robot task allocation problems efficiently. The Comparison with the previous study performed by Jose and Pratihar (2016) gives better result by avoiding the collisions successfully. In previous study they had divided the tasks into three parts where the first and second robot was assigned 25-35 locations for inspection and the remaining tasks was inspected by the third robot but in this current study a heuristic approach has been used and every robots has given equal priority for inspection of the tasks. The tasks have been arranged based on the increasing order of weight values where if any robot completes its previous task earlier than other robots then the nearest task will be assigned to it. Thus, by using the modified version of HPSO algorithm the completion time and the total fuel consumption has been reduced successfully by improving the efficiency up to 9.84% and 9.24% respectively as compare to the previous study.

REFERENCES

Binetti, G., Naso, D., & Turchiano, B. (2013). Decentralized task allocation for surveillance systems with critical tasks. *Robotics and Autonomous Systems*, *61*(12), 1653–1664. doi:10.1016/j.robot.2013.06.007

Caloud, P., Choi, W., Latombe, J. C., Le Pape, C., & Yim, M. (1990, July). Indoor automation with many mobile robots. *Proceedings of the IEEE International Workshop on Intelligent Robots and Systems, Towards a New Frontier of Applications* (pp. 67-72). IEEE. 10.1109/IROS.1990.262370

Coello, C. A. C., Pulido, G. T., & Lechuga, M. S. (2004). Handling multiple objectives with particle swarm optimization. *IEEE Transactions on Evolutionary Computation*, *8*(3), 256–279. doi:10.1109/TEVC.2004.826067

Davison, A. J., & Murray, D. W. (1998, June). Mobile robot localisation using active vision. *Proceedings of the European conference on computer vision* (pp. 809-825). Springer.

Dijkstra, E. W. (1959). A note on two problems in connexion with graphs. *Numerische Mathematik*, *1*(1), 269–271. doi:10.1007/BF01386390

Hart, P. E., Nilsson, N. J., & Raphael, B. (1968). A formal basis for the heuristic determination of minimum cost paths. *IEEE transactions on Systems Science and Cybernetics*, *4*(2), 100-107.

Huang, L., Ding, Y., Zhou, M., Jin, Y., & Hao, K. (2018). Multiple-Solution Optimization Strategy for Multirobot Task Allocation. *IEEE Transactions on Systems, Man, and Cybernetics. Systems*, 1–12. doi:10.1109/TSMC.2018.2847608

Hussein, A., Adel, M., Bakr, M., Shehata, O. M., & Khamis, A. (2014). Multi-robot task allocation for search and rescue missions. *Journal of Physics: Conference Series*, *570*(5), 052006. doi:10.1088/1742-6596/570/5/052006

Hussein, A., & Khamis, A. (2013, December). Market-based approach to multi-robot task allocation. *Proceedings of the 2013 International Conference on Individual and Collective Behaviors in Robotics (ICBR)* (pp. 69-74). IEEE. 10.1109/ICBR.2013.6729278

Jianping, C., Yimin, Y., & Yunbiao, W. (2009, November). Multi-robot task allocation based on robotic utility value and genetic algorithm. *Proceedings of the 2009 IEEE International Conference on Intelligent Computing and Intelligent Systems* (Vol. 2, pp. 256-260). IEEE. 10.1109/ICICISYS.2009.5357957

Jose, K., & Pratihar, D. K. (2016). Task allocation and collision-free path planning of centralized multi-robots system for industrial plant inspection using heuristic methods. *Robotics and Autonomous Systems*, *80*, 34–42. doi:10.1016/j.robot.2016.02.003

Ju, C., & Son, H. (2018). Multiple UAV systems for agricultural applications: Control, implementation, and evaluation. *Electronics (Basel)*, *7*(9), 162. doi:10.3390/electronics7090162

Lee, D. H. (2018). Resource-based task allocation for multi-robot systems. *Robotics and Autonomous Systems*, *103*, 151–161. doi:10.1016/j.robot.2018.02.016

Li, Z., Xu, B., Yang, L., Chen, J., & Li, K. (2009, June). Quantum evolutionary algorithm for multi-robot coalition formation. *Proceedings of the first ACM/SIGEVO Summit on Genetic and Evolutionary Computation* (pp. 295-302). ACM. 10.1145/1543834.1543874

Liu, C., & Kroll, A. (2012, April). A centralized multi-robot task allocation for industrial plant inspection by using A* and genetic algorithms. *Proceedings of the International Conference on Artificial Intelligence and Soft Computing* (pp. 466-474). Springer. 10.1007/978-3-642-29350-4_56

Liu, Y., & Bucknall, R. (2018). Efficient multi-task allocation and path planning for unmanned surface vehicle in support of ocean operations. *Neurocomputing*, *275*, 1550–1566. doi:10.1016/j.neucom.2017.09.088

Muhuri, P. K., & Rauniyar, A. (2017). Immigrants based adaptive genetic algorithms for task allocation in multi-robot systems. *International Journal of Computational Intelligence and Applications*, *16*(04), 1750025. doi:10.1142/S1469026817500250

Park, C. M., & Wang, G. N. (2009). Job allocation and scheduling in multi robotic tasks considering collision free operation. *Intelligent Automation & Soft Computing*, *15*(2), 249–261. doi:10.1080/10798587.2009.10643029

Parker, L. E. (1994, February). An experiment in mobile robotic cooperation. *Proceedings of the ASCE Specialty Conference on Robotics for Challenging Environments*, Albuquerque, NM. Academic Press.

Qian, B., & Cheng, H. H. (2018). Bio-Inspired Coalition Formation Algorithms for Multirobot Systems. *Journal of Computing and Information Science in Engineering*, *18*(2), 021010. doi:10.1115/1.4039638

Suemitsu, I., Izui, K., Yamada, T., Nishiwaki, S., Noda, A., & Nagatani, T. (2016). Simultaneous optimization of layout and task schedule for robotic cellular manufacturing systems. *Computers & Industrial Engineering*, *102*, 396–407. doi:10.1016/j.cie.2016.05.036

Wagner, G., & Choset, H. (2015). Subdimensional expansion for multirobot path planning. *Artificial Intelligence*, *219*, 1–24. doi:10.1016/j.artint.2014.11.001

Wang, J., Gu, Y., & Li, X. (2012). Multi-robot task allocation based on ant colony algorithm. *Journal of Computers*, *7*(9), 2160–2167. doi:10.4304/jcp.7.9.2160-2167

Yoon, S., & Kim, J. (2017). Efficient multi-agent task allocation for collaborative route planning with multiple unmanned vehicles. *IFAC-PapersOnLine*, *50*(1), 3580–3585. doi:10.1016/j.ifacol.2017.08.686

Zhou, X., Wang, H., Ding, B., Hu, T., & Shang, S. (2019). Balanced connected task allocations for multi-robot systems: An exact flow-based integer program and an approximate tree-based genetic algorithm. *Expert Systems with Applications*, *116*, 10–20. doi:10.1016/j.eswa.2018.09.001

Zhu, Z., Tang, B., & Yuan, J. (2017). Multirobot task allocation based on an improved particle swarm optimization approach. *International Journal of Advanced Robotic Systems*, *14*(3). doi:10.1177/1729881417710312

Chapter 10
Corrosion and Wear Behavior of Electroless Nickel Coatings

Suman Kalyan Das
https://orcid.org/0000-0001-7540-7954
Department of Mechanical Engineering, Jadavpur University, India

Supriyo Roy
https://orcid.org/0000-0001-8731-0068
Department of Mechanical Engineering, Halida Institute of Technology, India

Prasanta Sahoo
https://orcid.org/0000-0002-1538-0646
Department of Mechanical Engineering, Jadavpur University, India

ABSTRACT

This chapter describes how corrosion is the degradation of a material due to a reaction with its environment. Wear is another deteriorating phenomenon which erodes the contacting surfaces of any mechanical component gradually. Although corrosion and wear are two different phenomenon they are often interdependent and in most practical situations take place simultaneously. Fortunately, several surface modification methods exist which can protect the surface of metal against corrosion and wear. Among them, electroless nickel coatings is a technology where a coating is applied without the use of external current source. The primary advantages of electroless coatings include uniform coating and ability to coat non-conductive materials. Electroless nickel coatings possess excellent properties such as high hardness, good wear resistance and corrosion resistance. The corrosion and wear behavior of these coatings mainly depend on bath ingredients, deposition conditions, heat treatment temperature, etc.

DOI: 10.4018/978-1-7998-1831-1.ch010

INTRODUCTION

Corrosion is an aspect of the decay of materials by chemical or biological agents. Deterioration of materials by physical causes is not corrosion and is described as erosion or wear. Thus corrosion may be described as the surface wastage of metals that occurs when metals are exposed to reactive environments or in another way it is the destructive attack of a metal by chemical or electrochemical reaction with the environment. Wear in simple terms can be defined as the erosion of material from a solid surface by the action of another material which comes in contact with it. In most practical situations however, chemical attack accompanies physical deterioration and the combined result is termed as corrosive wear.

Corrosion and wear are two deteriorating phenomena which affect various devices and their service lives. Corrosion and wear cause environmental contamination, personal injuries and untimely shutdown of various machineries. Hence, no established industry can tolerate these deteriorating phenomena. Corrosion is frequently encountered in our daily life. The pipes which are used for water supply get corroded due to continuous flow of water. The iron rods placed inside the reinforced concrete of a building can also get corroded. The metal structure of bridge over the river is very much prone to corrosion. Marine sector is one of the worst affected by corrosion. Components of boats and ships sailing over the seas, onshore and offshore installations, etc. are all highly prone to corrosion due to their salt laden environment which accelerates the corrosion process. In power plants, the blades and bearings of the steam turbine, components of water pump are mostly affected due to corrosion. Again, wear occurs wherever two bodies come in contact and there is relative motion in between them. Wear is also encountered over a large practical domain ranging from the sole of a shoe to different sliding components in machinery. Wear is commonly associated with friction which results in huge energy losses in industries.

Corrosion and wear may lead to failure of infrastructures like, electrical and plumbing lines, chemical processing plants, power generation plants, and aircraft and marine equipment which are usually costly to repair. However undesired they might be, corrosion and wear cannot totally be eliminated; instead be controlled in a sustainable manner. The main motive behind controlling corrosion and wear are conservation, protection, economics and safety. The methods by which the surface of a metal can be protected against corrosion and wear are fall under the realms of surface engineering. Surface engineering refers to technologies that aim to design and modify the surface properties of components. Coatings are a well-established form of surface engineering. Several types of coating processes exist which provide protection of the metal surface against corrosion and wear. Among these, electroless nickel (EN) has emerged as a protective coating and has found widespread industrial acceptance due to its valuable properties (Sahoo and Das, 2011). Electroless nickel coatings are deposited chemically and well known for their deposit uniformity. These coatings are smooth and can be deposited virtually on any type of substrate. Moreover, the properties of electroless nickel coatings can be customised to suit a particular application by the incorporation of additional compatible elements.

ELECTROLESS COATING PROCESS

Electroless coating is based on chemical process in which the coating itself is catalytic to the reactions for the formation of next layer of coating. Hence, the reaction is autocatalytic and the whole process is executed without the aid of external electrical power which makes it different from the conventional electroplating process. In this method the metal which is to be coated (known as substrate) is submerged

Figure 1. Electroless coating setup

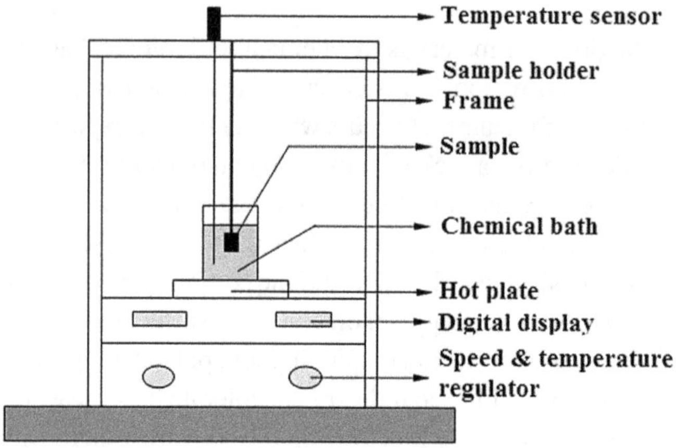

in a chemical solution (known as bath), kept at a certain temperature. The bath contains a source of metallic ions, reducing agent, complexing agent, stabilizer and other components. When the substrate is dipped in the bath it develops a potential and the reducing agent supplies the necessary electrons for the reduction of the metal ions and their deposition on to its surface. The reaction is controlled by the complexing agent and the stabilizer. A typical electroless coating setup is illustrated in Figure 1. The chemical and physical properties of an EN coating depend on its composition, which, in turn, depends on the formulation and operating conditions of the EN plating bath. The desirable properties can be varied by choosing different sets of pH level, bath temperature and bath composition.

Generally, there are three types of electroless coatings, viz. pure nickel, nickel alloy coatings and nickel composite coatings. Nickel alloy coatings may be binary, ternary or quaternary coatings depending on the number of alloying elements, viz. Ni-P, Ni-B, Ni-P-W, Ni-P-Cu, Ni-P-W-Cu, etc. Among these electroless Ni-P and Ni-B coatings are the most popular ones. Nickel composite coatings may be lubricating coating, wear resistance composite coatings and nano-composite coatings depending on the type of composite particles. Lubricating composite coatings consists of soft materials like PTFE (Ni-P-PTFE). Wear resistance composite coatings consist of hard particles like B_4C, SiC, TiO_2, Al_2O_3, etc. If the deposited particles are of nano size, the coating is classified as nano-composite coating.

CORROSION MECHANISM AND MEASUREMENT

Corrosion is associated with electrochemical reaction which is defined as a chemical reaction involving the transfer of electrons. Hence, corrosion measurement can be carried out by electrochemical techniques. Monitoring the current-potential relations under carefully controlled conditions can yield information on corrosion rates of coatings and films, their passivity, pitting tendencies and other important data. Nevertheless, conventional salt spray tests are also conducted like a corrosion endurance test.

When a metal specimen gets in contact with the corrosive medium, both reduction and oxidation processes occur on its surface. Ideally the specimen oxidizes (corrodes) and the medium (solvent) is reduced. The specimen must serve as both anode and cathode as both anodic and cathodic currents occur on the specimen surface. Corrosion is generally concerned with the anodic reactions that produce electrons and complementary cathodic reactions that consume electrons. Because there is a flow of electrons in the corrosion reaction, it can be measured and controlled electronically. Therefore, controlled electrochemical experimental methods can be used to characterize the corrosion properties of metals and metal components in combination with various electrolyte solutions. This may be achieved by using either a direct current (DC) or an alternating current (AC) source. The polarization testing instrument consists of a potentiostat/glavanostat, three electrodes, polarization cells and compatible software to store and plot data directly using a computer. In potentiostatic mode, a potentiostat/galvanostat will accurately control the potential of the auxiliary electrode (AE) against the working electrode (WE) so that the potential difference between the working electrode (WE) and the reference electrode (RE) is well defined, and correspond to the value specified by the user. In galvanostatic mode, the current flow between the WE and the CE is controlled. The potential difference between the RE and WE and the current flowing between the CE and WE are continuously monitored.

Most commonly used three-electrode potentiostat/galvanostat setup for corrosion measurement is shown in Figure 2. In this configuration, the current flows between AE and WE. The potential difference is controlled between the WE and the AE and measured between the RE and sample (S). Because the WE is connected with S and WE is kept at pseudo-ground (fixed, stable potential), by controlling the polarization of the AE, the potential difference between RE and WE is controlled all the time. The potential between the WE and AE usually is not measured. This is the voltage applied by the control amplifier and it is limited by the compliance voltage of the instrument. It is adjusted so that the potential difference between the WE and RE will be equal to the potential difference specified by the user.

There are several techniques to evaluate the corrosion behaviour of different materials based on their suitability. However, potentiodynamic polarization and electrochemical impedance spectroscopy are the most popular DC polarization methods which are employed to study the corrosion behaviour of electroless nickel coatings

Figure 2. Three-electrode potentiostat / galvanostat

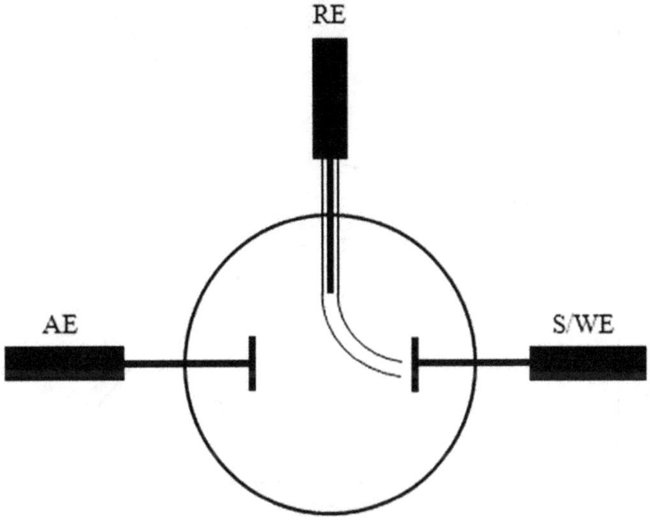

Potentiodynamic Polarization

Corrosion measurement by potentiodynamic polarization technique is conducted by application of current through electrolyte, so that the potential of the electrode can be varied over a wide range at a selected rate. In this case, corrosion environment is defined by the type of electrolyte. Generally, solutions of sodium chloride (NaCl), sulphuric acid (H_2SO_4) or Hydrochloric acid (HCL) are used as electrolyte. The method is based on cyclic sweep or cyclic polarization. In this method, during forward scan the potential is swept in anodic or noble direction until a chosen voltage is reached, then in reverse scans the potential starts sweeping in cathodic or active direction and terminated at another predetermined voltage. The important parameters considered are the position of the "anodic-to-cathodic" transition relative to the corrosion potential (E_{corr}), and the corrosion current density (I_{corr}). These parameters can be obtained from Tafel plot extrapolation. A typical Tafel plot is illustrated in Figure 3.

Electrochemical Impedance Spectroscopy

Electrochemical impedance spectroscopy is a recent tool in corrosion and solid-state laboratories that is slowly making its way into the service environment as units are decreased in size and become portable. The usefulness of impedance spectroscopy lies in the ability to distinguish the dielectric and electric properties of individual contributions of components under investigation. Electrochemical methods can be used to obtain insights into corrosion mechanisms and to establish the effectiveness of corrosion control methods, such as inhibition and coatings.

Impedance is the ratio of voltage and current. In electrochemical impedance spectroscopy (EIS), the response of an electrode to alternating potential signals of varying frequency is interpreted on the basis of circuit models of the electrode/electrolyte interface. The simplest circuit model for EIS analysis is shown in Figure 4. The circuit consists of the parallel combination of charge transfer resistance (R_{ct})

Figure 3. Tafel plot

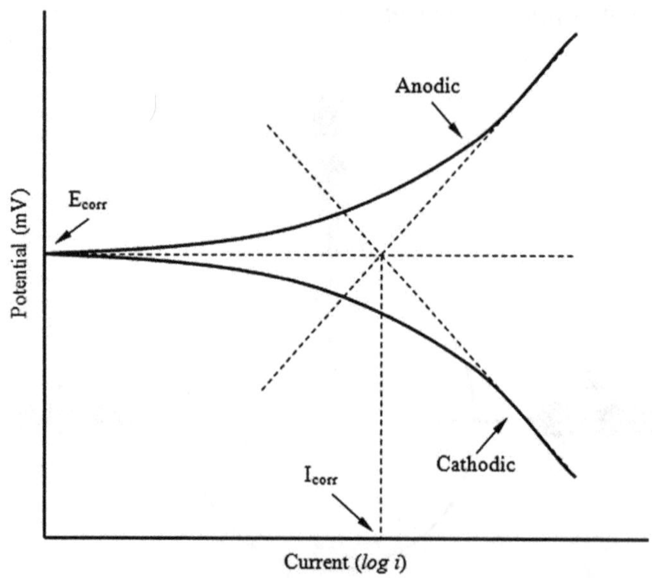

Figure 4. Simplest equivalent circuit of EIS

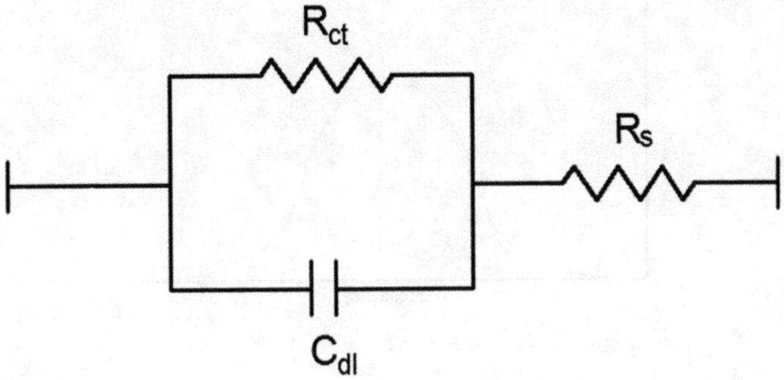

Figure 5a. Nyquist plot

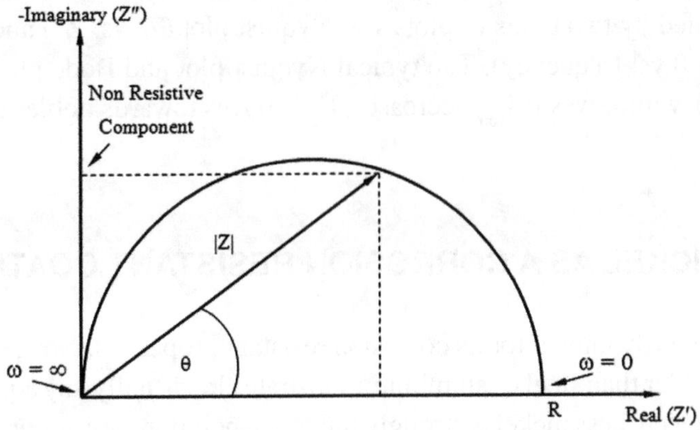

Figure 5b. Bode plot frequency vs. Impedance

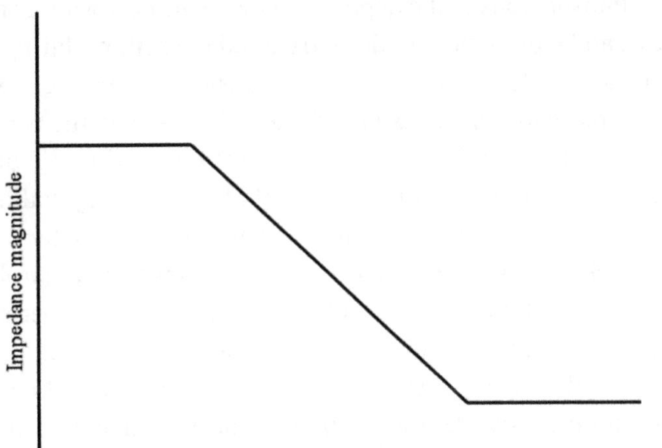

Figure 5c. Bode plot frequency vs. Phase angle

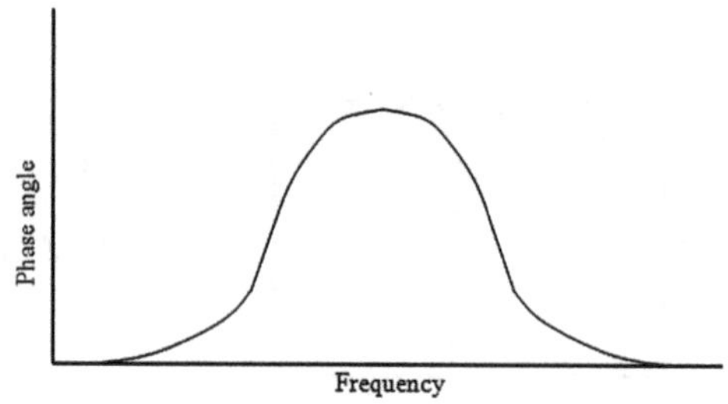

and double layer capacitance (C_{dl}) in series with solution resistance (R_s). As impedance is a complex number, it is generally expressed by the combination of real component (Z') and imaginary component (Z'') or by the frequency dependent modulus ($|Z|$) and the phase angle (θ). These impedance data can be graphically represented by two types of plots viz. Nyquist plot (Z' vs. Z'') and Bode plot (considering $|Z|$ vs. frequency or θ vs. frequency). The typical Nyquist plot and Bode plots are shown in Figure 5. The corrosion property improves if I_{corr} decreases, E_{corr} moves towards nobler direction, R_{ct} increases and C_{dl} decreases.

ELECTROLESS NICKEL AS A CORROSION RESISTANT COATING

Electroless nickel is primarily known for its corrosion resistant properties. Now, electroless nickel coatings are considerably nobler than steel or aluminium substrates in virtually any aqueous medium. Hence, corrosion resistance of electroless nickel is strongly linked to their porosity as corrosion reactions are set up at pores which extend down to the substrate. In general, these coatings have lower porosity and a more uniform thickness than the equivalent electroplated nickel and therefore provide better corrosion protection. However, corrosion behaviour of electroless nickel coatings is found to be largely dependent on their composition. Electroless Ni-P coating with high phosphorous content is found to possess higher corrosion resistance compared to coatings with lower phosphorous content (Mukherjee and Rajagopal 1992, Lu, Evans and Zangari 2004). Phosphorous in the deposit can make the corrosion potential increase and the corrosion current decrease, and it promotes anodic and cathodic reactions during the corrosion process, thereby increasing the anodic dissolution of nickel. The corrosion of electroless Ni-P coatings is mainly governed by the degree of amorphous state, extent of internal stress and the percentage of phosphorus content. The difference in corrosion resistance between electroless Ni-low P and Ni-high P coatings is in particular due to the difference in structure. Ni-low P possesses a crystalline and inhomogeneous structure whereas Ni-high P possesses an amorphous and homogeneous structure. Crystalline structure is more prone to corrosion attack at the grain boundaries which are absent in amorphous structure. Similar explanation could be given for electroless nickel coatings which normally experience a decrease in their corrosion resistance properties after heat treatment. Due to heat treatment, a microstructural transformation of the coating occurs and the coating turns crystalline resulting in the increase of grain boundaries which become active sites for corrosive reactions. Effect of heat treatment conditions on the corrosion resistance of electroless Ni-P coatings as studied by Rabizadeh, Allahkaram and Zarebidaki (2010) reveals that corrosion potential (E_{corr}) of Ni-P deposits is positively shifted from -0.376V to -0.25V with the increasing heat treatment temperature and reaches nobler potential at 600°C. On the other hand, corrosion current density (I_{corr}) reaches minimum value of 1×10^{-5} A/cm^2 when heat-treated at 400°C. The values of charge transfer resistance (R_{ct}), varies in the range of 13934 Ωcm^2 - 52903 Ωcm^2, and the maximum value reaches when heat treated at 400°C for 1hr. The capacitance value (C_{dl}) obtained for as-plated and heat treated electroless Ni-P coatings are in the order of 23-38 μF/cm^2. The lesser corrosion resistance at lower heat treatment temperatures can be attributed to the nano-crystalline nature having more grain boundaries and hence susceptible to the formation of larger number of corrosive sites. But with the increasing heat treatment temperatures from 200°C to 600°C coatings become denser and less porous resulting in the improvement of corrosion property. It is also reported that at 600°C and 400°C the alloys consist of two types of crystals of the nickel-rich phase and the Ni_xP_y phases. Due to presence of in-homogeneities the alloys are not continuous, which are active sites for corrosion attack. However, the grain growth at 400°C, triumph the formation of second phases and the corrosion resistance of the coating increases.

Electroless Ni-B coatings possess lower corrosion resistance compared to Ni-P coatings. But significant improvement in resistance is observed when incorporating phosphorous in the coating to form a Ni-P-B coating (Srinivasan et al. 2010). The lower corrosion resistance of electroless Ni-B coating may be attributed to the passivation films that form on Ni-B coated surfaces which are believed to be not as glassy or protective enough as those that form on high phosphorous electroless nickel coatings. In addition to that, phase boundaries present in Ni-B deposits might also be responsible for causing discontinuity of the passivation film, which are the preferred sites for the initiation of corrosion process (Narayanan and Seshadri 2004). Similar to Ni-P coating, increase of Boron contents results in the increase of corrosion resistance of Ni-B. To further enhance the applicability of Ni-B coating in corrosive environments, concepts of Ni-P/Ni-B duplex coatings have been introduced. In this type of coating, each of the layers is developed on top of the other. Duplex coatings show improved corrosion behaviour compared to the individual coatings. However, the coating with Ni-P as the outer layer is expected to have better resistance against corrosion.

Incorporation of certain elements viz. copper, tin, tungsten, etc. can improve the corrosion performance of electroless nickel coatings significantly. Roy and Sahoo (2012) studied the corrosion behavior of electroless Ni-P-W coatings. It is revealed that with a sodium tungsten concentration of 25g/l, the E_{corr} reaches a value of -278.71 mV and corresponding value of I_{corr} is 0.08 μA/cm^2. With 25 g/l sodium tungstate the optimum value of R_{ct} (13.54×10^5 Ωcm^2) and C_{dl} (0.4931×10^{-5} F) were obtained. A comparative study between Ni-P-W and Ni-P-Al_2O_3 was carried out by Hamdy, Shoeib, Hady and Salam (2007). The study revealed that presence of alumina in as-deposited condition improves the corrosion resistance of steel in 3.5% NaCl solution. With the increase of Al_2O_3 from 25 g/l to 75 g/l, the surface resistance increases from 0.60×10^4 Ω cm^2 to 7.00×10^4 Ω cm^2. Further increase of Al_2O_3 decrease the surface resistance and with 100 g/l alumina the surface resistance reaches a level of 3.50×10^4 Ω cm^2. Heat treatment decreases the corrosion resistance. With 75 g/l Al_2O_3, the corrosion resistance of the heat-treated specimen reaches a value of 3.30×10^4 Ω cm^2. This is due to the fact that heat treatment converts more corrosion-resistant alumina phase to less corrosion-resistant aluminium phosphate and hence, the surface resistance decreased sharply. But for ternary Ni-P-W alloy coatings in as-deposited condition with 5 g/l tungsten, the maximum surface resistance value obtained as 12.0×10^4 Ω cm^2 which is much higher than Al_2O_3 (7.00×10^4 Ω cm^2). Further increase of tungsten decreases the corrosion resistance. After heat-treatment the surface resistance decreased to be 10.0×10^4 Ω cm^2. This may be due to heat treatment that decomposes the high corrosion-resistant tungsten phosphide film to less protective nickel phosphide form and hence, the surface resistance to chloride attack decreased compared to the samples before heat treatment. Similar type of study on Ni-P-Cu was carried out by Liu et al. (2010). The result of polarization study reveals that the lowest current density of electroless Ni-P-Cu is 0.2μA/cm^2. Copper also enhances the corrosion resisting power of Ni-W-P coatings in both salty and acidic environment (Balaraju & Rajam, 2005). However, tin improves the anti-corrosion property of Ni-P only in brine solution. Small amount of tin addition actually promotes the formability of the amorphous phase in Ni-P coating (Zou et al., 2010). Hence, electroless nickel alloy coatings may be considered as corrosion protective coatings amongst EN coatings.

Assimilation of third particle is also found to improve the corrosion properties of basic Ni-P coatings. Hard particles are normally incorporated to increase the wear resistance of the coatings but not without affecting the corrosion properties. Araghi and Paydar (2010) conducted the polarization study of electroless Ni-P and Ni-P-B_4C coatings and compared the corrosion resistance characteristics. It was revealed that the binary Ni-P coating showed better corrosion resistance than Ni-P-B_4C composite coatings. The corrosion potential (E_{corr}) of Ni-P-B_4C composite coated specimen (-1.031 V) is more negative than that of Ni–P coated specimen (-0.808 V), proving the higher corrosion tendency of the composite coated alloy. Corrosion current densities obtained through the extrapolation of the linear portion of cathodic and anodic polarization curves show a decrement from 84 μA/cm^2 for composite coated specimen down to 7 μA/cm^2 for Ni-P coated alloy. However, it may be noted that the corrosion resistance of composite coated specimen is measured to be lower than that of Ni-P coated alloy which may be attributed to formation of micro-cracks on the surface due to impregnation of B_4C particles into the coating. Composite coating using PTFE displayed highest corrosion resistance by increasing the PTFE co-deposition which results in lower P content. It may be due to the fact that PTFE particles are inert with very low electrical conductivity and their co-deposition in the coating leads to more uniform surface with fewer defects. The study by Mafi and Dehghanian (2011) revealed that the lowest corrosion current density using PTFE with CTAB surfactant is 5×10^{-7} A/cm^2, compared to 8×10^{-6} A/cm^2 of Ni-P deposit. The charge transfer resistance values of Ni-P-PTFE lie in the range of 27000 Ωcm^2 to 40000 Ωcm^2, which is also higher than pure Ni-P coatings. Novakovic and Vassiliou (2009) studied and compared the electrochemical results obtained from polarization studies for electroless Ni-P and Ni-P-TiO_2 composite coatings in non-deaerated 3.5% sodium chloride aqueous solution. The study revealed that corrosion current density (I_{corr}) for the as-deposited electroless Ni-P coating is 6.81 μA/cm^2, whereas I_{corr} values for the Ni-P-TiO_2 composite coatings are in the range of 8.34–15.20 μA/cm^2 and this increased with the increase in TiO_2 particle incorporation. According to Liu and Zhao (2004), the corrosion resistance of any alloy depends on the rate of dissolution of the passive film apart from the amount of phosphorous present in a deposit. When the particles are co-deposited, it is believed that the formation of passive layer might be disturbed and this could be the reason for decrease in corrosion resistance for composite coatings. Compared to as-deposited coatings, vacuum heat-treated electroless Ni-P and Ni-P-TiO_2 composite coatings exhibit greater I_{corr}. It is because the amorphous alloys offer better resistance to corrosion attack than equivalent polycrystalline materials due to freedom from grain and grain boundaries. After vacuum heat treatment, a mixture of crystalline nickel and nickel-phosphides is created so areas of different corrosion potential are produced. This leads to the formation of minute active/passive corrosion cells and consequently to an accelerated attack by an aggressive medium.

It is generally seen that incorporation of composite material into basic Ni-P coatings decrease the corrosion resistance due to porosity of the coated surface. This porosity can be reduced by using nano sized hard composite particles. Rabizadeh and Allahkaram (2011) studied the corrosion characteristics of nano-SiO_2 incorporated Ni-P coatings and found that the nano-composite coatings showed better corrosion resistance than Ni-P coatings. The electrochemical results obtained from polarization studies for electroless Ni-P and Ni-P/nano-SiO_2 composite coatings in non-deaerated 3.5 wt.% sodium chloride solution showed that for nano-composite coatings, E_{corr} value (-288mV vs. Ag/AgCl) is nobler than Ni-P (-330.6 mV vs. Ag/AgCl). The I_{corr} value also improves from 0.416μA/cm^2 to 0.308μA/cm^2 due to incorporation of nano particles. From the EIS data, the charge transfer resistance and double layer capacitance for composite coatings (R_{ct} = 17169 Ωcm^2 and C_{dl} = 18.265 μF/cm^2) gives better result than Ni-P coatings (R_{ct} = 13934 Ωcm^2 and C_{dl} = 38.334 μF/cm^2). This seems to be logical, because even though the apparent area remains the same, the effective metallic area prone to corrosion is decreased considerably in the case of electroless composite coating.

Corrosion resistance of electroless nickel is also found to largely depend on the deposition conditions and the substrate. It is invariably found that electroless Ni-P coatings developed from acidic bath have superior corrosion performance compared to the same developed from alkaline baths. Coatings developed from alkaline baths have a tendency to develop pores which will initiate corrosion. Acidic baths tend to give pore free deposits and hence greater corrosion resistance. On the other hand, composition and structure of substrate has not yet been firmly proved to affect corrosion behaviour of electroless nickel. However, roughness of the substrate plays an important part of the final corrosion behaviour of the coating. Pits and asperities have a high tendency to act as pore initiation sites. Hence, it can be concluded that rougher the substrate surface, lesser the corrosion resistant is a coated sample for any deposit thickness.

WEAR MECHANISM AND MEASUREMENT

Wear as defined by Rabinowicz (1995) is the interaction between surfaces and more specifically the removal and deformation of material on a surface as a result of mechanical action of the mating surface. Corrosion may be included in wear phenomenon, but the damage is amplified and performed by chemical reactions. Wear is basically erosion of metal from the contacting surfaces when both the surfaces are in sliding motion. Based on the mechanism, wear may be classified as adhesive wear, abrasive wear, corrosive wear and fatigue wear.

When two flat surfaces come in sliding contact then adhesive wear takes place. In microscopic view most of the surfaces consists of asperities or peaks and valleys. A temporary bonding takes place when this type of surfaces comes in contact. During sliding of one surface over another a fragment of softer surface get shear off and get detached from its base material due to this adhesive bonding. Abrasive wear takes place due to entrapped wear particles within the sliding surfaces. There are generally two modes of abrasive wear, two-body abrasion and three-body abrasion. When hard surface slides over a soft surface, forming a groove in the soft material, called two-body abrasion. In case of three body abrasion, the hard worn out particles get entrapped between two sliding surfaces cause additional grooves in the softer material due to ploughing effect. If wear takes place in corrosive medium, it is called corrosive wear. In case of corrosive wear both the sliding and corrosion takes place simultaneously. Corrosive wear is important in many industries such as mining, mineral processing, chemical processing, etc. Fatigue wear of a material is caused by a cycling loading during sliding and rolling. Surface cracks are produced due to repeated loading and unloading. These cracks come out in the form of large fragments after a critical number of cycles. Besides these there are some minor forms of wear like, fretting, erosion, percussion.

The traditional means of measuring wear is by the volume or weight loss from a certain specimen after its interaction with another body. The difference of weight before and after the interaction is normally reported as the wear encountered. However, modern tribometers are equipped to report wear undergone as depth of wear. Instruments like pin-on-disc tribometers, multitribotesters, etc., make the specimen slide against a much harder conterface. The wear depth is captured by suitable displacement transducers and plotted real-time in a PC attached to the tribometer. Users can change the sliding speed, applied normal load and the duration of sliding and hence simulate the actual service conditions or any desired condition at which the test is planned.

WEAR RESISTANT CHARACTERISTICS OF ELECTROLESS NICKEL

Electroless nickel coatings are much appreciated for their superior hardness and wear resistance compared to their electrodeposited counterpart. The properties and microstructures of these coatings depend on the composition of the deposit. Wear resistances of electroless Ni-P and Ni-B are found to be proportional to the phosphorous and boron contents respectively. The post-deposition heat treatment also has a major influence over the wear resistance of the coating. Balaraju and Seshadri (1999) studied the wear characteristics of as-deposited as well as heat-treated electroless Ni-P coating. The coating contained 10-12% phosphorous and the heat treatment is carried out for 1 hour at a constant temperature of 400°C. Wear test is carried out with a disc-on-disc configuration where the counter disc was high carbon, high aluminium steel having hardness of 60 HRC. The test is carried out with a constant speed of 1000 rpm and loads were varied within the range of 20-60N. It is revealed that heat treated Ni-P samples have low specific wear rate compared to as-deposited coatings. This is believed to be due to very low mutual solubility of nickel phosphide and iron thus presenting a relatively incompatible surface. Electroless Ni-B coatings are preferred to Ni-P coatings for wear-based applications because of the higher hardness of the former in as-deposited condition which can be further increased with the application of a suitable heat treatment cycle. The wear mechanism in electroless nickel coating is primarily found to be either adhesive or abrasive or a combination of both. The columnar structure of electroless Ni-B coatings helps in retaining lubricants which is found to be useful in preventing adhesive wear (Srinivasan et al., 2010). In general heat treatment increases the hardness of electroless nickel coating and hence its wear resistance. The optimal heat treatment temperature for maximum hardness is found to lie around 350-400°C. However, at higher heat treatment temperature grain coarsening occurs resulting in softening of the coating and hence declined wear performance.

The wear property of electroless nickel coatings can further be improved by incorporating hard particles like B_4C, Al_2O_3, TiO_2, SiC, etc. Ebrahimian-Hosseinabadi, Azari-Dorcheh, and Vaghefi (2006) reported that wear resistance of electroless Ni-P-B_4C composite coatings under loads of 15, 30 and 60N is improved from 60 to 90% with increasing B_4C up to 25 vol. %. However, B_4C beyond 25 vol. % results in a decrease of wear performance which may be due to the B_4C particles which act as origins for developing and growth of cracks. More cracks develop and merge together forming larger cracks with increase of B_4C particles from 25 to 33 vol.%. Wear resistance of electroless Ni-P-SiC composite deposits has been studied by many researchers (Apachitei et al., 2002; Jiaqiang et al., 2006; Huang et al., 2003). The wear resistance increases with increase in SiC incorporation level. The dispersion strengthening effect of SiC particles and the strong Ni-P matrix enables such coatings to offer excellent wear resistance. Furthermore, the Ni-P matrix should possess high phosphorus content (>7 wt. %) in order

Figure 6a. Variation of wear of Ni-P
(Alirezaei et al., 2007)

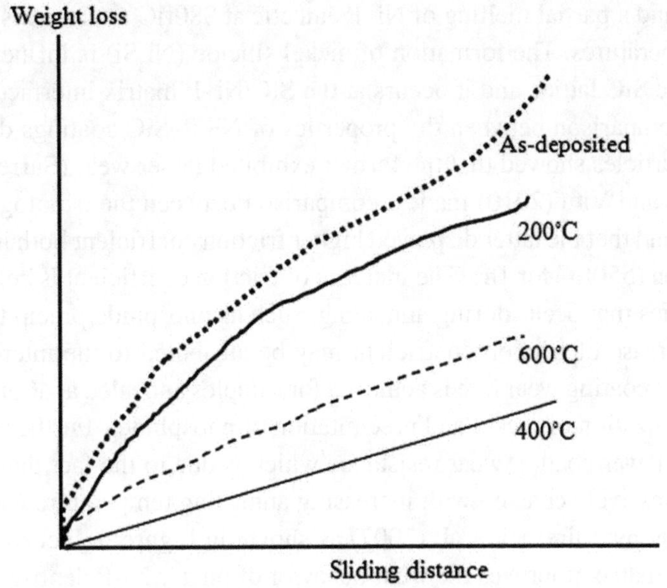

to support the SiC particle to provide superior wear resistance (in both as-deposited and heat-treated conditions). The influence of heat treatment on the wear resistance of electroless Ni-P-SiC composite deposits suggests that wear resistance increases with increase in heat-treatment temperature up to 350fiC

Figure 6b. Variation of wear of Ni-P-Al_2O_3
(Alirezaei et al., 2007)

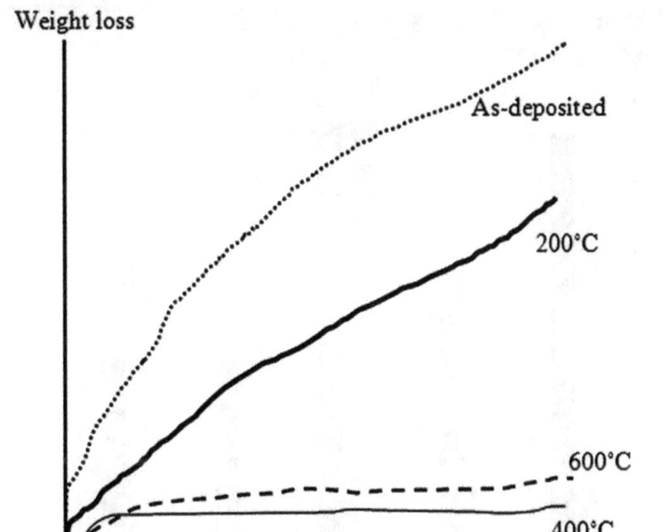

(Broszeit, 1982). Further increase in temperature (up to 600fiC) results in increase in the wear rate with the situation worsening at even higher temperatures (800fiC). The possible reaction between nickel and SiC at around 580fiC and a partial melting of Ni–P eutectic at 880fiC are responsible for the poor wear resistance at high temperatures. The formation of nickel silicide (Ni_3Si) is influenced by the diffusion of nickel atoms into the SiC lattice and it occurs at the SiC/Ni–P matrix interface upon heat-treatment at 500fiC for 1 h. A comparison between the properties of Ni–P–SiC coatings developed with macro and micro sized SiC particles showed that the former exhibited lesser wear (Sarret, Müller, and Amell, 2006). Novak, Vojtech and Vitu (2010) made a comparison between the tribological behavior of Ni-P and Ni-P-Al_2O_3 and found that the latter displayed higher friction coefficient both in as-deposited as well as heat treated condition (550fiC for 1h). The increase of friction coefficient is believed to be due to the structural heterogeneities that occur during annealing, such as phosphide precipitation and subsequent grain growth. This increase of friction coefficient may be attributed to the intense abrasive action of Al_2O_3 fibers. The lowest coating wear rate is achieved for samples annealed at 400fiC for 1h. This can be attributed to the crystallization of nickel and precipitation of phosphides. Further increase of annealing temperature leads to a lower coating wear resistance which is due to the fact that the nickel and phosphide crystallites progressively coarsen with increasing annealing temperature and time. Similar result was obtained previously by Alirezaei et al. (2007) as shown in Figure 6. Incorporation of SiO_2 nanoparticles in to the Ni-P matrix improves the wear behavior of binary Ni-P deposit as reported by Hazan et al. (2010). Heat treatment further improves the wear resistance. The effect of particle concentration also has significant impact on the wear behavior. The wear rate in terms of weight loss at different heat treatment temperatures with different SiO_2 concentration is shown in Figure 7.

Irrespective of the heat treatment temperature, the coating consisting of 3 g/l SiO_2 particle shows minimum weight loss, i.e. maximum wear resistance. In case of lubricating composite coatings, incor-

Figure 7. Variation of wear of Ni-P-SiO_2 at different condition
(Hazan et al., 2010)

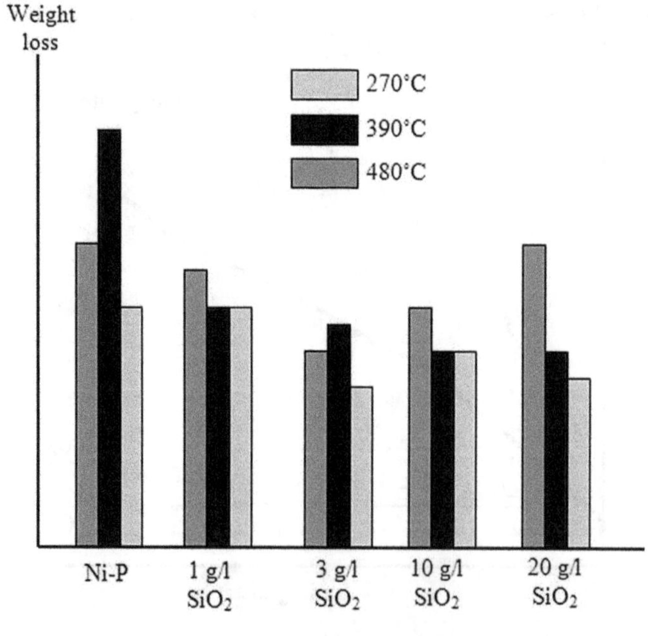

Figure 8. Variation of wear due to incorporation of MoS_2 and PTFE
(Mohammadi et al., 2011)

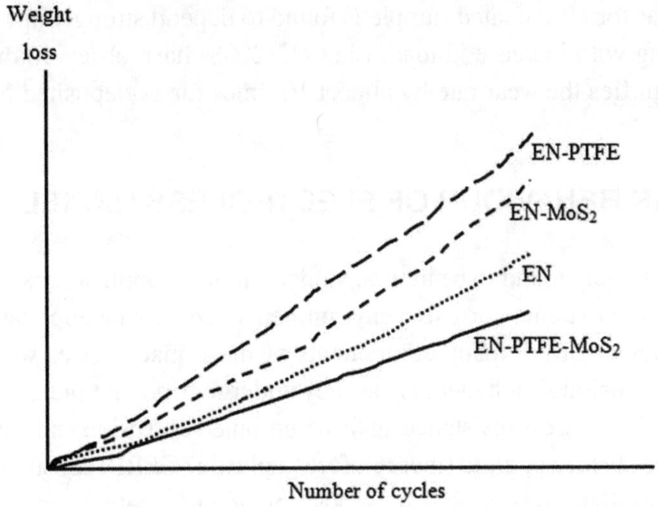

poration of lubricating particles into Ni-P matrix results in reduction of wear resistance. Mohammadi M and Ghorbani M (2011) studied the effect of incorporation of PTFE and MoS_2 particles in the Ni-P coatings. The study reveals the variation of wear rate in terms of weight loss as illustrated in Figure 8. It is evident from the figure that due to the negative effect of PTFE and MoS_2 particles the Ni-P-PTFE and Ni-P-MoS_2 coatings suffer more weight loss than basic Ni-P coating. But the Ni-P-PTFE-MoS_2 coating shows better wear resistance. The simultaneous co-deposition of the PTFE and MoS_2 particles into the nickel coating improves the wear characteristics of the coating by about 30% relative to binary electroless nickel coating. The heat treatment of the Ni-P–PTFE–MoS_2 coating at different temperatures increases the hardness of the coating which reaches maximum at 400fiC due to the formation of stable lubricant layer of Ni_3P phase.

Like composite coatings, inclusion of a third element like tungsten or copper proves to be beneficial in terms of wear characteristics of the coating. Insertion of tungsten could result in the solid solution strengthening of the nickel matrix and hence improve the wear resistance (Palaniappa and Seshadri 2008).

Recent trends include co-deposition of carbon nano tubes (CNT) with electroless Ni-P coatings. Wear performance of Ni-P-CNT is found to be much superior to coatings such as Ni-SiC and Ni-Graphite due to improved mechanical properties, self-lubrication and unique microstructure of the coatings. The factors which infiuence the CNT content in deposits are types of agitation, surfactants and CNT concentration in the plating bath. Ni–P–CNTs composite coating greatly increased the hardness and wear resistance with the increase of the content of carbon nanotubes in deposits. The examination of the wear surface revealed that Ni–P–CNT coatings encountered abrasive wear. Very few cracks, scales and very fine scratches appeared on the wear surface but at the same time some small bulges were observed and the abrasion grooves were discontinuous. These demonstrated that the CNTs dispersed in the nickel matrix block, or at least delayed the movement of dislocations in the metal nickel and thus inhibited plastic deformation, which improved the wear behaviour of the (Chen et al. 2006).

Based on the range of values reported by various researchers employing different methods of wear testing, wear behaviour of electroless nickel is believed to be a function of the test method used to test it.

Several test methods have been used for testing electroless nickel viz. pin-on-disc, block on disc, Taber abrader, scratch test, etc. The wear behaviour of electroless nickel coatings also depends on the testing parameters. Rate of wear for Ni-P coated sample is found to depend strongly upon the applied load and the same increases along with increase in load. Li et al. (2008) have observed that an increase of load from 1 N to 1.5 N, magnifies the wear rate by almost 10 times for as-deposited Ni-P.

CORROSIVE WEAR BEHAVIOUR OF ELECTROLESS NICKEL

Corrosion and wear are largely found to be interdependent in many applications. Many modern types of machinery are subjected to extreme corrosive environments viz. marine applications, chemical plants, etc. Due to the dynamics involved, many components of these machineries would require protection against both wear and corrosion simultaneously as they undergo tribo-chemical interactions. Electroless nickel coatings are found to possess resistance against combined corrosion and wear. These coatings are favourably applied to the surface of glass fiber reinforced plastic (GFRP) substrate for making blades of offshore wind turbine installations (Lee, 2008). Deposition of Ni-P coatings on a carbon fiber reinforced plastic (CFRP) substrate also increases its wear-corrosion resistance. This wear-corrosion resistance increased as coating thickness and phosphorous content increased, and the polishing condition decreased (Lee, 2009). Liu (2006) considered the effect of NaOH concentration in medium, the content of Cr_2O_3 in the coating and technology of heat treatment on corrosive wear resistance of a Cr_2O_3 composite electroless plating, which was compared with Ni-P coating. The corrosive wear of the coating was found to be linearly dependent on NaOH concentration, and the corrosive wear resistance of the composite electroless plating was superior to the Ni–P coating in any NaOH concentration. When the content of Cr_2O_3 in coating was 8%, the corrosive wear resistance of the composite electroless plating is the best, and the corrosive wear resistance is improved with the employment of proper technology of heat treatment.

CLOSURE

This chapter basically emphasises on the corrosion and wear characteristics of electroless nickel coatings. The recent developments of tribological and electrochemical behavior of electroless nickel alloys, composites and nano-coatings have been described. For the requirement of growing modern industries, new alloy can be deposited with the binary electroless nickel-phosphorous matrix to satisfy the desired criteria. Work remains to be done on improving the plating of different substrates that require thick and eco-friendly coatings for certain applications and at the same time lowering the chemical usage and cost. In addition, getting maximum efficiency from the bath is of vital importance in the near future. Electroless nickel composite has the potential to be wear resistant or self-lubricating coatings and other special applications like corrosion prevention. With the advent of automatic controllers and more advanced processing technology, poly-alloys and composites may become the next frontier of electroless nickel technology. The results of electroless nickel nano-coating have shown that these coatings could be deposited successfully in the Ni-P matrix by to prevent corrosion and wear with desired level of satisfaction. These electroless nickel-based nano-composite/nanocoatings have gained much attention in research community due to their ability to produce coatings that possess useful physical and mechanical properties. The properties of electroless nickel coatings are improved continuously by suitable surface treatments (post deposition heat treatment, laser treatment, vacuum thermal treatment etc.) and the incorporation of various alloying elements (copper, tungsten, etc.) and composite particles (SiC, TiO_2, etc.) to suit various applications. Hence, electroless nickel (EN) coatings have already proved their mettle and awareness needs to be created for their widespread application as an anti-corrosion and wear resistant coating.

REFERENCES

Alirezaei, S., Monirvaghefi, S. M., Salehi, M., & Saatchi, A. (2007). Wear behavior of Ni–P and Ni–P–Al_2O_3 electroless coatings. *Wear*, *262*(7-8), 978–985. doi:10.1016/j.wear.2006.10.013

Apachitei, I., Tichelaar, F. D., Duszczyk, J., & Katgerman, L. (2002). The effect of heat treatment on the structure and abrasive wear resistance of autocatalytic Ni-P and Ni-P-SiC coatings. *Surface and Coatings Technology*, *149*(2-3), 263–278. doi:10.1016/S0257-8972(01)01492-X

Araghi, A., & Paydar, M. H. (2010). Electroless deposition of Ni–P–B_4C composite coating on AZ91D magnesium alloy and investigation on its wear and corrosion resistance. *Materials & Design*, *31*(6), 3095–3099. doi:10.1016/j.matdes.2009.12.042

Balaraju, J. N., & Rajam, K. S. (2005). Electroless deposition of Ni-Cu-P, Ni-W-P and Ni-W-Cu-P alloys. *Surface and Coatings Technology*, *195*(2–3), 154–161. doi:10.1016/j.surfcoat.2004.07.068

Balaraju, J. N., & Seshadri, S. K. (1999). Preparation and characterization of electroless Ni-P and Ni-P-Si_3N_4 composite coatings. *Transactions of the Institute of Metal Finishing*, *77*(2), 84–86. doi:10.1080/00202967.1999.11871253

Broszeit, E. (1982). Mechanical, thermal and tribological properties of electro and chemodeposited composite coatings. *Thin Solid Films*, *95*(2), 133–142. doi:10.1016/0040-6090(82)90235-8

Chen, X. H., Chen, C. S., Xiao, H. N., Liu, H. B., Zhou, L. P., Li, S. L., & Zhang, G. (2006). Dry friction and wear characteristics of nickel/carbon nanotube electroless composite deposits. *Tribology International*, *39*(1), 22–28. doi:10.1016/j.triboint.2004.11.008

Ebrahimian-Hosseinabadi, M., Azari-Dorcheh, K., & Vaghefi, S. M. M. (2006). Wear behavior of electroless Ni-P-B_4C composite coatings. *Wear*, *260*(1-2), 123–127. doi:10.1016/j.wear.2005.01.020

Hamdy, A. S., Shoeib, M. A., Hady, H., & Abdel Salam, O. F. (2007). Corrosion behavior of electroless Ni-P alloy coatings containing tungsten or nano-scattered alumina composite in 3.5% NaCl solution. *Surface and Coatings Technology*, *202*(1), 162–171. doi:10.1016/j.surfcoat.2007.05.030

Hazan, Y., Zimmermann, D., Z'graggen, M., Roos, S., Aneziris, C., Bollier, H., ... Graule, T. (2010). Homogeneous electroless Ni–P/SiO_2 nanocomposite coatings with improved wear resistance and modified wear behavior. *Surface and Coatings Technology*, *204*(21-22), 3464–3470. doi:10.1016/j.surfcoat.2010.04.007

Huang, Y. S., Zeng, X. T., Annergren, I., & Liu, F. M. (2003). Development of electroless Ni–P-PTFE–SiC composite coating. *Surface and Coatings Technology*, *167*(2-3), 207–211. doi:10.1016/S0257-8972(02)00899-X

Jiaqiang, G., Lei, L., Yating, W., Bin, S., & Wenbin, H. (2006). Electroless Ni–P–SiC composite coatings with superfine particles. *Surface and Coatings Technology*, *200*(20-21), 5836–5842. doi:10.1016/j.surfcoat.2005.08.134

Lee, C. K. (2008). Corrosion and wear-corrosion resistance properties of electroless Ni-P coatings on GFRP composite in wind turbine blades. *Surface and Coatings Technology*, *202*(19), 4868–4874. doi:10.1016/j.surfcoat.2008.04.079

Lee, C. K. (2009). Structure, electrochemical and wear-corrosion properties of electroless nickel-phosphorus deposition on CFRP composites. *Materials Chemistry and Physics*, *114*(1), 125–133. doi:10.1016/j.matchemphys.2008.08.088

Li, Z., Chen, Z., Liu, S., Zheng, F., & Dai, A. (2008). Corrosion and wear properties of electroless Ni–P plating layer on AZ91D magnesium alloy. *Transactions of Nonferrous Metals Society of China*, *18*(4), 819–824. doi:10.1016/S1003-6326(08)60142-9

Liu, G., Yang, L., Wang, L., Wang, S., Chongyang, L., & Wang, J. (2010). Corrosion behavior of electroless deposited Ni–Cu–P coating in fiue gas condensate. *Surface and Coatings Technology*, *204*(21-22), 3382–3386. doi:10.1016/j.surfcoat.2010.03.056

Liu, T. (2006). Study on corrosive wear resistance of a Ni–P–Cr_2O_3 composite electroless plating with low-content phosphor in NaOH. *Run Hua Yu Mi Feng*, *4*, 90–91.

Liu, Y., & Zhao, Q. (2004). Study of electroless Ni–Cu–P coatings and their anti-corrosion properties. *Applied Surface Science*, *228*(1-4), 57–62. doi:10.1016/j.apsusc.2003.12.031

Lu, G., Evans, P., & Zangari, G. (2004). Investigations of the effect of chromate conversion coatings on the corrosion resistance of Ni-based alloys. *Journal of the Electrochemical Society*, *49*, 1461–1473.

Mafi, I. R., & Dehghanian, C. (2011). Comparison of the coating properties and corrosion rates in electroless Ni–P/PTFE composites prepared by different types of surfactants. *Applied Surface Science*, *257*(20), 8653–8658. doi:10.1016/j.apsusc.2011.05.043

Mohammadi, M., & Ghorbani, M. (2011). Wear and corrosion properties of electroless nickel composite coatings with PTFE and/or MoS_2 particles. *Journal of Coatings Technology and Research*, *8*(4), 527–533. doi:10.100711998-011-9329-y

Mukherjee, D., & Rajagopal, D. (1992). Electrodeposition of amorphous nickel alloys. *Metal Finishing*, *90*, 15–19.

Narayanan, T. S. N. S., & Seshadri, S. K. (2004). Formation and characterization of borohydride reduced electroless nickel deposits. *Journal of Alloys and Compounds*, *365*(1-2), 197–205. doi:10.1016/S0925-8388(03)00680-7

Novak, M., Vojtech, D., & Vitu, T. (2010). Influence of heat treatment on tribological properties of electroless Ni–P and Ni–P–Al_2O_3 coatings on Al–Si casting alloy. *Applied Surface Science*, *256*(9), 2956–2960. doi:10.1016/j.apsusc.2009.11.057

Novakovic, J., & Vassiliou, P. (2009). Vacuum thermal treated electroless Ni-P-TiO_2 composite coatings. *Electrochimica Acta*, *54*(9), 2499–2503. doi:10.1016/j.electacta.2008.12.015

Palaniappa, M., & Seshadri, S. K. (2008). Friction and wear behavior of electroless Ni-P and Ni-W-P alloy coatings. *Wear*, *265*(5–6), 735–740. doi:10.1016/j.wear.2008.01.002

Rabinowicz, E. (1995). *Friction and Wear of Materials*. New York: John Wiley and Sons.

Rabizadeh, T., & Allahkaram, S. R. (2011). Corrosion resistance enhancement of Ni–P electroless coating by incorporation of nano-SiO_2 particles. *Materials & Design*, *32*(1), 133–138. doi:10.1016/j.matdes.2010.06.021

Rabizadeh, T., Allahkaram, S. R., & Zarebidaki, A. (2010). An investigation on effects of heat treatment on corrosion properties of Ni-P electroless nano-coatings. *Materials & Design*, *31*(7), 3174–3179. doi:10.1016/j.matdes.2010.02.027

Roy, S., & Sahoo, P. (2012). Potentiodynamic Polarization Behaviour of Electroless Ni-P-W Coatings. *ISRN Corrosion*, *2012*, 914867. doi:10.5402/2012/914867

Sahoo, P., & Das, S. K. (2011). Tribology of electroless nickel coatings – a review. *Materials & Design*, *32*(4), 1760–1775. doi:10.1016/j.matdes.2010.11.013

Sarret, M., Müller, C., & Amell, A. (2006). Electroless Ni-P micro and nano-composite coatings. *Surface and Coatings Technology*, *201*(1-2), 389–395. doi:10.1016/j.surfcoat.2005.11.127

Srinivasan, K. N., Meenakshi, R., Santhi, A., Thangavelu, P. R., & John, S. (2010). Studies on development of electroless Ni-B bath for corrosion resistance and wear resistance applications. *Surface Engineering*, *26*(3), 153–158. doi:10.1179/174329409X409468

Zou, Y., Cheng, Y., Cheng, L., & Liu, W. (2010). Effect of tin addition on the properties of electroless Ni-P-Sn ternary deposits. *Materials Transactions*, *51*(2), 277–281. doi:10.2320/matertrans.MC200917

Chapter 11
Tribological Behavior, Machinability, and Optimization of EDM of AA-2014 Hybrid Composite Reinforced With SiC and Glass Particulates

Mohan Kumar Pradhan
https://orcid.org/0000-0002-2718-7181
Maulana Azad National Institute of Technology, Bhopal, India

ABSTRACT

In the present work, the mechanical, tribological behavior of the machinability using electro discharge machining on a hybrid reinforced with different compositions of SiC and glass particulates are fabricated through a stir casting route. The effect of composition with machine parameters on the responses was observed. The Taguchi L9 orthogonal array was used for conducting the experiments, and the optimal EDM machining condition was obtained by performing MULTI-MOORA method integrated with PCA method. However, the effect of load, speed, and composition was investigated for tribological properties. It has been observed that SiC and glass particulates, significantly affect the compressive strength and hardness. However, the impact energy reduces an increase in the brittle nature of the composite. The glass particles do not contribute to the tensile strength but significantly improve the tribological property. The composition and pulse current found to be the most influencing parameters, the composition (Al- 2014+2wt% glass+3 wt% SiC) has led to the best combination for machinability.

INTRODUCTION

Composite materials have been emerging materials over the last few decades, its applications have steadily been increased and show remarkable growth in market penetration. The properties of composites appear as per their constituent materials, their interaction, and distribution, which makes it possible to obtain

DOI: 10.4018/978-1-7998-1831-1.ch011

a unique combination of material properties. Hybrid composites are more advanced composites than conventional fiber-reinforced polymer composites. The hybridization may have more than one reinforcing phase and a single matrix phase or multiple matrices with single reinforcing phases and have better flexibility in comparison to fiber-reinforced composites. It usually contains a high modulus fiber with a low modulus fiber Vogelesang and Vlot (2000).

Aluminum alloys have many applications due to the advantage of having a high strength to weight ratio, however, low hardness and low wear resistance restricted their applications in specific areas. Aluminum metal matrix (Al-MMC) composites with particulate reinforcements are considered an encouraging solution to provide superior wear resistance to aluminum alloys. Addition of silicon carbide and alumina to aluminum alloys improves its wear resistance, similar to glass, granite, cerium dioxide, boron carbide, nickel alumina, aluminum diboride, aluminum nitride, and titanium dioxide are also reported as an active enhancer of the tribological properties of aluminum-based alloys. The wear resistance can be enhanced by increasing the reinforcement material because of the greater stiffness and strength of the reinforcement phase. In recent years, SiCp-Al MMCs have received exceptional consideration and have been the main interest of their manufacturing techniques, which influence the mechanical properties of the compounds Bunma et al. (2011). Metal matrix composites (MMC) have greatly improved properties, including high specific strength, specific modules, damping capacity, and excellent wear resistance compared to un-reinforced alloys. Composites with low density and low-cost reinforcement are also drawing increasing attention. Powder metallurgy, stirring and compression casting can be performed in several ways, of which casting under stirring is preferred for the preparation of composite material due to better adhesion. The ability of MMCs to continuously synthesize the reproducibility of the microstructure and MMCs capable of replacing traditional monolithic materials. Applications, where the prerequisite of lightweight in addition to energy savings are the dominant factors in design, the pre-eminence of MMC over conventional monolithic alloys, have been favored due to the recent development of manufacturing techniques to fabricate the components. The addition of the reinforcement in the composite increases the mechanical strength while maintaining the ductility considered by the matrix Lim et al. (1999).

COMPOSITE

A composite material obtained by two or more different materials, with different physical and electrical properties, predictors than the properties of the resulting material are superior. Usually comprised of two phases: the reinforcement phase, and the matrix phase, Vlot and Gunnink (2011). Each material has different physical, mechanical and chemical properties, its properties depend on the composition of the particles and, consequently, the level of dispersion of the mixed particles is a factor that determines the properties of the compounds. The factor that also affects the properties of composite materials is the interface between the host matrix and the particles. it is made of a base material called matrix, has low properties while being affordable, reinforced with different materials in the form of continuous fibers, short fibers or particles with mechanical, physical properties or chemical Helu et al. (2011).

CLASSIFICATION

Based on Matrix Material

Metal Matrix Composites

In MMC there will be at least two constituent parts, essentially a metal, the other may not necessarily being metal a ceramic or organic compound. When there are three or more materials are present, then it is called hybrid composite. Because MMCs have outstanding mechanical and tribological qualities and are hence regarded as potential engineering materials for many tribological applications. Hence, several research attempts are being made on the sliding mechanism of MMC reinforced with ceramic particles, whiskers or short fibers, and observed an improvement in wear and abrasion Alpas and Zhang (1992) and continuous attempts are increasingly being done.

Ceramic Matrix Composites (CMC)

The reinforcements used in ceramic matrix composites make it possible to improve the toughness of the combined material system the advantage is the high strength and Young's modulus of the ceramic matrix. The most common embodiment of reinforcement is a ceramic fiber of continuous length, with a modulus of elasticity generally slightly higher than that of the matrix. Ceramic fiber reinforcements not only increase the initial resistance of the composite structure to crack propagation but also allow the CMC to avoid the sudden brittle fractures characteristic of monolithic ceramics. This behavior is different from the behavior of ceramic fibers in polymer matrix composites and MMC, with fibers generally fracturing before the matrix due to the higher deformation capabilities of these matrices. Alumina (Al_2O_3), Silicon Carbide (SiC), Carbon Fibers (C) and Mullite (Al_2O_3-SiO2) are most often used for CMCs.

Polymer Matrix Composites (PMC)

A polymeric matrix composite is a composite material that is composed of a variety of small or continuous fibers. PMCs are used to transfer charges through fibers through the matrix. The main advantages of the PMCs are their lightweight, high rigidity and great resistance.

Based on Reinforcements

Particulate Composites

In this class of composite, the reinforcement is in the form of particles, dispersed in the matrix material. There are many forms of particulate composites. The particles used for reinforcement can be different forms such as chopped fibers, tiny particles (< 0.25 microns), flakes, hollow spheres or new materials such as Bucky beads or carbon nanotubes. In each case, the particles provide desirable material properties, and the matrix serves as a liaison for structural applications. Particulate composites do not usually have high tensile strength, unlike fibrous composites. The particles increase the hardness of the material, but do not have much effect on the strength of the composite. In many cases, particulate composites were used to improve performance at high temperatures. In other cases, these compounds were used to improve thermal and electrical properties.

Fibrous Composites

This matrix contains fibers as reinforcement for specific applications, as in aircraft parts, which requires greater strength per unit weight (specific resistance). This can be achieved with composites containing of a low density (and flexible) matrix reinforced with rigid fibers. The resistance depends on the length of the fiber and its orientation. The efficiency of load-transfer between the matrix and the fiber depends on the interfacial bond.

Laminate Composites

It is composed of two-dimensional layers that have a preferred direction of the force. These layers are stacked and cemented together as needed. The materials used in their manufacture include foils, cotton, paper, woven glass fibers, embedded in a plastic matrix, etc. Many laminar composites are used to increase corrosion resistance while maintaining low cost, high strength, or lightweight.

Advantages of Composites

Composites can be engineered, particularly to meet the current necessities of the society, they are high strength to weight ratio, resistance to wear, fatigue, and corrosion degradation, vibration damping, provides excellent thermal and electrical properties as well as high-temperature performance. Directional customization capabilities the fiber pattern can be used to adapt Improved bump resistance Composite panels are not damaged as easily as thin sheets. Composites offer better torsional stiffness sign high rotation speed, a reduced number of intermediate bearings and structural support elements.

Limitations of Composites

In spite of the several advantages, it has some limitations like Anisotropy, Non-homogenous, which makes it challenging to model and analyze, its cost makes their applications in such cases where their benefits exceed their costs. Machinability, Difficult in fabrication, they are subject to residual heat stress, because they are manufactured at high temperature and then cooled. Thus, making structures from such materials is difficult, time-consuming, and expensive.

Applications of Composites

Composites are the most widely used materials, due to the superior properties, adaptability to situations and with the other materials to meet specific objectives and to exhibit desirable properties. Surface transport and heavy transport offer a wide reach and great receptivity for design changes, materials, and processes, strength to weight ratio, rigidity, and cost-effectiveness, ease of availability of raw materials, reproducibility, resistance to handling by semi-skilled workers. The polyester resin with suitable fillers and reinforcements. The choice made of properties such as low cost, ease of design and production of functional parts, etc. Manufacture of automotive parts, glass, and sisal fibers are commonly used. The fatigue properties of the materials and their low weight, their ability to withstand the stress of engine heat and low-frequency road vibrations are features that favor composites in trucks and other heavy vehicles. Reinforced plastics are not too expensive for tooling and are now widely used for auto parts, indicator lights, traffic lights, and other accessories. Commercial aircraft applications are the most important uses of composites. Aircraft, unlike other vehicles, required more safety on weight and Glass-reinforced composites were used extensively in many automotive parts such as panels, bumpers, dashboards, and intake manifolds, etc.

Fabrication Techniques of Composite

There are several methods to manufacture composite components, the choice of a method depends on the material, the design of the piece and the end-use or application. Stir Casting is a liquid state technique of composite material's fabrication, the reinforcement is mixed in the molten metal of matrix material and stirring through mechanical stirrer to mix the reinforcement. This process of fabrication is economical and simple. the Distribution of reinforcement materials is a challenge of a uniform distribution of reinforcement particles exist in matrix material melt due to the density difference between the reinforcement particles and the matrix alloy melt, for the improvement of the mixing the mechanical stirrer is employed. the Wettability between the matrix material and reinforcement plays a vital role in the fabrication of MMC. The accomplishment of proper inclusion of reinforcement into the matrix material depends on the wettability, and the melt needs to wet the solid ceramic particles. It involves the surface chemistry and surface tension between the ceramic particles and molten metal. it can be improved by Increment in surface energies of the solid, reducing the surface tension of molten metal, decreasing the interfacial energy between the reinforcement particle and matrix alloy. The transient layer can be formed between the particles and the liquid matrix to enhance the interfacial strength by alloying certain elements. It is observed that the addition of Mg, Ca, Ti, or Zr can improve the wettability. Mg is widely used to promote the wetting of many ceramic particles such as SiC and mica. The mechanical properties of the composite are significantly influenced by the porosity, its size, and its distribution in a matrix. these Defects also affect the corrosion resistance of the molded part. It can be avoided by controlling process parameters of the speed of stirrer, holding times, and the size and position of the impeller. Chemical reactions between the reinforcement material and the matrix alloy Sometimes the chemical reaction takes place between reinforcement material and the matrix alloy, which is undesirable, and it should be avoided

LITERATURE REVIEW

Many research efforts have been made in the last few decades.

Madhu Kumar (2018) studied the Effect of Particle Size on the Mechanical Properties of Glass Particulate Reinforced Aluminium Alloy6061 MMC. the matrix had been prepared by particulate of size (75 pm, 88 pm, 105 pm, and 250 pm) by varying wt.% of 3%, 6%, 9%, and 12%. it was observed that with the increase in the wt.% of the reinforcement, hardness of composites increases. James et al. (2014) investigated machining and mechanical property of hybrid Al-MMC reinforced with SiC and TiB2. It was reported that the hardness increases with an increase in reinforcement SiC and TiB2 but a reduction in hardness up to 15 wt.% have been noticed. Sathish and Rathnaraj (2015) et al. studied the behavior of LM-6 by reinforcement of borosilicate glass particles with wt.% of 2.5%, 5%, 7.5% and 10%, fabricated through stir casting method the composite showed an improvement in toughness and hardness. Forquin et al (2003) et al. have studied the effect of aluminum reinforcement on the dynamic fragmentation of SiC ceramics. Manikonda et al. (2018) et al studied the wear of silica carbide particles reinforced in the aluminum matrix. Al 2014 is selected for the matrix material. The tests were performed on the Al2104 with-out reinforcement and the MMC with 5wt% SiC reinforcement. The hardness of the base material Al 2014 is 60.31 BHN and after introducing reinforcement of 5wt% SiC the hardness increased to 65.01 BHN. Aribo et al. (2017) et al. has studied the behavior with the varying percentage of snail-shell-ash (SSA) and SiC reinforcement. the reinforcement up to 7.5wt% SiC + 7.5wt% SSA gives maximum hardness but dropped at 10wt% SiC + 10wt% SSA. Srivyas and Charoo (2018) Explain the strengthening of a density of the dislocations. Kumar et al. (2011) et al. has investigated the effect of Glass and SiC particulates on the workability of Aluminium (Al) through the powder metallurgy route. The Al powder of 80 pm particle size was used. Al, glass and SiC particles size were investigated microscopically. The Powder was prepared by mixing the SiC and glass of particle size of 60, 90, 120 and 150 pm with varying wt.% with the aluminum in a ball mill to obtain a homogeneous powder blend. The weight percentage taken for SiC and Glass are 0%, 4%, 8%, 12% and 16%. the influence of the particle size and reinforcement % of glass and SiC on the behaviour of the aluminum hybrid composite specifically the workability, when subjected to triaxial stress state, has been investigated. Sathish and Rathnaraj (2015) also show in his study that by addition of borosilicate glass in aluminum alloy tensile strength improved. In his study, he found the increment in tensile strength up to the maximum value of 147 MPa at 7.5wt% reinforcement and decline after that due to the agglomeration of borosilicate glass. James et al. (2014) et al. found that the tensile strength increased by 20% by the addition of the SiC to the base metal but the addition of TiB2 decreases 50-60% of strength.

Wear and friction cannot be conquered completely but with the help of lubrication or by making the smooth surfaces, it can be reduced to some extent (Wilson and Alpas (1997), Deuis et al. (1997)). On introducing the reinforcements of Sic and glass the wear rate is influenced. James et al. (2014) et al. shows the TiB2 particle increases the wear resistance, the wear of SiC10%-TiB2 0% has 20% more wear than SiC 10%-TiB2 2.5% specimen. Manikonda et al. (2018) studied the coefficient of friction with a varying load of 3 kg, 6 kg and 9 kg. with the increase in sliding distance the coefficient of friction also increases for both the plain Al2014 and the Al MMC with 5wt% SiC reinforcement, however, the coefficient of friction decreases for the higher load as introducing the reinforcement. it is found that the wear loss decreases after reinforcement at higher loads. Umanath et al. (2013) studied the wear behaviour of Al6061-T6 reinforced with SiC and Al oxide (Al2O3) for the factors like applied Load, rotational speed, reinforcement volume fraction and counterface hardness influence the wear property of the composite was considered in which the reduction in wear is found at low applied load, high counter face hardness, low rotational speed and high-volume fraction. Sharma et al. (1998) have conducted the wear test on zinc- Al alloy composite. The ZA-27 is chosen for the matrix material and the glass reinforcement is introduced in it. The short glass fibers with a weight percentage of 1% to 5% in step of 2% is taken as reinforcements. The reduction in wear rate with an increment of reinforcement by 43% at 20 N, 46% at 40 N, and 83.2% at 60 N have been noticed. Sahin and Murphy (1996) had investigated the wear behaviour of Al alloy reinforced with SiC coated, uni-directional boron fabricated by metal infiltration technique and found the reduced wear rates in comparison with the matrix alloy. Dwivedi et al. (2012) investigated the effect of CNC lathe process parameter on surface roughness while machining of A356 alloy with 5wt% SiC particulate MMC. A decrement in surface roughness with an increase in cutting speed, while surface roughness increases by an increase in depth of cut and feed. Ravindran et al. (2013) studied reinforcement of graphite (Gr) and investigated the tribological and mechanical property of the composite under dry sliding condition and found that by introducing the SiC particle in the composite the wear resistance of the composites improved, besides the hardness. Furthermore, it is seen that by increasing applied load and sliding distance the wear loss of the hybrid composites decreases due to which the low friction coefficient and low wear loss can be attained at high sliding speeds. The improved tribological property of the composite is achieved at 5 wt.% Gr and 20wt% SiC. also, it is found that by reinforcing hard material SiC and the soft material graphite reformed the tribological property. The current information about the tribological behaviour of particulate MMC based on Al alloys is inferred. The recent study differentiates abrasive wear to adhesive wear although the mechanism of abrasion also operates in other types of wears, and vice versa (Dante (2015), Zamzam (1991), Rohatgi (1992), Larsen-Basse (1990)). Kumar et al. (2017) et al. analyzed experimentally the wear rate of PMC fabricated by glass fiber reinforced epoxy resin filled with Al powder. Resin LY556 is used as matrix material, E-glass fiber as reinforcement and Al powder 200mesh as a filler material. The composite is prepared by mixing the chopped glass fiber having aspect ratio 10:1 with Al powder; HY-951 is used as curing agent for epoxy LY-556. The composite is prepared with reinforcement % as (60% Epoxy + 20% Glass fiber + 20% Al powder), (70% Epoxy + 10% Glass fiber + 20% Al powder) and (80% Epoxy + 10% Glass fiber + 10% Al powder). The pin-on-flat test apparatus is used to check wear property with varying loads and speed. The result shown in the below figure revealed that the specimen having combination of 80% Epoxy + 10% Glass fiber + 10% Al powder has maximum wear rate and the specimen of 60% Epoxy + 20% Glass fiber + 20% Al powder has lowest wear rate. It shows as the percentage of the glass fiber and respectively. Chen and Hoshi (1992) machined on SiC whisker composite of Al alloy with the reinforcement of 18 vol% of SiC and found that the carbide tool used in the rotary tool has more wear resistance than the PCD tool. Behera and Sutradhar (2012) on his machinability test of LM6/SiCp MMC observed the effect of machining parameters. The surface roughness and cutting forces are strongly influenced by the depth of cut and the cutting speed at a constant

Davim and Antonio (2001) got the optimizing value of machining parameter on turning and drilling Al matrix composite with SiC reinforcement of particle size 20 mm, the cutting tool used for machining was PCD insert. The optimal values are found to be for turning cutting velocity 350m/min, feed 0.1mm/rev, cutting time 19min whereas for drilling cutting velocity 50m/min, feed 0.05mm/rev, cutting time 10min. Davim (2002) studied the machinability of aluminum metal-matrix composites (A356/SiC/20p) with brazed PCD insert and Chemical Vapour Deposition (CVD) tool and compares the results. The turning operation is done on the MMC and the tool wear is observed for both inserts and it is found that the CVD diamond insert suffers more severe flank wear rapidly than the PCD insert. El-Gallab and Sklad (1998) studied the effect of machining parameters like depths of cut, cutting speeds and feed rates on the surface integrity of aluminum composite (Al/20%SiC). The Dry high-speed turning is performed on the specimen with polycrystalline diamond (PCD) tools and surface roughness was measured. by increasing feed and cutting speed, the roughness of the surface improves, but by increasing the DOC, it decreases. Kilickap et al. (2005) investigated the tool wear and surface roughness on turning the 5% SiCp reinforced Al MMC. The coated, uncoated TiN grade cutting tool K10 used for the turning operation at different cutting speeds of 50, 100 and 150 m / min and a feed rate of 0.1, 0.2 and 0.3 mm / turn. The abrasion wear on the flank side of the tool. It can also be seen that the cutting speed is the determining parameter of the wear of the tool. As the cutting speed increases, tool wear increases. The feed rate is a determining parameter after the cutting speed and it is found that when it increases, the wear of the tool also increases.

Many studies have been performed on the Al alloy composites since long and are still continuing. Al extends the advantage of high strength, light in weight therefore for the popularity of Al in the engineering field. The studies of Al composite focus on the fabrication route and the effect of the reinforcement on the property of the base Al alloy, to improve the properties. It is found that Limited researches have been performed on Al-2014, though there is a wide range of application on the aerospace industries. Thus, Mechanical as well as Tribological properties of this alloy becomes very important and to improve these properties researches are progressing. The manufacture of this compound through the stir casting route has not yet been done. The study in the area of Machinability of the Al-2014 compound is limited and to meet this gap, the current work is attempted.

The objective of the present work is to fabricate the Aluminum-2014 alloy composite reinforced with SiC and Glass particulates to improve the Tribo-Mechanical properties and to the machinability of the composite on performing a die-sinking operation on EDM. The study of different electrical discharge machining (EDM) parameter and its effect on the composite has been done.

MATERIAL PREPARATION

Matrix Material

Aluminium-2014

Al constituting more than 8% of the central mass of the Earth. It is also the third most common chemical element on our planet after oxygen and silicon. Aluminum offers a rare combination of valuable properties. One of the advantageous properties of Al is that it is very light in weight and offers good strength as well as corrosion resistant. In this research work Al-2014 is used. It is the strongest among all the available aluminum alloys. It is a copper-based alloy of Al. It contains 3.9%-5% of copper with other elements. The composition of Al-2014 is given in Table 1

Table 1. Chemical composition of Aluminium-2014

Chemical Composition	Weight Percentage
Copper:	3.9 - 5
Magnesium	0.2 - 0.8
Iron	0.7
Chromium	0.1
Manganese	0.4 - 1.2
Silicon	0.5 - 1.2
Titanium	0.15
Zinc	0.25
Aluminum	Balance

Reinforcement Materials

Silicon Carbide

SiC is an extremely hard and crystalline compound of silicon and carbon. Its chemical formula is SiC. Since the end of the 19th century, SiC has been an important material for cutting tools, grinding wheels and abrasive paper. It is widely used in heating elements and refractory coatings for industrial furnaces as well as in the wear-resistant parts like in rocket motor, pumps, turbines, etc. In this research, the particle size of 420pm of SiC is used as reinforcement material.

Glass Particulates

Glass is an amorphous or non-crystalline solid derived from the Latin word glaesum. The first synthetic glass was discovered in Egypt, around 3,500 years ago. Generally, glass can be defined as an inorganic melt which has been cooled to a rigid state without crystallization. The glass behavior is similar to that of amorphous polymers because glass does not have a melting point or a distinct freezing point. Today, there are about 750 types of glasses available commercially. Commercial glass mainly manufactured in the world is soda-lime glass (63-74% sand, 12-16% soda and limestone 7-14%). Chemically composition of glass particulates is given in Table 2.

FABRICATION OF COMPOSITE

Preparation of Composite Through Stir Casting Route

The Electric stir casting furnace was used to fabricate the composite as shown in Figure 1. It consists of a stirrer mechanism having three blades at an angle of 120°. The stirrer mechanism is operated by an electric motor whose speed can be varied between 400 — 700 rpm shown in Figure 2. The Al alloy is melted in a crucible having a capacity of 2 kg in the furnace up to the temperature of 850° *C* for

Table 2. Chemical composition of glass

Chemical Composition	Weight Percentage
SiO_2	64 - 70
K_2O	0-3
B_2O	2-5
ZnO	1-5
Na2O	8-13
MgO	1-4
CaO	3-7
AI_2O_3	3-6
TiO_2	0-3

the time period of 1.5 - 2 hrs. to melt completely. To remove the impurities from the melt the flux was added to the melt. Before adding the reinforcement, it is preheated at the temperature of 200°C and it is added to the melt having 650° C while stirring the molten alloy through stirrer mechanism at the speed of 600 rpm for 10 - 15 min. Before pouring the molten metal into the Die cavity, the graphite solution is applied to the internal cavities of the die to avoid adhesive action between molten metal and the die surface. After that taking out the molten metal in the crucible, and it is poured into the pre-heated Die cavities. After solidification of the molten material in the cavity, it is taken out. For the sake of simplicity, the different composition of composite material coded as A1(Al-2014+1.5wt% glass+3.5wt% SiC), A2(Al- 2014+2wt% glass+3 wt.% SiC) and A3(Al-2014+2.5wt% glass+2.5wt%SiC).

Specimen Preparation

The specimen was prepared by further machining of the cast material in lathe machine as well as other machining operations. The various test specimen was prepared included Compressive Test specimen, Impact test Specimen and for the EDM operation. All the specimens were prepared according to the ASTM standard. The compression test specimen is also prepared according to the ASTM E9 having a dimension of 15mm diameter and 30mm height as shown in Figure 3. Impact test specimen having a length of 55mm and cross-section of 10mm x 10mm and V-notch of 45° shown in Figure 4. The wear test was performed on the cylindrical specimen has 15 mm diameter 100 mm length

Figure 1. Electric stir casting furnace

Figure 2. Stirrer mechanism of the furnace

TESTING MACHINES USED

Compression Testing Machine

Compression test was performed in the compression testing machine having capacity of 2000kN shown in Figure 5. The test was performed on the specimen of 15mm diameter and 30mm height, kept between the upper plates and lower plate, the load was applied to the specimen up to failure point and reading was taken from the display

Wear Testing Machine

Wear test was conducted on the pin on disc wear testing machine developed at MANIT, Bhopal. Weight loss was measured with the electronic weighing machine as shown in Figure 7. The temperature change displayed in the machine using a thermocouple. The test was performed at varying load of 1 kg, 2 kg, and 3 kg with the variation of speed at 30, 40 and 50 rpm. The wear testing machine is shown in Fig. 8.

Hardness Testing Machine

The Hardness test was performed on the Brinell Hardness Tester machine with 750 kg of load and the indenter of Ball having 5mm of diameter. Figure 9 shows the Brinell Hardness Tester.

Figure 3. Specimen for compression test

Figure 4. Specimen for impact test

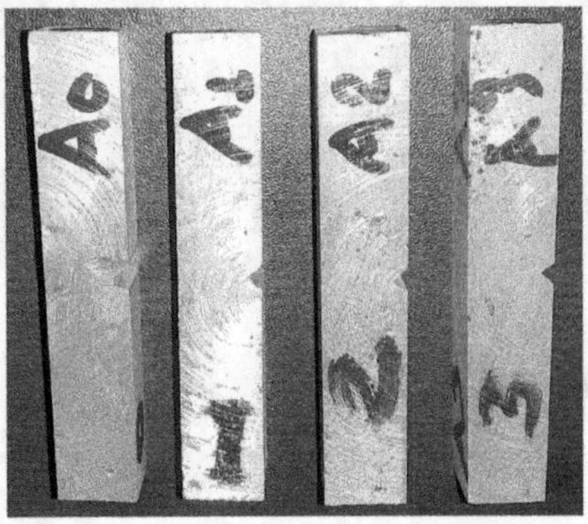

Figure 5. Impact testing machine

Figure 6. Impact testing machine

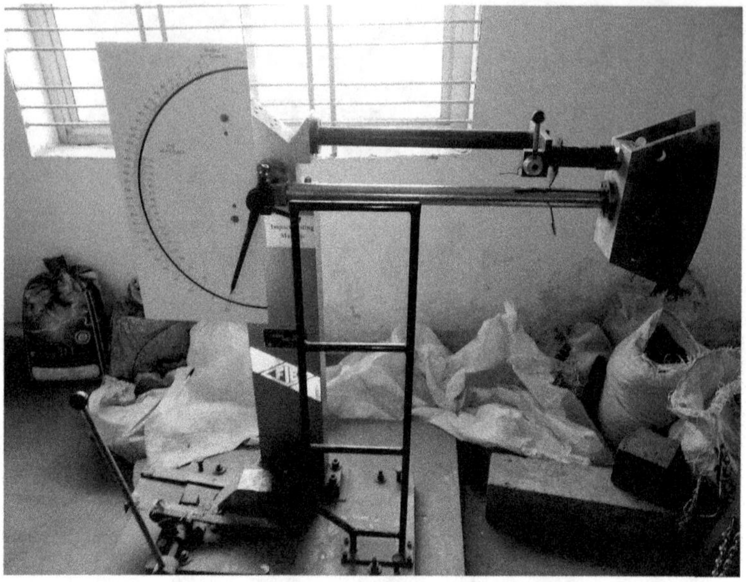

Figure 7. Electronic weighing machine

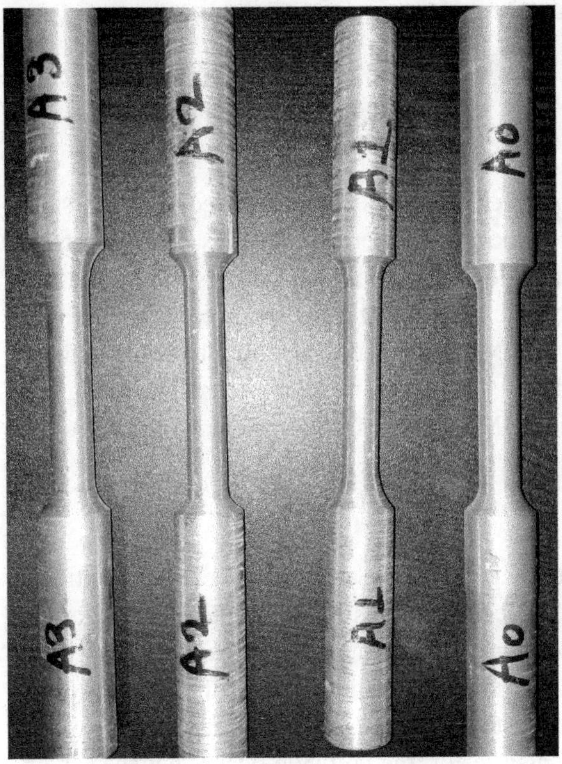

Figure 8. Pin-on-disc wear testing machine

Figure 9. Brinell Hardness Tester

Study of Microstructure

The mechanical properties of composite materials are influenced by their microstructure. properties like strength, ductility, corrosion resistance, toughness, and wear resistance, etc. are greatly influenced. The glass particles and SiC reinforced in the matrix material plays an important role in the morphology of the composite.

Figure 10 shows the EDMed surface of the A1with parameter combination of 3 A, 100 (µs) and 40 V, the surface has dark patches that could be caused by the burning of Glass Particulates. Similarly, Figure 11 is for A1 with 7 A, Ton = 150 (µs) and Voltage = 50 V, in this parameter such dark patches are not present. Figure 12 shows the machining surface when parameters having current = 11 A, Ton = 200(µs) and Voltage = 60 V and in these parameters, no black spots observed somehow decolourisation of surface occurred. Figure 13 shows the machining surface of A2 composition when having current = 3 A, Ton = 150 (µs) and voltage = 60 V, and at this parameters no significant spots observed neither decolourisation of surface occurred, when the parameters vary to current = 7, Ton = 200 and voltage = 40 V, the machined surface is seen in Figure 14, the number of black spots increases. Figure 15 shows the microstructure of A2 when current increases to 11 A, Ton reduce to 100 (µs) and voltage increase to 50V in this the surface decolourisation occurs but no significant changes in several black spots observed. Figure 16 showing the surface condition of A3 composite when Current = 3 A, Ton = 200(µs) and Voltage = 50V and the surface is free from the black spots this indicates no burning of glass particulates and neither the decolourisation occurs. When current increases to 7 A, Ton reduces to 100 and voltage increases to 60, the microstructure with these parameters are shown in Figure 17 and black spots observed and for Current = 11 A, Ton = 150 (μs) and Voltage = 40 V the microstructure shown in Figure 18.

ELECTRICAL DISCHARGE MACHINING

EDM is a non-conventional manufacturing process in which desired shape is created with the spark obtained by electrical energy. it capable of machining any kind of material as long as it is conductive, hence preferred for the difficult-to-machine materials and alloys. The complex contours and cavities can be machined. High precision and good surface quality can be obtained, the aerospace parts and automotive industry, and surgical components (Pradhan 2010). EDM has been the best-established machining option to replace the other traditional machining operations such as milling, drilling, and grinding, etc. in many manufacturing industries around the world. The EDM machine set up is shown in the Figure19.

Working Principle

The material removal takes place in the workpiece due to the rapid cyclic current between the electrodes submerged in a dielectric medium (Pradhan, 2012). The electric field is established between them by carrying the proper interelectrode gap and voltage. The electric field intensity is increased by increasing the gap voltage due to which the power of the electric field dielectric strength as a result, it breaks-down an allow current to flow between them. The electrons emitted from the tool flows at very high speed towards the workpiece through the dielectric medium, as a result, the collision occurs between them, results in ionization of di-electric molecules. Therefore, more positive ions and electrons generated as the result of the collision and the concentration increased due to the cyclic process. The Avalanche motion of electrons took place due to the less electrical resistance of the plasma and could be seen in the form of a spark. So, the thermal energy of the spark is obtained from electrical energy. The high-speed electrons and ion then strike the workpiece, the kinetic energy of electrons and ions during impact with the work surface and the tool would be converted into thermal energy or heat flux. Such localized heat

Table 3. List of Figures of EDMed surface.

S. No.	Composition	Current (Ip)	Ton(μs)	Voltage (V)	Figures
1	A1	3	100	40	Figure 10
2	A1	7	150	50	Figure 11
3	A1	11	200	60	Figure 12
4	A2	3	150	60	Figure 13
5	A2	7	200	40	Figure 14
6	A2	11	100	50	Figure 15
7	A3	3	200	50	Figure 16
8	A3	7	100	60	Figure 17
9	A3	11	150	40	Figure 18

Figure 10. Microstructure of EDMed surface of Al-2014+1.5wt% glass+ 3.5wt% SiC composite

Figure 11. The microstructure of the EDMed surface of Al-2014+1.5wt% glass+3.5wt% SiC

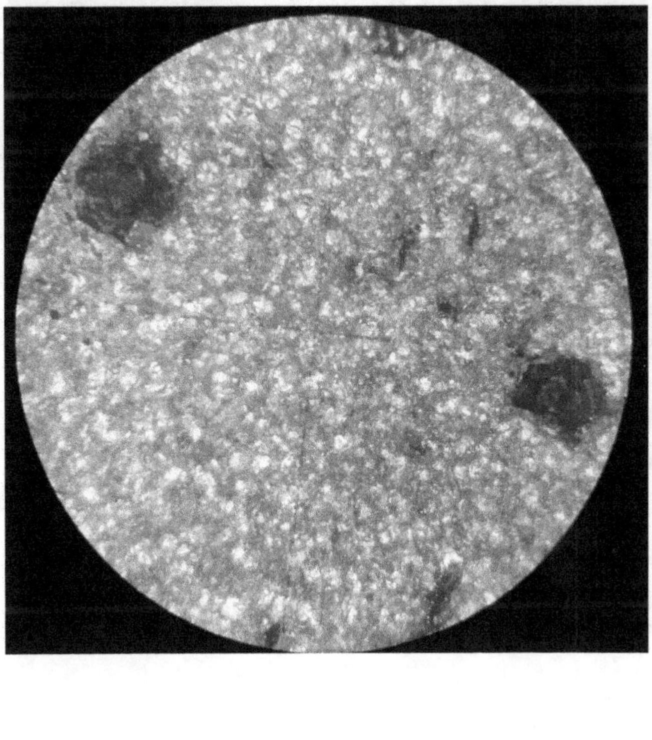

Figure 12. The microstructure of the EDMed surface of Al-2014+1.5wt% glass+3.5wt% SiC

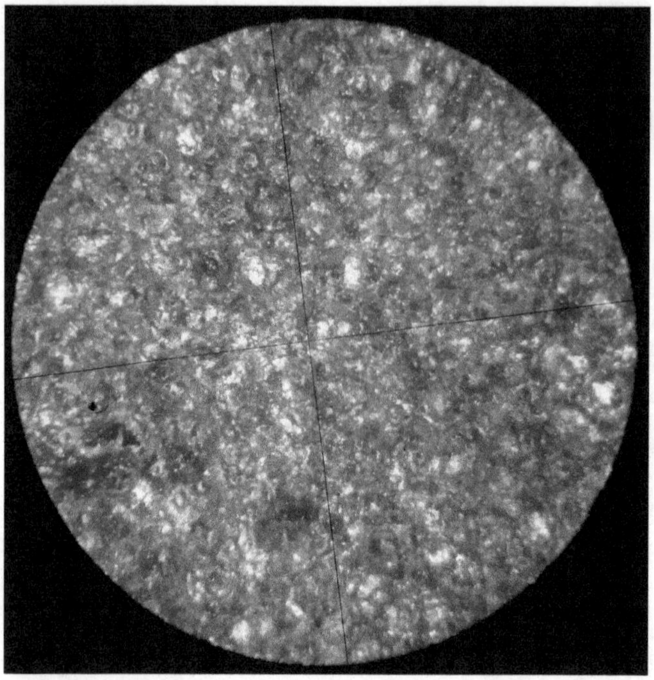

Figure 13. Microstructure of EDMed surface of Al-2014+2wt% glass + 3wt% SiC

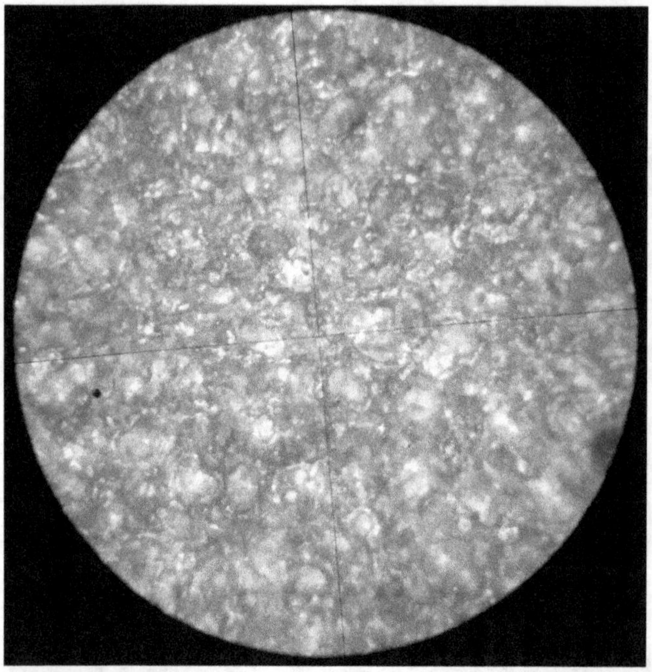

Figure 14. Microstructure of EDMed surface of Al-2014+2wt% glass+3 wt.% SiC

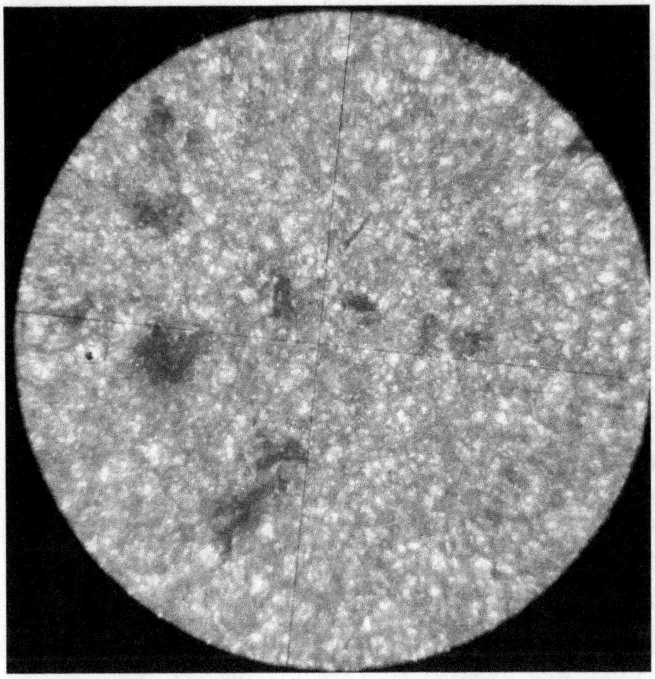

Figure 15. Microstructure of EDMed surface of Al-2014 + 2wt% glass + 3 wt.% SiC

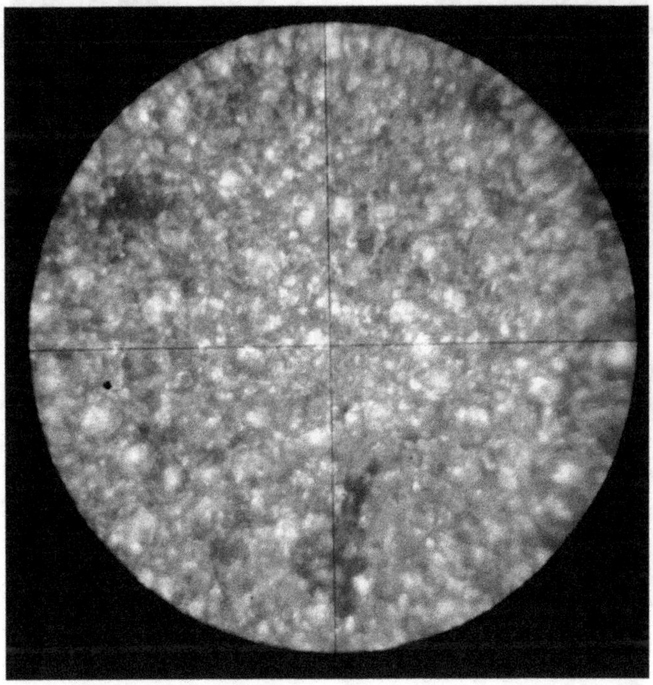

Figure 16. Microstructure of EDMed surface of Al-2014 + 2.5wt% glass + 2.5 wt.% SiC

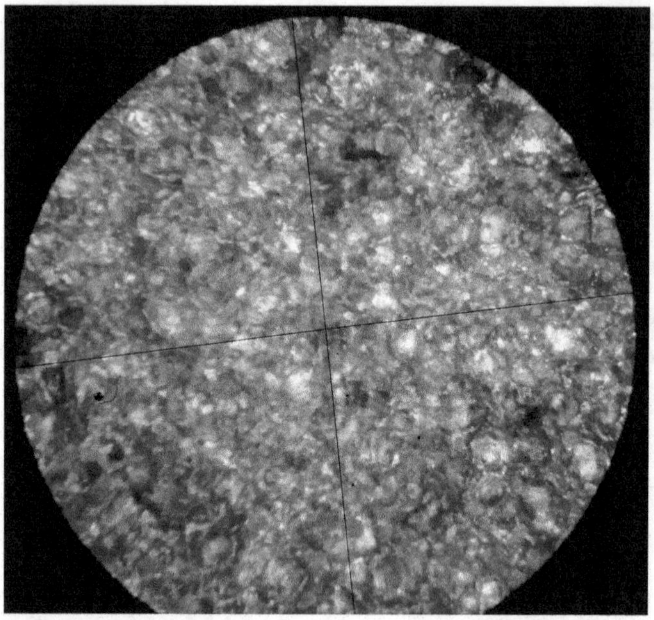

Figure 17. Microstructure of EDMed surface of Al-2014 + 2.5wt% glass + 2.5 wt.% SiC

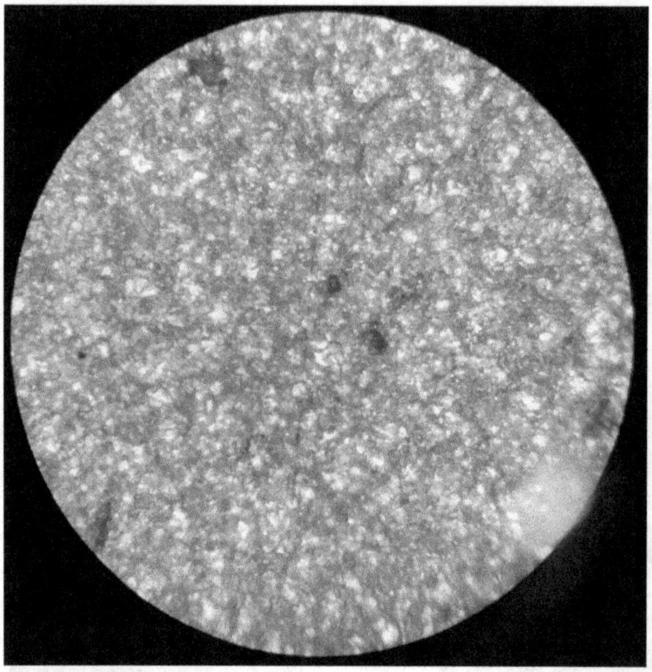

Figure 18. Microstructure of EDMed surface of Al-2014 + 2.5wt% glass + 2.5 wt.% SiC

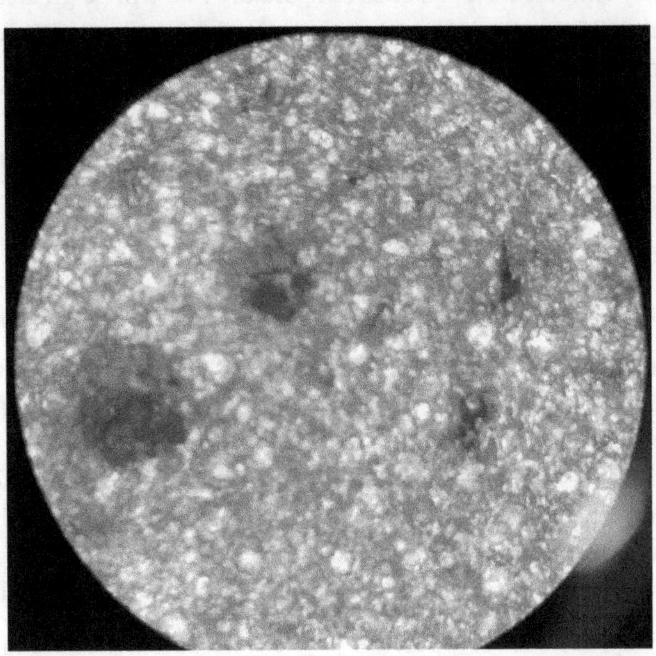

Figure 19. Electrical discharge machine

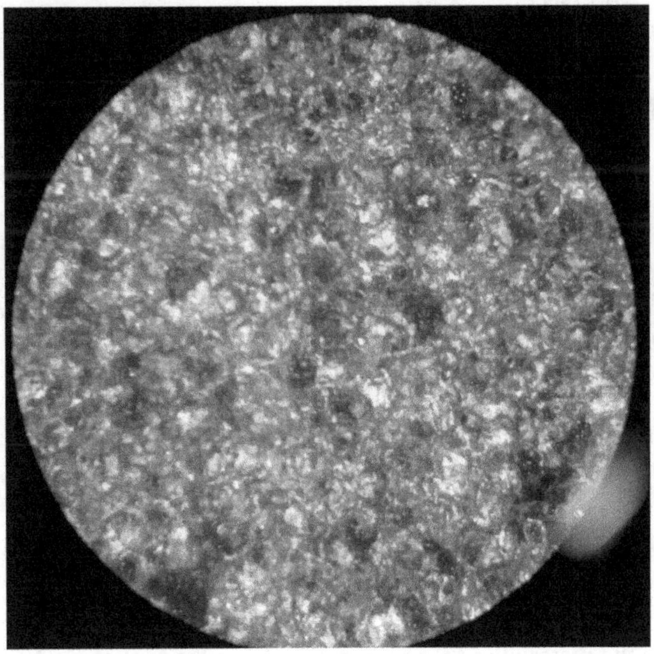

flux leads to an extremely extreme instantaneous temperature rise will be greater than 10,000oC, leads to melt and even vaporize the material. The crater is formed when the molten metal is eroded by the dielectric and the generation of shock wave or pressure, a result of collapsing plasma channel when the potential difference is removed.

Process Parameters

Electrical Parameters

- **Peak Current:** The peak current is the maximum available current for each pulse of the power supply. It is the amount of power used in discharge machining, measured in amperage units, and the major machining parameter in EDM. The current increases to a predefined level at each pulse expressed as a peak current (Pradhan and Singh (2019).
- **Pulse Duration and Pulse Interval:** The cycle consists of an on-time and off-time period, expressed in units of microseconds. The on-time is the time period in which work is done due to which these pulses and the number of cycles per second (frequency) become significant. The amount of energy delivered during on-time directly influences the material removal rate and can be controlled by the length of the on-time and peak amperage.
- **Gap Voltage:** It is the voltage between the tool and workpiece maintained at a certain distance referred as Discharge Gap or" Spark Gap". The tool servo-mechanism controls responsively the working gap to the predefined value. An appropriate" Gap" is needed to generate sparks between an electrode and a workpiece.

EDM PERFORMANCE MEASURES

Material Removal Rate (MRR)

This is the rate of material removed from the workpiece during EDM machining, measured by the volume removed per unit time (mm3 / min). Weight reduction of workpiece is the difference in weight of workpiece before and after machining, which is measured by digital weighting machine.

$$MRR = \frac{W_i - W_f}{\rho \times t} \tag{1}$$

Where Wj = weight of workpiece before machining Wf = weight of workpiece after machining p = density of tool T = time (min)

Tool Wear Rate:

It is the rate at which the material of the tool removes from the tool and calculated by measuring the weight loss of the tool given in equation 2.

$$TWR = \frac{W_i - W_f}{\rho \times t} \tag{2}$$

Where W_j = weight of tool before machining, Wf = weight of tool after machining p = density of tool T = time (min). For measurement of the weight before and after machining, a digital micro-weighing balance was used.

Radial overcut:

In the EDM process, the cavity formed by the electrode is slightly larger than the electrode. The excess dimension calculated as half of the difference in the dimension of the cavity diameter and the dimension of the electrode diameter is called Radial overcut (ROC).

It is always desirable to achieve minimum ROC Calculated from equation 3

$$ROC = \frac{D_h - D_t}{2} \tag{3}$$

Where, D_h = Diameter of the hole created in workpiece D_t = Diameter of tool

DESIGN OF EXPERIMENT (DOE)

The design of the experiment allows designers to concurrently determine the individual and interactive effects of the responses. In this work, the Taguchi method has been used. The value of this loss function is then transformed into a signal-to-noise ratio (S / N). The experiments were conducted in Die-sinking operation on Al-2014, Al-2014 + 1.5wt% glass + 3.5wt% SiC, Al-2014 + 2wt% glass + 3 wt.% SiC, Al-2014 + 2.5wt% glass + 2.5wt% SiC to investigate and optimization machinability in terms of variation in MRR, TWR, and ROC. The factors that affect die-sinking operation on an EDM machine is shown in Table 4. The control factors are considered in this work are the composition, current, ton, and voltage. Factors with their levels are presented in Table 5.

Table 4. Factors that affect die-sinking operation

Control Factors	Noise Factors
Current	Vibration
Ton	Raw material variation
Voltage	Machine Condition
Composition	Temperature
Flushing Pressure	Strongly more important
Tau	Operator Skill

Table 5. Control factors and their levels

Factors & Units	Levels		
	1	2	3
Composition	A1	A2	A3
Current (Ip)	3	7	11
Ton(μs)	100	150	200
Voltage (V)	40	50	60

OPTIMIZATION TECHNIQUES

Getting the optimal solution out of the similar options needed to adopt strategic options, Various options are related to the actual set of institutions, making the process of choosing the best of the best choices delicate and require some insight to it. To do this, some qualitative and quantitative factors should be selected, which can help in differentiation in options. This set of factors is collectively called a criterion. There are options in comparison to the criteria selected for an optimal selection. Therefore, MCDM is the composition of a group of many criteria, an option of an option or a comparison of them in one way or another.

WEIGHTED SUM METHOD

In this method, the options are compared by assigning scores and generation of standard values using these scores for considered options (Pradhan 2016). The result will be as good, better and best. The parameters are measured according to the severity of each. The sum of all these weights should be 1. Each option is evaluated in relation to each attribute Janic and Regini (2002). A summarized score of each option is calculated by the 7 equation.

Table 6. L9 orthogonal array

S. No.	Composition	Current (Ip)	Ton (^s)	Voltage (V)
1	A1	3	100	40
2	A1	7	150	50
3	A1	11	200	60
4	A2	3	150	60
5	A2	7	200	40
6	A2	11	100	50
7	A3	3	200	50
8	A3	7	100	60
9	A3	11	150	40

$$SS_i = \sum_{i=1}^{m} w_i x_{ij} \quad (7)$$

where

SS_i = Summated Score of ith alternative for each of the n alternative.
W_j = Weight of the jth criteria for each of the m criteria.
Xij = Score of the ith alternative with respect to the jth criteria.
n = Number of alternatives.

Analytic Hierarchy Process (AHP)

Analytical Hierarchy Process is an MCDM method which is based on the theory of priorities, deals with the problem involves the complexity of many criteria/alternatives. The ability of this method to integrate data and expert judgment in the model in a logical way to give a scale for the measurement of the intangible assets and setting of priorities to address the interdependence of elements makes this method a valuable contribution to the MCDM area. Comparative judgment involves a two-level process, first, an A (N × N) type matrix is formed in which the relative comparison of another option is performed. So here N is the number of alternatives and Similarly, M metrics are formed. So, basically, this level of comparison consists of one of the criteria (one with respect to each criterion). The entries in these matrices are made according to the initial AHP measurement scale, from 1 to 9, as shown in the following Table 7.

This value is essentially chosen by the decision-maker defining the relative value (importance) of an alternative with respect to the other, by maintaining a fixed criterion, that is to say by prioritizing the alternatives according to individual criteria. After the matrix, corresponding to each criterion has been made, it is time to build the normalized matrix from these. The final step involves the ranking of the alternatives according to the particular criteria by forming eign vector. In the eign vector, the largest

Table 7. Scale of the AHP

Degree of Importance	Definition
1	Equal importance (no preference)
2	Intermediate between 1 and 3
3	Moderately more important
4	Intermediate between 4 and 5
5	Strongly more important
6	Intermediate between 5 and 7
7	Very strongly important
8	Intermediate between 7 and 9
9	Extremely important
1/2 1/3 1/4 1/5 1/6 1/7 1/8 1/9	Reciprocal of 2, 3, 4, 5, 6, 7, 8 and 9

value is ranked higher than the smallest, then the smallest. The ranking of all the alternatives with every possible criterion can be done by recursive application of the same sequential steps. The accuracy of the rank can be checked following equation of Consistency Ratio.

$$CR = \frac{CI}{RI} \tag{8}$$

Where, CR = Consistent ratio shows whether the results are consistent or not. CI = Consistency Index. RI = Random Index.

Technique for Order of Preference by Similarity to Ideal Solution (TOPSIS)

The TOPSIS method was revealed in 1981 by Yoon and Hwang. The central idea of this method is that the selected option is necessarily the shortest distance of the ideal solution and the farthest from the negative-ideal solution in a geometric judgment. Ranking of alternatives is based on this method the high-grade is the examination of both subjective and objective factors is performed in the TOPSIS (Pradhan and Tiwari 2018, Pradhan2018).

VIKOR Method

Serafim Opricovic developed this method that could deal with contradictory and incompatible standards. Recognize that the agreement is acceptable to resolve conflicts. The decision-maker seeks a solution closer to the ideal and in accordance with all the rules, the options are evaluated (Gangil, 2016).

Preference Ranking Organization Method for Enrichment Evaluations (PROMETHEE)

This method is developed in the 1980s. It is an outranking method in which ranking of a finite number of alternatives is done on the basis of a finite number of criteria. First of all, it is comparing the rank of the alternatives on each criterion. PROMETHEE make the preferential function to link the difference of preference between a set of alternatives on each criterion (Pradhan 2019)

ELECTRE (Elimination Et Choix Traduisant La Realite/ Elimination and Choice Expressing Reality)

This method is stated by Bernard Roy, while applied for the first time in 1965, the ELECTRE method considered of choosing the best action (s) from a given set of actions. An ELECTRE application has two main segments: First, building one or more external links that aim to compare each pair of actions. Second, use a procedure which develops the recommendations acquired in the first phase. The types of recommendation based on the problem being addressed: sorting, choice, or ranking (Gangil et al., 2017).

MULTI-MOORA (Multi-Objective and Optimization on the Basis of Ratio Analysis)

The MOORA (Multi-Objective and Optimization on the basis of the Ratio Analysis) method has been implemented in this work. The MULTI-MOORA method consists of three steps, ranking using Ratio System, Reference Point Approach, and Full Multiplicative Form, the last step is to rank based on dominance approach.

Ratio System of MOORA

The steps for the Ratio System of MOORA is given below.

Step 1 Formation of Decision matrix X where *xj* is the value of ith (i = 1, 2, ..., m) alternative based on j^{th} (j = 1, 2, ..., n) criterion.

$$\left[X\right] = \begin{bmatrix} x_{11} & x_{12} & \cdots & x_{1n} \\ x_{21} & x_{22} & \cdots & x_{2n} \\ \vdots & \vdots & \ddots & \vdots \\ x_{m1} & x_{m2} & \cdots & x_{mn} \end{bmatrix}$$

Step 2 Normalizing the Decision matrix using equation 9

$$x_{ij}^{*} = \frac{x_{ij}}{\sqrt{\sum_{j=1}^{n} x_{ij}^{2}}} \tag{9}$$

Step 3 Formation of Weighted normalized decision matrix using equation 10. Where w_j is the weight of the jth criterion.

$$V_{ij} = W_j \times x_j \tag{10}$$

Step 4 Obtaining Final preference values (y^*) by using equation 11 where j = 1, 2, ..., g shows the criteria to be maximized and j = g +1, g + 2, ..., n shows the criteria to be minimized.

$$y_i^{*} = \sum_{j=1}^{g} v_{ij} - \sum_{j=g+1}^{n} v_{ij} \tag{11}$$

Step 5 Alternatives are ranked by ranking y^* in descending order.

THE REFERENCE POINT APPROACH OF MOORA

In the reference point approach, the first three steps will be the same as MOORA's Ratio System. The other steps can be summarized as

Step 1 Formation of decision matrix like in the first step of the Ratio system.
Step 2 Normalizing decision matrix using equation 9.
Step 3 Formation of weighted decision matrix using equation 10.
Step 4 Determination of Reference points *(rj)* for each criterion. Reference points are determined by choosing the highest values for maximization criteria and minimum value for minimization criteria.
Step 5 Calculation of distance between the alternatives and the best alternatives are chosen by using Tchebycheff Min- Max metric given in equation 12.

$$\min_j\left\{\max_j\left(w_j r_j - v_{ij}\right)\right\} \tag{12}$$

as per equation 12, firstly, the maximum distance values of each of the alternatives for the reference points are determined, then these values within, the minimum value selected as the best option. The final ranking of the alternatives is obtained by ranking the maximum distance values in ascending order. Full Multiplicative Form of MOORA following are the steps for the Full Multiplicative Form of MOORA.

Step 1. Formation of decision matrix like in the first step of the Ratio system.
Step 2. Determination of Multiplicative ranking index U_i for each alternatives using equation13.

$$U_i = \frac{A_i}{B_i} \tag{13}$$

where $A_{i}=\prod_{j=1}^{g} x_{ij}^{w_j}$, *for* i^{th} alternative for the criteria to be maximized (j =1,2, ..., g) and $B_{i}=\prod_{j=g+1}^{g} x_{ij}^{w_j}$, for i^{th} alternative for the criteria to be minimized (j= $g+1$, $g+2$, ..., n).

Step 3. Alternatives are ranked by ranking U_i values in descending order.

MULTI-MOORA

The MULTI-MOORA method is also based on the theory of dominance also aims to reach a final ranking from the three rankings, with the ratio system, the reference point approach and the complete multiplicative form. In this method, the final ranking, that is why it is a robust method.

PRINCIPAL COMPONENT ANALYSIS (PCA)

PCA is a mathematical tool that helps the conversion of a set of observations of possibly correlated variables into a set of values of uncorrelated variables. Pearson (1901) invented it, and further developed by Hotelling (1933), later on, it used widely as a tool to investigate data analysis as well as the predictive model (Pradhan, 2013). PCA is performed by decomposition of the eigenvalue of data covariance matrix or singular value of a data matrix. It helps in finding patterns in data, and presentations The PCA has the advantage that once the pattern in data found, it can be compressed by minimizing the number of dimensions, without much loss of information

Design of Experiment Using Taguchi Approach

The experiments performed on the Al composite to study the effects of various machining parameters like Ip, Ton, V and composition on MRR, TWR and ROC. For this experiment three-level control factors chosen as discussed in the previous section. The quality characteristic considered for the study i.e. MRR, TWR and ROC is calculated as according to the Taguchi approach. Using L9 orthogonal array, 9 experimental runs performed as shown in Table 8.

METHOD TO DETERMINE THE WEIGHTS

The decision-maker decides the weights of the different attributes considering the comparative importance based on the preference on the attributes. In MCDM methods like TOPSIS, Gray relational analysis, SAW, etc. needed the information about the relative importance of attributes. The principal component analysis method can be used effectively to determine weights. Table 9shows the weight of the attributes.

Table 8. Responses of EDM machining

S. No.	Composition	Current (Ip)	Ton	Voltage (V)	MRR (mm3/ min)	TWR (mm3 / min)	ROC (mm)
1	A1	3	100	40	14.54	0.009	0.164
2	A1	7	150	50	40.22	0.071	0.221
3	A1	11	200	60	62.71	0.201	0.304
4	A2	3	150	60	13.83	0.009	0.168
5	A2	7	200	40	41.28	0.162	0.259
6	A2	11	100	50	63.94	0.019	0.272
7	A3	3	200	50	11.77	0.023	0.197
8	A3	7	100	60	31.26	1.307	0.256
9	A3	11	150	40	69.28	0.205	0.297

Table 9. Eigenvectors and Contribution

Variable	PC1	PC2	PC3	Contribution
MRR	0.674	0.299	-0.676	0.454276
TWR	0.221	-0.954	-0.202	0.048841
ROC	0705	0.013	0.709	0.497025

OPTIMIZATION USING MULTI-MOORA METHOD

Ratio System of MOORA

The steps are followed and the decision matrix is formed. After the formation of the decision matrix, it is normalized using equation 9 and the normalized decision matrix is shown in Table 11. Weights are obtained using the PCA method shown in Table 9. Weighted normalized matrix is formed by multiplying the corresponding weights of the attributes and ranking of the alternatives are done by obtaining final preference value using equation 11 and arranging it in descending order shown in Table 12. The ranks of the alternatives by ratio system of MOORA is given in Table 13.

Reference Point Approach of MOORA

In this method reference points of each alternative are obtained from the weighted normalized decision matrix. For determining the reference points for maximization criteria, the highest value is selected and for the minimization criteria, the minimum value is selected, given in the last row of Table 14. Then the distance is calculated between the reference point and each alternative as shown in Table 14 and the maximum value of the distance is selected, given in the last column of Table 14. By using the Tcheby-cheff Min-Max metric, the best alternatives are chosen and the final ranking is done by arranging the maximum distance value in increasing order. The ranking of the alternatives is presesnted in Table 15.

Full Multiplicative Form of MOORA

In this form of MOORA, multiplicative ranking index U_i is determined for each alternative by using equation 13, the values are shown in the last column of Table 16. The ranking of the alternatives is done by arranging U_i in descending order. Ranking of the alternatives is shown in table 17.

Table 10. Eigen analysis of the correlation matrix

Principal Component	Eigenvalue	Explained Variations (%)
FIRST	1.9664	65.5
SECOND	0.992	30.30
THIRD	0.0424	0.014

Table 11. Normalized Decision matrix

S. No.	MRR (mm3/min)	TWR (mm3/min)	ROC (mm)
1	0.10942	0.00007	0.22522
2	0.30266	0.00376	0.30380
3	0.47186	0.02983	0.41790
4	0.10410	0.00006	0.23048
5	0.31064	0.01932	0.35535
6	0.48114	0.00040	0.37345
7	0.08857	0.02326	0.27035
8	0.23520	1.26568	0.35145
9	0.52132	0.03112	0.40850

*Table 12. Weighted Normalized Decision matrix and y**

S. No.	MRR (mm3 /min)	TWR (mm3/min)	ROC (mm)	y*
1	0.0497	0.0000	0.1119	-0.0894
2	0.1375	0.0002	0.1510	-0.0887
3	0.2144	0.0015	0.2077	-0.1118
4	0.0473	0.0000	0.1146	-0.0931
5	0.1411	0.0009	0.1766	-0.1135
6	0.2186	0.0000	0.1856	-0.0863
7	0.0402	0.0000	0.1344	-0.1161
8	0.1068	0.0618	0.1747	-0.1880
9	0.2368	0.0015	0.2030	-0.0970

Table 13. Ranking of alternatives with Ratio System of MOORA

Composition	Current (Ip)	Ton (µs)	Voltage (V)	Rank
A2	11	100	50	1
A1	7	150	50	2
A1	11	200	60	3
A2	3	150	60	4
A2	7	200	40	5
A2	11	100	50	6
A3	3	200	50	7
A3	7	100	60	8
A3	11	150	40	9

Table 14. Reference Point Approach of MOORA

Alternatives	MRR (mm3 /min)	TWR (mm3 /min)	ROC (mm)	MAX DISTANCE
1	0.1871	0.0618	0.0958	0.1871
2	0.0993	0.0616	0.0567	0.0993
3	0.0225	0.0604	0.0000	0.0604
4	0.1895	0.0618	0.0931	0.1895
5	0.0957	0.0609	0.0311	0.0957
6	0.0183	0.0618	0.0221	0.0618
7	0.1966	0.0618	0.0733	0.1966
8	0.1300	0.0000	0.0330	0.1300
9	0.0000	0.0603	0.0047	0.0603
REFERENCE POINT	0.2368	0.0618	0.2077	

Table 15. Ranking of alternatives with Reference Point Approach of MOORA

Composition	Current (Ip)	Ton (μs)	Voltage (V)	Rank
A2	11	100	50	1
A3	11	150	40	2
A2	7	200	40	3
A1	11	200	60	4
A1	7	150	50	5
A3	7	100	60	6
A1	3	100	40	7
A2	3	150	60	8
A3	3	200	50	9

Table 16. Full Multiplicative Form of MOORA

MRR (mm3 /min)	TWR (mm3/min)	ROC (mm)	Ai	Bi	Ui
14.54	0.009	0.164	1.738	0.324	5.36
40.22	0.071	0.221	2.143	0.415	5.16
62.71	0.201	0.304	2.349	0.512	4.59
13.83	0.009	0.168	1.72	0.327	5.26
41.28	0.162	0.259	2.155	0.467	4.61
63.94	0.019	0.272	2.359	0.431	5.47
11.77	0.023	0.197	1.663	0.371	4.49
31.26	1.307	0.256	2.035	0.514	3.96
69.28	0.205	0.297	2.398	0.506	4.74

Table 17. Ranking of alternatives with full multiplicative form of MOORA

Composition	Current (Ip)	Ton (μs)	Voltage (V)	Rank
A2	11	100	50	1
A1	3	100	40	2
A2	3	150	60	3
A1	7	150	50	4
A3	11	150	40	5
A2	7	200	40	6
A1	11	200	60	7
A3	3	200	50	8
A3	7	100	60	9

Table 18. Ranking of alternatives with MULTI-MOORA approach

Alternatives	Ratio System	Reference Point	Full-Multiplicative	Final Rank
1	3	7	2	2
2	2	5	4	4
3	6	4	7	7
4	4	8	3	3
5	7	3	6	6
6	1	1	1	1
7	8	9	8	8
8	9	6	9	9
9	5	2	5	5

Table 19. Final rank of alternatives

Composition	Current (Ip)	Ton (μs)	Voltage (V)	Rank
A1	3	100	40	2
A1	7	150	50	4
A1	11	200	60	7
A2	3	150	60	3
A2	7	200	40	6
A2	11	100	50	1
A3	3	200	50	8
A3	7	100	60	9
A3	11	150	40	5

FINAL RANKING OF ALTERNATIVES USING MULTI-MOORA

In this stage, the final ranking is obtained by using the dominance theory of MULTI- MOORA. The three ranks obtained from the Ratio System, Reference Point Approach and Full Multiplicative Form are finalized into one final ranking shown in Table 18. The final ranking is given in Table 19.

Mechanical Behavior

The reinforcement particles act as a barrier for the movement of dislocations as the result, mechanical properties influenced by it. So, to check the effect of glass and SiC particles, a number of the mechanical tests has been carried out as discussed below.

Compressive Strength

Test of Compressive strength was performed on the pure Aluminum-2014 and found to be 116 kN, while the composite of composition Al-2014 + 1.5wt% glass + 3.5wt% SiC has the compressive strength of 118.6 kN, Al-2014 + 2wt% glass + 3wt % SiC is 126.5 kN, and Al-2014 + 2.5wt% glass + 2.5wt% SiC has 104.5 kN of compressive strength. The composition of Al-2014 + 2wt% glass + 3wt% SiC gives the maximum compressive strength. It can be seen from the result that by reinforcing the glass and SiC particles, the compressive strength increases due to the glass and SiC particles act as a barrier for the movement of dislocations, but at the composition of Al-2014 + 2.5wt % glass + 2.5wt% SiC it reduces. It may be caused by the agglomeration of glass particles and a reduction in SiC particles. Figure 20 shows the Compressive Strength of Composite.

HARDNESS

The Hardness test was performed on the Brinell Hardness Tester machine with 750 kg of load and the indenter of Ball having 5 mm of diameter. Result shows, the pure Al-2014 has 96.975 BHN and the reinforced composite A1 = 97.2 BHN, A2= 98.15 BHN, and A3= 91.2 BHN as presented in Figure 21. it can be observed that A2 has maximum hardness but it decreases A3. It may be due to the agglomeration of glass particles as well as due to having minimum SiC content.

Impact Test

The Izod impact test had been conducted on pure aluminum-2014 and the composite reinforced with glass particulates and SiC. Al-2014 has an impact energy of 5 joules while the A1 = 4 joules, A2 = 3 joules and A3 = 5 joules exhibited in Figure 22. The increase in impact energy in the A3 structure may be the result of a compromised value. The impact energy cannot improve due to an increase in the fragile nature of the composite. As reinforcing the glass particles increases the brittleness of the composite because of an amorphous structure. It does not contribute to impact strength.

Figure 20. Compressive strength of composite

Figure 21. The hardness of composite

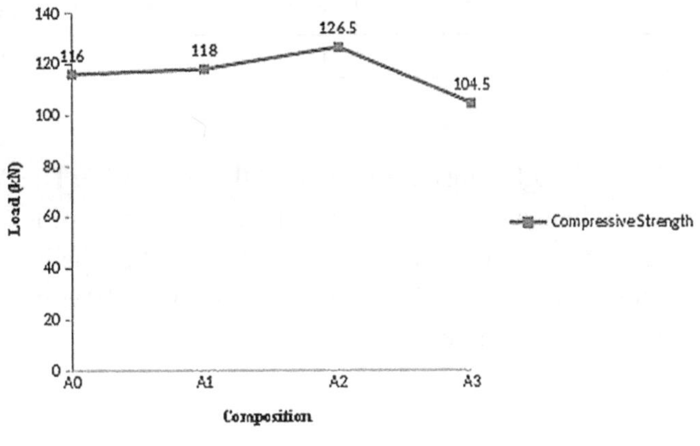

TRIBOLOGICAL BEHAVIOR

Wear test had been conducted on the wear testing machine (Fig-8), and Taguchi design of experiment using L9 orthogonal array has been applied for the same and presented in Table 20. The input-variables chosen for the test was load (1kg, 2kg, and 3kg), speed (30, 40, and 50 rpm), and the composition (A1, A2, and A3). Weight loss measured in electronic weighing machine and temperature was recorded using a thermocouple.

Figure 22. Impact energy of composite

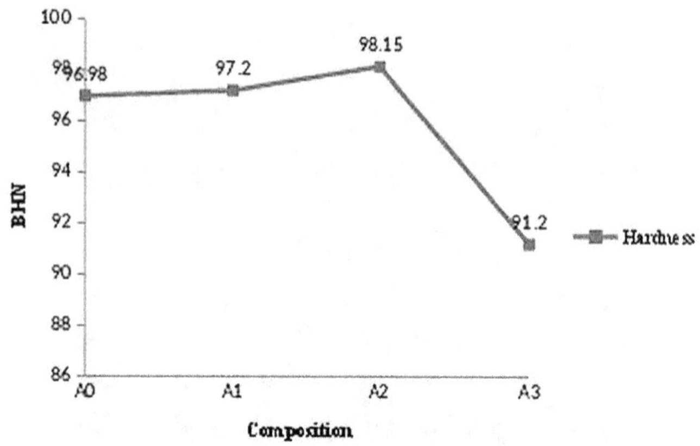

Table 20. L9 orthogonal array for the wear test

Composition	Load (kg)	Speed (rpm)	Weight Loss (mg)
A1	1	30	21
A1	2	40	34
A1	3	50	65
A2	1	30	45
A2	2	40	16
A2	3	50	26
A3	1	40	79
A3	2	50	119
A3	3	30	471

Figure 23 shows the effect of the different composition of the composite. The A1 composition has the minimum weight loss amongst them and A3 has the maximum. the effect of load and speed on weight loss of the composite also depicted, the weight loss also increases when the increase in load. Variation in speed affects the wear rate, on increasing the speed from 30 to 40 RPM the weight loss decreases, but when increased from 40 to 50 it increases.

MACHINABILITY

The machining operation was made up to 1 mm of the depth using copper electrode and MRR, TWR and ROC were calculated. The value of Tau kept constant at 20. Electrode gap was maintained at 0.1 mm. The Table 21 shows the input-variables and responses of the EDM machining.

Figure 23.

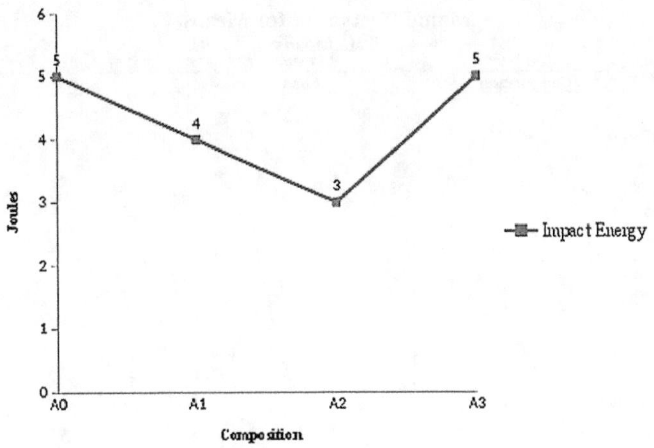

Table 21. L9 orthogonal array for EDM machining

S. No.	Composition	Current (Ip)	Ton (μs)	V	MRR (mm3/min)	TWR (mm3/min)	ROC (mm)
1	A1	3	100	40	14.542	0.009	0.164
2	A1	7	150	50	40.224	0.071	0.221
3	A1	11	200	60	62.710	0.201	0.304
4	A2	3	150	60	13.834	0.009	0.168
5	A2	7	200	40	41.284	0.162	0.259
6	A2	11	100	50	63.943	0.019	0.272
7	A3	3	200	50	11.771	0.023	0.197
8	A3	7	100	60	31.258	1.307	0.256
9	A3	11	150	40	69.283	0.205	0.297

Effect on MRR

The effect of input-parameters on MRR are depicted in figure 24. It can be observed that the MRR is maximum at the composition of A2. When the current increase, MRR also increases and at Ip = 11 A, MRR reaches to the maximum values on increasing Ton from 100 to 150, MRR increase but from 150 to 200 it decreases and the maximum value is at 150. The gap voltage also affects the MRR, at 40 it is maximum and decreases when the voltage increased to 50 and from 50 to 60 it again decreases.

Effect on TWR

Effect of Ip, Ton, Voltage, and composition on TWR are depicted in figure 25, the maximum TWR is active in the composition of A2 while it is maximum at A3. For TWR it is beneficial to have the minimum value. When current increases from 3 to 7 A, the current increases from 7 to 11 A.

Figure 24.

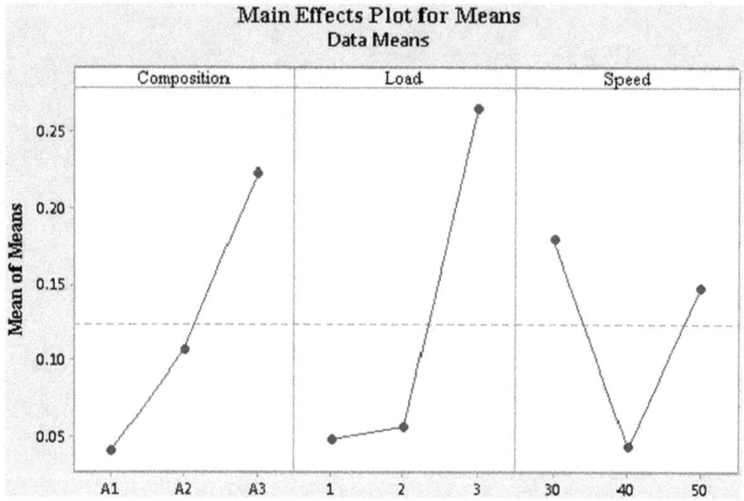

Figure 25.

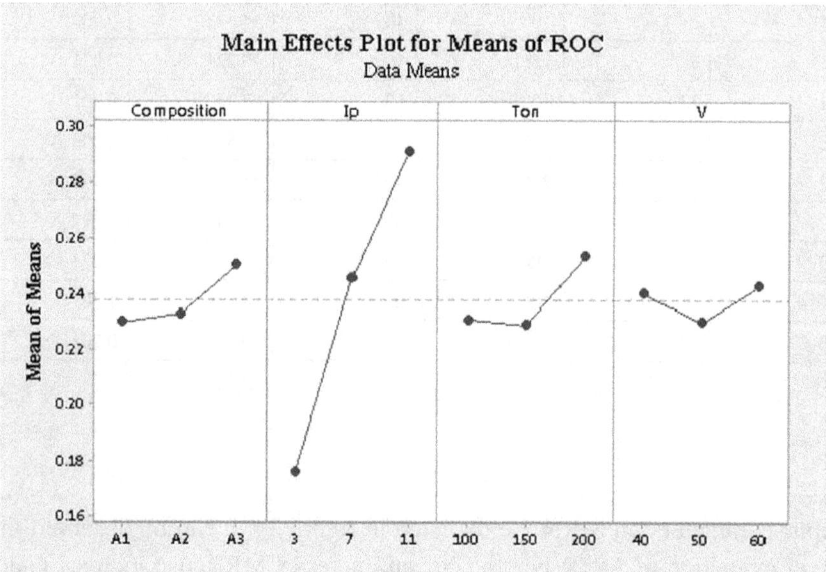

Effect on ROC

Effect of Ip, Ton, Voltage, and composition on ROC are depicted in figure 26. The minimum value of ROC is at the composition of A1, on increasing the current value of ROC increases and maximum at 11 A while minimum at 3A For Ton, the minimum value is at 150 μs. When the voltage increases from 40 to 50 V, the ROC decreases and reaches a minimum value at 50 V and again increases to the maximum value at 60 V.

Figure 26. Main effect plot for the mean of ROC

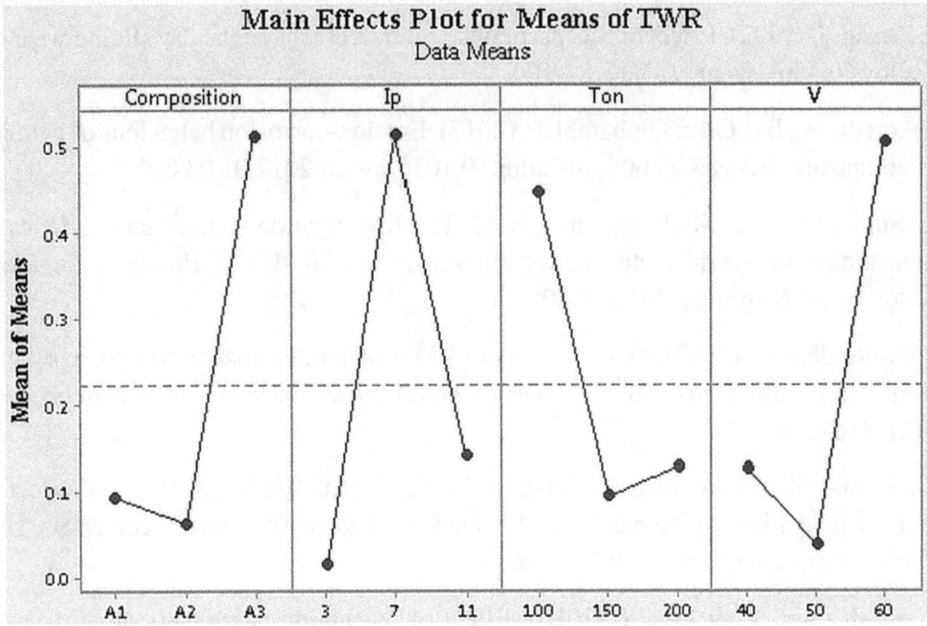

CONCLUSION

In this work, the investigation of the tribo-mechanical properties of the fabricated AA-2014 hybrid composite reinforced with SiC and Glass particulates have been conducted. Various mechanical tests have been performed to investigate the effects on the properties of the aluminum-2014 alloy on reinforcing the SiC and Glass particulates. SiC particles are widely used for the abrasives due to their high Hardness value, Glass particulates also provide good Hardness value, as well as they are, anti-corrosion and chemical resistant. Following conclusions are drawn from the experimental studies: The Compressive strength of aluminum alloy increases by reinforcing the SiC and glass particulates. However, the Impact energy decreases with the addition of reinforcements. Hardness is found to increase with the inclusion of reinforcement, A2 composition of the composite has a higher BHN value. The optimum machinability condition for the EDM machining of AA-2014 composite is found at A2 composition, pulse current 11A, Ton 100, voltage 50. Pulse current is found to be the most influencing parameter. MRR and accuracy in terms of ROC, increase with the increase in the pulse current. However, the TWR and ROC decrease with decrease in Pulse on time with the selected parametric range. composition A1 has a minimum wear rate, this may be due to the effect of SiC particles as the volume fraction of SiC in the A1 composition is more. However, at the low load and speed leads to the lower wear rate.

REFERENCES

Alpas, A. & Zhang, J. (1992). Effect of SiC particulate reinforcement on the dry sliding wear of aluminium-silicon alloys (a356). *Wear*, *155*(1), 83-104.

Aribo, S., Fakorede, A., Ige, O., & Olubambi, P. (2017). Erosion-corrosion behaviour of aluminum alloy 6063 hybrid composite. *Wear*, *376*, 608–614. doi:10.1016/j.wear.2017.01.034

Babu, T. M., Sugin, M. A., & Muthukrishnan, N. (2012). Investigation on the characteristics of surface quality on machining of hybrid metal matrix composite (Al-SiC-B4C). *Procedia Engineering*, *38*, 2617–2624. doi:10.1016/j.proeng.2012.06.308

Behera, R., & Sutradhar, G. (2012). Machinability of LM6/SiCp metal matrix composites with tungsten carbide cutting tool inserts. *Journal of Engineering and Applied Sciences (Asian Research Publishing Network)*, *7*(2), 216–221.

Benham, A., Kumar, K. A., Vettivel, S., Pandian, M. S., & Paul, C. G. (2018). The effect of matrix treatment and SiCp addition on the mechanical behaviour of glass fibre reinforced epoxy composites. *Materials Today: Proceedings*, *5*(2), 7029–7036.

Bhushan, R. K., Kumar, S., & Das, S. (2010). Effect of machining parameters on surface roughness and tool wear for 7075 al alloy SiC composite. *The International Journal of Advanced Manufacturing Technology*, *50*(5-8), 459-469.

Bunma, M., Subarporn, P., Bobuangern, R., & Patthannkitdamrong, T., Thuan- wong, T., and Patcharawit, T. (2011). Process parameter-microstructure- mechanical property relations of SiCp-reinforced aluminum composites produced by powder-injection casting. *Journal of Metals, Materials, and Minerals*, *21*(2), 85–93.

Chawla, N., Deng, X., & Schnell, D. (2006). Thermal expansion anisotropy in extruded SiC particle reinforced 2080 aluminum alloy matrix composites. *Materials Science and Engineering A*, *426*(1-2), 314–322. doi:10.1016/j.msea.2006.04.054

Chen, P., & Hoshi, T. (1992). High-performance machining of SiC whisker- reinforced aluminium composite by self-propelled rotary tools. *CIRP Annals*, *41*(1), 59–62. doi:10.1016/S0007-8506(07)61152-4

Dante, R. C. (2015). *Handbook of friction materials and their applications*. Woodhead Publishing.

Davim, J. P. (2002). Diamond tool performance in machining metal-matrix composites. *Journal of Materials Processing Technology*, *128*(1-3), 100–105. doi:10.1016/S0924-0136(02)00431-4

Davim, J. P., & Antonio, C. C. (2001). Optimisation of cutting conditions in the machining of aluminium matrix composites using a numerical and experimental model. *Journal of Materials Processing Technology*, *112*(1), 78–82. doi:10.1016/S0924-0136(01)00551-9

Deuis, R., Subramanian, C., & Yellup, J. (1997). Dry sliding wear of aluminium composites—A review. *Composites Science and Technology*, *57*(4), 415–435. doi:10.1016/S0266-3538(96)00167-4

Dwivedi, S., Kumar, S., & Kumar, A. (2012). Effect of turning parameters on surface roughness of a356/5% SiC composite produced by electromagnetic stir casting. *Journal of Mechanical Science and Technology*, *26*(12), 3973–3979. doi:10.100712206-012-0914-5

El-Gallab, M., & Sklad, M. (1998). Machining of Al/SiC particulate metal matrix composites: Part ii: Workpiece surface integrity. *Journal of Materials Processing Technology*, *83*(1-3), 277–285. doi:10.1016/S0924-0136(98)00072-7

Forquin, P., Tran, L., Louvigne, P.-F., Rota, L., & Hild, F. (2003). Effect of aluminum reinforcement on the dynamic fragmentation of SiC ceramics. *International Journal of Impact Engineering*, *28*(10), 1061–1076. doi:10.1016/S0734-743X(03)00034-4

Gangil, M., & Pradhan, M. K. (2016). Optimization the machining parameters by using VIKOR Method during the EDM process of Titanium alloy.

Gangil, M., Pradhan, M. K., & Purohit, R. (2017). Review on modelling and optimization of electrical discharge machining process using modern Techniques. *Materials Today: Proceedings*, *4*(2), 2048–2057.

Gao, H., Liu, X., Zhang, S., & Qi, J. (2018). Synergistic effect of glass fibre and al powder on the mechanical properties of glass-ceramics. *Ceramics International*, *44*(13), 15167–15175. doi:10.1016/j.ceramint.2018.05.155

Helu, M., Vijayaraghavan, A., & Dornfeld, D. (2011). Evaluating the relationship between use phase environmental impacts and manufacturing process precision. *CIRP Annals*, *60*(1), 49–52. doi:10.1016/j.cirp.2011.03.020

Hosking, F., Portillo, F. F., Wunderlin, R., & Mehrabian, R. (1982). Composites of aluminium alloys: Fabrication and wear behaviour. *Journal of Materials Science*, *17*(2), 477–498. doi:10.1007/BF00591483

James, S. J., Venkatesan, K., Kuppan, P., & Ramanujam, R. (2014). Hybrid aluminium metal matrix composite reinforced with SiC and TiB2. *Procedia Engineering*, *97*, 1018–1026. doi:10.1016/j.proeng.2014.12.379

Janic, M., & Reggiani, A. (2002). An application of the multiple criteria decision making (mcdm) analysis to the selection of a new hub airport. *European Journal of Transport and Infrastructure Research*, *2*(2).

Kilickap, E., Cakir, O., Aksoy, M., & Inan, A. (2005). Study of tool wear and surface roughness in machining of homogenised SiC-p reinforced aluminium metal matrix composite. *Journal of Materials Processing Technology*, *164*, 862–867. doi:10.1016/j.jmatprotec.2005.02.109

Kumar, D., Loganathan, C., & Narayanasamy, R. (2011). Effect of glass in an aluminum matrix on workability and strain hardening behavior of powder metallurgy composite. *Materials & Design*, *32*(4), 2413–2422. doi:10.1016/j.matdes.2010.12.008

Kumar, T. S., Shivashankar, G., Dhotey, K., & Singh, J. (2017). Experimental study wear rate of glass fibre reinforced epoxy polymer composites filled with aluminium powder. *Materials Today: Proceedings*, *4*(10), 10764–10768.

Larsen-Basse, J. (1990). Role of microstructure and mechanical properties in abrasion. *Scripta Metallurgica et Materialia*, *24*(5), 821–826. doi:10.1016/0956-716X(90)90119-2

Lim, S., Gupta, M., Ren, L., & Kwok, J. (1999). The tribological properties of Al-Cu/SiCp metal-matrix composites fabricated using the rheocasting technique. *Journal of Materials Processing Technology*, *89*, 591–596. doi:10.1016/S0924-0136(99)00067-9

Madhukumar, U. (2018). Characterization of glass particulate reinforced aluminium alloy6061 metal matrix composites. *Materials Today: Proceedings*, *5*(8), 7604–7608.

Manikonda, R. D., Kosaraju, S., Raj, K. A., & Sateesh, N. (2018). Wear behavior analysis of silica carbide-based aluminum metal matrix composites. *Materials Today: Proceedings*, *5*(9), 20104–20109.

Patil, N. A., Mulik, S. S., Wangikar, K. S., & Kulkarni, A. P. (2018). Characterization of glass laminate aluminium reinforced epoxy-a review. *Procedia Manufacturing*, *20*, 554–562. doi:10.1016/j.promfg.2018.02.083

Pradhan, M. K. (2010). *Experimental investigation and modelling of surface integrity, accuracy and productivity aspects in EDM of AISI D2 steel* [Doctoral dissertation].

Pradhan, M. K. (2012). *Process simulation, modelling, and estimation of the temperature and residual stresses electrical discharge machining of AISI D2 steel*. ISCI.

Pradhan, M. K. (2013). Estimating the effect of process parameters on MRR, TWR and radial overcut of EDMed AISI D2 tool steel by RSM and GRA coupled with PCA. *International Journal of Advanced Manufacturing Technology*, *68*(1-4), 591–605. doi:10.100700170-013-4780-9

Pradhan, M. K. (2018). Optimization of EDM process for MRR, TWR and radial overcut of D2 steel: A hybrid RSM-GRA and entropy weight-based TOPSIS approach. *International Journal of Industrial and Systems Engineering*, *29*(3), 273–302. doi:10.1504/IJISE.2018.093047

Pradhan, M. K., Gangil, M., & Kumar, P. R. (2016, December). Application of MCDM for Optimization of EDM of LM6 Silicon Carbide Boron Carbide Hybrid Composite. *Proceedings of the 6th International and 27th All India Manufacturing, Technology, Design, and Research conference (AIMTDR 2016)* (pp. 557-562). Academic Press.

Pradhan, M. K., & Singh, B. (2019). *Machinability and Multi-response Optimization of EDM of Al7075/SIC/WS2 Hybrid Composite Using the PROMETHEE Method. In Optimization for Engineering Problems* (pp. 39–75). John Wiley and Sons.

Pradhan, M. K., & Tiwari, A. (2018). Modeling and Optimization of End Milling Parameters on Ti-6Al-4V Titanium Alloy using TLBO Algorithm and TOPSIS Algorithm. *International Journal of Machining and Machinability of Materials*, *20*(6), 513–535. doi:10.1504/IJMMM.2018.096379

Ravindran, P., Manisekar, K., Rathika, P., & Narayanasamy, P. (2013). Tribological properties of powder metallurgy-processed aluminium self-lubricating hybrid composites with SiC additions. *Materials & Design*, *45*, 561570.

Rohatgi, P. (1992). Friction and wear of metal-matrix composites. In ASM Handbook (Vol. 18, p. 801). ASM.

Saaty, T. L. (2008). Decision making with the analytic hierarchy process. *International Journal of Services Sciences*, *1*(1), 83–98. doi:10.1504/IJSSCI.2008.017590

Sahin, Y., & Murphy, S. (1996). Wear performance of aluminium alloy composites containing unidirectionally-oriented silicon carbide coated boron fibres. *Wear*, *197*(1-2), 248–254. doi:10.1016/0043-1648(96)06959-1

Sathish, S., & Rathnaraj, J. D. (2015). Experimental studies of the mechanical and microstructure properties of lm6 metal matrix composite with borosilicate glass reinforcement. *International Journal of Applied Engineering Research*, *10*(11), 10063–10069.

Sharma, S., Satish, B., Girish, B., Kamath, R., & Asanuma, H. (1998). Dry sliding wear of short glass fibre reinforced zinc-aluminum composites. *Tribology International*, *31*(4), 183–188. doi:10.1016/S0301-679X(98)00020-6

Srivyas, P. D., & Charoo, M. (2018). Role of reinforcements on the mechanical and tribological behavior of aluminum metal matrix composites-a review. *Materials Today: Proceedings*, *5*(9), 20041–20053.

Umanath, K., Palanikumar, K., & Selvamani, S. (2013). Analysis of dry sliding wear behaviour of Al6061/SiC/Al2O3 hybrid metal matrix composites. *Composites. Part B, Engineering*, *53*, 159–168. doi:10.1016/j.compositesb.2013.04.051

Vlot, A., & Gunnink, J. W. (2011). *Fibre metal laminates: an introduction*. Springer Science & Business Media.

Vogelesang, L. B., & Vlot, A. (2000). Development of fibre metal laminates for advanced aerospace structures. *Journal of Materials Processing Technology*, *103*(1), 1–5. doi:10.1016/S0924-0136(00)00411-8

Wilson, S., & Alpas, A. (1997). Wear mechanism maps for metal matrix composites. *Wear*, *212*(1), 41–49. doi:10.1016/S0043-1648(97)00142-7

Zamzam, M. (1991). Abrasive wear of aluminum-matrix composites. *Metal*, *45*(3), 250–254.

Chapter 12
Temperature Gradient-Based Laser Bending of Full Plates and Plates With Cutout

Paramasivan Kalvettukaran
https://orcid.org/0000-0003-0851-5730
Jadavpur University, India

Sandip Das
Jadavpur University, India

Sundar Marimuthu
The Manufacturing Technology Centre, UK

Dipten Misra
Jadavpur University, India

ABSTRACT

The laser bending process, also called the laser forming process, consists of irradiating the surface of a sheet or a plate by means of a moving laser beam with a predefined scanning strategy to generate the desired shape through thermally induced residual stress. This chapter presents the mechanisms of a laser bending process and the technological aspects concerning laser v-bending of rectangular AISI 304 plates for full plates and plates with a central cutout at its middle to highlight the process fundamentals and how processing affects the final bending angle of the workpieces. Laser bending involving plates with a cutout will have numerous applications for car bodies, such as front and rear panels where bending is required to be performed on panels with cutout geometries. To investigate the effects of shape and size of the cutout on temperature distribution, stress distribution, and final bending angle, different shapes such as circular, ellipse, rectangular, and square, as well as different dimensions of cutouts have been chosen.

DOI: 10.4018/978-1-7998-1831-1.ch012

INTRODUCTION

In recent years, laser beam forming (LBF) has emerged as a new and promising technique in diverse industries, such as aerospace, automotive, microelectronics and shipbuilding. The reasons for the ever-increasing applications of laser forming over conventional techniques are some typical advantages of the process. Specifically, this method requires no mechanical contact and hence is noise free from noise, vibration and frequent repair and maintenance of the system. It offers a versatile and flexible manufacturing process by way of generating different complex shapes (Hennige, 2000; Gollo et al., 2015) (e.g., sinusoidal, dish-shape, saddle, pillow, etc.) and micro-bending for precise adjustment (Dirscherl et al., 2006; Hu et al., 2011), without using special tools and dies, with a controllable moving beam. In addition, laser forming has attracted considerable attention over flame bending since the former can produce a steep thermal gradient in thin materials with high reflectivity and high thermal conductivity. Moreover, material degradation, caused by laser forming, is less than that by flame heating, as the laser beam produces a smaller and narrower heat affected zone (HAZ).

In laser bending process, the laser beam is traversed across the sheet surface; the scanning path depends on the target shape. In the simplest case, the beam delivery may be at a single point. For V-bending, it may follow a straight line across the whole part as shown in Figure 1. For complex shapes, the scanning path may involve combination of radial, circular, archimedean or other trajectories (Venkadeshwaran et al., 2010; Safari et al., 2015). Moreover, during the scanning process, the beam power, the beam diameter and the scanning speed may need to be varied as per the process requirement. Multipass scanning following the same path may be used to achieve large bending (Kant & Joshi, 2015). The workpiece thickness is the most important parameter as it directly controls the temperature gradient along the thickness direction (Dixit et al., 2015). The increase in the thickness leads to the change in the bending mechanism from the one to another. The bending angle is approximately inversely proportional to the square of the workpiece thickness. Moreover, the microstructures of laser irradiated material varies depends on the plate thickness. The grain size is smaller for smaller strip thickness since the temperature of the top surface is higher which implies that thinner the strip, the harder becomes the laser irradiated surface (Fetene et al., 2017).

Different mechanisms of laser bending processes are postulated. These are: Temperature gradient mechanism (TGM), Buckling mechanism (BM) and upsetting mechanism (UM) (Geiger and Vollertsen, 1993). The temperature gradient mechanism is applicable for thick plates with low thermal conductivity, high scanning speed and small beam diameter, setting up a high temperature gradient across the thickness of the plate. For a thin plate with a high thermal conductivity, when a laser beam of large diameter with a low scanning speed traverses the plate, the heated region throughout the plate thickness assumes a more or less uniform temperature and the plate is subjected to a compressive stress. When this compressive stress exceeds the critical load for buckling, it buckles, particularly if there is a certain amount of pre-stress in the plate. Moreover, if the stress in the plate, exceeds the yield limit, a permanent deformation will set in the plate (Marya & Edwards, 2000; Vollertsen et al., 1995). On the other hand, for a thick plate with a high thermal conductivity, when it is subjected to a laser beam of large diameter with a low scanning speed, a large amount of in plane compressive force builds up, causing a stress level exceeding the yield point and this may cause a shortening of the plate with a local bulging of the plate (Vollertsen, 1998). Moreover, laser irradiation of blind hole bottoms can lead to bending away from the laser beam, since the stresses caused by upsetting deformation effects are induced in the sheet metal below the neutral axis (Woizeschke et al., 2019). All these mechanisms are discussed in detail in the

Figure 1. Schematic diagram of laser forming process

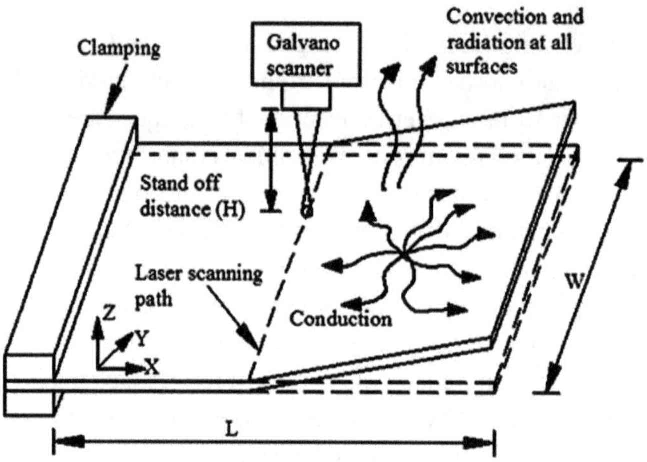

proposed book chapter. Moreover, the TGM being the most prevalent mechanism for laser bending, a numerical study involving the evolution of stress and deformation in the plate during the laser bending process is planned to be presented in the proposed book chapter.

In present research, laser bending of AISI 304 is studied through experimental investigation and numerical modeling. First, the mechanism of laser bending process is presented which includes the evolution of stress and deformation in the plate during the laser bending process. Then the experimental investigations on V-bending of rectangular AISI 304 plates for full plates are presented. Finally, the Laser bending involving plates with different size and shape cutouts and the developed mathematical models using response surface methodology are presented.

MECHANISM OF LASER BENDING PROCESS

Finite Element Simulation of the Laser Forming Process

A finite element modeling for the laser forming process of AISI 304 steel is carried out to understand the mechanism of laser forming process. AISI 304 sheet of $100 \times 50 \times 1.5$ mm^3 is used to study the laser bending process. Simply supported boundary conditions are applied at the fixed end of the plate, and the other end is kept free. The laser beam scanning path passes through the middle of the plate along the y axis as shown in Figure 2. COMSOL MULTIPHYSICS® has been used to compute the temperature distribution and deformation of sheet metal.

The input parameters used are, laser power 300 W, scanning speed 10 mm/s, spot diameter 2 mm, and plate thickness 1.5 mm (along z-axis). Figure 2 shows the schematic view of the geometry of workpiece and coordinate system. Initial temperature of the plate is considered as 300 K. To study the process of laser bending, two cross-sectional views of the workpiece are taken, one along the X-Z plane, across the scanning path, and another along the Y-Z plane, passing through the scanning path. The cut-sections are shown in Figures 3(a-b). The temperature, stress and displacement distributions are captured (a) from the front view on the X-Z plane passing through the centre of the plate width, and (b) from the right side

Figure 2. Schematic view of the geometry of the workpiece and coordinate system

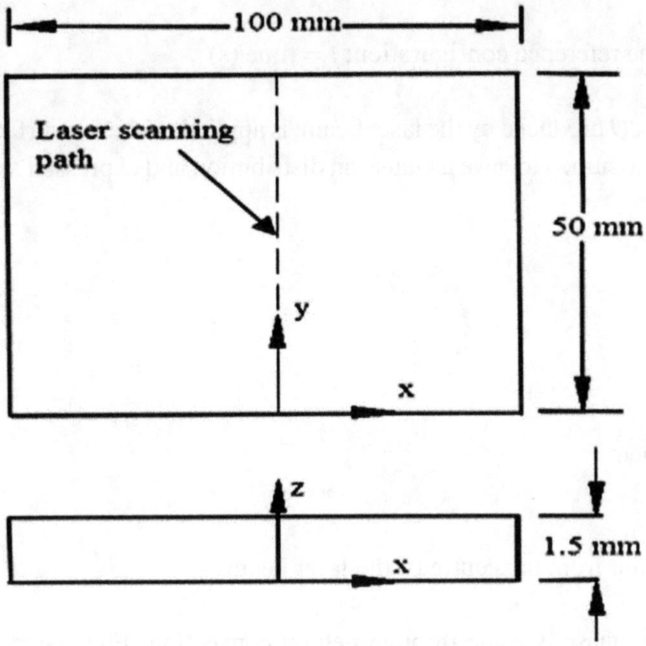

view on the Y-Z plane which crossing the centre of the plate length. The distributions are monitored on both cross sections during the period of laser beam scanning and during the cooling phase as well.

In the present work, thermal linear elastic material model is used which combines linear elastic material with thermal expansion. During laser forming, the transient temperature field is calculated based on the mechanism of the heat conduction. The governing equation for heat conduction (Eq. 1) within the specimen can be written as follows:

$$\rho c_p \frac{\partial T(r,t)}{\partial t} = k_{xx}\frac{\partial^2 T}{\partial x^2} + k_{yy}\frac{\partial^2 T}{\partial y^2} + k_{zz}\frac{\partial^2 T}{\partial z^2} \quad (1)$$

where:

ρ = material density (kg/m^3)
c_p = specific heat (J/kg.K)

Figure 3. Cut-sections for monitoring of temperature, stress and displacement distributions

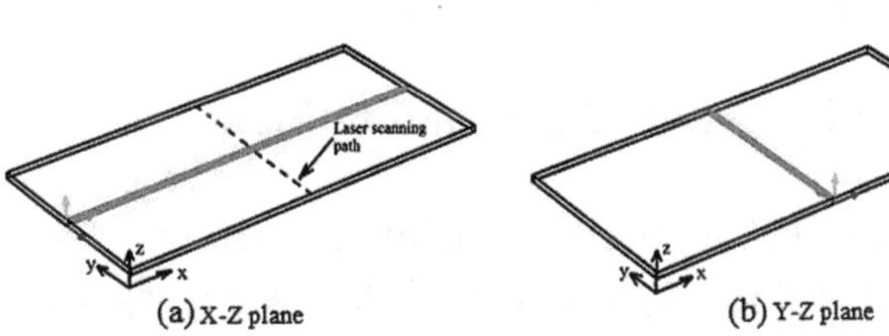

k_{xx}, k_{yy} and k_{zz} = thermal conductivities (W/m.K)
$T(r,t)$ = temperature (K)
r = coordinate (m) in the reference configuration; t = time (s)

The moving heat flux Q produced by the laser beam, is applied on the top surface of the sheet metal. In this work, laser beam is assumed to have a Gaussian distribution and expressed as follows in Equation 2:

$$Q = \frac{2\mathrm{AP}}{\grave{\mathrm{A}}\mathrm{R}^2}\exp\left(-\frac{2r^2}{R^2}\right) \tag{2}$$

Where:

A = absorption coefficient
P = laser power (W)
R = laser beam radius (m)
r = distance (m) of a point from the centre of the laser beam.

The material cooling phase is made through natural convection (Eq. 3) and radiation (Eq. 4) from its surfaces exposed to ambient air. The convection boundary condition can be expressed as follows:

$$q_{conv} = h\left(T_s - T_o\right) \tag{3}$$

where:

h = heat transfer coefficient (W/m^{2}K)
T_s = sheet metal surface temperature (K)
T_0 = ambient temperature (K)

The radiation boundary condition can be expressed as follows:

$$q_{rad} = \varepsilon\sigma_{bol}\left(T_{s^4} - T_{o^4}\right) \tag{4}$$

where:

ε = emissivity
σ_{bol} = Stefan Boltzmann constant (5.6703×10^{-8} W/m^{2}K^4)

Transient Temperature, Stress and Deformation Fields of a Rectangular Plate

In this section, to get a better insight of the mechanism of bending process, the temperature, stress distributions and displacement fields on the X-Z plane at y= 25 (half-way along the scan line) and Y-Z plane at x=50 (along the scan line) are presented at different time intervals covering Figure 4 through Figure 15. The plate behaviour during the laser bending process has been studied during three vital stages: (i) at the early stage of heating, (ii) the intermediate portion of the plate, (iii) the farthest edge of the plate and during the cooling phase. The monitored distributions on the planes are discussed hereunder. The displacement of nodes lying on the X-Z and Y-Z plane are presented with arrow plot (representing vector quantity) to understand the plate deformation direction and plate thickening.

Figures 4 (a-d) show the temperature, stress and displacement distribution on the X-Z plane when the laser beam is fully inside the plate with the beam centre having a coordinate of (50, 1, 1.5) (at t = 0.7 s). It can be seen from Figure 4 (a) that there is no change in temperature; however, the top surface of X-Z plane attains a small amount of tensile stress due to counter bending as can be seen from Figure 4 (b). Figures 5 (b) and (c) show the x and y-components of stress distribution on the Y-Z plane. The

Figure 4. Temperature distribution, stress distribution and displacement field on the X-Z plane (t = 0.7s)

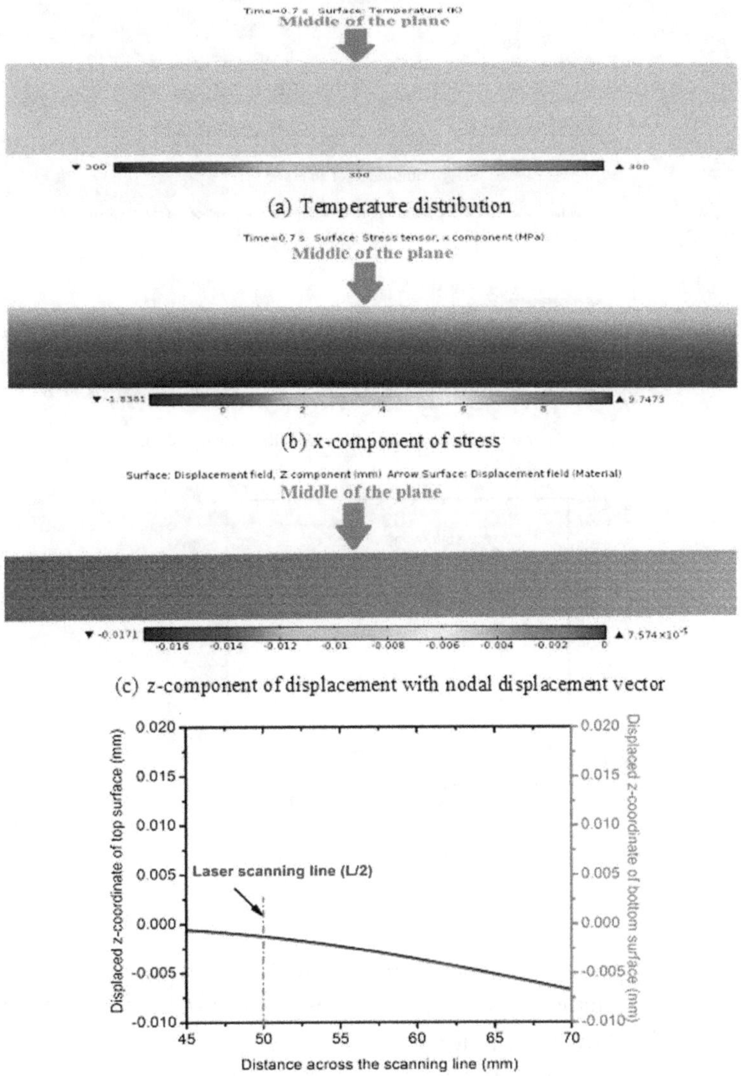

(a) Temperature distribution

(b) x-component of stress

(c) z-component of displacement with nodal displacement vector

(d) Displacement across the scanning path

Figure 5. Temperature distribution, stress distribution and displacement field on the Y-Z plane (t = 0.7s)

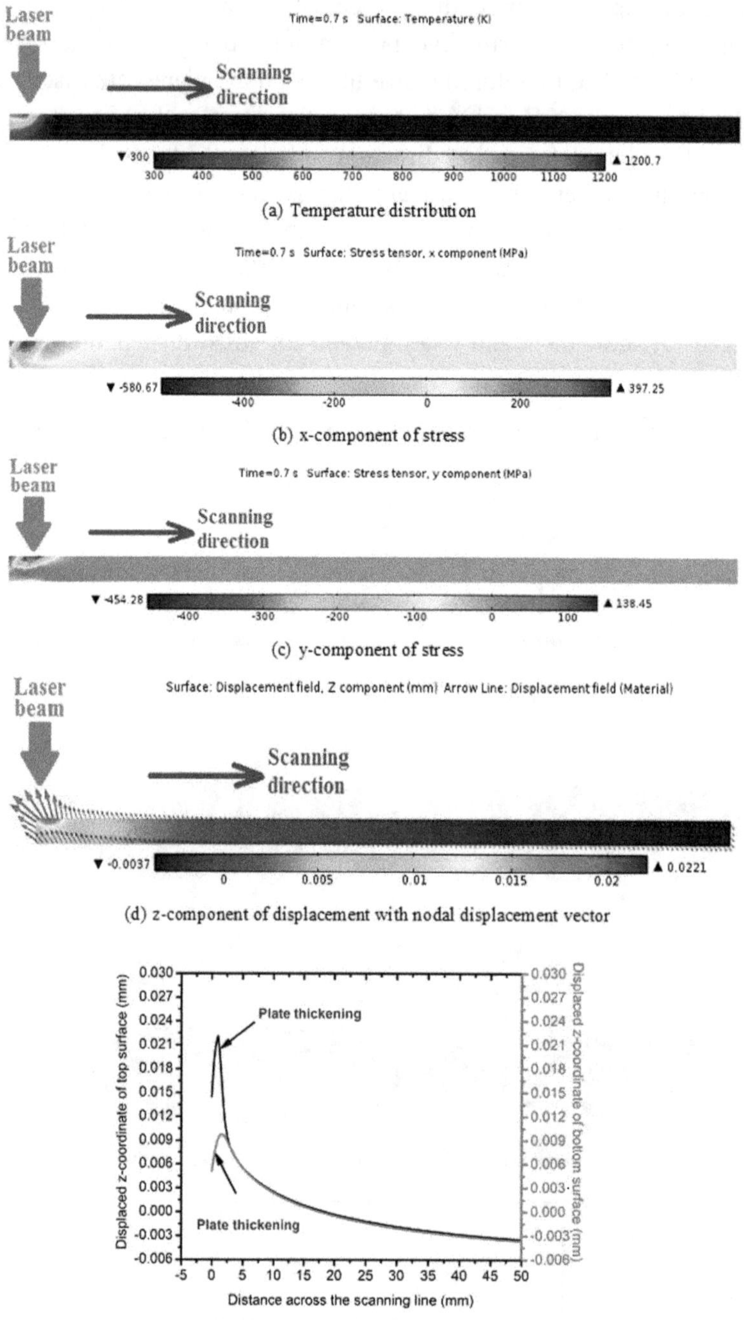

positive stress values representing the red region indicate tensile stresses, and the negative stress values represented by the blue region indicate compressive stresses. The bottom surface of X-Z plane develops a compressive stress. As shown in Figure 4 (c), at this time, the plate experiences a displacement towards +ve x-direction and downward direction of z-axis as well. In Figure 4 (d), x axis indicates the

Figure 6. Temperature distribution, stress distribution and displacement field on the X-Z plane (t = 0.8s)

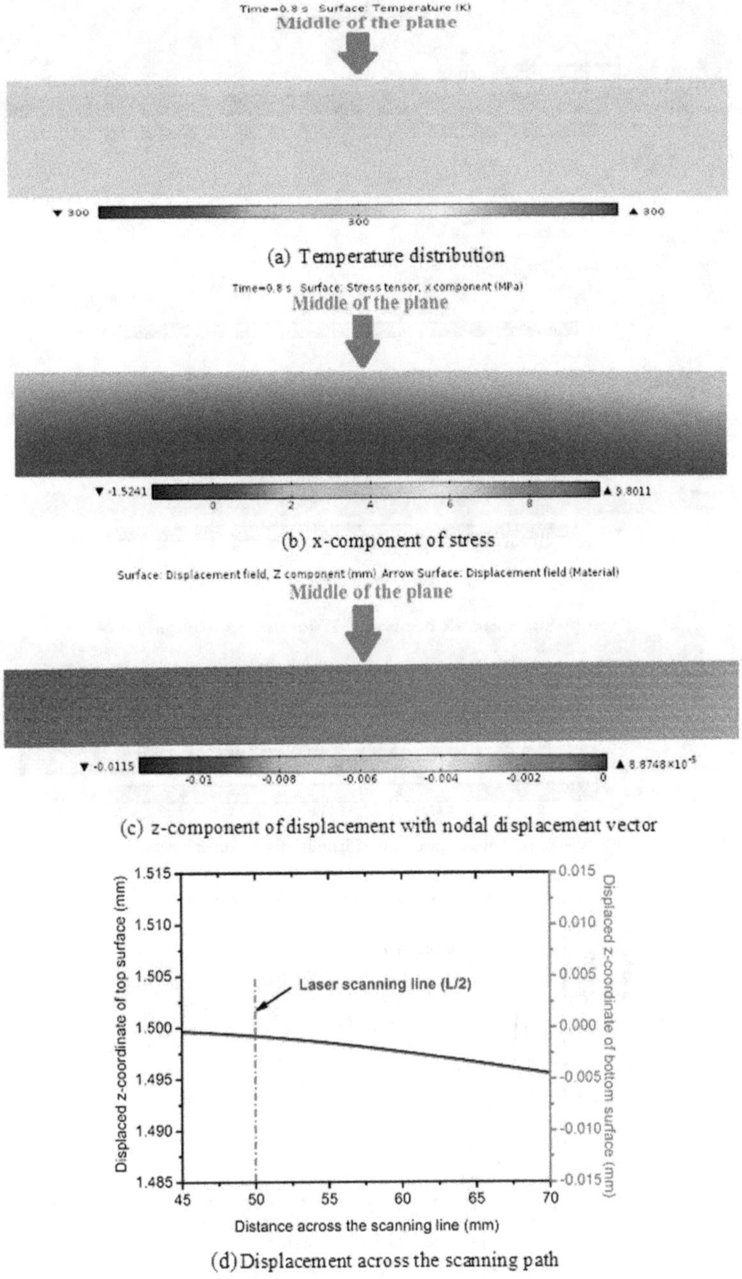

(a) Temperature distribution

(b) x-component of stress

(c) z-component of displacement with nodal displacement vector

(d) Displacement across the scanning path

distance on the X-Z plane, left side y axis indicates the z-displacement at the top surface of X-Z plane from initial state and right side y axis indicates the z-displacement at the bottom surface of X-Z plane from initial state. The magnitude of counter-bending (about 6.7 μm) involved at this time can be seen from Figure 4 (d).

The temperature distribution, x and y-components of stress distribution, z-component of displacement and displacement vector of nodes lying on the Y-Z plane along the scanning path when the centre of the

Figure 7. Temperature distribution, stress distribution and displacement field on the Y-Z plane (t = 0.8s)

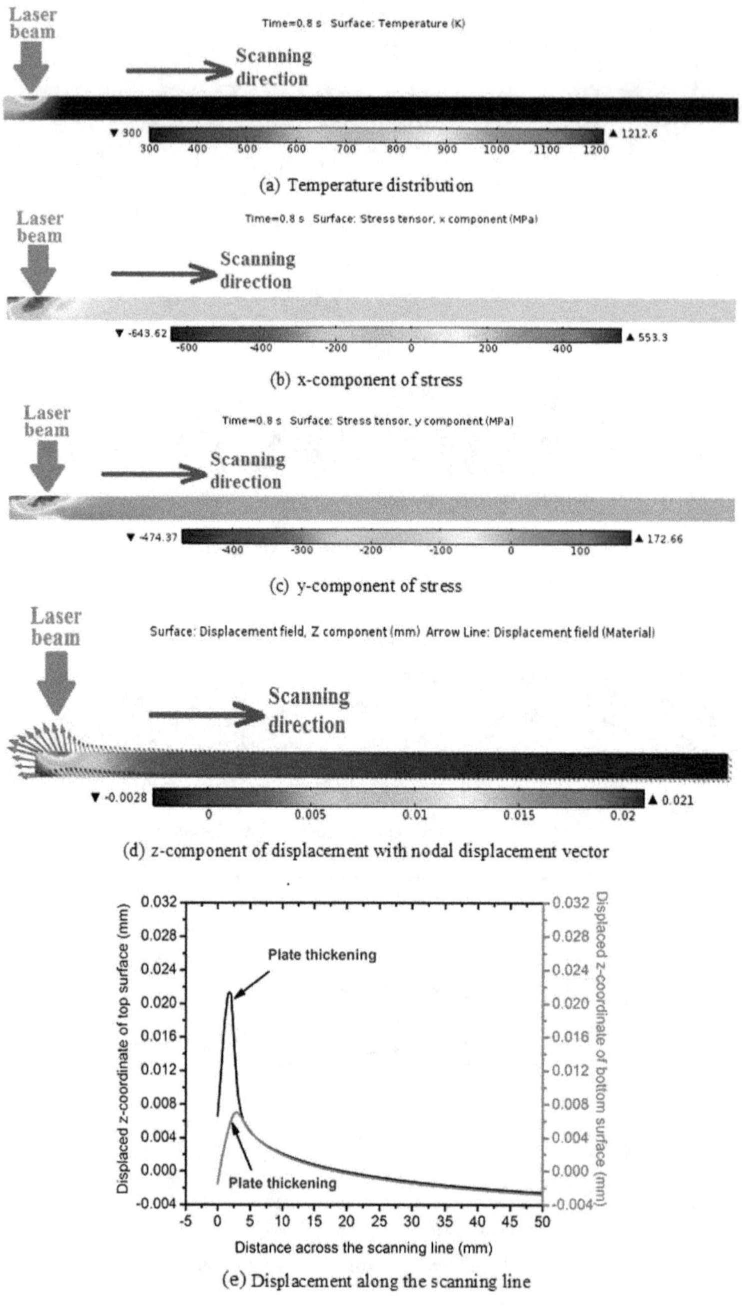

(a) Temperature distribution

(b) x-component of stress

(c) y-component of stress

(d) z-component of displacement with nodal displacement vector

(e) Displacement along the scanning line

laser beam occupies a coordinate of (50, 1, 1.5) (at t = 0.7 s) are shown in Figures 5 (a-d). As can be seen in Figure 5 (a) that the temperature at the top heated surface increases further and it is propagating to the surrounding zone through conduction. Figures 5 (d) and (e) show the z-displacement of the nodes on the Y-Z plane. It is indicating that due to further rise in temperature and stresses at the heated and surrounding region, the plate attains further bulging at the heated region and plate displacement occurs

Figure 8. Temperature distribution, stress distribution and displacement field on the X-Z plane (t = 2.98s)

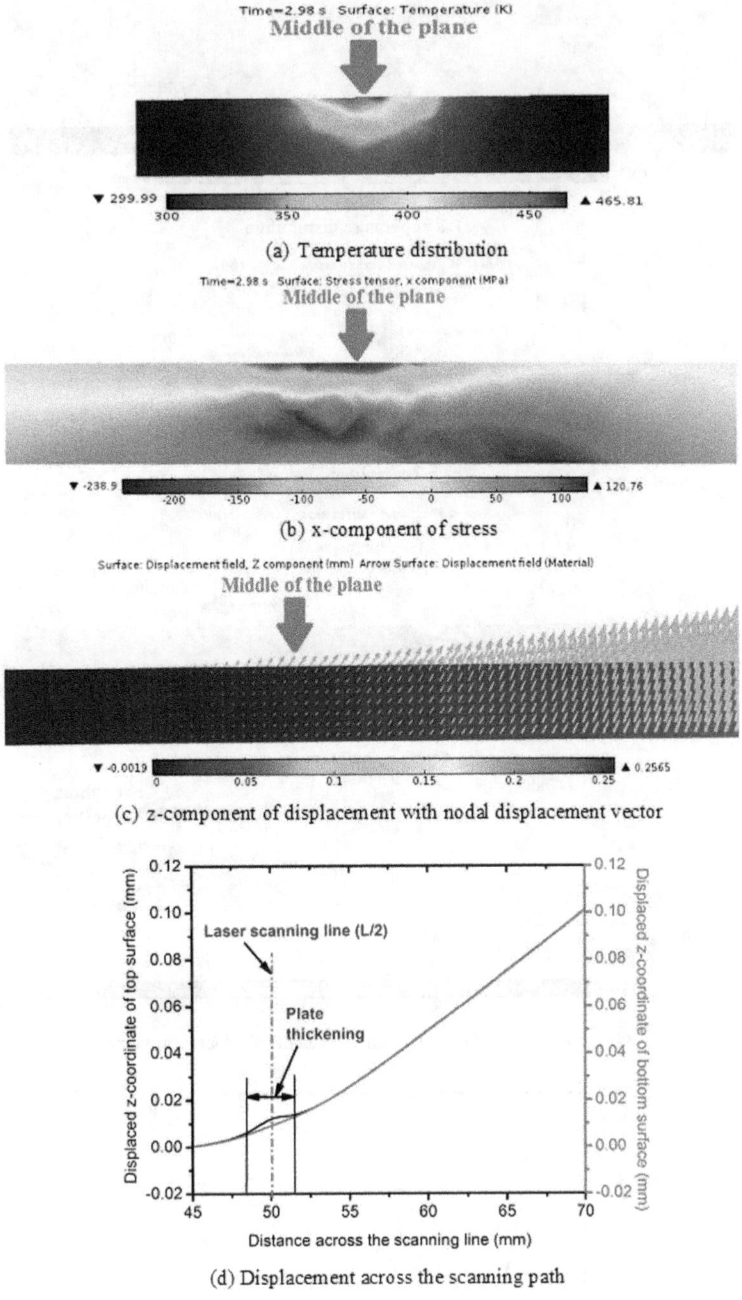

towards to the negative direction of laser irradiation (-ve Y axis). In Figure 5 (e), x axis of this graph indicates the distance on the Y-Z plane and y axis indicates the z-displacement at the top and bottom surface of Y-Z plane.

Figures 6 (a-d) show the temperature, stress and displacement distributions on the X-Z plane when the centre of the laser beam is at (50, 2, 1.5) (for t = 0.8 s). It can be seen from Figures 6 (a-c) that, there

Figure 9. Temperature distribution, stress distribution and displacement field on the Y-Z plane (t = 2.98s)

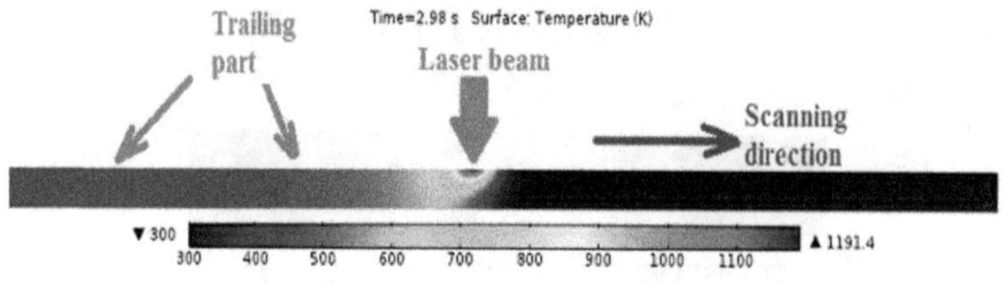

(a) Temperature distribution

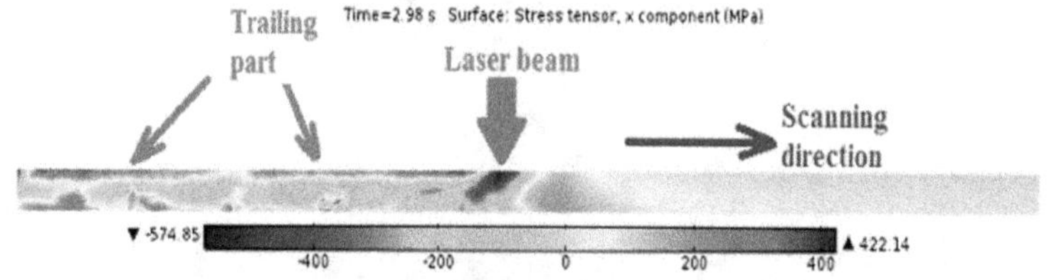

(b) x-component of stress

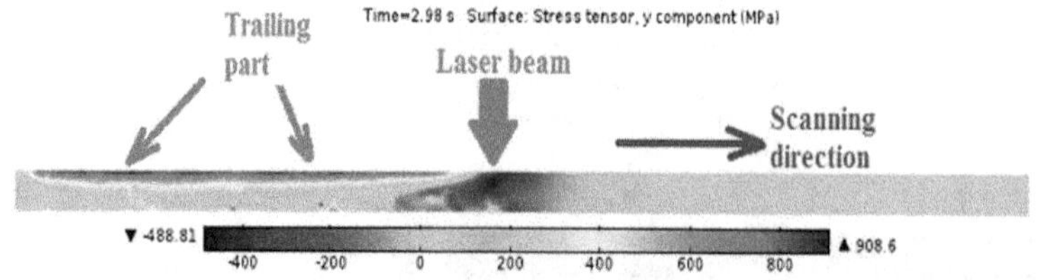

(c) y-component of stress

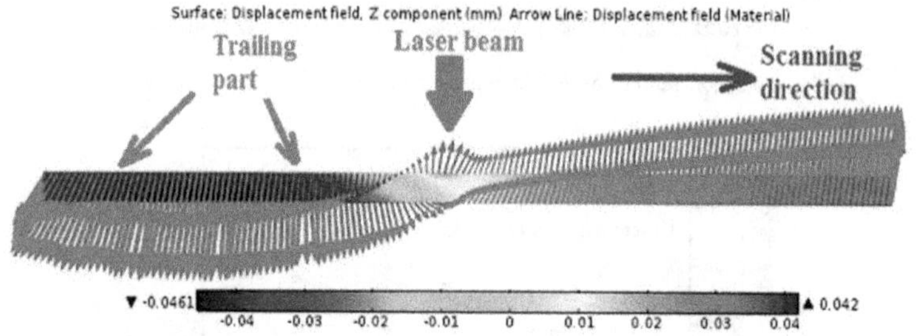

(d) z-component of displacement with nodal displacement vector

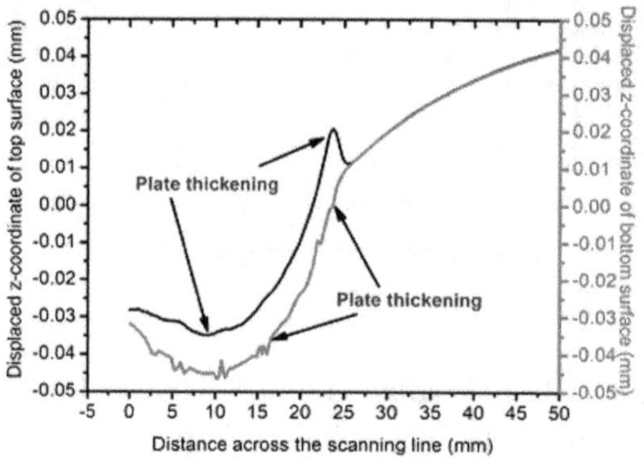

(e) Displacement along the scanning line

Figure 10. Temperature distribution, stress distribution and displacement field on the X-Z plane (t = 3.12s)

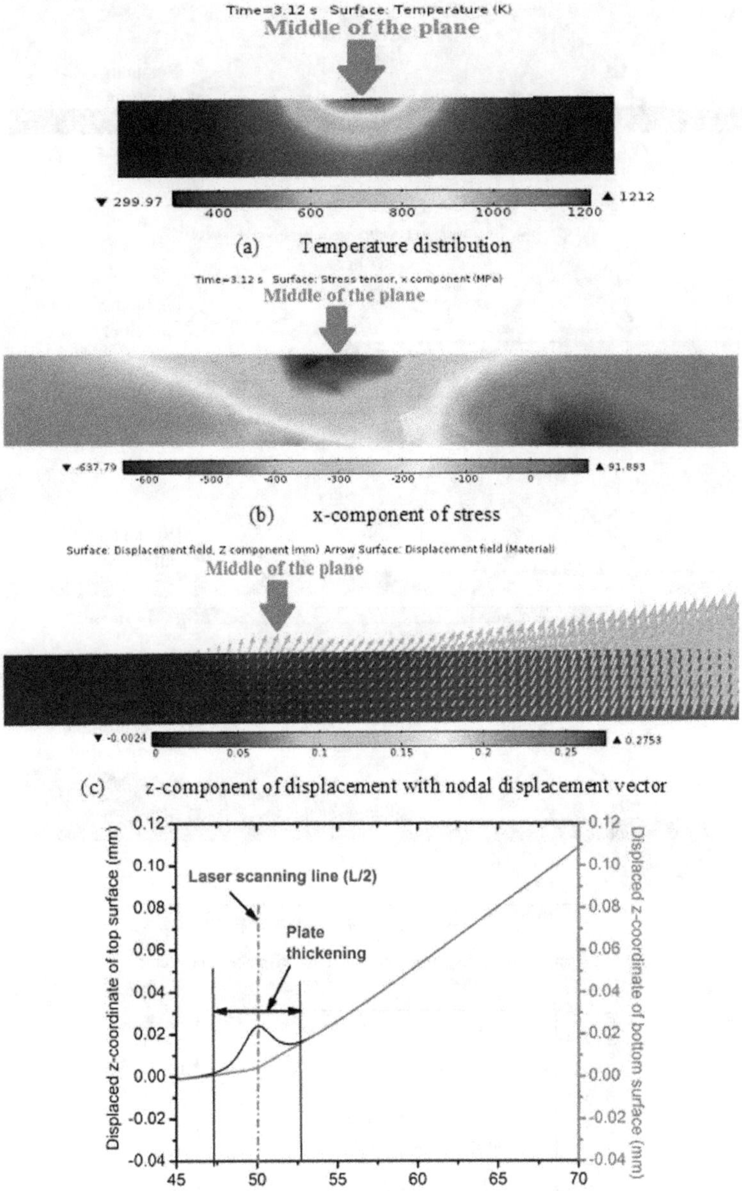

is not much change in temperature, stress and displacement on the X-Z plane compared to the time step at t = 0.7 s discussed above. However, it is noticed from figure 6 (d) that, the plate has started to bend towards to the laser beam. The temperature distribution, x and y-components of stress distribution, z-component of displacement and displacement vector of nodes situated on the Y-Z plane along the scanning path are shown in Figures 7 (a-d) at t = 0.8 s when the centre of the laser beam is at (50, 2, 1.5). It is observed from the results of numerical simulation that when the laser beam approaches the point,

Figure 11. Temperature distribution, stress distribution and displacement field on the Y-Z plane (t = 3.12s)

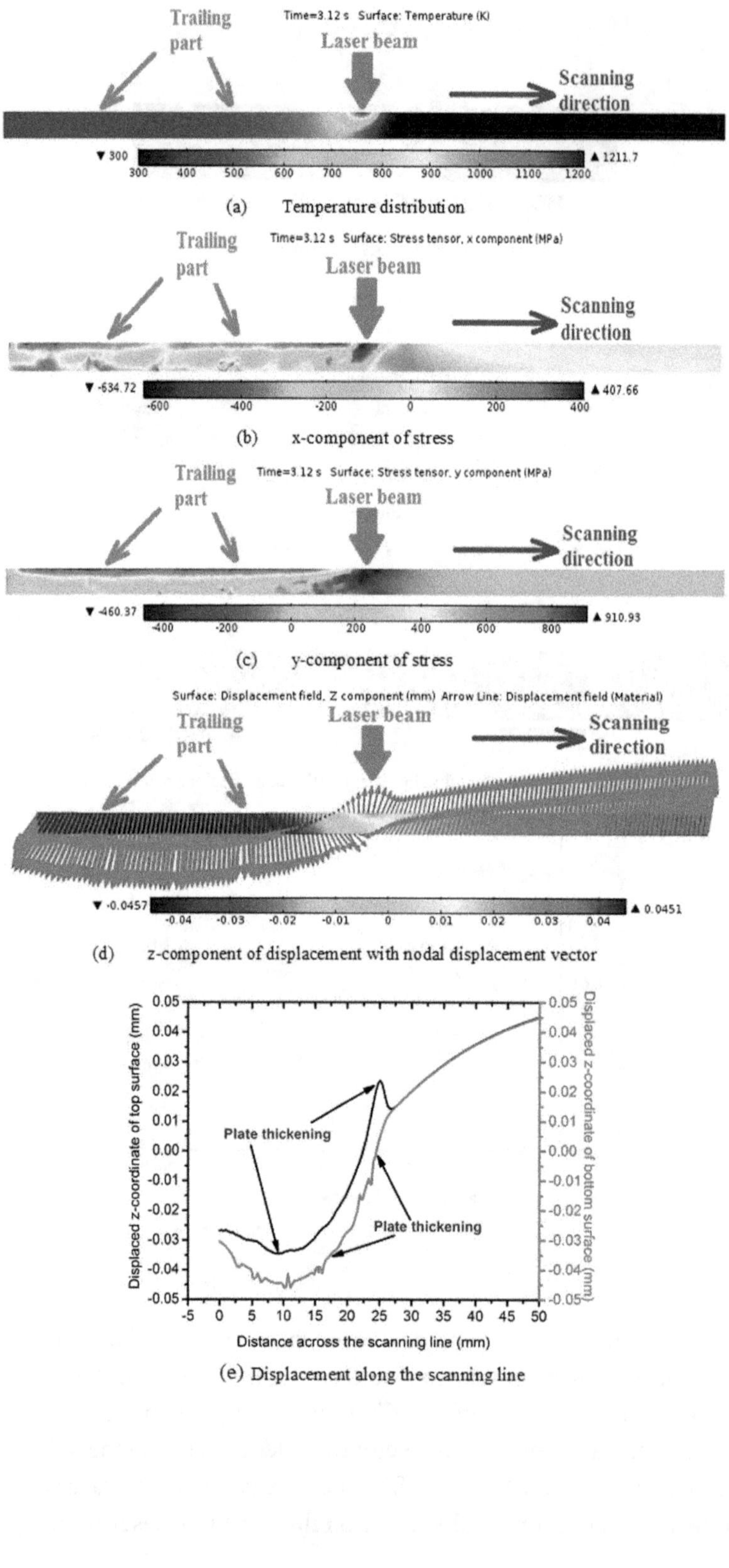

(a) Temperature distribution

(b) x-component of stress

(c) y-component of stress

(d) z-component of displacement with nodal displacement vector

(e) Displacement along the scanning line

Figure 12. Temperature distribution, stress distribution and displacement field on the X-Z plane (t = 5.6s)

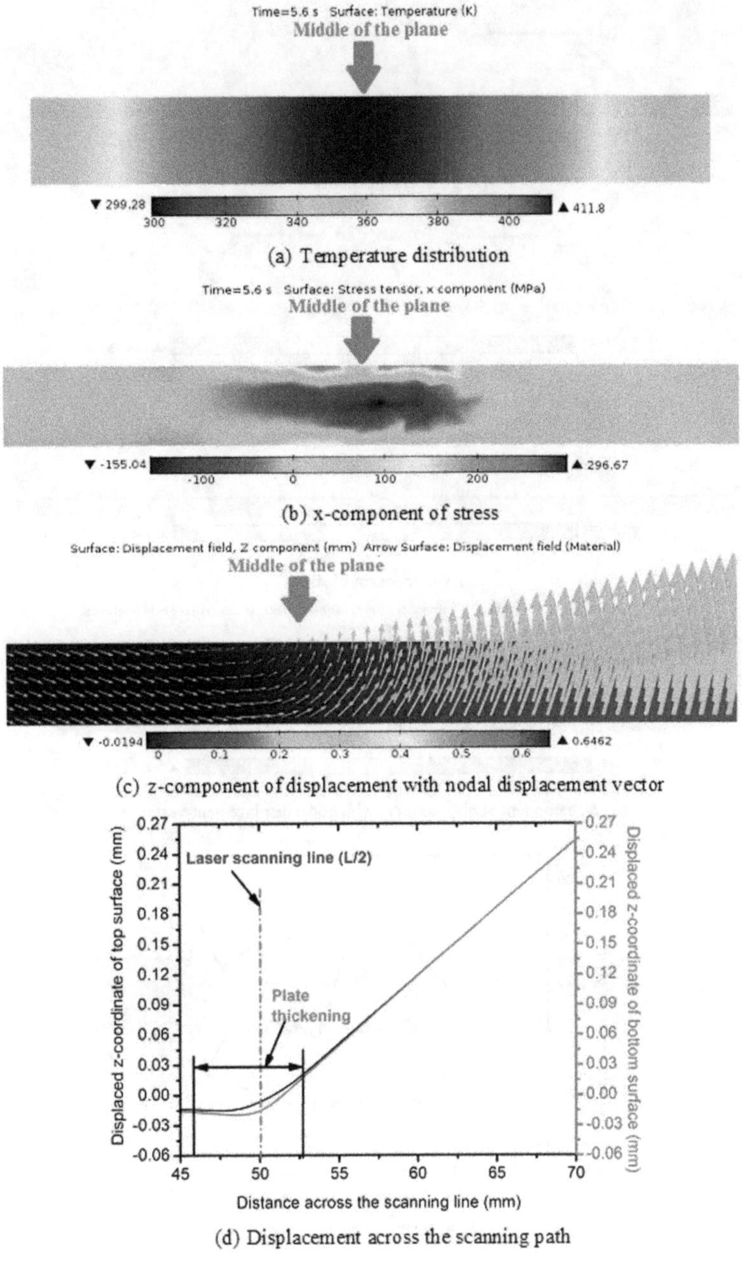

the heated region of the plate develops further amount of compressive stress due to thermal expansion of a deeper and wider zone surrounding the beam. It is also noted that the compressive stress starts to decrease at the heated region of the plate due to rapid cooling of the region, and it changes to tensile stress from compressive due to the contraction of the material. Figures 7 (d) and (e) show z-displacements along the scanning path showing thickening of the plate and changes in plate dimension (width wise) produced by local heating.

Figure 13. Temperature distribution, stress distribution and displacement field on the Y-Z plane (t = 5.6s)

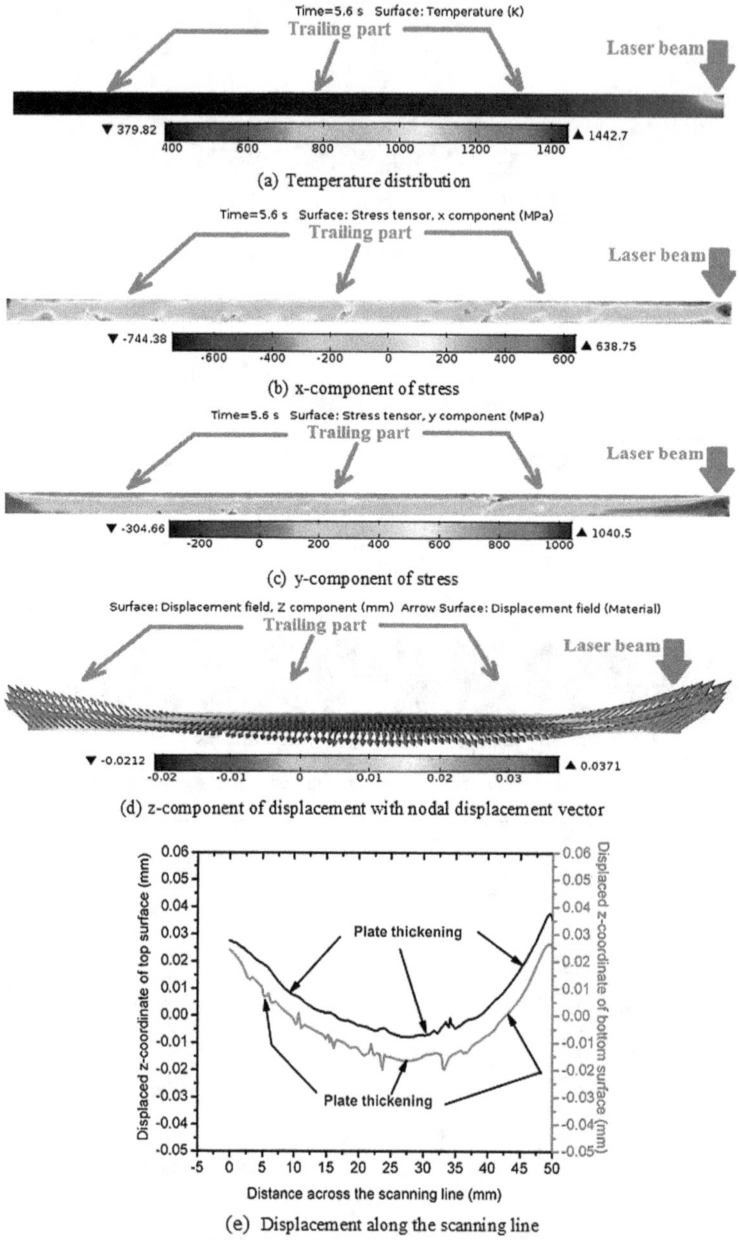

Temperature, stress, displacement distributions and z-displacement of the top and bottom surface across the scanning line on the X-Z plane are presented in Figures 8 (a-d). It is observed from figure 8 (a) that when the centre of the laser beam is positioned at (50, 23.8, 1.5) (t = 2.98 s), temperature starts to rise rapidly at the top surface of X-Z plane due to conduction so that the top surface of X-Z plane alters from tensile to compressive stress due to thermal expansion of the zone surrounding the beam. It is also noted that the bottom surface of the tensile stress on the X-Z plane starts to decline due to the temperature gradient along the plate thickness. The plate z-displacement along with plate thickening is

Figure 14. Temperature distribution, stress distribution and displacement field on the X-Z plane (t = 59.9s)

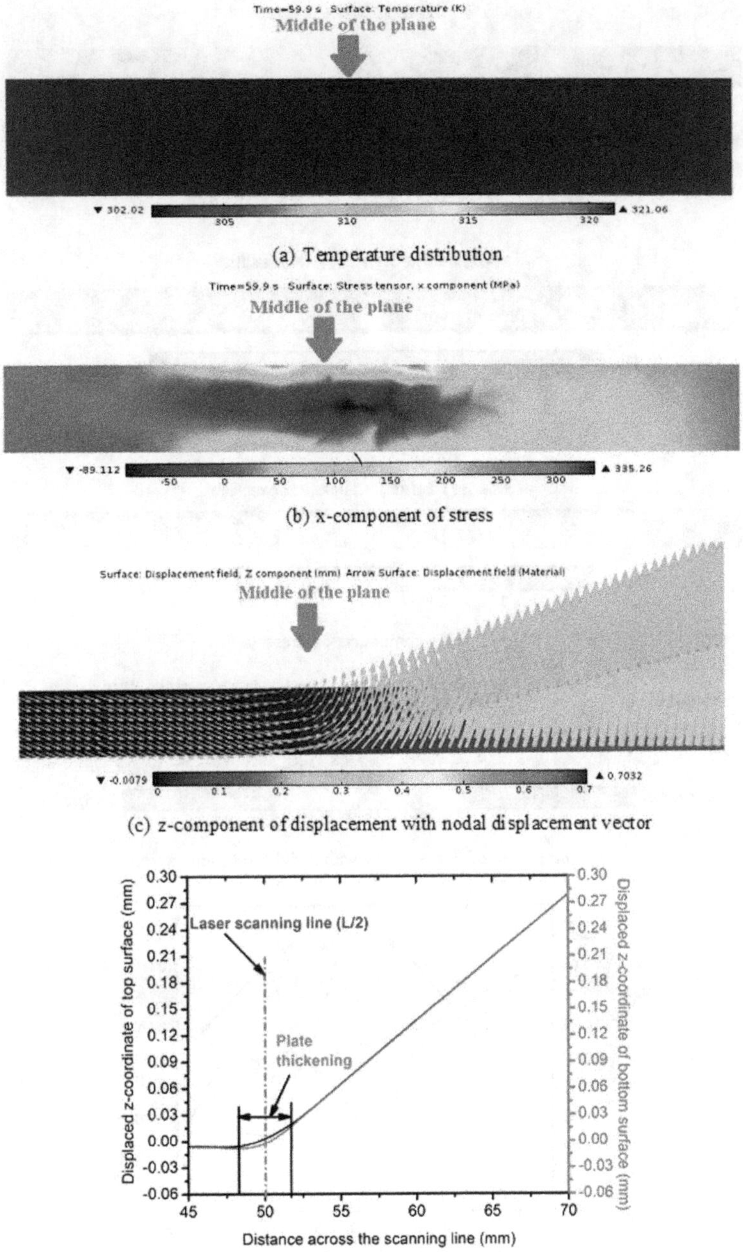

(a) Temperature distribution

(b) x-component of stress

(c) z-component of displacement with nodal displacement vector

(d) Displacement across the scanning path

shown in Figures 8 (c) and (d). Due to effect of conduction, the selected X-Z plane is preheated even before laser beam reaches the X-Z plane. Therefore, the top surface of the X-Z plane attains a bulging.

Figures 9 (a-e) show the temperature, stress distributions and z-displacement of the top and bottom surfaces along the scanning line on the Y-Z plane. Figures 9 (d) and (e) show the z-displacement along the scanning path with the plate thickening and increment in plate geometry because of local heating.

Figure 15. Temperature distribution, stress distribution and displacement field on the Y-Z plane (t = 59.9s)

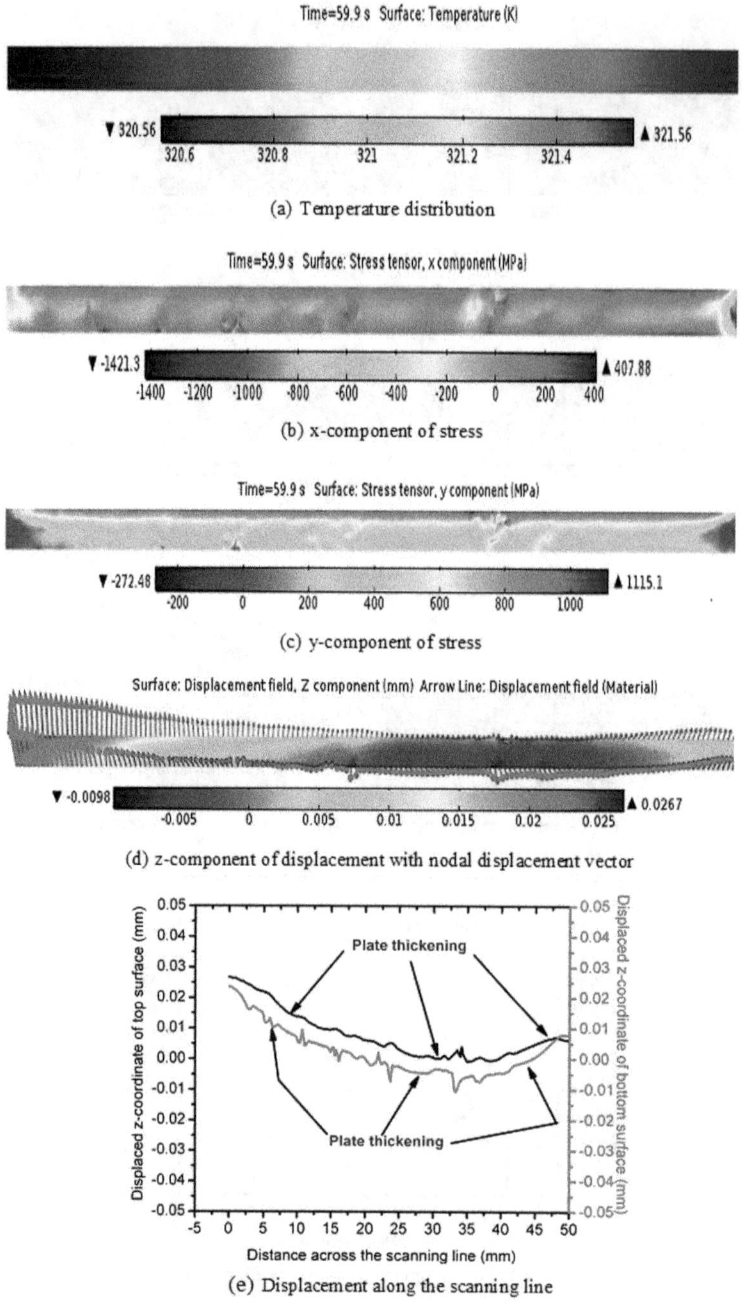

Figure 10 (a) shows the temperature distribution on the X-Z plane. At t = 3.12 s when the centre of the laser beam reaches the top surface of the X-Z plane, the temperature attains a maximum. Figure 10 (b) shows the x-component of stress distribution corresponding to the temperature distribution (Figure 10 (a)). In that, the maximum compressive stress is produced at the top surface of X-Z plane due to occurrence of maximum temperature. At this time, the compressive stress of 200 MPa develops at the bottom surface of the X-Z plane due to increment in temperature at the bottom surface of the X-Z plane. It is also shown that the induced stress gradient through the plate thickness which occurs due to thermal expansion of the material in the heated zone. Figure 10 (c) reveals that the z-displacement across the scanning path with the plate thickening produced by local heating. As can be seen in Figures 10 (c) and (d) that plate thickening occurred only at the laser scanning path, and there is no uneven thickening at the regions towards to the free end of the plate.

Temperature and stress distributions and z-displacement of the top and bottom surface along the scanning line on the Y-Z plane are presented in Figures 11 (a-e). Figures 11 (d) and (e) show the z-displacement of Y-Z plane with the plate thickening along the scanning path and increment in plate geometry.

Figures 12 (a-d) illustrate the temperature, stress and displacement distributions on the X-Z plane when the centre of the laser beam is at farthest edge (50, 32.8, 1.5) of the plate (at t = 5.6 s). It is seen from the figures that the temperature and stresses of X-Z plane decreases further due to cooling.

As shown in Figure 13 (a), a much higher temperature is obtained at the end of the scan line compared to that at the beginning as well as that at the middle part of the plate. The reason for this is that heat from the incident laser beam and the heat stored by the material behind the beam flows into the cold region ahead of the beam, and as the beam reaches the edge the heat flowing ahead of the beam cannot travel any further and, consequently, a heat buildup occurs near about the farthest edge of the plate. As the temperature of the plate increases, result in higher compressive stresses which can be seen from Figure 13 (b). The plate geometry also alters towards to the scanning line direction (+ve Y) when the laser beam is heating at the farthest edge of the plate as shown in Figure 13 (d). Figure 13 (e) shows the plate thickening along the scanning line produced by laser irradiation.

Figures 14 (a-d) illustrate the temperature and stress distributions on the X-Z plane after sufficient time to cool the plate, at t ≈ 60 s. It is seen from the figures that the temperature on the X-Z plane decreases near to ambient. As a result, the tensile stress at the top and bottom surfaces on the X-Z plane increases. Meanwhile, the compressive stress on the X-Z plane also decreases which can be seen from Figure 14 (b). Figures 14 (c) and (d) show the z-component of displacement of the top and bottom surface of X-Z plane.

Figures 15 (a-e) show the temperature, stress and displacement distributions on the Y-Z plane after sufficient cooling time, t ≈ 60 s. As we know that the temperature starts rising rapidly when the laser beam approaches the beginning edge of the plate. Afterwards, the temperature reaches a steady until the beam approaches the farthest edge of the plate. During heating, peak temperature is attained at the farthest point on the scanning line. Afterwards, the temperature starts to drop once laser beam leaves the plate as shown in Figure 15 (a). It can be seen from the figure, after cooling, the temperature drops to near ambient along Y-Z plane of the plate at t ≈ 60 s. Figures 15 (d) and (e) show the plate geometry changes and plate thickening along the scanning line produced by laser beam irradiation.

To observe the plate geometry changes when the laser beam is irradiating along the mid-length of the plate, the variation of x-component of displacement with time at a point having a coordinate (100, 25, 1.5) on the plate free edge top surface is monitored and shown in Figure 16 (a). Similarly, the variation of y and z-component of displacements are monitored with time at the same coordinate as shown in Figures 16 (b) and (c).

From Figure 16 (a), it can be seen that the dimension of the plate increases in the plate length direction (+ve X). The reason for this is that during heating process the solid particles in the plate will heat up and it does expand by thermal expansion. Once the heat which has been given to the plate is dissipated, the plate dimension starts to decline due to contraction as shown in Figure 16 (a). From Figure 16 (b), it can be seen that the plate moving towards to the laser scanning direction (+ve Y) until the laser beam irradiates the half width of the plate. Afterwards, the plate starts to move in opposite direction to the laser scanning direction (-ve Y) during rest of the heating. During the cooling phase, the plate moves towards to the laser scanning direction (+ve Y) with very slow rate due to contraction. Figure 16 (c) shows the variation of z-component of displacement with time during the heating and cooling process. It can be seen from the figure that the z-component of displacement increases during the laser heating process and remains fixed during the cooling time. Detailed explanation of the bending mechanism during the laser forming process can be found in the published research article of (Paramasivan et al., 2019; Paramasivan, 2019).

EXPERIMENTAL AND FINITE ELEMENT ANALYSIS OF V-BENDING OF RECTANGULAR FULL PLATES

Laser forming of metal sheets for full plates as well as plates with cutout is studied in order to produce V-shaped formed parts. In this section, experimental and numerical investigations are undertaken to study the bending behaviour of AISI 304 steel sheet using a low power Nd:YVO4 laser source. A series of experiments have been performed to study the effects of process parameters on bending angle. In ad-

Figure 16. Displacement at different components at top surface of the plate edge

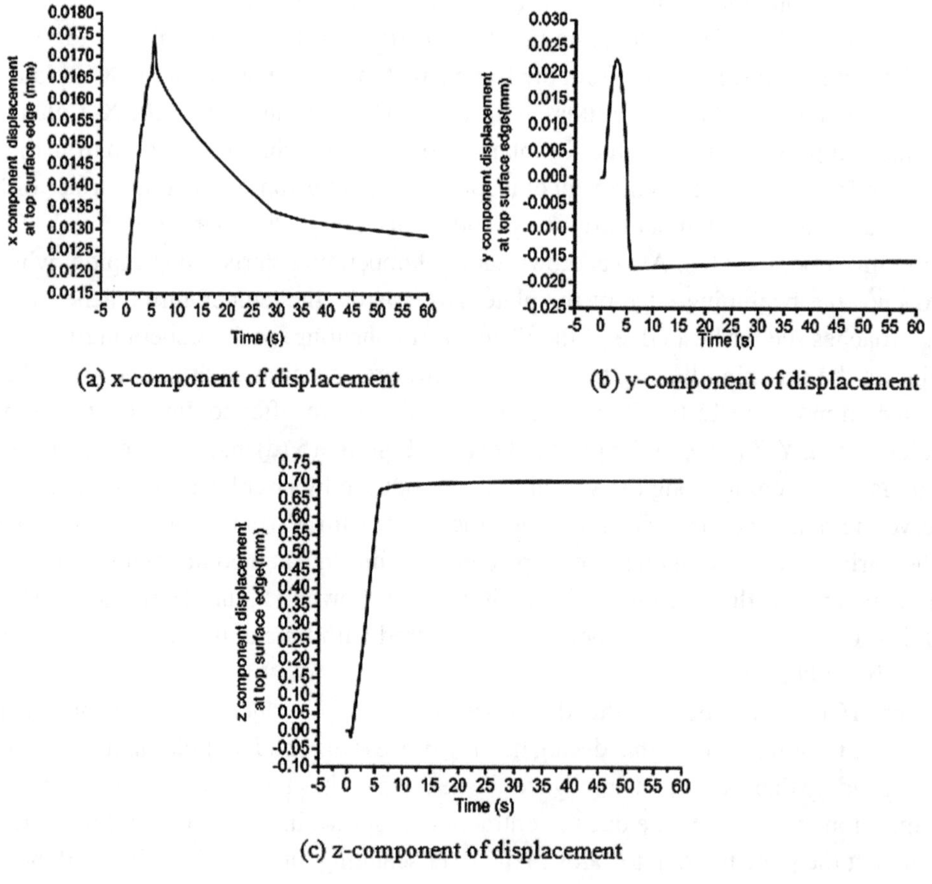

dition, the effect of number of scans on relative variation of bending angle (RBAV) along the free edge is also investigated. A metallurgical investigation is also conducted on laser formed AISI 304 specimens to study the effect of laser forming on surface morphology.

Experimental Procedure

AISI 304 sheet (70mm × 40mm × 0.8mm) is chosen in this study. The sample is mounted on a table, with one end fixed and the other end free. The defocused laser beam is traversed over it along the line 30 mm away from the fixed end of the plate. For all the experiments, the number of passes and the inter pass time are maintained constant at 20 and 12 seconds, respectively. Figure 17 shows the photographic view of the experimental set-up used in this study. The system is equipped with a 12 W Nd: YVO4 laser. The laser beam is focused to a diameter of 50 μm using a 195mm telecentric lens. A laser displacement sensor (Make: Micro-Epsilon, Germany, Model: ILD 1300, Series: optoNCDT 1300) with an accuracy of 4 μm is used to measure the Z-direction displacement at the free end of the specimen during the process. The laser parameters and the specimen dimensions, used in the present set of experiments, are listed in Table 1.

Effect of Scanning Speed on Bending Angle

Figure 18 shows the effect of scanning speed on bending angle at different power. Bending angle decreases with increase in scanning speed.

As scanning speed increases, the interaction time between the work piece and laser beam decreases, resulting in reduced absorption of heat by the work piece, lower peak temperature and lower plastic stress. Thus, the amount of bending decreases with increase in scanning speed.

Figure 17. Experimental set-up for laser bending

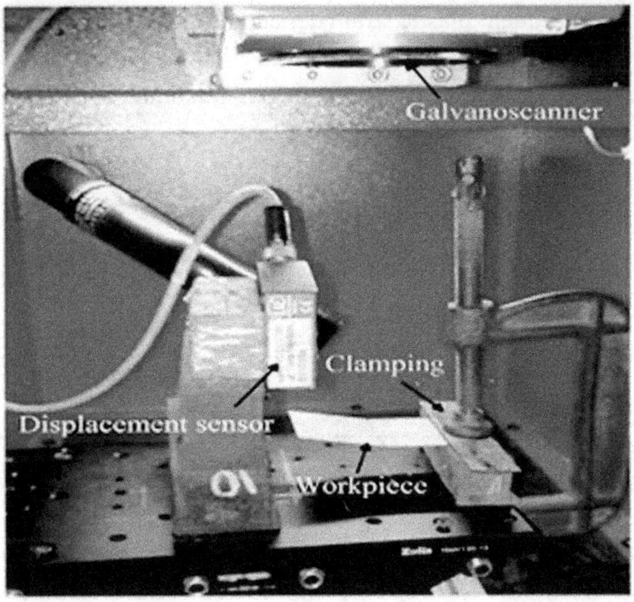

Figure 18. Effect of scanning speed on bending angle at different power (No. of pass= 20)

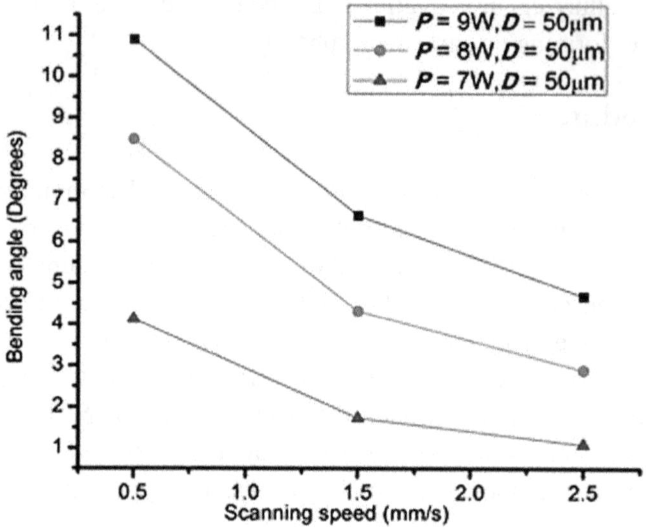

Effect of Operating Distance on Bending Angle

An effect of operating distance (the distance of the scan line from the fixed edge) on bend angle has received very little attention. In real time applications, the beam is expected to move not necessarily along the midline of the sheet.

It is expected that the amount of bending changes with operating distances. Figure 19 shows the variation of bending angle with operating distance. There is a reduction in bend angle when the operating distance is closer to the free edge. In the present case, there is only internal restraint offered by the bulk of unheated mass surrounding the heated zone and there is no external constraint, as the plate is fixed along one edge only. Hence, in the present case, in absence of external constraint, the magnitude of the final tensile stress and the consequent plastic strain depend mostly on the quantum of mass surrounding the heated zone. As the scan line moves closer to the free edge, the amount of mass unaffected by laser heating increases and hence, lesser internal restraint is generated and lesser bending is achieved.

Table 1. Laser and specimen parameters used in experiments

Laser power	W	7-9
Scanning speed	mm/s	0.5-2.5
Laser beam diameter	μm	50
Number of scans	--	20
Specimen length	mm	70
Specimen width	mm	40
Specimen thickness	mm	0.8

Effect of Number of Passes on Bending Angle

Multiple laser scans are necessary to obtain the required magnitude of bending angle for most real-life applications. Figure 20 shows the variation of the bend angle with number of passes or laser scans. It can be observed that cumulative bend angle increases almost linearly with the number of scans, with the rate of increase decreasing at higher number of passes, due to work hardening. The extent of work hardening is evident from the graph of incremental bending angle in each pass.

Effect of Number of Passes on RBAV

To evaluate the variation of bending angle along the free edge, termed as edge effect, bending angle at nine equivalently spaced positions along the free edge of the sheet is measured. The edge effect is calculated according to Eq. (5).

$$RBAV = \frac{\Delta\alpha}{\bar{\alpha}} = \frac{\alpha_{max} - \alpha_{min}}{\sum_{i=1}^{n} \alpha_i} \quad (5)$$

RBAV is the relative variation of bending angle, $\Delta\alpha$ is the difference between maximum bending angle $\left(\alpha_{max}\right)$ and minimum bending angle $\left(\alpha_{min}\right)$ along the free edge of the sheet. $\bar{\alpha}$ is the arithmetic average of the bending angles at the equivalently spaced sampling points. A low value of RBAV is desirable, since it implies a low value of difference between the maximum and minimum bending angle, $\Delta\alpha$ and/or a high value of average bending angle, $\bar{\alpha}$.

Figure 19. Effect of operating distance

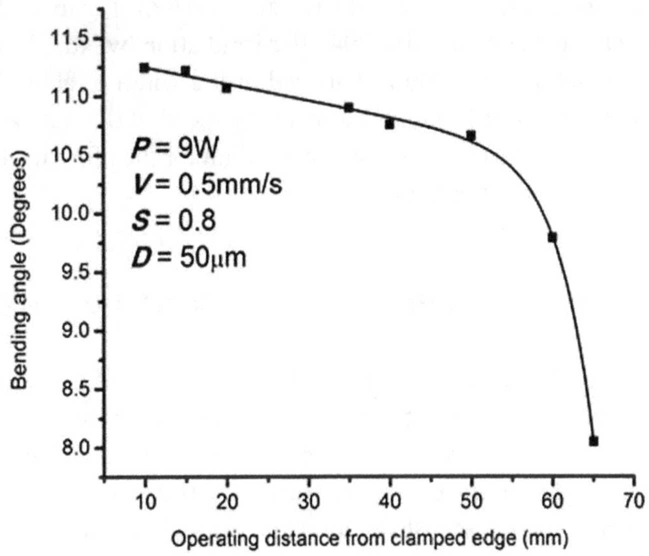

Figure 20. Effects of number of scan

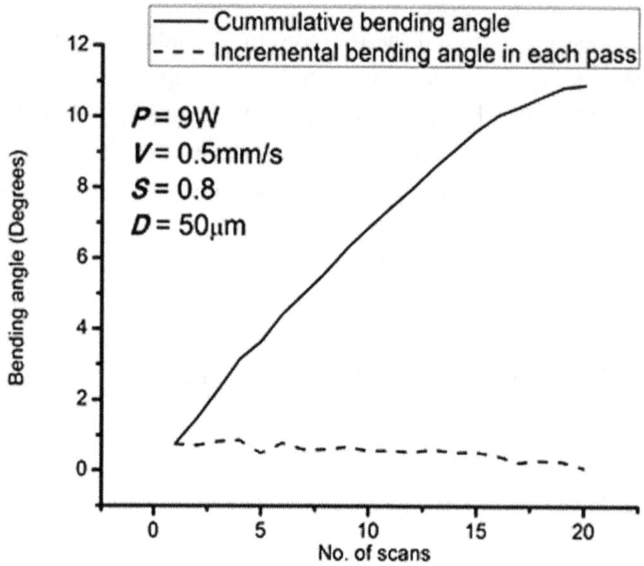

As can be seen from Figure 21, that with increase in number of laser passes, the average bending angle $\bar{\alpha}$ increases, the difference between the maximum and the minimum bending angle $\Delta\alpha$ first increases and then decreases; the net effect on RBAV is that it decreases monotonically with the number of laser scans.

Heat Affected Zone of AISI 304

A metallurgical investigation is conducted on laser formed AISI 304 specimens to study the effect of laser forming on the micro-structure and heat affected zone (HAZ). Figures 22 (a) and (b) show the unaffected zone and heat affected zone of AISI 304 after irradiation by Nd:YVO4 laser. From this microstructure, it can be seen that the substrate is damaged over a width of about 80 μm due to the high temperatures experienced during laser forming at a scanning speed of 0.5 mm/sec, laser powers of 9 W and beam diameter of 50 μm. The narrow HAZ of about 80 μm for the aforementioned parameters may be attributed to the small diameter of the laser beam.

V-BENDING OF RECTANGULAR PLATES WITH CENTRAL CUTOUT

In this section, laser forming of rectangular plates with a central cutout at its middle is presented. To investigate the effects of shape and size of the cutout on temperature distribution, stress distribution and final bending angle, different shapes such as circular, ellipse, rectangular and square, and different dimensions of cutouts are chosen. Investigation of the effects of variation of the dimensions of the cutouts on the thermo-mechanical stress distribution and the final bending is carried out and detailed results are presented in this section.

Figure 21. Effect of number of scans: a) variation of bending angle along the free edge, b) average bending angle, c) difference between maximum and minimum bending angles, d) relative variation of bending angle

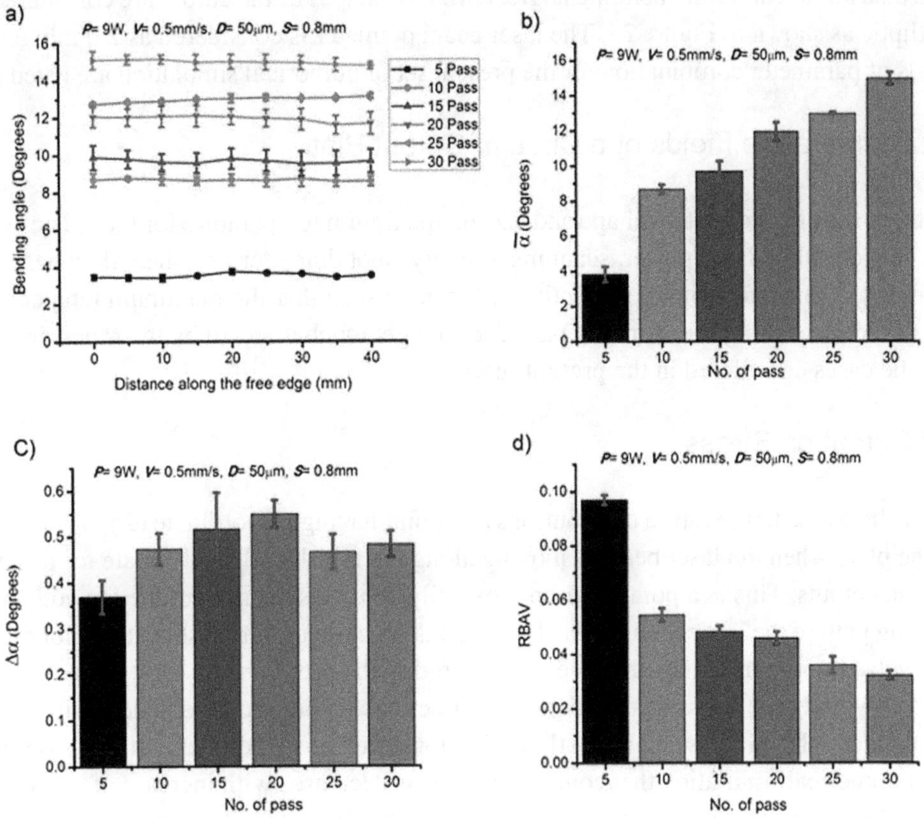

Figure 22. SEM images of the heat affected zone: a) view of unaffected and affected zones

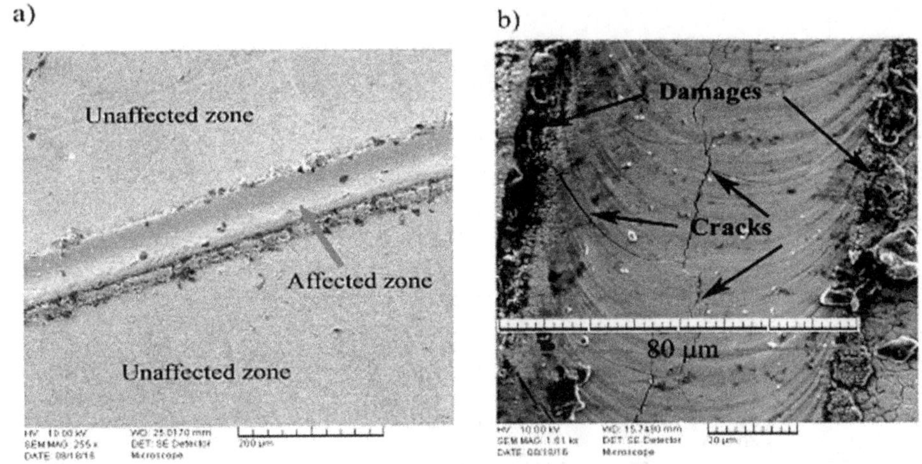

Laser Forming of Rectangular Plates with Circular/Elliptic Cutout

Ten number of cutout plates are considered to investigate the effect of shape and size of the cutouts on temperature distribution and final bending angle. Different shapes of the cutout are considered, viz., circular and elliptic as shown in Figure 23. The laser beam diameter is considered as 2.0 mm for the entire study. Details of parameter combination for the present set of numerical simulation are listed in Table 2.

Transient Temperature Fields of a Circular Cutout Plate

Figure 24 shows the effect of cutout shape and size on maximum temperature for four plates with cutout of different shapes, while laser power, scanning velocity, spot diameter and sheet thickness are kept at 250 W, 15 mm/s, 2 mm and 1.5mm, respectively. It can be seen that the maximum temperature on the top surface of the plate along the scanning path does not get much affected by the shape and size of the cutouts, for the cases considered in the present research.

Effects of Cutout on Stress

Figure 25 (a) shows the temperature distributions at a point having a coordinate (25, 48, 1.5) on the top surface of the plate when the laser beam is moving along the mid-length of the plate for different shape and size of the cutouts. This is a point, chosen arbitrarily, which is near about the top edge of the plate on the scanning path of the laser beam. It can be seen that the temperature distributions for all shape and size cutouts gets almost similar, for the cases considered in the present study.

Figure 25 (b) shows the stress distribution in X-direction at a point on the plate having a coordinate (25, 48, 1.5) which falls on the scanning path (x=25) for different shape of cutouts. It is observed from the results of numerical simulation that compressive stress decreases with increase of cutout dimension in Y-direction, as can be observed from the results of the 20×10 to 10×20 elliptic cutout plates.

Effects of Cutout on Bending Angle

Figure 26 (a) shows the variation of bending angle with time at the top surface of the plate free end at middle having a coordinate (50, 25, 1.5) for different shape of cutouts. It can be seen that under the

Table 2. Cutout details of the 50 × 50 × 1.5 AISI 304 plates and process parameters

Cutout Shape	Cutout Dimensions (mm)	Laser Power P (W)	Scanning Speed V (mm/s)
Circular (D)	10 20 30 40	250, 300, 350	15
Elliptic (axb)	10×20 20×10 20×30 30×20 30×40 40×30	250, 300, 350	15

Figure 23. Schematic diagrams showing the few numbers of work pieces and irradiation path for numerical simulation: (a) plate with Ø 20 mm circular cutout, (b) plate with 30 × 20 mm² elliptic cutout

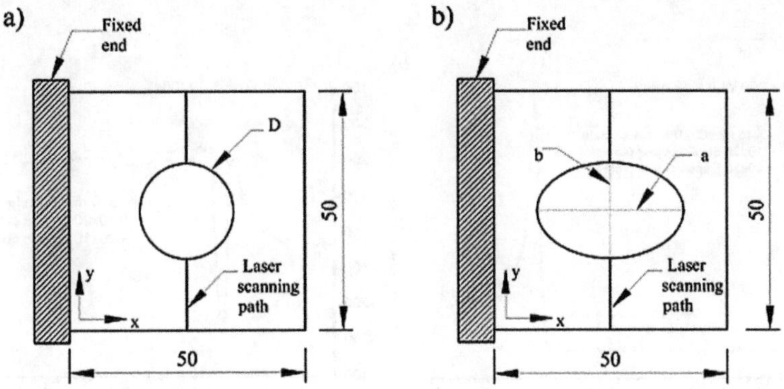

conditions of same process parameters, the bending angle is varying depending upon the shape of the cutouts. The maximum bending angle is achieved for a full plate without cutout followed by circular cutout and square cutout. The reason for this is that sample without cutout has more heating time, de-

Figure 24. Variation of maximum temperature with different cutout shapes and sizes: (a) different shapes of cutout, (b) circular cutout plate, (c) elliptic cutout plate (laser power = 250W, scanning speed = 15 mm/s, plate thickness = 1.5 mm, beam diameter = 2 mm)

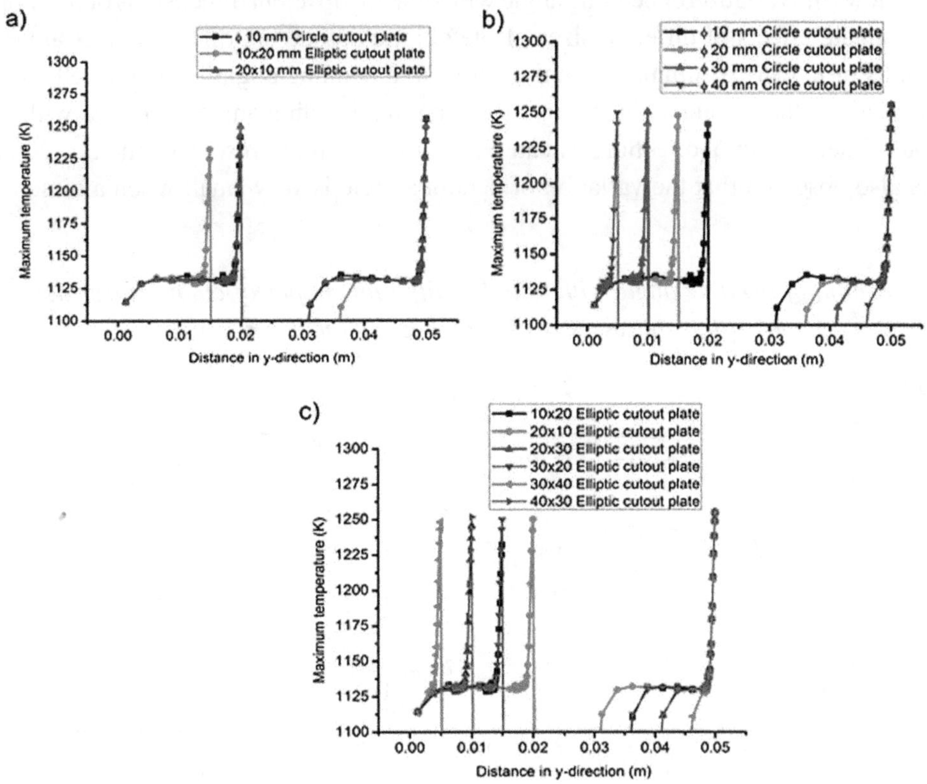

Figure 25. Variation of (a) temperature and (b) x-component of stress with time for different shape and size of cutouts (laser power = 250W, scanning speed = 15 mm/s, plate thickness = 1.5 mm, beam diameter = 2 mm)

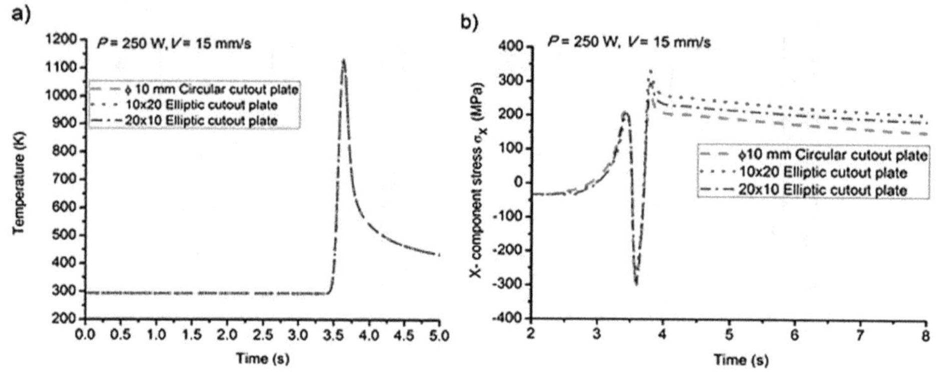

livering maximum amount of energy. So, more amounts of thermal stress and plastic strain develop in the region of the scanning path compared with the cutout plates. However, the bending angle difference between the different cutout shape plates is very small under the same process parameters. When cutout dimension increases the bending angle is reduced, as shown in Figure 26 (b).

As the scan-wise dimension of the cutout increases, the interaction time between the work piece and laser beam decreases, resulting in reduced absorption of heat by the work piece. Thus, the amount of bending angle decreases with an increase in scan-wise dimension of the cutout.

Figure 27 shows the variation of bending angle with time for different dimensions of the elliptic cutout plate. To investigate effects of different sizes of the elliptic cutout, six different sizes are considered. From the results of numerical simulation, unique trends in bending angle are identified while varying the dimensions along the X- and Y-directions. It is seen that bending angle decreases with increase of elliptic cutout dimension in Y-direction, as can be realised from the results of the 30×20 and 30×40 cutouts. It is also observed that the variation of bending angle is very small when a dimension of the

Figure 26. Variation of bending angle with time for different shapes of cutout (laser power = 250W, scanning speed = 15 mm/s, plate thickness = 1.5 mm, beam diameter = 2 mm)

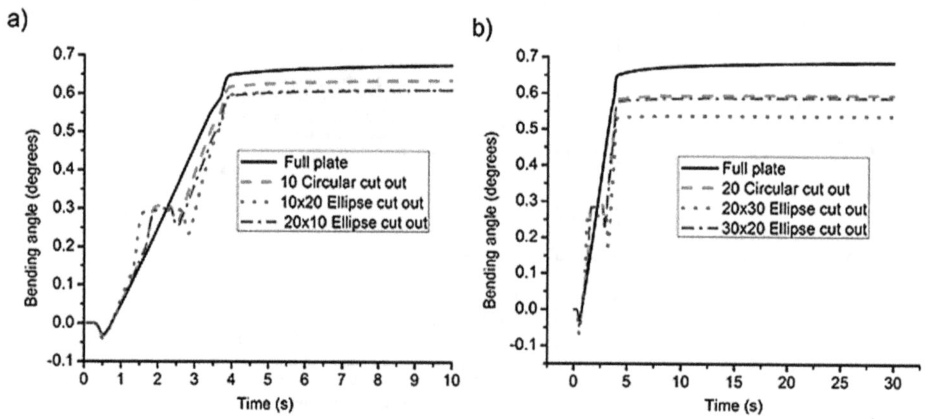

elliptic cutout is changed in the X-direction, as can be seen from the results of the 10×20 and 30×20 cutout plates. This is because, as the dimension of the elliptic cutout is more in the scanning path direction, there is lesser interaction between the laser beam and the plate material.

Figure 28 shows the effects of laser power on bending angle for four different shapes of cutout stainless steel plate while scanning speed is kept at 15 mm/s. It is seen that bend angle increases with an increase in laser power. When the length of the cutout along the scanning direction increases, bending angle decreases when process parameters are kept constant. On the other hand, when the length of the cutout across the scanning direction is altered, it does not have much effect on bending angle. In general, under the conditions of same parameters, the amount of bending angle does not change much with different shapes of the cutout.

Prediction of Bending Angle Using Response Surface Methodology

Though simulation of laser bending of AISI 304 plate with a central cutout at its middle using finite element model provides the insight into laser bending, it demands huge computation time and resource. Therefore, Statistical techniques are used to develop a mathematical model based on simulation results to predict bending angle. Developing an empirical relation capable of predicting the bend angle in a given parameter range will help to utilise the benefits of laser bending in the manufacturing industries. The process parameters (low actual and high actual) and their notation and units are presented in Table 3.

The final regression equation (Eq. 6) for bending angle in terms of actual factors as determined by Design expert software is shown below:

Figure 27. Variation of bending angle with time for elliptic cutout plates (laser power = 250W, scanning speed = 15 mm/s, plate thickness = 1.5 mm, beam diameter = 2 mm)

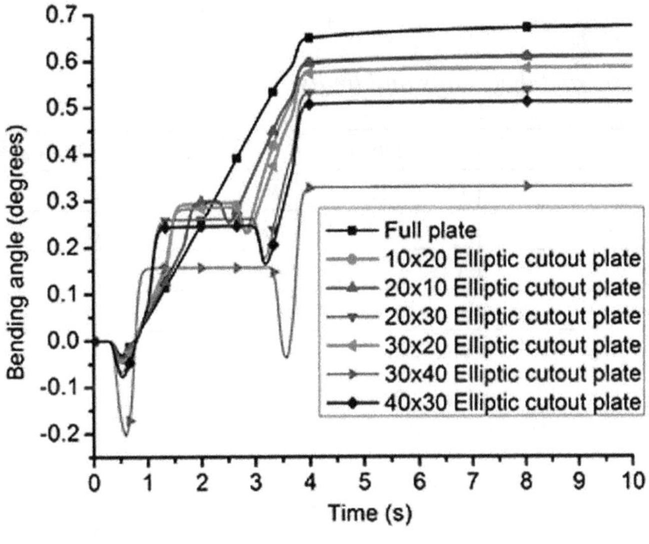

Figure 28. Effect of laser power on bending angle for different shape and size of the cutout plates. (scanning speed = 15 mm/s, plate thickness = 1.5 mm, beam diameter = 2 mm)

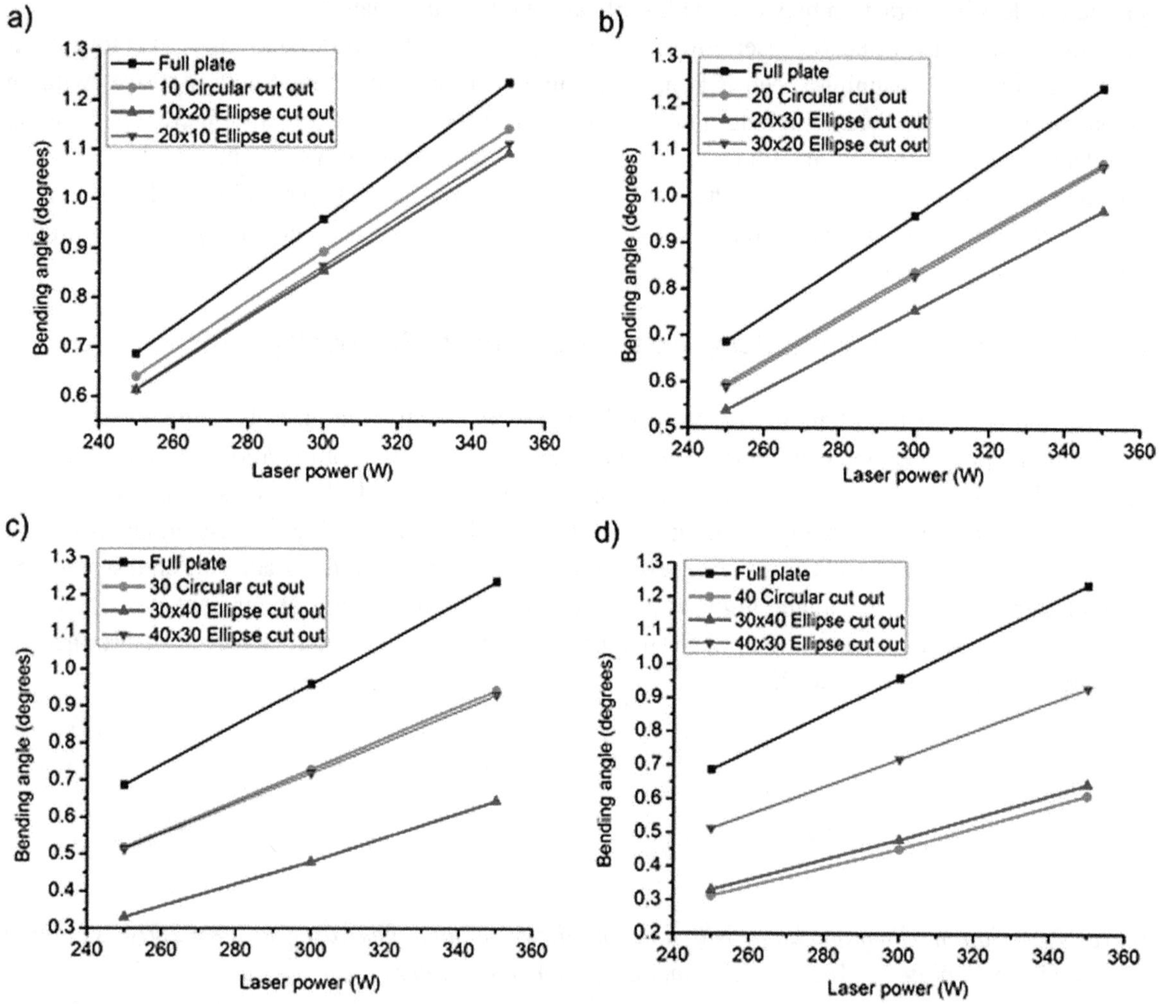

Table 3. Process parameters and their units and limits

Parameter	Notation	Unit	Low Actual	High Actual
Laser power	*P*	W	275	375
Scanning speed	*V*	mm/s	10	20
Cutout dimension in x-direction	*X*	mm	15	35
Cutout dimension in y-direction	*Y*	mm	15	35

$$\begin{aligned}\alpha_b &= 0.46916 + 2.04905\times10^{-3}\times P - 0.066625\times V - 9.42853\times10^{-3}\times X + 0.011769\times Y \\ &-1.04894\times10^{-4}\times P\times V + 1.15609\times10^{-6}\times P\times X - 2.30896\times10^{-5}\times P\times Y \\ &+1.60462\times10^{-5}\times V\times X + 3.55841\times10^{-4}\times V\times Y - 3.55719\times10^{-5}\times X\times Y \\ &+4.78233\times10^{-6}\times P^2 + 2.01319\times10^{-3}\times V^2 + 1.11022\times10^{-4}\times X^2 - 3.09635\times10^{-4}\times Y^2\end{aligned} \tag{6}$$

Laser Forming of Rectangular Plates With Rectangular/Square Cutout

Laser bending of a rectangular AISI 304 plate with a rectangular/square cutouts are investigated in this section. The numerical simulation conditions are listed in Table 4. The deformation of sheet metal has been obtained from numerical simulations and discussed below.

Effect of Width of the Cutout on Bending Angle

Figure 29 shows the variation of bending angle with time at the mid-point of the free end of the plate having coordinates (100, 25, 1.5) for four different widths (w = 10, 20, 30, 40 mm) of the cutout while length of the cutout (l) is kept constant at 10 mm. It is seen that bending angle decreases with increase in width of the cutout.

As a result, counter bending increases with width of cutout. When the laser beam enters the cutout, bending angle does not change. When laser beam re-enters the plate through the other edge of cutout, the plate initially undergoes the same amount of counter bending, and the bending angle increases in the later part of the beam traverse. After the laser beam leaves the work piece, a steady bending angle is obtained as evident from these Figures.

Table 4. Cutout details of the 100 × 50 × 1.5 AISI 304 plates and process parameters

Specimen Conditions	Dimension of Cutout(l×w)	Laser Power P (W)	Scanning Speed V (mm/s)	Beam Diameter (mm)
Full plate (L× W× T) without cutout		250, 300, 350	10, 15, 20	2
Plate with cutout	10 × 10 10 × 20 10 × 30 10 × 40 20 × 10 20 × 20 20 × 30 20 × 40 30 × 10 30 × 20 30 × 30 30 × 40 40 × 10 40 × 20 40 × 30 40 × 40	250, 300, 350	10, 15, 20	2

L-length of plate; W- width of plate; T- thickness of plate; l- length of cutout; w-width of cutout

Figure 29. Bending angle versus time for different widths (w) of cutout and laser powers: (a) laser power = 250 W, scanning speed = 10 mm/s, (b) laser power = 300 W, scanning speed = 10 mm/s, (c) laser power = 350 W, scanning speed = 10 mm/s

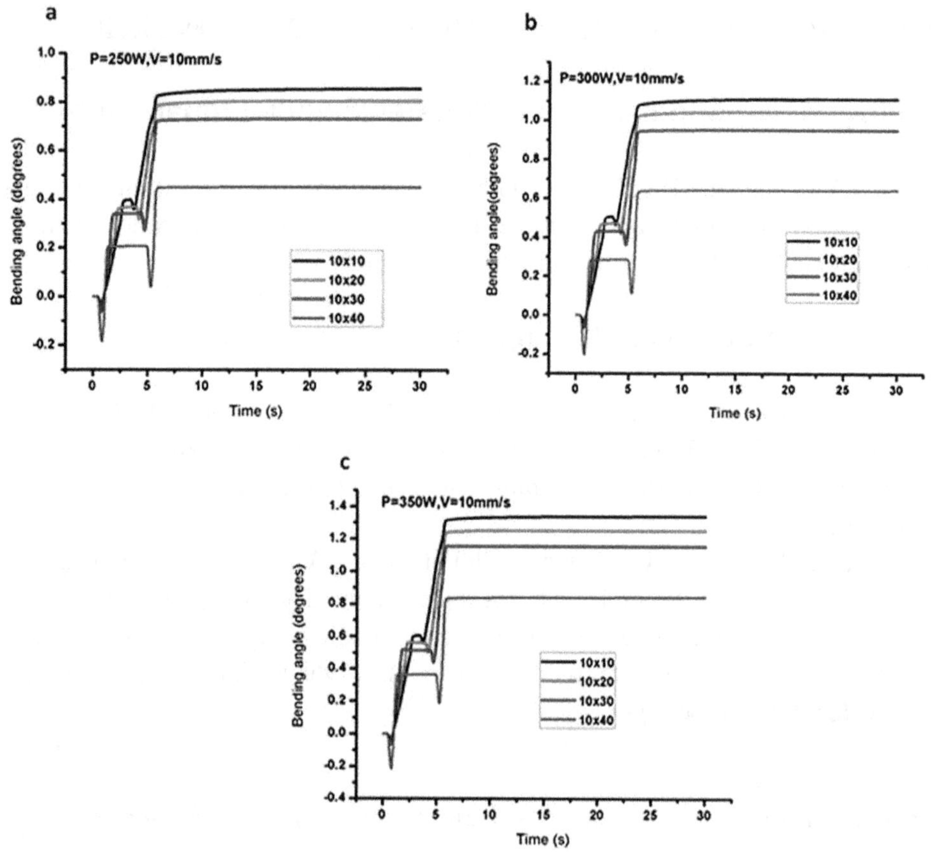

Effect of Length of the Cutout on Bending Angle

Figure 30 shows the effect of changing the length of the cutout with a constant width on the bending angle under different laser powers and scanning speeds. It is observed that the variation of bending angle is very small when length of the cutout is changed from 10 mm to 40 mm, while the width of the cutout is kept constant at 10 mm. From the results it can be seen that as the length of the cutout is increased, keeping the width constant, there is no significant variation in bending angle.

Effect of Cutout at Different Locations on Plate

To investigate effect of cutout at different position in the plate, the same dimension of the cutout is used in this study. Eight numbers of cutout plates and one full plate (without cutout) are considered. All work pieces have the same length (100 mm), width (50 mm) and thickness (1.5 mm). The cutout dimension is taken as 10 mm × 10 mm (length × width). The other process parameters are: laser power 250 W,

Figure 30. Variation of bending angle with time for different laser powers and cutout dimensions

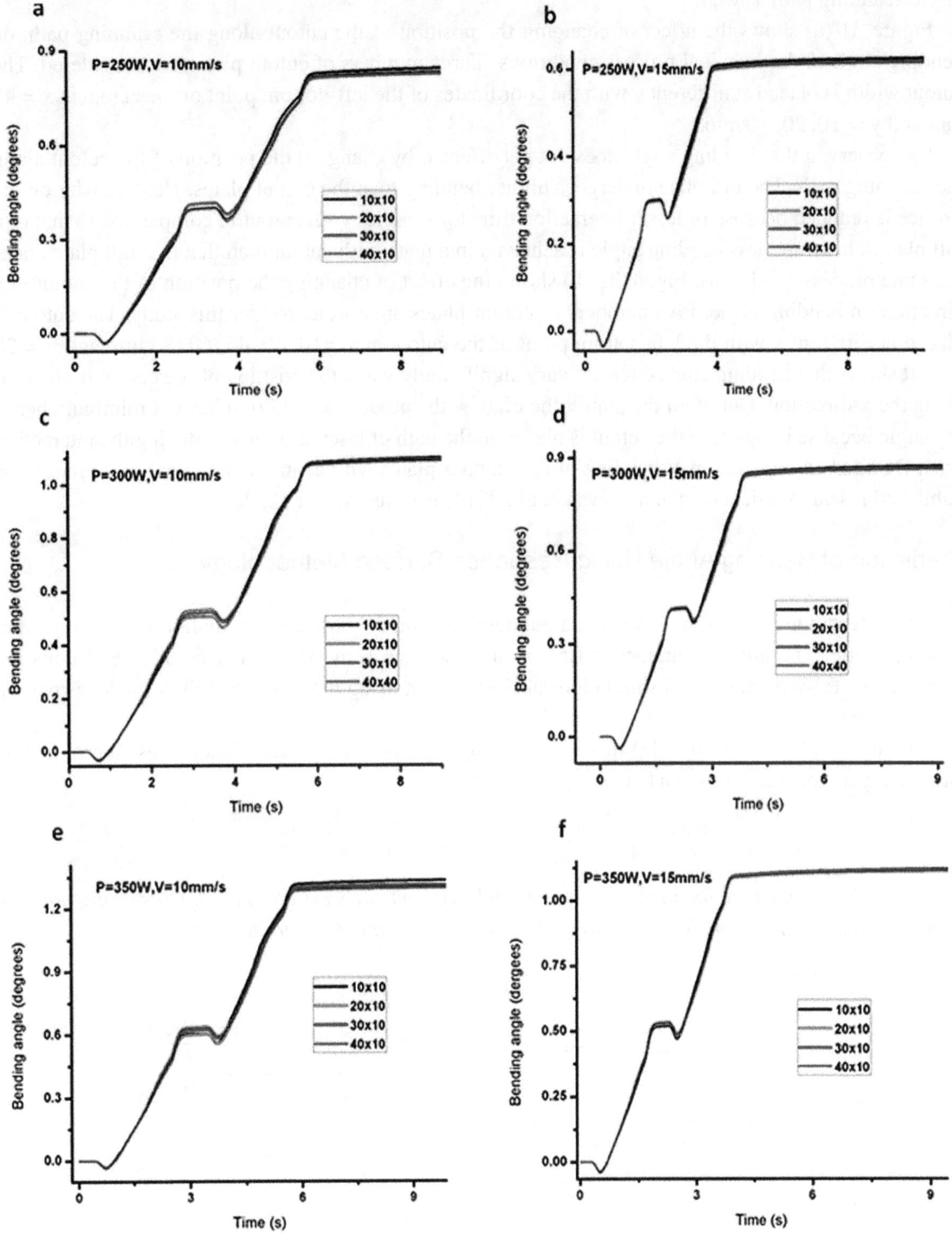

velocity 10 mm/s, spot diameter 2 mm. The cutout position is varied along the scanning path and outside of the scanning path as well.

Figure 31 (a) shows the effect of changing the position of the cutout along the scanning path, on bending angle under identical process conditions. Three numbers of cutout plates are considered. The cutout width is placed at different y with the coordinates of the left-bottom point of the cutout at x = 45 mm and y = 10, 20, 30 mm.

It is observed that bending angle does not get affected by changing the position of the cutout along the scanning path. The full plate undergoes higher bending than the cutout plates. Heat transfer on the surface is reduced because of lesser interaction time for samples with cutout in comparison to that of a full plate. Therefore, less bending angle is achieved in a plate with cutout than that in a full plate under the same process conditions. Figure 31 (b) shows the effect of changing the position of the cutout in x direction on bending angle. Five numbers of cutout plates are considered for this study. The cutout is placed at different x with the left-bottom point of the cutout at x =10, 30, 45, 60, 80 mm and y = 20 mm. It shows that bending angle does not vary significantly when the position of the cutout is changed along the x-direction. Out of all the plates, the plate with cutout at x = 45 mm has got minimum bending angle because in this case the cutout is placed in the path of laser scanning, reducing the interaction time. Detailed explanation of V-bending of rectangular plates with central cutouts can be found in the published research article of (Paramasivan et al., 2014; Paramasivan et al., 2015).

Prediction of Bending Angle Using Response Surface Methodology

Statistical techniques are used to develop a mathematical model based on simulation results to predict bending angle. The process parameters (low actual and high actual) and their notation and units are presented in Table 5. Numerical simulations are carried out using the developed FEA model as per the design layout.

The final mathematical model (Eq. 7) for bending angle in terms of actual factors as determined by Design expert software is shown below:

Figure 31. Effect of cutout location on bending angle for same process parameter: (a) varying position of cutout along scanning path (y direction), (b) varying position of cutout in x-direction

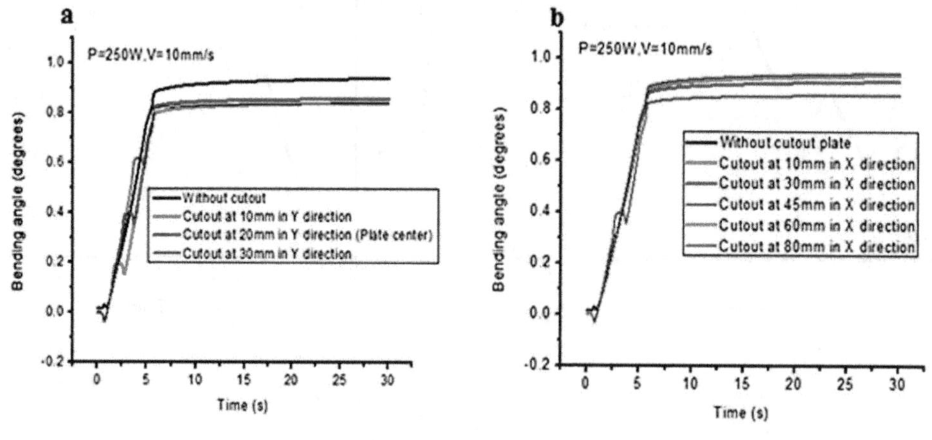

Table 5. Process parameters and their units and limits

Parameter	Notation	Unit	Low Actual	High Actual
Laser power	*P*	W	275	375
Scanning speed	*V*	mm/s	10	20
Cutout dimension in x-direction	*X*	mm	15	35
Cutout dimension in y-direction	*Y*	mm	15	35

$$\begin{aligned}\alpha_b &= -0.10949 + 4.31367\times10^{-3}\times P - 0.052200\times V - 2.06924\times10^{-3}\times X + 0.016146\times Y\\ &-1.06297\times10^{-4}\times P\times V - 7.67447\times10^{-6}\times P\times X - 3.08705\times10^{-5}\times P\times Y\\ &+1.50649\times10^{-4}\times V\times X + 4.67978\times10^{-4}\times V\times Y - 1.57584\times10^{-5}\times X\times Y\\ &+2.20973\times10^{-6}\times P^2 + 1.35620\times10^{-3}\times V^2 + 2.27744\times10^{-5}\times X^2 - 4.08603\times10^{-4}\times Y^2\end{aligned} \quad (7)$$

CONCLUSION

Laser forming process of AISI 304 steel is discussed for V-shaped formed rectangular plates with or without cutout. In the first section of presented chapter, mechanism of the process is monitored and discussed through the numerically computed the temperature, stress and displacement distributions on the two cross sections of the plate. In the second part of the chapter, V-bending of metal sheets for full plates is studied using a low power Nd:YVO4 laser source. It is revealed that bending angle increases with laser power and decreases with increase in scanning speed. Moreover, bending angle increases with the number of scans and relative variation of bending angle (RBAV) decreases drastically when the number of laser pass increases. However, the increment of bending angle decreases with successive scans due to work hardening of the material. V-bending of metal sheets for cutout plates is presented in the final part of the chapter. It is observed that bending angle decreases by increasing the width of the cutout in laser scanning path with the same process parameters due to the reduced interaction time between the work piece and the laser beam. Bending angle does not depend much on the cutout dimension across the scanning path. Moreover, the position of the cutout along the scanning path and across the scanning path has no substantial effect on the bending angle. The developed regression models for V-bending of the cutout plates has good predictability for evaluating the bending angle within the parameter space considered.

REFERENCES

Dirscherl, M., Esser, G., & Schmidt, M. (2006). Ultrashort pulse laser bending. *Journal of Laser Micro Nanoengineering*, *1*(1), 54–60. doi:10.2961/jlmn.2006.01.0011

Dixit, U. S., Joshi, S. N., & Kant, R. (2015). Laser forming systems: A review. *International Journal of Mechatronics and Manufacturing Systems*, *8*(3/4), 160–205. doi:10.1504/IJMMS.2015.073077

Fetene, B. N., Kumar, V., Dixit, U. S., & Echempati, R. (2018). Numerical and experimental study on multi-pass laser bending of AH36 steel strips. *Optics & Laser Technology*, *99*(1), 291–300. doi:10.1016/j.optlastec.2017.09.014

Geiger, M., & Vollertsen, F. (1993). The mechanisms of laser forming. *CIRP Annals*, *42*(1), 301–304. doi:10.1016/S0007-8506(07)62448-2

Gollo, M. H., Nadi, G., Mehdi, M., & Abbaszadeh, M. (2015). Experimental and numerical study of spiral scan paths on cap laser forming. *Journal of Laser Applications*, *27*(1), 0120023.

Hennige, T. (2000). Development of irradiation strategies for 3D-laser forming. *Journal of Materials Processing Technology*, *103*(1), 102–108. doi:10.1016/S0924-0136(00)00392-7

Hu, Y., Yao, Z., & Xu, X. (2011). Experimental investigation on the bending deformation of thin 1060 pure aluminum sheet by laser peen forming. *Proceedings of the international conference on shot peening*, South Bend, IN (pp. 341-346). Academic Press.

Kant, R., & Joshi, & S.N. (2013). Finite element simulation of multi-pass laser bending process with forced cooling. *Proceedings of the International Conference on Precision, Meso, Micro and Nano Engineering*, Kerala, India (pp. 772-777). Academic Press.

Marya, M., & Edwards, G. R. (2000). Factors affecting the laser bending of Ti-6Al-2Sn-4Zr-2Mo. *Journal of Laser Applications*, *12*(4), 149–159. doi:10.2351/1.521926

Paramasivan, K. (2019). *Study of laser forming of sheet metal with and without cutout* [Doctoral dissertation]. Jadavpur University, Kolkata, India.

Paramasivan, K., Das, S., Marimuthu, S., & Misra, D. (2015). Thermo-mechanical Simulation of the Laser Forming of AISI 304 Stainless Steel Square Plate with a Central Cutout. *Lasers in Engineering*, *32*(5-6), 393–410.

Paramasivan, K., Das, S., Marimuthu, S., & Misra, D. (2019). Numerical Study of the Mechanism of Laser Forming Process. *International Journal of Laser Science*, *1*(3-4), 347–382.

Paramasivan, K., Das, S., & Misra, D. (2014). A study on the effect of rectangular cutout on laser forming of AISI 304 plates. *International Journal of Advanced Manufacturing Technology*, *72*(9), 1513–1525. doi:10.100700170-014-5695-9

Safari, M., & Farzin, M. (2015). Experimental investigation of laser forming of a saddle shape with spiral irradiating scheme. *Optics & Laser Technology*, *66*, 146–150. doi:10.1016/j.optlastec.2014.09.003

Venkadeshwaran, K., Das, S., & Misra, D. (2010). Finite element simulation of 3-D laser forming by discrete section circle line heating. *International Journal of Engineering Science and Technology*, *2*(4), 163–175. doi:10.4314/ijest.v2i4.59284

Vollertsen, F. (1998). *Forming, sintering and rapid prototyping, Handbook of the Eurolaser Academy* (pp. 357–453). London: Chapman & Hall. doi:10.1007/978-1-4615-5297-0_6

Vollertsen, F., Komel, I., & Kals, R. (1995). The laser bending of steel foils for micro-parts by the buckling mechanism - A model. *Modelling and Simulation in Materials Science and Engineering*, *3*(1), 107–119. doi:10.1088/0965-0393/3/1/009

Woizeschke, P. (2019). Approach for bidirectional laser bending of sheet metal with one-sided accessibility. *CIRP Annals*, *68*(1), 277–280. doi:10.1016/j.cirp.2019.04.098

Chapter 13
Modelling and Analysis of Functionally-Graded Cracked Beams Subjected to Static and Dynamic Loadings

B. Panigrahi
https://orcid.org/0000-0002-1721-9127
Birla Institute of Technology and Science, Pilani, India

Goutam Pohit
Jadavpur University, India

ABSTRACT

In last few decades, continuous efforts have been put by the researchers on developing the field of structural systems considering the applicability and enhancing material properties of the structural elements. In this regard functionally graded materials (FGMs) have gained tremendous popularity and have been the active area of research interest for researchers in recent decades. In this chapter, a detailed method of analysis for FGM cracked beam subjected to static and dynamic loadings is presented. Method based on Timoshenko beam theory and Ritz method is successfully implemented to get the displacement and natural frequency of cracked beam subjected to static as well as dynamic loadings.

INTRODUCTION

In last few decades, continuous efforts have been put by the researchers on developing the field of structural systems considering the applicability and enhancing material properties of the structural elements. In this regard functionally graded materials (FGMs) have gained tremendous popularity and have been the active area of research interest for researchers in recent decades. Functionally graded materials (FGMs) are inhomogeneous composites obtained by mixing two or more constituent materials continuously following a given volume fraction. As a result, material properties become a function of position and a

DOI: 10.4018/978-1-7998-1831-1.ch013

continuous variation from one surface to another can be achieved. FGMs have much more advantages and now been used in various engineering sectors such as space vehicles, aerospace, automotive, biomedical, optics, electronics, and military applications, specifically in adverse operating conditions where uneven thermal gradients, chemical and mechanical loads are present. Concept of FGM was first introduced in Japan in 1980s, since then continuous efforts have been focused on developing and studying various aspects of FGM structures. In the light of advantages which FGM brings to the structural systems, it becomes necessary to study the behaviour of such structures subjected to static and dynamic loadings. For a healthier structure there are numerous amounts of works which can be found in open literatures. However, there are very limited works focused on showing the behaviour of cracked FGM structures. It is quite obvious that FGM structures responds differently with respect to the homogenous isotropic structures if there is a flaw such as cracks in the structures. These aspects need to get explored for various static and dynamic loading conditions.

Structures respond quite differently if there is a flaw such as cracks in the structures. Study of such cracked structures gained popularity in 1970s. Thereupon, most of the works were experimental based and concluded with some empirical and semi-empirical theories. In recent times due to the computational advancements various numerical and analytical studies on vibration behaviour of cracked beams with isotropic as well as FGM properties have been performed. Although various researchers have investigated the stress and strain behaviour near the crack tip on structural members Irwin (1957), the first work on the vibration monitoring of cracked structures is reported by Dimarogonas (1970), later by Pafelias (1974) in a project by General Electric Company. In this work, influence of cracks on the dynamic response of cracked rotors is explained by compliance measurement technique. Dimarogonas and Papadopoulos (1983) used the Paris energy equation to obtain local flexibility due to crack. Authors also obtained it experimentally. Subsequently, using these data, bending vibration of a cracked shaft is investigated. Gudmundson (1982) presented an equation to relate the change of natural frequencies with that of change in severity of cracks using a perturbation technique for transversely cracked bar problem. Author compared the results with the experimental and finite element results. In a later study Gudmundson (1983) proposed two different methods to determine static flexibility matrix, which infact represents the crack. One method was based on the integration of the stress intensity factors and other method was based on finite element techniques. Authors used these techniques for vibration analysis of edge cracked cantilever slender beams. Christides and Barr (1984) proposed a very famous crack model which is not based on fracture mechanics. The theory based on the assumption of an exponential decay in stress when moves away from crack location. Several empirical relations between the stress intensity factors and function of the crack depth can be found in several handbooks by Tada et al. (1985). Papadopoulos and Dimarogonas (1987) modeled an open crack which leads to similar behaviour to that of a rotor with dissimilar moments of inertia along two perpendicular directions. Subsequently, authors investigated the bending-extension coupling effect on a rotating shaft. Wauer (1990) published a brief review on the dynamic behaviour of cracked rotors. Author explained the necessity and the evolution on the field of cracked rotors in an extensive manner. Krawczuk and Ostachowicz (1992a) modeled a closing crack to investigate the influence of such cracks on the transverse vibration response of a cracked beam. Flexibility is obtained as the function of time as it varies with the opening and closing of crack. In connection to that first the variation of crack depth is modeled as a time dependent function. In first half of time period depth goes on increasing and decreasing with time, on the other half of time period crack depth remains constant as it vibrates as an intact beam. In another study Krawczuk (1993a) hhhconsidered an open non propagating crack in a plate to analyze the free vibration characteristics using a finite element technique. It is assumed that the presence of crack influences the stiffness only and the mass remains unchanged. In a subsequent article Krawczuk and Ostachowicz (1993b) investigated the effect of open crack on the forced vibration response of a beam. Finite element technique is used to analyze the dynamic response of cracked beam under vertical loads. Ostachowicz and Krawczuk (1993c) modeled an open crack by means of beam finite element and presented a bending-tensional coupled vibration analysis of a cracked rotor. Ostachowicz and Krawczuk (1991a) considered two types of cracks, double-sided crack, for the cyclic loading conditions, and single-sided crack, for fluctuating loading condition. Influence of such open cracks on the natural frequencies of beam vibration is examined. Papaeconomou and Dimarogonas (1989) explained the local flexibility induced due to presence of surface crack and subsequently used this to obtain a compliance matrix for the behavior of the beam in the vicinity of the crack. Effect of these parameters on the vibration behaviour of crack beams is investigated. Krawczuk (1992) modelled a non-propagating crack by incorporating an additional flexibility matrix, derived by using fracture mechanics approach, in the flexibility matrix of the corresponding healthier beam. Petroski (1981, 1984) used the equivalent slot approach to solve vibration problems of damaged beams, modeling the crack with a local member of reduced cross-section. Chondros and Dimarogonas (1980) studied the effect of cracks in welded structure and performed analytical as well as experimental investigation on dynamic response of such structures. In the subsequent work Chondros et al. (1998a) and Chondros (2001) used the cracked model of a continuous flexibility, found with fracture mechanics methods using the displacement field in the vicinity of the crack to determine its dynamic response. Dimarogonas (1996) published a review of the state of the art of vibration-based methods for testing cracked structures. Dentsoras and Dimarogonas (1983) studied the fatigue crack growth on resonating members of distributed mass. Authors investigated the deviations in resonance due to crack propagation. Crack propagation induces additional flexibility to the system and as a result shifts the resonance phenomenon considerably. Silva and Gomez (1990b) proposed an experimental technique in order to investigate the size and shape of crack. Crack detection in a beam like structure is done considering the modal analysis and vibration behaviour via an experimental technique. Kikidis and Papadopoulos (1992b) investigated the effect of slenderness ratio on the dynamic response of cracked beams. Authors used Euler Bernoulli's and Timoshenko beam theory and compared how the cracked thick beam response is quite different with that of thin beam when two different beam theories are employed. Qian et al. (1990c) used the fracture mechanics approach to derive the element stiffness matrix with the presence of crack in a beam. Subsequently using this matrix a finite element model is developed to study the eigenfrequency of cracked beam vibration for different crack severity. Krawczuk and Ostachowicz (1995) investigated Eigen frequency of a cantilever beam made of graphite fibre reinforced polyimide with transverse open crack. Crack is modelled as mass less spring, flexibility of the crack is found by using fracture mechanics theory and castigliaono's theorem. Problem is also solved by FEM approach. Nandwana and Maiti (1997a) measured the natural frequency for detection of the location and size of a crack in stepped cantilever beam using a rotational beam model. Relationship between stiffness of spring and crack location is found for at least three vibration modes. Point of intersection of the three curves gives the crack location and corresponding stiffness. Crack size is then computed using the standard relation between stiffness and crack size. Nandwana and Maiti (1997b) extended the method for crack detection in stepped cantilever beams. The crack is represented as a rotational spring, and the method involves obtaining plots of its stiffness with crack location for any three natural modes through the characteristic equation. The point of intersection of the three curves gives the crack location. Chati et al. (1997c) investigated the nonlinear effects due to opening and closing of crack on a cantilever beam. Modal analysis is performed for a transverse cracked cantilever beam. Nonlinearity is modelled as a piecewise linear system and idea of 'bilinear frequency' is used. A perturbation technique is used to detect nonlinear characteristics. Doebling et al. (1998b) published another extensive literature review of the state of the art of vibration-based damage identification methods. Kisa et al. (1998c) analyzed vibration characteristics of cracked Timoshenko beam using FEM and component mode synthesis. Flexibility matrix including interaction forces is derived and then problem is solved using fracture mechanics theory. Yokoyama and Chen (1998d) analyzed free vibration of uniform Euler-Bernoulli beam with single edge crack using a modified line spring model. Natural frequencies and mode shapes are obtained for various crack parameters. Takahashi (1999) studied vibration and stability analysis of a non-uniform cracked Timoshenko beam subjected to a tangential follower force distributed over the centre line by use of transfer matrix approach. Effects of cross section, span and stiffness of intermediate supports and Crack parameters on frequency and flutter loads are investigated. Kisa and Brandon (2000) developed a finite element scheme for computing the eigen system for a cracked beam for different degrees of closure. They integrated the finite element method, the component mode synthesis method and the linear elastic fracture mechanics theory for modeling the cracked structures. Sinha et al (2002) developed a simplified approach to modeling cracks in beams undergoing transverse vibration. The modeling approach uses Euler-Bernoulli beam elements with small modifications to the local flexibility in the vicinity of cracks. This crack model is then used to estimate the crack locations and sizes, by minimizing the difference between the measured and predicted natural frequencies via model updating. This simplified crack model allows the location and damage extent to be estimated directly. Nahvi and Jabbari (2005a) developed an analytical, as well as experimental approach to the crack detection in cantilever beams by vibration analysis. An experimental setup is designed in which a cracked cantilever beam is excited by a hammer and the response is obtained using an accelerometer attached to the beam. To avoid non-linearity, it is assumed that the crack is always open. To identify the crack, contours of the normalized frequency in terms of the normalized crack depth and location are plotted. Hsu (2005b) studied effect of sampling point number, crack parameter, axial loading, and foundation stiffness on natural frequency of a clamped-free and hinged-hinged single edge cracked Euler- Bernoulli beam by solving an eigen value problem derived numerically using differential quadrature method. Wu and Shih (2005c) investigated the dynamic instability and non-linear response of the cracked plates subjected to periodic excitation. Galerkin's method is used to reduce the governing equation of a cracked plate to a time-dependent Mathieu equation. The incremental harmonic balance (IHB) method is applied to solve the non-linear temporal equation of motion and analyze the dynamic instability. Marur and Prathap (2005d) employed three simplifications, which are not consistent with the governing variational principles. First one is equivalent quasi-linearization technique, which is shown to have a reduced non-linear stiffness. The second, where 'neglect of in plane displacements' takes place, is seen to register an excessive non-linear stiffness. Thirdly, when both these simplifications are introduced together, they produce results closer to those of variationally correct ones. They highlighted the necessity of formulating this class of problems in variationally correct and consistent manner. Chasalevris and Papadopoulos (2006a) studied the dynamic behaviour of a cracked beam with two transverse surface cracks. Each crack is characterised by its depth, position and relative angle. A local compliance matrix of two degrees of freedom, bending in the horizontal and the vertical planes is used to model the rotating transverse crack in the shaft and is calculated based on the available expressions of the stress intensity factors and the associated expressions for the strain energy release rates. Bikri et al. (2006b) investigated geometrically non-linear free vibrations of a clamped–clamped beam containing an open crack. Authors introduced a new function called ''cracked beam functions'', which satisfy the natural and geometrical end conditions, as well as the inner boundary conditions at the crack location. Iterative solution of a set of non-linear algebraic equations is obtained numerically. Then, an explicit solution is derived and proposed as an alternative procedure. Lin and Chang (2006c) developed an analytical method for the dynamic analysis of a cracked cantilever beam subjected to a concentrated moving load. The crack is modeled as a rotational spring with sectional flexibility. Two Individual sub beams are assumed to follow Euler-Bernoulli's beam theory. Considering the compatibility requirements on the crack, the relationships between these two spans are obtained. By using the analytical transfer matrix method, Eigen solutions of this cracked system are obtained explicitly. The forced responses are obtained by the modal expansion theory using the determined Eigen functions. Loya et al. (2006d) obtained natural frequencies for bending vibrations of Timoshenko cracked beams. The beam model promotes discontinuities in both transverse as well as rotational displacements. The differential equations for the free bending vibrations are established and then solved individually for each segment with the corresponding boundary conditions and the appropriated compatibility conditions at the cracked section. The problem is also solved by the perturbation method. Sadettin [46] studied the free and forced vibration analysis of a cracked beam in order to identify the crack in a cantilever beam. Single- and two-edge cracks are evaluated. Kisa and Gurel (2007a) proposed a numerical technique applicable to analyze the free vibration characteristics of uniform and stepped cracked beams with circular cross section. In this approach the beam is detached into parts from the crack section. These substructures are joined by using the flexibility matrices taking into account the interaction forces, derived by virtue of fracture mechanics theory as the inverse of the compliance matrix found with the appropriate stress intensity factors and strain energy release rate expressions. Aydin (2007b) presented analytical method to determine the vibrational frequencies and mode shape functions of axially-loaded Timoshenko beams with an arbitrary number of cracks. The local compliance induced by a crack is described by a mass less rotational spring model. A set of boundary conditions are used as initial parameters to define the mode shape of the segment of the beam before the first crack. Using this, the remaining set of boundary conditions and the mode shape function of vibration of the beam containing multiple cracks are determined. Viola et al. (2007c) investigated the changes in the natural frequencies and modal response introduced by the presence of a crack on an axially loaded uniform Timoshenko beam. A new method based on the coupling of dynamic stiffness matrix and line-spring element is introduced to model the cracked beam. The application of the theory is demonstrated by two examples of bending–torsion coupled beams with different end conditions, for which the influence of axial force, shear deformation and rotary inertia on the natural frequencies is studied. Demir and Mermertas (2008a) investigated the natural frequencies of annular plates with circumferential open cracks by using finite element method. In the event of the selection of appropriate half sector angle, taking into account the closeness of the geometries of the sector and trapezoidal element, the flexibility matrix of the sector element with crack was obtained by the flexibility matrix of the trapezoidal element. Also, because of small half sector angles, the circumferential cracks in the elements were assumed to be straight cracks. Yang et al. (2008b) presented an analytical study on the free and forced vibration of inhomogeneous Euler–Bernoulli beams containing open edge cracks. The beam is subjected to an axial compressive force and a concentrated transverse load moving along the longitudinal direction. The rotational spring model is used to model the crack causing sectional flexibility. The forced response is determined by employing modal series expansion technique. Ke et al. (2009a) studied the post buckling response of beams made of functionally graded materials containing an open edge crack based on Timoshenko beam theory and von Karman nonlinear kinematics. Ritz method is employed to derive the nonlinear governing equations which are solved using Newton -Raphson method. Effect of bending-extension coupling is also studied. Kitipornchai et al. (2009b) modeled the functionally graded Timoshenko beam as two sub beam connected by a rotational mass less spring. Using Ritz method followed by a direct iterative process fundamental nonlinear frequencies are obtained. Effect of bending extensional coupling for functionally graded material is shown which has not any significant effect on isotropic materials. Resaee and Hassannejad (2011a) proposed a new approach to free vibration analysis of a beam with a breathing crack based on mechanical energy balance method. They modelled the crack opening and closing phenomenon as a fatigue crack. Local stiffness changes at the crack location are considered to be a nonlinear amplitude-dependent function and it is assumed that during one half cycles, the frequencies and mode shapes of the beam vary continuously with time. Yan and Yang (2011b) presented an analytical study on the forced flexural vibration of functionally graded beams with open edge cracks under a combined action of an axial compressive force and a concentrated transverse load moving along the longitudinal direction. The cracked section is modelled as a rotational spring whose sectional flexibility is calculated through fracture mechanics. The forced vibration response is determined by employing modal series expansion technique. Zheng and Ji (2012a) has analysed a crack beam by representing a cracked beam as an un-cracked beam with equivalent reduced sections around the cracks. Wei et al. (2012b) introduced an analytical method for solving the free vibration of cracked FGM beams with axial loading using rotational beam model. Based on the transfer matrix method, then the recurrence formula is developed to get the Eigen value equations of free vibration of FGM beams. Due to the decrease in the determinant order, the developed method is simpler and more convenient to analytically solve the free vibration problem of cracked FGM beams. Mazanoglu (2015a) studied forward and backward problem of cracked beam. In forward problem drop in natural frequencies caused by cracks are studied. On the other hand by knowing the connection between crack depth and natural frequency drop, inverse problem is carried out, which helps significantly in crack detection process. Wang and Wang (2015b) investigated the free vibration response of cracked nanobeams. Influence of surface energy and transverse shear deformation on frequencies for fundamental as well as higher mode of vibration of cracked nanobeams is studied. Yang et al. (2015c) studied the free vibration response of edge cracked FGM beams based on Euler-Bernoulli beam theory and a continuous stiffness beam model. Crack modelling is established by treating the cracked FGM beam as an intact beam with continuously varying bending stiffness along its longitudinal direction. Batihan and Kadioglu (2016) modeled the crack as an elastic rotational spring. Using different beam theories dynamic analysis of cracked beams with an Elastic Foundation is investigated. Capozucca and Magagnini (2017a) examined the natural frequencies of vibration of cracked RC beam using a finite element technique as well as experimentally. Khiem and Hang (2017b) proposed an exact expression in the frequency domain for the response of a multiple-cracked beam subjected to a moving harmonic load. The obtained solution is used first for response analysis of the beam in dependence on the load speed, frequency, and crack parameters. Then, based on the solution a procedure is developed for detecting multiple cracks in a beam from the measured frequency response. Liu et al. (2017c) proposed a new cracked cantilever beam model to study the vibration of a cantilever beam with a slant edge crack. An equivalent stiffness model is proposed by dividing the beam into numerous uniform independent thin pieces along its width. The crack is assumed to be distributed along the width of the beam as a straight line and a parabola. The methodology proposed in this work can also be extended to model a crack with an arbitrary curve. Effects of crack depths on the non-dimensional equivalent stiffness at the crack section of the cracked cantilever beam are studied.

It is evident from the literatures that various works have been focused on various aspects of FGM and cracked beams individually. However, in this chapter the objective is to combine all the modelling techniques for cracks and FGM beams into a single chapter and subsequently proposing two different model for static and dynamic analysis of such FGM structures.

Method of Analysis

In the present section a detailed method of analysis for FGM cracked beam subjected to static and dynamic loadings is presented. Firstly, the FGM models are discussed. Subsequently crack modelling and method of analysis is presented in detail. In the present work, the FGM modelling is accomplished by assuming the continuous variation in the material properties in the thickness direction of a beam. In order to achieve these properties, two different gradation models are taken into consideration. In model 1, an exponential variation in the material properties is assumed along the thickness of the beam. In model II, gradation is chosen by a power law considering two different constituents. In figure (1), cracked FGM beams are shown. Length, width, thickness, crack location and crack depths are represented as L, b, h, L_1 and a, respectively. Origin of the coordinate frame is chosen at the mid surface of the left fixed end. As shown in figure (1), downward z direction from mid surface is considered as positive. In figure (1a), properties such as Young's modulus (E), mass density (ρ) are represented by P_1 at top surface, whereas P_2. at bottom surface of the beam. Gradation relations for model I and model II are give in equation (1).

$$\begin{aligned} &Model\,I : P(z) = P_1 \sqrt{k} e^{\left(\frac{z}{h}\ln(k)\right)} \\ &Model\,II : P(z) = \left(P_2 - P_1\right)\left(\frac{z}{h} + \frac{1}{2}\right)^k + P_1 \end{aligned} \tag{1}$$

In equation (1), k be material property index. It is to be noted here that FGM should be selected from different groups of the constituents, Material 1 and Material 2 as in model II of present work. This gives a more physical insight to the material distribution. However, exponential gradation relation is also a type of gradation. Although it has a constraint $\frac{E_2}{E_1} = \frac{\rho_2}{\rho_1}$, various authors have used this exponential gradation relation due to its simplicity in modelling. In this regard, we have considered one another FGM model as mentioned in equation (1) as model II., where P(z) is the properties of the beam, which varies in thickness direction. At the bottom surface (z = h/2), P(z) = P_2 (ceramic properties) and at top surface

Figure 1. a) FGM beam model with a crack (b) Crack modeled as rotational spring.

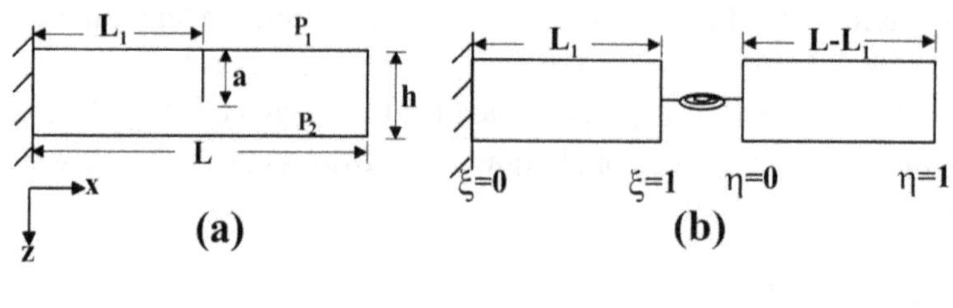

Figure 2. Variation in properties along depth (a) Model I (b) Model II.

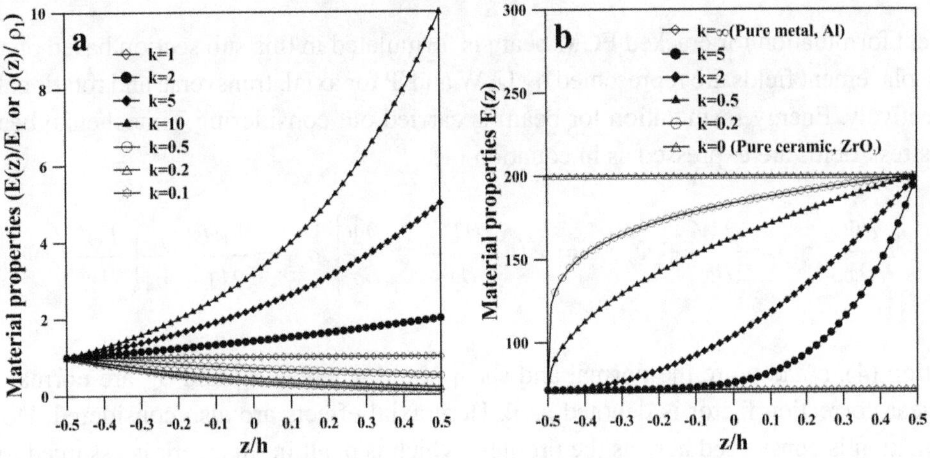

(z = -h/2), P(z)= P_1 (metal at top surface). Variations in the properties in thickness direction for both the models considering various material indices (k) are shown in figure (2).

Crack is modeled as proposed by Panigrahi and Pohit [65] assuming a mass less rotational spring at the cracked section and the stiffness of rotational spring (K_t) is obtained as inverse of the flexibility (1/G) induced due to crack. Locally induced flexibility due to crack is derived using fracture mechanics approach and is provided in Equation (1)

$$G = \int_0^{\varsigma} \frac{72\pi(1-\nu^2)\varsigma F^2\left(\varsigma\right)}{E(z)h^2} d\varsigma \tag{2}$$

In, equation (2), υ, ς and E are the poison's ratio, crack depth ratio (a/h) and modulus of elasticity, respectively. F (ς) represents the crack correction factor which is given by Kitipornchai et al. [53] and Panigrahi and Pohit [65] as in equation (3)

$$\begin{aligned} &F(\varsigma) = 1.150 - 1.662\varsigma + 21.667\varsigma^2 - 192.451\varsigma^3 \\ &+909.375\varsigma^4 - 2124.310\varsigma^5 + 2395.83\varsigma^6 - 1.031.750\varsigma^7 \ \text{for k} = 1 \\ &F(\varsigma) = 1.762 + 0.3\varsigma - 24.225\varsigma^2 + 170.387\varsigma^3 \\ &-554.549\varsigma^4 + 1009.567\varsigma^5 - 971.767\varsigma^6 + 395.919\varsigma^7 \ \text{for k} = 2 \\ &F(\varsigma) = 1.910 - 2.752\varsigma - 4.742\varsigma^2 + 146.776\varsigma^3 \\ &-770.750\varsigma^4 + 1947.830\varsigma^5 - 2409.170\varsigma^6 + 1177.980\varsigma^7 \ \text{for k} = 0.2 \\ &F(\varsigma) = 0.650 - 0.859\varsigma + 12.551\varsigma^2 - 72.627\varsigma^3 \\ &+267.910\varsigma^4 - 535.236\varsigma^5 + 545.139\varsigma^6 - 211.706\varsigma^7 \ \text{for k} = 5 \end{aligned} \tag{3}$$

Energy Formulation of Statically Loaded FGM Cracked Beam

Mathematical formulation for cracked FGM beam is formulated in this sub section based on the energy method. Displacement fields are represented by U, W and Ψ for axial, transverse and rotational displacement, respectively. Energy formulation for beam is carried out considering Timoshenko beam theory. Strain and stress fields are expressed as in equation (4)

$$\varepsilon_{xx} = \frac{\partial U}{\partial x} + z\frac{\partial \Psi}{\partial x},\ \gamma_{xz} = \frac{\partial W}{\partial x} + \Psi, \sigma_{xx} = E(z)\left[\frac{\partial U}{\partial x} + z\frac{\partial \Psi}{\partial x}\right], \sigma_{xz} = \frac{\kappa E(z)}{2\left(1+\upsilon\right)}\left(\frac{\partial W}{\partial x} + \Psi\right) \tag{4}$$

In equation (4), ε and γ are the normal and shear strain whereas σ_{xx} and σ_{xz} are normal and shear stresses. Shear correction factor is denoted as κ. Here axial effects are also considered. However, no axial deformation is considered here as the problem which is dealt in this work is assumed to be in the linear range and no additional stretching is considered. Therefore, in the context of present work, axial deformation can be neglected and all the subsequent formulations are carried out without considering the axial deformation. Potential energy of the problem can be derived as mentioned in equation (5).

$$U = \frac{1}{2}\int_0^L \int_{\frac{-h}{2}-C}^{\frac{h}{2}-C} \int_{-b/2}^{b/2} \left\{E(z)z^2\left(\frac{\partial \Psi}{\partial x}\right)^2 + \frac{\kappa E(z)}{2\left(1+\nu\right)}\left(\frac{\partial W}{\partial x} + \Psi\right)^2\right\} dx dy dz + \frac{1}{2}K_t\left(\Delta\Psi_{x=L_1}\right)^2 \tag{5}$$

Stiffness parameters are considered as

$$K_1 = \int_{\frac{-h}{2}-C}^{\frac{-h}{2}-C} E(z)z^2 dz, K_2^b = \int_{\frac{-h}{2}-C}^{\frac{-h}{2}-C} \frac{\kappa E(z)}{2\left(1+\nu\right)} dz \tag{6}$$

In equations (5) and (6), C be the distance between mid plane and neutral axis and given by

$$C = \int_{-h/2}^{h/2} zE(z)dz \Bigg/ \int_{-h/2}^{h/2} E(z)dz\,.$$

On the substitution of parameters from equation (6) to equation (5), equation (5) reduces to

$$U = \frac{b}{2}\int_0^L \left\{K_1\left(\frac{\partial \Psi}{\partial x}\right)^2 + K_2\left(\frac{\partial W}{\partial x} + \Psi\right)^2\right\} dx + \frac{1}{2}K_t\left(\Delta\Psi_{x=L_1}\right)^2 \tag{7}$$

It has been concluded in the crack modeling by Panigrahi and Pohit (2016) that first mode of fracture dominates the total strain energy of the system and hence it can be assumed that at the crack location, there is continuity in transverse (W) direction whereas rotational displacement is discontinuous. Therefore, two different sub beams in either side of the crack is assumed and entire span of the beam is divided into two parts. Displacement fields are named with subscript 1 and 2 for sub beams. Non dimensional scheme is followed as mentioned in equation (8)

$$\xi = \frac{x}{L_1}\ ,\ \eta = \frac{x - L_1}{L - L_1}, w = \frac{W}{h}, \psi = \Psi, \alpha = \frac{L_1}{h}, \alpha_1 = \frac{L - L_1}{h}, \beta = \frac{L - L_1}{L_1}$$

$$\{k_1, k_2\} = \left\{ \frac{K_1}{\int_{\frac{-h}{2}-C}^{\frac{-h}{2}-C} E(z)dzh^2}, \frac{K_2}{\int_{\frac{-h}{2}-C}^{\frac{-h}{2}-C} E(z)dz} \right\}. \tag{8}$$

Potential of load for the concentrated load F acting transversely at x_f distance from the left fixed end can be expressed as $V = FW_{x=x_f}$. Dividing equation (7) and load potential (V) by

$$\int_{\frac{-h}{2}-C}^{\frac{-h}{2}-C} E(z)dzh^2 \Big/ L - L_1$$

non-dimensional form of energy equations (U* and V*) for different sub beams can be rewritten as provided in equations (9) and (10).

$$U^* = \frac{\beta}{2}\int_0^1 \left\{ k_1\left(\frac{\partial \psi_1}{\partial \xi}\right)^2 + k_2\left(\frac{\partial w_1}{\partial \xi}\right)^2 + k_2\alpha^2\psi_1^2 + k_2\alpha\frac{\partial w_1}{\partial \xi}\psi_1 \right\} d\xi$$

$$+\frac{1}{2}\int_0^1 \left\{ k_1\left(\frac{\partial \psi_2}{\partial \xi}\right)^2 + k_2\left(\frac{\partial w_2}{\partial \xi}\right)^2 + k_2\alpha^2\psi_2^2 + k_2\alpha\frac{\partial w_1}{\partial \xi}\psi_2 \right\} d\eta + \frac{1}{2}k_t^*\left(\Delta\psi\right)^2 \text{ where } k_t^* = \frac{L - L_1}{\int_{\frac{-h}{2}-C}^{\frac{-h}{2}-C} E(z)dzh^2}K_t \tag{9}$$

Normalised load potential is given as

$$V^* = \frac{Fw_{\xi p}\left(L - L_1\right)}{\int_{\frac{-h}{2}-C}^{\frac{-h}{2}-C} E(z)dzbh} \tag{10}$$

Energy functional can be obtained as

$$\pi = (U^* + V^*) \tag{11}$$

Selection of the Trial Functions

In the present work, displacement fields are assumed by choosing the suitable admissible coordinate functions. Therefore, by definition, displacement functions automatically satisfy kinematic boundary conditions as they are chosen from the class of admissible functions. For the sake of simplicity these functions are chosen as polynomials depending on the axial coordinates. Since these class of admissible functions do not automatically satisfy the compatibility conditions arising due to presence of cracks, additional set of conditions are required to be imposed on the displacement functions to take care of the discontinuities in the displacement fields due to crack. As mentioned earlier that mode II and mode III fracture has insignificant influence hence neglected for the present formulation, restrains on discontinuity for transverse displacement fields automatically vanishes. Only restrain persists is on the rotational discontinuity at the crack location as it arises due to mode I fracture. In order to satisfy this criterion, there must be a discontinuity in the rotational displacement and at the crack section rotational displacement should satisfy the moment compatibility condition. Here, the trial functions for clamped-clamped beam are derived as mentioned by Panigrahi and Pohit [65] again for the ready reference. For clamped-clamped beam moment (M_x) can be derived as

$$M_x = \int_{-h/2}^{h/2} \sigma_{xx} z dz = \int_{-h/2}^{h/2} E(z) z^2 \left(\frac{\partial \Psi}{\partial x} \right) dz = \left(K_1 \right) \frac{\partial \Psi}{\partial x} \tag{12}$$

Discontinuity in the rotational displacement can also be written in the form of bending moment at that point as

$$M_{|x=L_1} = K_t \left(\Psi_{2|x=L_1} - \Psi_{1|x=L_1} \right) \tag{13}$$

Comparing equations (12) and (13) at the crack location one can obtain the required compatibility criterion. In equation (14), this condition is mentioned in the non dimensional form as

$$k_t^* \left(\psi_{2|\eta=0} - \psi_{1|\xi=1} \right) = \left(k_1 \right) \frac{\partial \psi}{\partial \eta} |_{\eta=0} \tag{14}$$

Imposing the crack compatibility equation derived in equation (14) on the kinematic admissible coordinate functions, displacement fields for a clamped-clamped cracked FGM beam are obtained as mentioned in equation (15)

$$w_1 = \sum_{j=1}^{j=N} B_j \xi^{j+1}\left(1-\xi\right) + \xi \sum_{j=1}^{j=N} b_j,\ \ w_2 = \sum_{j=1}^{j=N} b_j \left(1-\eta\right)^{j+1}$$
$$\psi_1 = \sum_{j=1}^{j=N} C_j \xi^{j+1}\left(1-\xi\right) + \xi \sum_{j=1}^{j=N} \left\{1 + \left(j+1\right)\frac{k_1}{k_t^*}\right\} c_j,\ \ \psi_2 = \sum_{j=1}^{j=N} c_j \left(1-\eta\right)^{j+1} \tag{15}$$

In equation (15), B_j, b_j, C_j and c_j are the unknown coefficients, which need to get determined in the subsequent stages in order to obtain the complete solution of the displacement fields. Substituting the trial functions from equations (15) into equation (11) and optimizing the energy functional with respect to unknown coefficients (B_j, b_j, C_j and c_j), set of algebraic equations can be derived as

$$[K]\{c\} = \{f_v\} \tag{16}$$

In equation (16), K be the stiffness matrix, c is vector of unknown coefficients and f_v is the force vector. Equation (16), can be easily solved now using matrix inversion technique.

Solution of Dynamic Problem

In the present work, dynamic response of the FGM cracked beam is also obtained for initial excitation (free vibration). Method of analysis is an extended version of the static problem mentioned earlier. However, the displacement fields for this case is expressed in variable separable for considering the spatial functions as in equation (15) and a temporal function. Since the problem here is linear, it would be convenient for the computational purpose to assume a simple harmonic motion and temporal function for both transverse and rotational displacement fields are assumed as $\exp(i\Omega t)$, where Ω be the natural frequencies for free vibration of FGM cracked beam and t be the time. Therefore, considering the maximum kinetic energy, temporal terms can be eliminated. Kinetic energy expressions in non-dimensional form can be derived similar to the potential energy expression and can be obtained as mentioned in equation (17)

$$T_{k\max}^* = \frac{\omega^2}{2}\left[\int_0^1 \left(m_1\psi_1^2 + w_1^2\right) d\xi + \beta \int_0^1 \left(m_1\psi_2^2 + w_2^2\right) d\eta\right] \tag{17}$$

Various parameters of equation (17) are described here in equation (18).

$$\omega = \Omega \sqrt{L_1\left(L - L_1\right) \frac{\int_{\frac{-h}{2}-C}^{\frac{h}{2}-C} \rho(z)dz}{\int_{\frac{-h}{2}-C}^{\frac{h}{2}-C} E(z)dz}},\ m_1 = \frac{M_1}{\int_{\frac{-h}{2}-C}^{\frac{h}{2}-C} \rho(z)dz h^2},\ M_1 = \int_{\frac{-h}{2}-C}^{\frac{h}{2}-C} \rho(z)z^2 dz \tag{18}$$

In equation (18), ρ be the mass density of the material. Energy functional for the free vibration problem can now be expressed as mentioned below

$$\pi_d = \left(U^* - T^*_{k\max}\right) \tag{19}$$

Optimizing equation (19), with respect to unknown coefficients similarly as mentioned for static problem, one can derive the set of algebraic equations in the form of eigen value problem as given in equation (20)

$$[K]\{c\} - \omega^2[M]\{c\} = \{0\} \tag{20}$$

K, and c in the equations are same as mentioned in equation (16), whereas M be the mass matrix. Solution of equation (20) for natural frequencies and unknown coefficients can be easily obtained using any eigenvalue solver. In the present work QR algorithm is adopted to find the eigenvalues and eigenvectors of the problem. By further processing, natural frequencies are extracted from the eigenvalues and modeshapes are obtained from eigenvectors.

Results and Discussions

Methodology presented above is implemented through a in code developed MATLAB. Firstly, a convergence study is performed to obtain the minimum numbers of polynomial terms required to completely specify the displacement fields. It is found that eight numbers of terms are sufficient enough for accurate results. Therefore, all the analysis performed here is carried out by taking eight numbers of terms. To validate the results obtained by the present methodology, similar analysis is also performed in a finite element package ANSYS. Geometric and material properties for beam are taken as mentioned in table 1 throughout the analysis.

A comparison of the natural frequencies of free vibration of FGM beams with model I and II with finite element results are presented in Table 2. It is found that the results obtained using proposed method matches quit well with that of finite element results. Geometric and material properties of beam for validation purpose are taken as E_1 = 70 GPa, ν = 0.33, ρ_1 = 2780 kg/m^3, L/h = 16, h = 0.1 m, ς = 0.2, κ = 5/6. For subsequent analysis in this chapter, these properties are chosen as mentioned in table 1.

Table 1. Geometric and material properties of the FGM beams

E_1	69GPa
Mass density of metal	2789 Kg/m^3
E_2	200GPa
Mass density of ceramic	5500 Kg/m^3
L	0.4m
h	0.01
b	0.05

Table 2. Comparison of present linear results with FEM results (Modal Analysis)

Crack Location (L1)	First modeFrequency (rad/sec)		Error (%)	Second modeFrequency (rad/sec)		Error (%)	Third modeFrequency (rad/sec)		Error (%)
	Present	ModalAnalysis (FEM)		Present	ModalAnalysis (FEM)		Present	ModalAnalysis (FEM)	
0.2	1288.2	1280	0.63	3455.6	3433.8	0.63	6462.9	6463.3	0.006
0.4	1299.2	1286.8	0.96	3401.7	3398.5	0.09	6385.9	6403.7	-0.27
0.6	1285.8	1278.4	0.58	3406.2	3400	0.18	6473.9	6464	0.10
0.8	1276.8	1272.7	0.32	3455	3431.8	0.67	6360.9	6376.2	-0.2

Figure 3 shows the static displacement response of FGM beams with a load 100N applied at the mid of a C-C beam. It can be concluded here that the displacement increases as crack depth increases. This is due to the additional flexibility which is induced by the crack.

Figure 3. Static response of cracked FGM beams for model II FGM; (a) k=infinity; (b) k=5; (c) k=2; and (d) k=0.2.

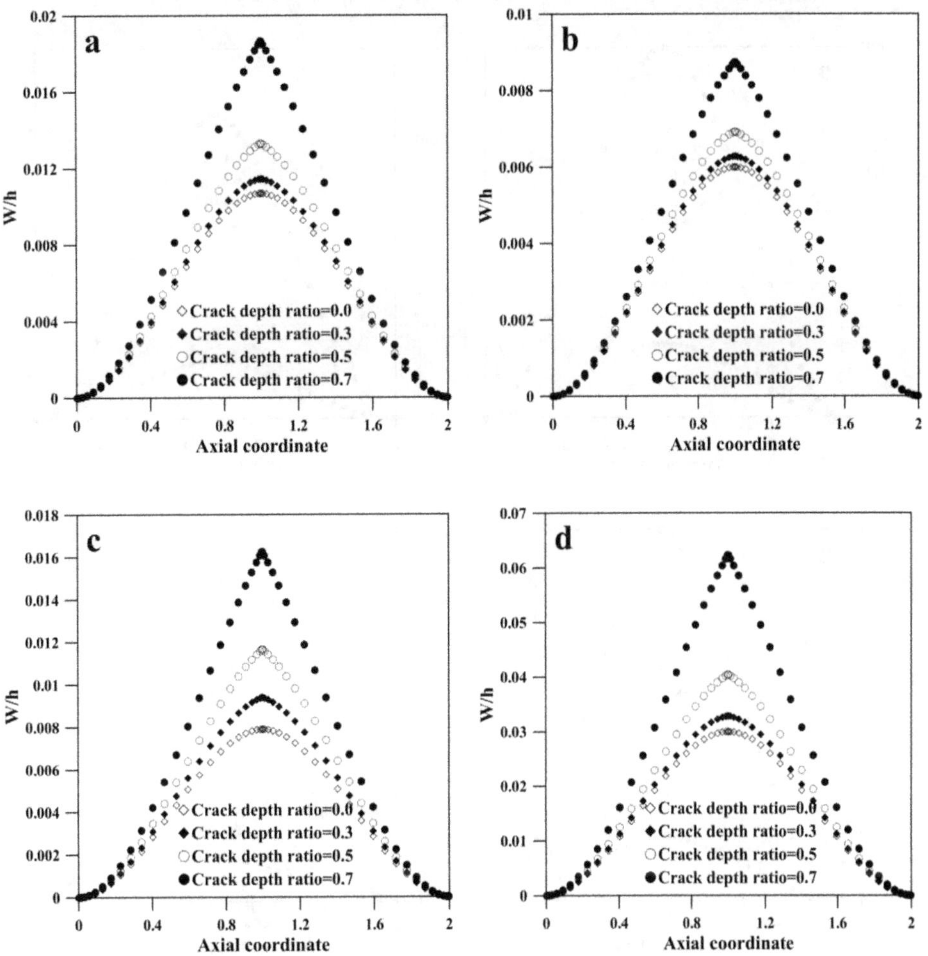

In Figure 4, response of model I FGM beams for different values of material indices are compared for various crack depth ratio.

Variation in the maximum displacement when the crack location is varied can be observed in figure 5. It can be observed that maximum displacement is obtained when the crack is located at the mid beam. Crack severity is maximum when it is situated at the mid of a C-C beam and hence it induces maximum flexibility to the beam causing maximum displacement. It can also be noted that displacement is maximum for k = 0.2.

Variation in the natural frequency of vibration for Model II FGM beams are shown in figure 6. A significant reduction in the natural frequency can be obtained. It can be observed that the linear frequencies decrease significantly with increasing crack depths. Cracks introduce an additional local flexibility to the system and stiffness gets reduced. A crack also influences the mass properties but decrement in natural frequencies with increment in crack depth signifies that influence of cracks on mass is insignificant compared to influence on stiffness. It can also be observed that influence of cracks is significantly different for different material indices. It can be observed that the FGM beam with k = 0.2 shows a significantly higher reduction in the natural frequencies compared to the FGM beam with k = 5.

Figure 4. Static response of model I FGM beams with F=100N at mid pint; (a) crack depth ratio=0; (b) crack depth ratio=0.3; (c) crack depth ratio=0.5; (d) crack depth ration=0.7.

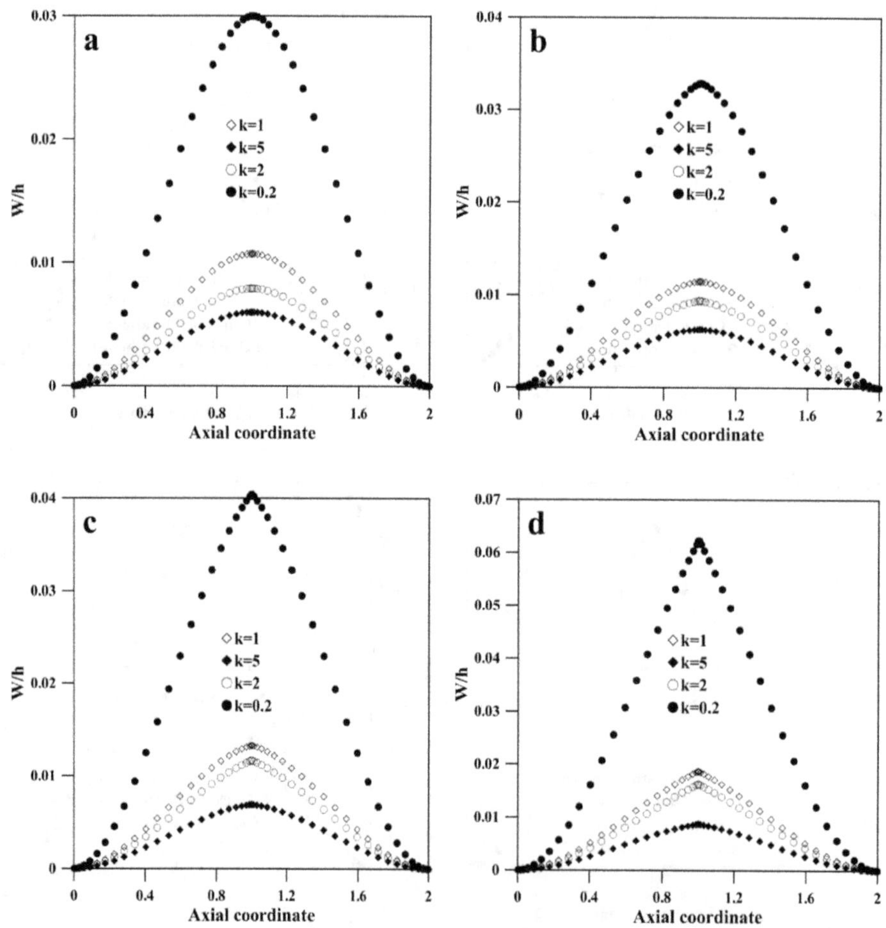

Figure 5. Variation in maximum displacement vs. crack location.

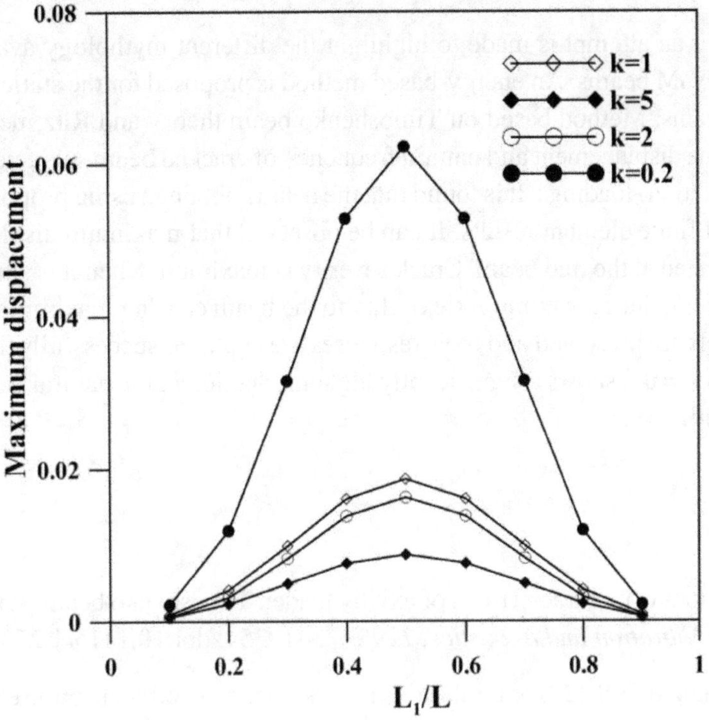

Figure 6. Variation in natural frequency of vibration vs. crack depth ratio.

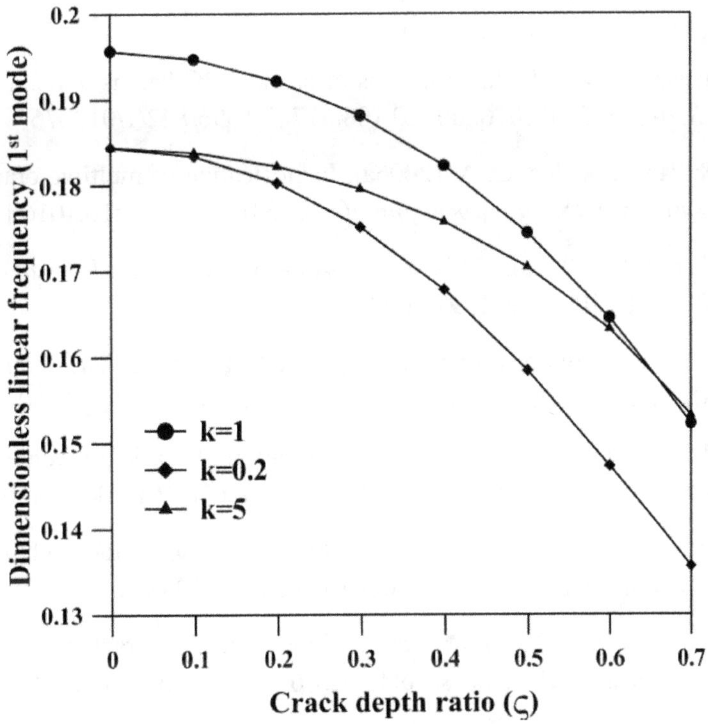

CONCLUSION

In the present chapter an attempt is made to highlight the different mythology available to model and analyze the cracked FGM beams. An energy-based method is proposed for the static and dynamic analysis of the cracked beams. Method based on Timoshenko beam theory and Ritz method is successfully implemented to get the displacement and natural frequency of cracked beam subjected to static as well as dynamic (initial excitation) loadings. It is found that the results obtained using proposed method matches quit well with that of finite element results. It can be observed that maximum displacement is obtained when the crack is located at the mid beam. Crack severity is maximum when it is situated at the mid of a C-C beam and hence it induces maximum flexibility to the beam causing maximum displacement. Two different FGM models are presented and their responses are captured successfully. It has been observed that FGM beam with $k = 0.2$ shows a significantly higher reduction in the natural frequencies compared to the FGM beam with $k = 5$.

REFERENCES

Aydin, K. (2007c). Vibratory characteristics of axially loaded Timoshenko beams with arbitrary number of cracks. *Journal of Vibration and Acoustics*, *129*(3), 341–354. doi:10.1115/1.2731411

Batihan, A. C., & Kadioglu, F. S. (2016a). Vibration analysis of a cracked beam on an elastic foundation. *International Journal of Structural Stability and Dynamics*, *16*(05), 1–18. doi:10.1142/S0219455415500066

Bikri, K. E., Benamar, R., & Bennouna, M. (2006b). Geometrically nonlinear free vibration of clamped-clamped beams with an edge crack. *Computers & Structures*, *84*(7), 485–502. doi:10.1016/j.compstruc.2005.09.030

Capozucca, R., & Magagnini, E. (2017a). Analysis of cracked RC beams under vibration. *Journal of Physics: Conference Series*, *842*, 012076. doi:10.1088/1742-6596/842/1/012076

Chasalevris, A. C., & Papadopoulos, C. A. (2006a). Identification of multiple cracks in beams under bending. *Mechanical Systems and Signal Processing*, *20*(7), 1631–1673. doi:10.1016/j.ymssp.2006.03.008

Chati, M. R., & Mukharjee, S. (1997c). Modal analysis of a cracked beam. *Journal of Sound and Vibration*, *207*(2), 249–270. doi:10.1006/jsvi.1997.1099

Chondros, T. G. (2001). The continuous crack flexibility model for crack identification. *Fatigue & Fracture of Engineering Materials & Structures*, *24*(10), 643–650. doi:10.1046/j.1460-2695.2001.00442.x

Chondros, T. G., & Dimarogonas, A. D. (1980). Identification of cracks in welded joints of complex structures. *Journal of Sound and Vibration*, *69*(4), 531–538. doi:10.1016/0022-460X(80)90623-9

Chondros, T. G., Dimarogonas, A. D., & Yao, J. A. (1998). Continuous cracked beam vibration theory. *Journal of Sound and Vibration*, *215*(1), 17–34. doi:10.1006/jsvi.1998.1640

Christides, S., & Barr, A. D. S. (1984a). One-dimensional theory of cracked Bernoulli-Euler beams. *International Journal of Mechanical Sciences*, *26*(11-12), 639–648. doi:10.1016/0020-7403(84)90017-1

Demir, A., & Mermertas, V. (2008a). Natural frequencies of annular plates with circumferential cracks by means of sector element. *Engineering Fracture Mechanics*, *75*(5), 1143–1155. doi:10.1016/j.engfracmech.2007.04.016

Dentsoras, A., & Dimarogonas, A. D. (1983c). Resonance controlled fatigue crack propagation in a beam under longitudinal vibrations. *International Journal of Fracture*, *23*(1), 15–22. doi:10.1007/BF00020154

Dimarogonas, A. D. (1970). *Dynamic response of cracked rotors*. Schenectady, NY: General Electric Co.

Dimarogonas, A. D. (1996). Vibration of cracked structures: The state of the art review. *Engineering Fracture Mechanics*, *55*(5), 831–857. doi:10.1016/0013-7944(94)00175-8

Dimarogonas, A. D., & Papadopoulos, C. A. (1983a). Vibration of cracked shafts in bending. *Journal of Sound and Vibration*, *91*(4), 583–593. doi:10.1016/0022-460X(83)90834-9

Doebling, S. W., Ferrar, C. R., & Prome, M. B. (1998a). A summary review of vibration-based damage Identification methods. *The Shock and Vibration Digest*, *30*(2), 91–105. doi:10.1177/058310249803000201

Gudmundson, P. (1982). Eigen frequency changes of structures due to cracks, notches or other geometrical changes. *Journal of the Mechanics and Physics of Solids*, *30*(5), 339–353. doi:10.1016/0022-5096(82)90004-7

Gudmundson, P. (1983b). The dynamic behavior of slender structures with cross-sectional cracks. *Journal of the Mechanics and Physics of Solids*, *31*(4), 329–345. doi:10.1016/0022-5096(83)90003-0

Hsu, M. H. (2005b). Vibration analysis of edge-cracked beam on elastic foundation with axial loading using the differential quadrature method. *Computer Methods in Applied Mechanics and Engineering*, *194*(1), 1–17. doi:10.1016/j.cma.2003.08.011

Irwin, G. R. (1957). Analysis of stresses and strains near the end of a crack transversing a plate. *Journal of Applied Mechanics*, *24*, 361–364.

Ke, L. L., Yang, J., & Kitipornchai, S. (2009a). Post buckling analysis of edge cracked functionally graded Timoshenko beams under end shortening. *Composite Structures*, *90*(2), 152–160. doi:10.1016/j.compstruct.2009.03.003

Khiem, N. T., & Hang, P. T. (2017b). Analysis and identification of multiple-cracked beam subjected to moving harmonic load. *Journal of Vibration and Control*, 1–20.

Kikidis, M. L., & Papadopoulos, C. A. (1992c). Slenderness ratio effect on cracked beam. *Journal of Sound and Vibration*, *155*(1), 1–11. doi:10.1016/0022-460X(92)90641-A

Kisa, M., Brandon, J., & Topcu, M. (1998b). Free vibration analysis of cracked beams by a combination of finite elements and component mode synthesis methods. *Computers & Structures*, *67*(4), 215–223. doi:10.1016/S0045-7949(98)00056-X

Kisa, M., & Brandon, J. A. (2000). The effects of closure of cracks on the dynamic of a cracked cantilever beams. *Journal of Sound and Vibration*, *238*(1), 1–18. doi:10.1006/jsvi.2000.3099

Kisa, M., & Gurel, M. A. (2007b). Free vibration analysis of uniform and stepped cracked beams with circular cross sections. *International Journal of Engineering Science*, *45*(2-8), 364–380. doi:10.1016/j.ijengsci.2007.03.014

Kitipornchai, S., Ke, L. L., Yang, J., & Xiang, Y. (2009b). Nonlinear vibration of edge cracked functionally graded Timoshenko beams. *Journal of Sound and Vibration*, *324*(3-5), 962–982. doi:10.1016/j.jsv.2009.02.023

Krawczuk, M. (1992b). Finite Timoshenko-type beam element with a crack. *Engineering Transactions*, *40*, 229–248.

Krawczuk, M. (1993a). Natural vibrations of rectangular plates with a through crack. *Archive of Applied Mechanics*, *63*, 491–504.

Krawczuk, M., & Ostachowicz, W. M. (1992a). Parametric Vibrations of beam with crack. *Archive of Applied Mechanics*, *62*(7), 463–473. doi:10.1007/BF00810237

Krawczuk, M., & Ostachowicz, W. M. (1993b). Transverse natural vibrations of a cracked beam loaded with a constant axial force. *Journal of Vibration and Acoustics*, *115*(4), 428–524. doi:10.1115/1.2930381

Krawczuk, M., & Ostachowicz, W. M. (1995). Modelling and vibration analysis of a cantilever composite beam with a transverse open crack. *Journal of Sound and Vibration*, *183*(1), 69–89. doi:10.1006/jsvi.1995.0239

Lin, H. P., & Chang, S. C. (2006c). Forced responses of cracked cantilever beams subjected to a concentrated moving load. *International Journal of Mechanical Sciences*, *48*(12), 1456–1463. doi:10.1016/j.ijmecsci.2006.06.014

Liu, J., Shao, Y. M., & Zhu, W. D. (2017c). Free vibration analysis of a cantilever beam with a slant edge crack. *Journal of Mechanical Engineering Science*, *231*(5), 823–843. doi:10.1177/0954406216631006

Loya, J. A., Rubio, L., & Fernandez-Saez, J. (2006d). Natural frequencies for bending vibration of Timoshenko cracked beams. *Journal of Sound and Vibration*, *290*(3-5), 640–653. doi:10.1016/j.jsv.2005.04.005

Marur, S. R., & Prathap, G. (2005d). Non-linear beam vibration problems and simplifications in finite element models. *Computational Mechanics*, *35*(5), 352–360. doi:10.100700466-004-0622-9

Mazanoglu, K. (2015a). A novel methodology using simplified approaches for identification of cracks in beams. *Latin American Journal of Solids and Structures*, *12*(13), 2460–2479. doi:10.1590/1679-78251723

Nahvi, H., & Jabbari, M. (2005a). Crack detection in beams using experimental modal data and finite element model. *International Journal of Mechanical Sciences*, *47*(10), 1477–1497. doi:10.1016/j.ijmecsci.2005.06.008

Nandwana, B. P., & Maiti, S. K. (1997a). Modelling of vibration of beam in presence of inclined edge or international crack for its possible detection based on frequency measurements. *Engineering Fracture Mechanics*, *58*(3), 193–205. doi:10.1016/S0013-7944(97)00078-7

Nandwana, B. P., & Maiti, S. K. (1997b). Detection of the location and size of a crack in stepped cantilever beams based on measurements of natural frequencies. *Journal of Sound and Vibration*, *203*(3), 435–446. doi:10.1006/jsvi.1996.0856

Ostachowicz, W. M., & Krawczuk, M. (1993c). Coupled torsional and bending vibrations of a rotor with an open crack. *Archive of Applied Mechanics*, *62*, 191–201.

Ostachowitz, W. M., & Krawczuk, M. (1991). Analysis of the effect of cracks on the natural frequencies of a cantilever beams. *Journal of Sound and Vibration*, *150*(2), 191–201. doi:10.1016/0022-460X(91)90615-Q

Pafelias, T. (1974). Dynamic behavior of a cracked rotor. General Electric Co.

Panigrahi, B., & Pohit, G. (2016b). Nonlinear modelling and dynamic analysis of cracked Timoshenko functionally graded beams based on neutral surface approach. *Journal of Mechanical Engineering Sciences, 230*(9), 1486-1497.

Papadopoulos, C. A., & Dimarogonas, A. D. (1987). Coupled longitudinal and bending vibrations of a rotating shaft with an open crack. *Journal of Sound and Vibration*, *117*(1), 81–93. doi:10.1016/0022-460X(87)90437-8

Papaeconomou, N., & Dimarogonas, A. (1989). Vibration of cracked beams. *Computational Mechanics*, *4*, 130–137.

Petroski, H. J. (1981). Simple static and dynamic models for the cracked elastic beam. *International Journal of Fracture*, *17*(4), 71–76. doi:10.1007/BF00036201

Petroski, H. J. (1984b). Stability of a crack in a cantilever beam undergoing large plastic deformation after impact. *International Journal of Pressure Vessels and Piping*, *16*(4), 285–295. doi:10.1016/0308-0161(84)90019-X

Qian, G. L., Gu, S. N., & Jiang, J. S. (1990c). The dynamic behavior and crack detection of a beam with a crack. *Journal of Sound and Vibration*, *138*(2), 233–243. doi:10.1016/0022-460X(90)90540-G

Rezaee, M., & Hassannejad, R. (2011a). A new approach to free vibration analysis of a beam with a breathing crack based on mechanical energy balance method. *Guti Lixue Xuebao*, *24*, 185–194.

Sadettin, O. (2007a). Analysis of free and forced vibration of a cracked cantilever beam. *NDT & E International*, *40*(6), 443–450. doi:10.1016/j.ndteint.2007.01.010

Silva, J. M. M., & Gomez, A. J. M. A. (1990b). Experimental dynamic analysis of cracked free-free beams. *Experimental Mechanics*, *30*(1), 20–25. doi:10.1007/BF02322697

Sinha, J. K., Friswell, M. I., & Edwards, S. (2002). Simplified models for the location of cracks in beam structures using measured vibration data. *Journal of Sound and Vibration*, *251*(1), 13–38. doi:10.1006/jsvi.2001.3978

Tada, H., Paris, P. C., & Irwin, G. R. (1985). *The Stress Analysis of Cracks Handbook*. Hellertown PA: Del Research Corporation.

Takahashi, I. (1999). Vibration and stability of non-uniform cracked Timoshenko beam subjected to follower force. *Composite Structures*, *71*(5), 585–591. doi:10.1016/S0045-7949(98)00233-8

Voila, E., Ricci, P., & Aliabadi, M. H. (2007d). Free vibration analysis of axially loaded cracked Timoshenko beam structures using the dynamic stiffness method. *Journal of Sound and Vibration*, *304*(1-2), 124–153. doi:10.1016/j.jsv.2007.02.013

Wang, K., & Wand, B. (2015). Timoshenko beam model for the vibration analysis of a cracked nano- beam with surface energy. *Journal of Vibration and Control*, *21*(12), 2452–2464. doi:10.1177/1077546313513054

Wauer, J. (1990a). Cracked rotor Dynamics: A state of the art survey. *Applied Mechanics Reviews*, *43*, 13–17. doi:10.1115/1.3119157

Wei, D., Liu, Y., & Xiang, Z. (2012b). An analytical method for free vibration analysis of functionally graded beams with edge cracks. *Journal of Sound and Vibration*, *33*(7), 1686–1700. doi:10.1016/j.jsv.2011.11.020

Wu, G. Y., & Shih, Y. S. (2005c). Dynamic instability of rectangular plate with an edge crack. *Composite Structures*, *84*(1-2), 1–10. doi:10.1016/j.compstruc.2005.09.003

Yan, T., & Yang, J. (2011b). Forced vibration of Edged – Cracked functionally graded beam due to a transverse moving load. *Procedia Engineering*, *14*, 3293–3300. doi:10.1016/j.proeng.2011.07.416

Yang, E. C., Zhao, X., & Li, Y. H. (2015b). Free Vibration Analysis for Cracked FGM Beams by Means of a Continuous Beam. *Shock and Vibration*, *2015*, 1–13.

Yang, J., Chen, Y., Xiang, Y., & Jia, X. L. (2008b). Free and forced vibration of cracked inhomogeneous beams under an axial force and a moving load. *Journal of Sound and Vibration*, *312*(1-2), 166–181. doi:10.1016/j.jsv.2007.10.034

Yokoyama, T., & Chen, M. C. (1998c). Vibration analysis of edge-cracked beams using a line–spring model. *Engineering Fracture Mechanics*, *59*(3), 403–409. doi:10.1016/S0013-7944(97)80283-4

Zheng, T., & Ji, T. (2012a). An approximate method for determining the static deflection and natural frequency of a cracked beam. *Journal of Sound and Vibration*, *331*(11), 2654–2670. doi:10.1016/j.jsv.2012.01.021

Chapter 14
Corrosion Protection of Construction Steel

Arkadeb Mukhopadhyay
https://orcid.org/0000-0002-4942-722X
Birla Institute of Technology, Mesra, India

Sarmila Sahoo
https://orcid.org/0000-0001-7000-3758
Heritage Institute of Technology, India

ABSTRACT

Reinforced concrete is one of the most versatile materials for construction. In spite of this, the performance is limited by corrosion, cracking, and spalling of the steel rebars. The steel embedded in the concrete is protected by a passive film from the corrosive attack of chlorides, carbon dioxide, and sulphates. As the concentration of chlorides, carbon dioxide, or sulphates increases above a certain threshold value at the concrete rebar interface, the passive film breaks and leads to a severe increase in the corrosion rate. Further, dynamic loading and the temperature of the surroundings also affect the durability of the reinforcements. The rebar may be protected from such a corrosion attack by the suitable selection of material, improving the concrete quality and tailoring its composition or application of protective coatings. The present chapter highlights and summarizes the different grades of steel for their high corrosion resistance. Further, surface engineering and application of corrosion resistance coatings for the prevention of corrosion of construction steel rebars has been also discussed elaborately.

INTRODUCTION

The most readily available and economic material for construction is reinforced concrete. One of the major causes of loss of durability of concrete structures is due to corrosion of the embedded steel rebars. Corrosion is a chemical attack of aggressive species that deteriorates the mechanical strength of the construction steel and rebars. Hence, it is harmful and at times may lead to catastrophic failure also. Corrosion at the steel-concrete interface has been a field of extensive research since decades. This is one of

DOI: 10.4018/978-1-7998-1831-1.ch014

the main reasons for limited life span of any concrete member and a consequent increase in maintenance cost of reinforced structures. It causes spalling, cracking of the concrete and thus reduces its strength. Early failure of reinforced and pre-stressed concrete structures leads to very high cost of maintenance, restoration and replacement worldwide (Bertolini, 2008).

The corrosion mechanism at the interface of steel and concrete is somewhat different from the normal corrosion of steel. There are many factors which are important for governing steel corrosion in concrete. The steel inside the concrete is normally corrosion resistant due to the formation of a passivation layer of oxide in the highly alkaline atmosphere i.e. often >13 (Sagüés et al., 2003). This happens mainly because of $Ca(OH)_2$ generated due to hydration of cement (Koleva et al., 2006). For steel to remain corrosion resistant this passive state needs to be maintained. Further, the concrete should be free from porosity which is mostly unavoidable since it is a matter of fact that no concrete is fully free of porosity (Shi & Ming, 2017). So, when any corrosive solution (Cl^-, CO_2^{2-}, SO_4^{2-}) penetrates the concrete and their concentration at the interface increases above a certain critical level, it starts reducing the alkalinity and then de-passivates the steel. In fact, the defects in concrete play a dual negative role i.e. loss of stability of passive films due to reduction of $Ca(OH)_2$ and accumulation of higher chloride species through the defects (Shi & Ming, 2017). Once the steel embedded in concrete de-passivates, it is corroded at a severe rate due to harsh atmosphere (Söylev & Richardson, 2008; Ožbolt et al., 2011). Once corrosion of the rebar occurs, cracking and damage of the concrete soon follows. Since the corrosion products of steel are quite voluminous, it results in swelling and generates stresses of sufficient magnitude to cause tensile failure of the concrete (Yeomans, 2004). Further expansion leads to delamination and spalling of pieces of concrete from the surface. The corrosion behavior of structural steel and rebars is also dependent on the external load as well as temperature (Deus et al., 2012; Shi et al., 2017). Therefore, corrosion of reinforcement steel necessities remedial action to increase the life of the structure or in certain cases complete replacement of the element. Globally, the cost associated with such replacements is expected to increase at an alarming rate (Yeomans, 2004).

Numerous researches have been done to reduce steel corrosion in concrete. Either the composition of concrete and surroundings is tailored or a suitable material is selected. The rebars may be also protected from pitting corrosion by application of coatings. The present chapter summarizes the corrosion phenomenon at the concrete-rebar interface and steps taken for prevention of corrosion mainly by suitable material selection or surface engineering. Future trends and scopes for improvement of corrosion resistance of reinforcement steel are also discussed. The present work would be also beneficial for adopting a suitable course of action for improving corrosion behavior of steel and leading to economic benefits in the long run.

CORROSION PREVENTION BY SUITABLE SELECTION OF MATERIAL AND IMPROVING THE STEEL QUALITY

Improving the reinforcement steel quality can reduce its possibility of corrosion. Several investigations have confirmed that stainless steel is much superior to carbon steel in its ability to resist chloride-initiated corrosion when embedded in concrete (Ping et al., 1996; Treadaway et al., 1989; Flint & Cox, 1988). So far most of the stainless steel used as reinforcement has been of the austenitic types (AISI 304 and 316), which are most readily available (Duarte et al., 2014) and have shown to have a 5-10 times higher chloride tolerance compared to carbon steel reinforcements (Ping et al., 1996). Duplex stainless steel, which has a ferritic/austenitic microstructure, combines a high material strength with increasing resistance to corrosion due to high molybdenum content (Duarte et al., 2014). Electrochemical studies in simulated seawater conditions (chloride attack) revealed the fact that both AISI 304 and 316 steels remained passive for most of the time with AISI 316 steel showing the highest corrosion resistance (Duarte et al., 2014). In fact, duplex stainless steel was also seen to be passive with the formation of a protective oxide layer. Duplex stainless steel with lower Ni demonstrated lower corrosion resistance in presence of chlorides and pitting corrosion was detected in the rebars over a long time period (Alonso et al., 2019). Molybdenum enhances resistance to initiation and propagation of pitting and crevice corrosion.

Stainless steel rebars are produced with the strength and dimensional properties to meet the requirements of structural concrete codes of practice. The British Standard Specification is BS 6744:1986 austenitic stainless-steel bars and lists alloys such as 304, 304L, 316, 316L, etc. Austenitic steel is resistant to corrosion in concrete with very high chloride content, and is the recommended material for rebars. Luo et al. (2017) examined the electrochemical and passivation behavior of 316L stainless steel in chlorinated simulated concrete pore solution. It was seen that the solution pH had a significant influence. Formation of molybdates in the outer layer of passive film further enhanced the corrosion resistance. The presence of high Cr/Fe ratio in the passive layer also resulted in nobler corrosion potential.

Castro et al. (2003) studied two different austenitic stainless steels (304LN and 316LN grades) as corrosion resistant materials for reinforcement in concrete. The chemical composition of both grades is shown in Table 1. Both the steels were subjected to hot rolling as well as cold rolling and subjected to electrochemical tests. Polarization resistance measurements, Tafel curves and cyclic potentiodynamic polarization curves at room temperature were determined in 3 wt.% NaCl solution using a potentiostat. Table 2 shows the corrosion potential of 304LN and 316LN steel (hot as well as cold rolled). Based on the corrosion potentials, cold rolled 304LN has nobler corrosion potential.

Criado et al. (2011) investigated the corrosion behavior of low-nickel austenitic stainless steel, AISI 304 austenitic stainless steel and carbon steel. A saturated calcium hydroxide solution with pH 12–13 and different NaCl concentrations (0.0, 0.4, 1.0, 2.0, 3.0 and 5.0 wt.%), was used to study corrosion

Table 1. Chemical composition (wt.%) of the 304LN and 316LN steels

Steel Designation	C	Si	Mn	Cr	Ni	Mo	N	Cu	S	P
304LN	0.023	0.38	1.49	18.55	8.77	0.21	0.18	0.18	0.022	0.027
316LN	0.030	0.36	1.31	17.62	11.70	2.87	0.20	0.10	0.004	0.023

(Castro et al., 2003)

Table 2. Corrosion potential of austentitic stainless steels after hot and cold rolling

Austenitic Grades	E_{corr} (mV)
316LN (hot rolled)	-154
304LN (hot rolled)	-139
316LN (cold rolled)	-153
304LN (cold rolled)	-132

(Castro et al., 2003)

behavior in a wide range of experimental conditions. All the experiments were performed at room temperature. Low nickel stainless steel and AISI 304 stainless steel exhibited lower corrosion current density in comparison with carbon steel and recommended as corrosion resistant construction steel rebars. It was also concluded that austenitic microstructures were more corrosion resistant than ferritic structures. Austentitic and duplex stainless steel were also observed to exhibit greater corrosion resistance in artificial sea water in comparison with carbon steel.

Cr modified HRB400 steel rebars were exposed to artificially simulated concrete pore solution (saturated $CaOH_2$) and electrochemical tests were carried out by Liu et al. (2015). The Cr modified steel exhibit higher corrosion resistance and it was seen that with an increase in Cr, formation of oxides of Cr take place. But, the primary constituents of the films in simulated concrete pore solution were that of iron oxides. Further, lower corrosion current density and higher impedance was observed for the Cr modified HRB400 steel rebars. Tian et al. (2019) investigated the corrosion resistance of Cr modified low alloy steel reinforcement exposed at marine splash zones over a period of two years. Here again it was concluded that Cr modified low alloy steels had higher corrosion resistance in marine environment. This was attributed to the fact that Cr had the capability to improve corrosion protection and buffering ability of the rust layer against aggressive species. Cr helps in increasing the Fe^{2+} and OH^- within the oxide films and also improves the compactness of the rust layer.

Carbonation also affects the corrosion resistance of steel rebars and formation of passive films. Electrochemical impedance spectroscopy and potentiodynamic polarization study was carried out for carbon steel as well as Cr modified low alloy steel in carbonated cement extract solution (Liu et al., 2017). In non-carbonated cement extract solution, there was no significant variation in the electrochemical parameters. But in carbonated solution, pitting potential was observed to shift in more negative direction as pH of the solution decreased. On the other hand, an increase in Cr content improved the passivation and pitting potential. The corrosion potential became nobler and corrosion current density decreases with an increase in Cr in low alloy steels.

The long-term corrosion performance over a period of 2 years was observed for stainless steel reinforcement in mortar exposed to chloride environment (Serdar et al., 2013). Electrochemical impedance spectroscopy at open circuit potential revealed that the corrosion behavior was divided into two phases: (a) passive phase and (b) pitting propagation phase. Austentitic stainless steel was seen to have a higher corrosion resistance over long term in comparison with steel types having low Ni, high N and Mn. Therefore, it may be concluded that Austentitic stainless steels are preferred for their higher corrosion resistance as well as mechanical strength as construction steel and reinforcement rebars. Cr modification also improves corrosion resistance of steel rebars in carbonated as well as non-carbonated concrete pore solution. Nevertheless, as a matter of fact, the environment friendliness of Cr is still under question. In fact, recent studies have also revealed the superior corrosion resistance of thermomechanical treated (TMT) rebars in comparison with conventional steel rebars (Torbati-Sarraf & Poursaee, 2018).

SURFACE ENGINEERING OF STRUCTURAL STEEL AND REBARS – APPLICATION OF COATINGS

Corrosion resistant coatings protect metal components against degradation due to moisture, salt spray, oxidation or exposure to a variety of environmental or industrial chemicals. Anti-corrosion coatings allow for added protection of metal surfaces and act as a barrier to inhibit the contact between chemical compounds or corrosive materials. Therefore, for active barrier protection of rebars, surface engineering and application of corrosion resistant coatings are discussed subsequently.

Enamel Coatings

In recent years, enamel coatings have been applied to reinforcing steel for corrosion protection and bond enhancement with concrete (Yan et al., 2012). Tang et al. (2012a; 2012b; 2013) studied the corrosion behavior of enamel-coated reinforcing steel bars in 3.5 wt.% NaCl solution. Grade 60 smooth reinforcing steel bar with a diameter of 13 mm was used in this study. Three types of enamel coatings were used: pure enamel (PE), mixed enamel (ME), and double enamel (DE). The typical chemical composition of the pure enamel is shown in Table 3. The mixed enamel consisted of 50% pure enamel and 50% calcium silicate by weight that was obtained by mixing enameling frit with Portland cement. The double enamel is a two-layer system with an inner pure enamel layer and an outer mixed enamel layer. Corrosion behavior of these enamel-coated reinforcing steel bars in 3.5 wt.% NaCl solution was evaluated by open-circuit potential, electrochemical impedance spectroscopy and potentiodynamic polarization. The passive current density, i_{pas}, was used as a parameter to measure the corrosion resistance of the steel samples in the passive state. A low passive current density indicates a high corrosion resistance. Figure 1 shows the passive current density of uncoated steel bars is higher than the enamel coated steel bars, indicating good protection of enamel coatings in the passive state. Among the three enamel coatings, the double enamel has the lowest passive current density and the mixed enamel coating has the highest passive current density. Both pure and double enamel coatings can significantly improve corrosion resistance, while the mixed enamel coating offers very little protection. In another study by Tang and Chen (2015), PE and DE coatings were again suggested for design of reinforced concrete structures and their corrosion protection. In the same study, coated deformed rebars exhibited lower corrosion resistance in comparison with smooth rebars. The PE and DE coatings were observed to be more sensitive to coating thickness in comparison with ME coatings with interconnected pores.

Steel bonds mortar seven times more strongly when coated with 50/50 enamel (Yan et al., 2012). Enamel coatings also increase the bond strength of deformed rebar in concrete by approximately 15% (Wu et al., 2012). But the bond strength between a smooth steel rod and mortar can be increased by as much as seven times when the rod is coated with a mixture of 50% (by weight) enamel and 50% calcium silicate (CS) particles. This increase in bond strength is due to the increased surface roughness of the coating and chemical bonding of the embedded CS particles to the surrounding mortar (Yan et al., 2012).

Epoxy Coatings

Epoxy coating is a good non-metallic coating used to prevent corrosion. Studies show that epoxy coating is a good method to prevent reinforcing steel corrosion (Kobayashi & Takewaka, 1984; Pul & Sprinkel,

Table 3. Chemical composition of pure enamel

Materials	SiO_2	B_2O_3	Na_2O	K_2O	CaO	CaF_2	Al_2O_3	ZrO_2	CoO	MnO_2	NiO
Wt%	44.0	19.3	15.8	2.8	0.0	4.7	4.6	5.3	0.9	1.5	1.0

(Tang et al., 2012a)

Figure 1. Passive current density of differently coated rebars
(Tang et al., 2012a)

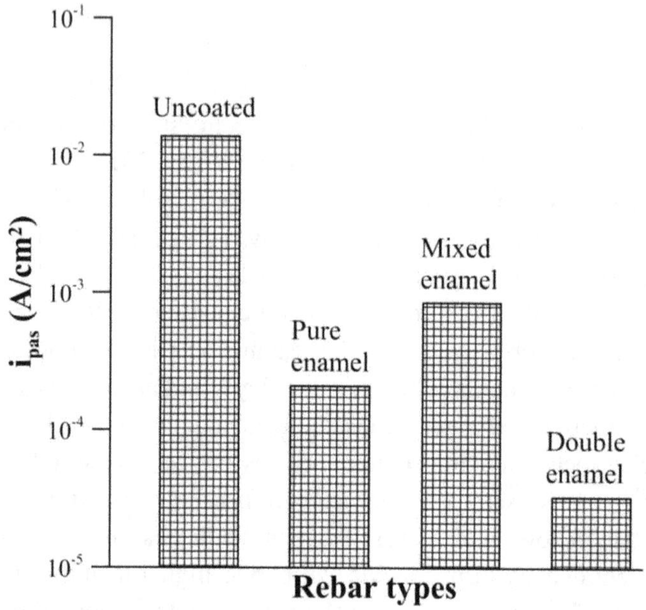

1998; Saravanan et al., 2007). Fusion-bonded, epoxy-coated reinforcement was first used in 1973 in reinforced concrete bridge deck construction to combat the severe effects of deicing salts in North America (Smith & Virmani, 1996). Its use is now varied and widespread, and epoxy coating is clearly the dominant coated reinforcement in general use. Epoxy coatings provide excellent corrosion protection to steel. The coating has very low permeability to Cl^-, is highly chemically resistant, and is essentially unaffected in many environments. In fact, intact epoxy coatings offer higher resistance to corrosion in comparison with enamels (Tang et al., 2012b). It protects the steel by isolating the metal completely from the local environment. Even though epoxy coatings provide enhanced corrosion resistance, their ability in very aggressive environments has been in question due to its tendency of getting damaged naturally and during transportation. Hence careful handling is necessary to minimize abrasion and mechanical damage.

Erdoğdu et al. (2001) examined the corrosion performance of undamaged and damaged epoxy coatings (1% and 2%) in chloride environment. It was seen that the corrosion current density was negligible for the coated rebars and regardless of exposure conditions, the undamaged epoxy coatings exhibited excellent corrosion resistance (Figure 2). With micro-damage present, corrosion occurs also locally without rust accumulation. Even with damage of 2% of the coating shows significant reduction in corrosion activity as can be seen in Figure 2. However, Yeomans (1994) noted that at damaged portions and unrepaired cut ends of bars, localized corrosion occurs and to a same extent as that of steel without coatings. Further, corrosion also proceeded under the coatings leading to their subsequent detachment (Yeomans, 1994). Manning (1996) studied the history of epoxy-coated reinforcement in North America from its development in the early 1970s to its status in early 1995. He found that all results indicated that due to defects occurring during handling and inside mortar itself long term performance of the epoxy coatings are not guaranteed.

In this regard, self-healing coatings with autonomous healing mechanism for damaged coatings were introduced by Weishaar et al. (2018). In general, epoxy coatings are applied to rebars by fusion bonding. But the self-healing coatings were deposited by utilizing electrostatic forces and powdered epoxy is sprayed onto heated reinforcement. The powder melts and a coating is formed on the surface of the reinforcement. Along with epoxy resin, 10% by wt. microencapsulated tung oil was introduced to form self-healing coatings. Post coating deposition, they were intentionally damaged and subjected to accelerated corrosion testing. The self-healing coatings were reported to demonstrate excellent corrosion resistance under damaged condition. The corrosion protection ability of epoxy coatings may be also significantly improved by incorporation of indole butyric acid (IBA)-modified gypsum along with epoxy resin (Rahman et al., 2019).

Therefore, it may be summarized that even though epoxy coatings provide high corrosion resistance, their applicability is restricted due to damages caused in them during handling and transportation and requires careful handling. Such problems may be overcome by the self-healing coatings or inclusion of IBA-modified gypsum in epoxy resin.

Metallic Coatings

Metallic coatings may be classified as sacrificial and non-sacrificial. Sacrificial coatings provide sacrificial protection to the base metal for example galvanizing of steel. Here the coating dissolves in preference to the base metal and provides protection even if the coating is delaminated at some point, exposing

Figure 2. Corrosion current density of rebars in slabs exposed to 3% sodium chloride with damaged (1-2%) and undamaged epoxy coating along with bare rebar
(Erdoğdu et al., 2001)

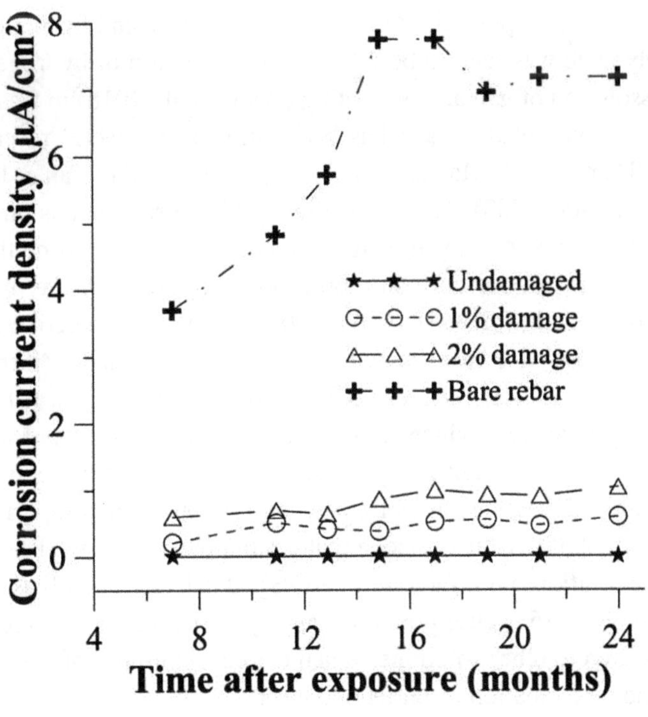

the base metal. The non-sacrificial type of coating provides barrier or shielding effect only. In case the coating gets delaminated, there is no protection of the base metal leading to accelerated corrosion since the coating is nobler in comparison with steel. Galvanizing has been widely used for corrosion protection of rebars and construction grade steels which provides sacrificial protection. On the other hand, electroless nickel coatings are being recently investigated for providing barrier protection to rebars due to their high corrosion resistance. Both the processes are discussed subsequently.

Galvanizing

Hot dip galvanization has been widely accepted and used for corrosion prevention of reinforced steel (Yeomans, 2004). Hot dip galvanization results in a metallurgical bond with the steel and resulting in high adhesion. Thus, strong and adherent coatings are formed. Zinc provides barrier protection from aggressive agents as well as sacrificial protection. In case of use of galvanized steel as reinforcement in concrete, the external layer of zinc passivates to form a protective layer (Andrade & Macias, 1988; Bertolini, 2008). But still there are chances of loss of adherence of the coating during the first contact days of zinc with concrete due to evolution of hydrogen on the rebar surface during cement hydration which releases hydroxyl ions into the concrete pore solution (Arenas et al., 2006). Due to this a chromate conversion was suggested which was later discarded because of its toxicity. In such a case, cerium salts may be chosen as environment friendly replacement. Arenas et al. (2006) observed that galvanized steel with cerium conversion coatings have higher corrosion resistance in saturated $Ca(OH)_2$ with 0.5 M NaCl for atleast 5 days after which galvanized steel without conversion coatings exhibit lower corrosion current density (Figure 3). Finally, it was concluded that cerium conversion coatings improve the corrosion resistance of galvanised steel in alkaline media, in absence or chlorides containing solutions, for short immersion times. In a recent study by Bellezze et al. (2018) nitrite was seen to be a promising inhibitor of hydrogen evolution but it was seen to be less effective than chromate. Nitrite decreases the hydrogen evolution by moving the corrosion potential to more positive values and formation of passive layers. In this regard, sodium molybdate was seen to be effective for reduction of hydrogen evolution leading to suppression of rapid dissolution of galvanized coating (Wang et al., 2018) in freshly prepared concrete.

The composition of the passivating film has been studied by several researchers particularly in $Ca(OH)_2$ as well as NaOH and KOH (Macias & Andrade, 1987a; 1987b; Tan & Hansson, 2008; Blanco et al., 1984; Macias & Andrade, 1983). The passivation of zinc also depends on the pH of the solution and on concentration of Ca^{2+} ions. A pH value higher than 13.3 results in formation of coarse corrosion products and this is unable to provide corrosion protection. In the pH range of 12.2 to 13.3, calcium hydroxyzincate (CHZ) is formed (Tan & Hansson, 2008) which has protective action. But galvanized steel may provide effective protection from corrosion attack of chlorides only if the CHZ is compact and continuous and the remaining coating thickness is high (Andrade & Macias, 1988). Andrade & Macias (1988) also observed less protective products at pH values lower than 12.2 and such corrosion products include ZnO and $Zn(OH)_2$. Even though the chloride ions may promote localized attack by breaking the CHZ, zinc and galvanized steel has higher corrosion resistance in comparison with bare steel. In a recent study by Zheng et al. (2018) it was seen that continuous protective layers were formed on the galvanized coating in the solutions with saturated $Ca(OH)_2$ at pH 12.5 as well as in solutions containing either KOH or NaOH at pH 13.15. Better protective ability of the passive film formed could be seen in saturated $Ca(OH)_2$ compared to other solutions. When the concentration of Ca^{2+} at pH 12.5 was lower, there was no reaction and no formation of protective layer.

The interaction between the galvanized steel and concrete matrixes obtained with different cements has also been widely studied. It has been observed in most research works that zinc passivation occurs with the formation of a layer of CHZ (Tan & Hansson, 2008; Bertolini, 2008). Farina and Duffó (2007) investigated the performance of pure zinc and galvanized steel in solutions simulating carbonated concrete. The influence of chloride content was also investigated. Potentiodynamic polarization results indicated that in presence of high levels of carbonates and bicarbonates, the corrosion resistance is improved. Galvanized steel remained passive even after carbonation. The corrosion rate was also of the same order of magnitude as found in non-carbonated concrete. On the other hand, Roventi et al. (2013) found out that carbonation destroys the CHZ and leads to the formation of amorphous products such as $CaCO_3$ and hydrozincite (HZ) which are also protective but lesser in comparison with CHZ. Further, carbonation also results in a degradation of threshold limit of chloride concentration which results in severe corrosion (0.29 wt. %) of galvanized steel in comparison with non-carbonated concrete (chloride threshold limit of 1.3-2 wt. %). On addition of $CaCl_2$ to concrete mix, it was seen that the resistance of galvanized steel to corrosion from chloride attacks was almost same in carbonated and non-carbonated concrete (Gonzalez & Andrade, 1982). Thus, the effect of chloride ions on corrosion resistance of galvanized steel post carbonation seems to be unclear.

Bellezze et al. (2006) studied the corrosion resistance of reinforcements galvanized with Zn–Ni–Bi and with Zn–Ni–Sn–Bi (modified baths), with respect to reinforcements galvanized with the Zn–Pb (traditional bath) and to that of a pure zinc rod. Concrete specimens that are cylindrical in structure were manufactured. They were reinforced with galvanized bars and with a pure zinc rod. Two different types of cement with different alkali content were used to study the effect of alkalinity of the concrete

Figure 3. Variation of the current density with time for galvanized steel (GS) and with cerium conversion coatings (CeGS) in $Ca(OH)_2$ saturated solution
(Arenas et al., 2006)

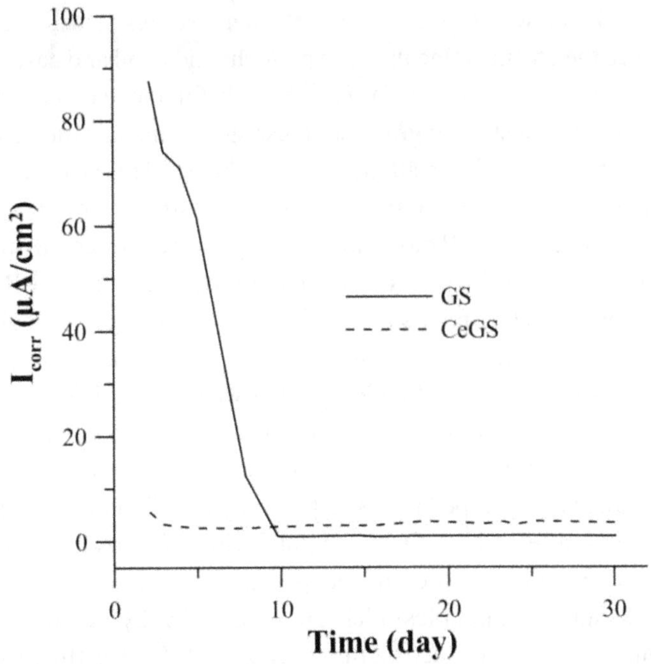

matrix on the behaviour of the reinforcements. After the air curing period, the concrete specimens were submitted to wet–dry exposure both in tap water and in a 5% sodium chloride solution. The corrosion rate and the corrosion potential of the bars were monitored during the air curing period and during the wet–dry exposure. The galvanized bars exhibited higher corrosion resistance compared to pure zinc rod since alloying elements was present in the coatings. The bars galvanized with Zn–Ni–Sn–Bi were not found to be suitable in highly alkaline concrete matrix and rather they show higher corrosion resistance in presence of chlorides. Further, localized corrosion was observed to start earlier in low alkali concrete matrix with higher penetration of chlorides. From the foregoing discussion it may be also clearly deduced that galvanizing has been widely investigated by researchers for corrosion protection by surface modification of steel rebars. The corrosion resistance is largely dependent on the zinc passivation process. Due to the sacrificial nature of the coatings, galvanizing results in higher ability of the coatings to protect rebars in comparison with epoxy coatings (Yeomans, 1994).

Electroless Nickel Coatings

Electroless means without the use of electricity and it has several advantages over conventional electroplating. The coating deposition is uniform and with superior corrosion resistance and tribological characteristics with one main advantage being the ability to be deposited over a wide variety of substrates (Sahoo & Das, 2011; Sudagar et al., 2013). Over the years, electroless nickel (EN) has evolved into a huge family of functional coatings with wide range of applications (Loto, 2016). The Ni-P alloy coatings appear quite economical and technically feasible for corrosion protection of steel (Elsener et al., 2008; Sun et al., 2018). The latest developments in deposition technologies and bath formulations have made it possible to produce amorphous/microcrystalline, flawless coating having low, medium and high phosphorous (Loto, 2016). The corrosion resistance generally improves with an increase in phosphorus content (Omar et al., 2018). Inclusion of W or Cu also improves the corrosion resistance of the coatings (Mukhopadhyay & Sahoo, 2019; Duari et al., 2018). Surface morphology of electroless Ni-P and its alloys are shown in Figure 4. A few micrometer thick of such coatings provide very effective protection against corrosion and hence the coatings find usage in petrochemical, oil and gas, chemical, food processing industries, etc. (Mukhopadhyay et al., 2018). The chloride tolerance of such coatings is comparable with expensive materials such as austenitic grade stainless steels and hastalloy (Singh & Ghosh, 2006). Nickel is reported to develop passive film in alkaline environments. This is attributed to the precipitation of $Ni(OH)_2$ at the metal/electrolyte interface that makes the substrate immune to attack (Wronkowska, 1993). Thus, EN coatings especially Ni-P based alloys are preferred for corrosion protection of steel.

Singh and Ghosh (2006) studied the effect of deposition of electroless Ni-P coatings with 8% and 16% by weight of phosphorus on steel rebars and their ability to provide corrosion resistance in simulated concrete pore solution. This solution was prepared by dissolving analytical reagent grades of 8.33 g/l NaOH+2.0 g/l CaO+3.36 g/l KOH in double distilled water. Also 1 M chloride ions were added. In order to study the effect of heat treatment on the corrosion resistance performance of the coatings, some specimens were heat treated in electric furnace at 600°C for 15 min. To determine the corrosion rate of the coatings, electrochemical impedance spectroscopy and direct current polarization resistance techniques were used. It was observed that Ni-P coatings with 8% phosphorus exhibited a very stable corrosion rate in comparison with 16% by weight phosphorus or the bare steel. The corrosion rate of mild steel (MS) and medium phosphorus electroless nickel plating i.e. 8% by weight of phosphorus (MEPEN) may be observed in Figure 5. Scanning electron microscopy and X-ray diffraction studies were used to study the kinetics and mechanism of corrosion. Passive films of Ni_2O_3 and Ni_5P_2 were observed in case of MEPEN coatings leading to enhanced corrosion resistance. Further, heat treatment was concluded to have a deleterious effect on the corrosion resistance of the coatings.

In a recent study, Mukhopadhyay & Sahoo (2019) investigated the corrosion resistance behaviour of Fe – 600 grade steel recommended for rebars by Bureau of Indian Standards to pitting attack of chlorides. For this, electrochemical tests such as potentiodynamic polarization and electrochemical impedance spectroscopy were carried out. Further, the rebars were coated with electroless Ni-P as well as more corrosion resistant variants such as Ni-P-W and Ni-P-Cu coatings. It was seen that in 3.5% NaCl, the coated rebars exhibited a nobler corrosion potential in comparison with the bare rebars. The corrosion current density also improved significantly as can be seen in Table 4. The Nyquist plots obtained from electrochemical impedance spectroscopy results also indicated a big semi-circular loop for the coated rebars in the high frequency region indicating a remarkable improvement in corrosion resistance. Further, scanning electron microscopy revealed that the bare steel was severely affected by the chloride attack and severe cracking could be observed and has been presented in Figure 6. Whereas, the corroded surfaces of EN coated rebars were devoid of any cracks as can be seen in Figure 7.

From the foregoing discussion, it may be clearly observed that EN coatings have the capability of providing barrier protection to rebars from pitting attack of chlorides and simulated concrete pore solution. But still literature regarding the same is scarce and the corrosion protection ability of the coatings to attack of sulphates and carbon dioxide require further investigation since they lead to spalling of rebars (Mukhopadhyay & Sahoo, 2019). If the coatings provide barrier protection to rebars in such aggressive environments, then the applicability of the EN alloys may be extended to the construction industry as well.

Figure 4. Scanning electron microscopy images of as-deposited electroless (a) Ni-P (b) Ni-P-W and (c) Ni-P-Cu coatings
(Mukhopadhyay & Sahoo, 2019)

Figure 5. Corrosion rate vs time plot for MS and MEPEN exposed in simulated pore solution with 1 M Cl⁻ (Singh & Ghosh, 2006).

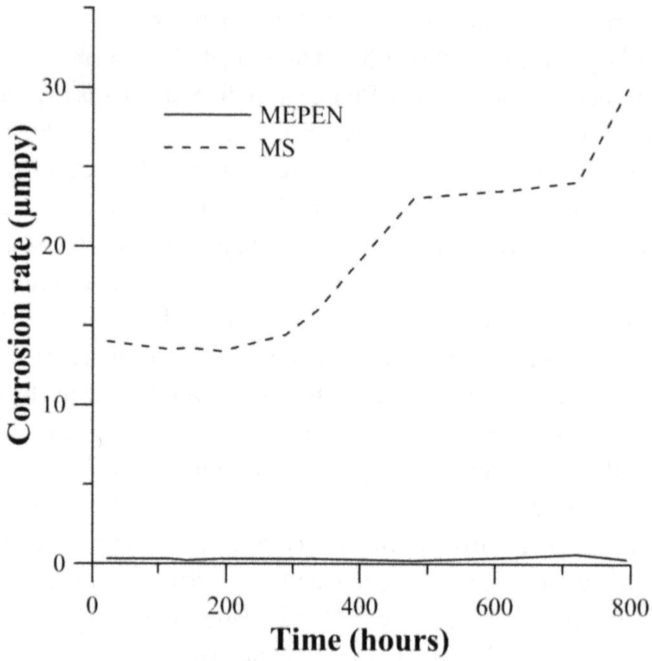

Duplex Coatings

Amongst coated rebars, either epoxy coated rebars or galvanized rebars are favored for corrosion resistance. But the corrosion protection ability of epoxy coating depends on the integrity of the coatings. It has been discussed previously that damage in epoxy coatings may lead to failure of the rebars and corrosion propagation from under the coatings and subsequent detachment. Moreover, during transit, there are chances of the coating getting damaged and hence proper handling is necessary. To overcome such problems, galvanized as well as epoxy (duplex) coated rebars may be used as reinforcement in concrete. In this regard, Dong et al. (2012) proposed epoxy/zinc duplex coating system. In this, the galvanized rebar coated with a thick epoxy coating, was developed as a protective material for use in the corrosive environments such as in ocean and seawater. It was observed that in epoxy coatings with damage, the corrosion was serious in the damaged area. While in case of the duplex coating, even though there was

Table 4. Potentiodynamic polarization results of uncoated and EN coated rebars in 3.5% NaCl

	E_{corr} (mV)	i_{corr} (µA/cm²)
Steel	-653	11.7
Ni-P	-436	8.2
Ni-P-W	-391	0.5
Ni-P-Cu	-356	0.78

(Mukhopadhyay & Sahoo, 2019)

Figure 6. Scanning electron microscopy image of uncoated rebar post corrosion in 3.5% NaCl *(Mukhopadhyay & Sahoo, 2019)*

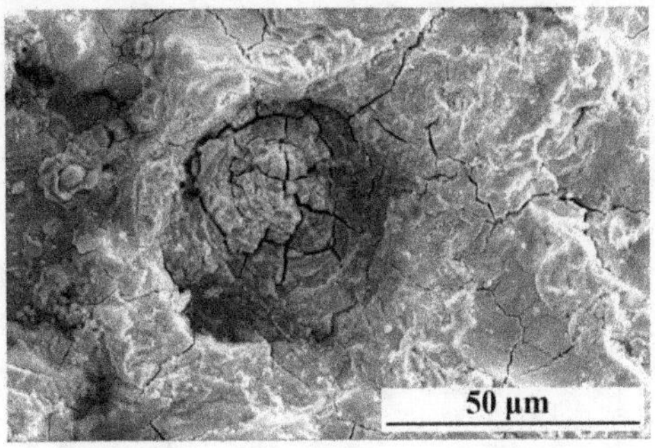

some damage, there was sufficient protection from corrosion. Thus, the galvanized coating provides effective protection when the epoxy coating suffers any damage. The corrosion potential and corrosion current density of the duplex coatings may be observed in Figure 8 and Figure 9 respectively. It was finally concluded that for long term protection in ocean water, the duplex coatings exhibit a lower corrosion potential and corrosion current density compared to epoxy coating or galvanized rebar.

Conversion Coatings

Even though galvanizing results in enhanced corrosion protection of steel rebars, they are rather energy consuming and costly (Jalili et al., 2009). Furthermore, above pH values of 13.3, the corrosion may increase exponentially. Also, epoxy coated rebars are costly when compared to bare rebars. They suffer from disadvantages such as reduction of bond strength with concrete especially when smooth round rebars are used. In this regard, Jalili et al. (2009) investigated the corrosion performance of environment friendly inorganic conversion coatings namely Zn-Ni-Mn phosphate, Zn-Ni phosphate and Zn-Ca phosphate. Zn-Ni phosphate coatings demonstrated highest corrosion resistance with a nobler corrosion potential and minimum corrosion current whereas the bond strength was the highest for Zn-Ca phosphate. Galvanized reinforcement steel with cerium conversion coatings provides an effective corrosion resistance for short immersion time in chloride solution by inhibiting hydrogen evolution on the galvanized coating during

Figure 7. Scanning electron microscopy image of Ni-P coated rebar post corrosion in 3.5% NaCl *(Mukhopadhyay & Sahoo, 2019)*

Figure 8. Corrosion potential (E_{corr}) of duplex coated rebars in concrete over long time (Dong et al., 2012)

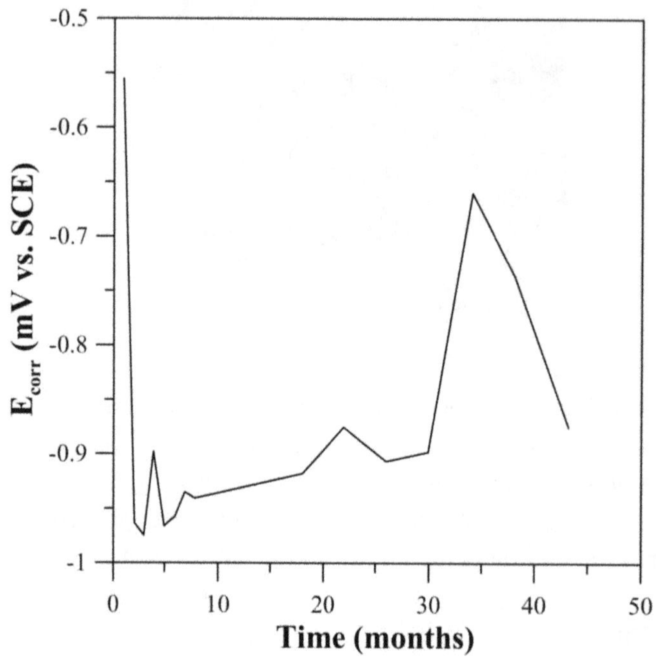

the initial stages as may be observed in Figure 3 (Arenas et al., 2006). Phytic acid conversion coating on steel has the ability to decrease the corrosion rate by 50-80% (Xiong et al., 2018). The conversion coating was prepared by treating steel in Zn^{2+} doped phytic acid solution at 20°C. The combination of phytic acid conversion coating and passivation products restrain chloride attack in alkaline environment. Passivation mainly occurs in the holes of phytic acid conversion coating layer. This further leads to compaction of the layer and further increase in corrosion resistance. Above a treatment time of 7.5 min in Zn-phytic acid solution, the corrosion resistance degrades due to the formation of cracks in the coating layer. Five different types of molybdate-phosphate conversion coating were deposited on steel rebars for corrosion protection by changing molybdate-phosphate ratio in the conversion bath (Verbruggen et al., 2019). Electrochemical studies were carried out in simulated concrete pore solution. The conversion coatings with higher phosphate in comparison with molydates provide active corrosion protection but lack barrier properties. By appropriately tuning the molybdate to phosphate ratio, conversion layers with active as well as barrier protection could be achieved.

Polypyrrole Coatings

The corrosion resistance of conducting polypyrrole coatings deposited on low carbon steel rebars in carbonated simulated concrete pore solution was investigated by Munot et al. (2018). Polypyrrole was synthesized and applied with a paint coating. Linear polarization technique was utilized to determine the corrosion rates of uncoated and coated rebars in 3.5% NaCl. The corrosion rate of coated rebar was significantly lower (0.1 mpy) compared to uncoated rebars (5 mpy). A shift of 106 mV in nobler direction was also observed for the coated rebars. However, as the chloride content increased in carbonated pore

Figure 9. Corrosion current density (I_{corr}) of duplex coated rebars in concrete over long time (Dong et al., 2012)

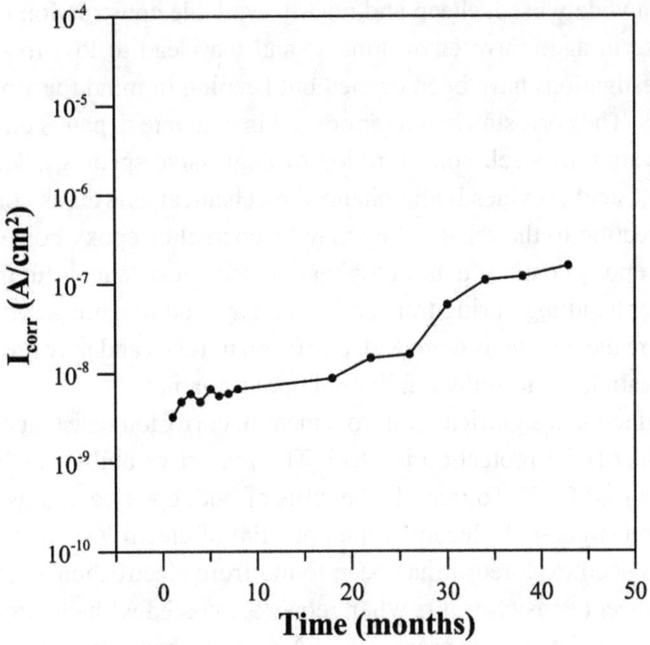

solution, the corrosion resistance of polypyrrole coated steel decreased. The ability of polypyrrole coatings for corrosion protection (chloride induced) of tunnel lining with pre-existing cracks and sustained loading was reported by Hu et al. (2019). When subjected to accelerated corrosion, polypyrrole coating delayed the reinforcement corrosion and led to lesser mass loss of steel (up to 50% lower mass loss). The transverse and longitudinal crack widths were also observed to be smaller for polypyrrole coatings.

FUTURE RESEARCH DIRECTIONS

Surface engineering has led to the achievement of enhanced corrosion resistance in rebars especially galvanizing. In this regard, the corrosion protection ability of epoxy coatings as well as duplex coatings has been also explored. But from the ongoing discussion, it may be also observed that even though electroless nickel coatings have high ability to provide active barrier protection to steel, their potential for corrosion prevention of rebars has remained unexplored. Future research works may be carried out to extend the applicability of electroless nickel coatings to corrosion prevention of reinforcement steel. Their corrosion resistance in simulated concrete pore solution has been also promising. In future, research works may be also directed towards the investigation of the corrosion behavior of electroless nickel coatings in environments containing sulphates or carbon dioxide since such species lead to spalling and cracking of rebars.

CONCLUSION

Reinforced concrete is a widely used, cheap and readily available construction material. But rebar corrosion restricts their use in aggressive environments and may lead to loss in structural integrity and high costs. Several investigations have been carried out keeping in mind the strength as well as corrosion resistance of rebars. The corrosion behavior of steel in concrete depends on several factors namely concrete quality, passivation of steel, concentration of aggressive species, alkalinity of the concrete, etc. Austentitic stainless steel provides both enhanced mechanical as well as corrosion characteristics. To provide barrier protection to the rebars, they may be enamel or epoxy coated. In comparison with enamel coatings, intact epoxy coatings exhibit higher corrosion resistance. But the problem with epoxy coatings is that improper handling during transportation may lead to damage of the coatings. In such a case, at the portion where the coating is damaged, corrosion initiates and may propagate under the intact coating portion. This again leads to sudden failure of epoxy coatings.

Galvanizing has resulted in a significant improvement in corrosion resistance of reinforcement steel rebars. Zinc provides sacrificial protection to steel. The protective ability again depends on the zinc passivation and formation of CHZ. To reap the benefits of both epoxy coatings as well as galvanizing duplex coatings has been suggested. Recently, the potential of electroless nickel plating for corrosion protection of rebars has been explored. It has been found from electrochemical tests that a significant decrease in corrosion current density occurs when rebars are coated with electroless nickel coatings. A nobler corrosion potential has been also observed for EN coated rebars. Recent studies have also reported the suitability of conversion coatings and polypyrrole coatings as prospective candidates for corrosion reduction of rebars.

REFERENCES

Alonso, M. C., Luna, F. J., & Criado, M. (2019). Corrosion behavior of duplex stainless steel reinforcement in ternary binder concrete exposed to natural chloride penetration. *Construction & Building Materials*, *199*, 385–395. doi:10.1016/j.conbuildmat.2018.12.036

Andrade, M. C., & Macias, E. A. (1988). Galvanized reinforcements in concrete. In A. D. Wilson, J. W. Nichols, & H. J. Prosser (Eds.), *Surface Coatings* (pp. 137–154). New York: Elsevier Applied Science. doi:10.1007/978-94-009-1351-6_5

Arenas, M. A., Casado, C., Nobel-Pujol, V., & de Damborenea, J. (2006). Influence of the conversion coating on the corrosion of galvanized reinforcing steel. *Cement and Concrete Composites*, *28*(3), 267–275. doi:10.1016/j.cemconcomp.2006.01.010

Bellezze, T., Malavolta, M., Quaranta, A., Ruffini, N., & Roventi, G. (2006). Corrosion behaviour in concrete of three differently galvanized steel bars. *Cement and Concrete Composites*, *28*(3), 246–255. doi:10.1016/j.cemconcomp.2006.01.011

Bellezze, T., Timofeeva, D., Giuliani, G., & Roventi, G. (2018). Effect of soluble inhibitors on the corrosion behaviour of galvanized steel in fresh concrete. *Cement and Concrete Research*, *107*, 1–10. doi:10.1016/j.cemconres.2018.02.008

Bertolini, L. (2008). Steel corrosion and service life of reinforced concrete structures. *Structure and Infrastructure Engineering*, *4*(2), 123–137. doi:10.1080/15732470601155490

Blanco, M. T., Andrade, C., & Macias, A. (1984). SEM Study of the Corrosion Products of Galvanized Reinforcements Immersed in Solutions in the pH Range 12·6 to 13·6. *British Corrosion Journal*, *19*(1), 41–48. doi:10.1179/000705984798273524

Castro, H., Rodríguez, C., Belzunce, F. J., & Canteli, A. F. (2003). Mechanical properties and corrosion behaviour of stainless steel reinforcing bars. *Journal of Materials Processing Technology*, *143*, 134–137. doi:10.1016/S0924-0136(03)00393-5

Criado, M., Bastidas, D. M., Fajardo, S., Fernández-Jiménez, A., & Bastidas, J. M. (2011). Corrosion behaviour of a new low-nickel stainless steel embedded in activated fly ash mortars. *Cement and Concrete Composites*, *33*(6), 644–652. doi:10.1016/j.cemconcomp.2011.03.014

Deus, J. M., Freire, L., Montemor, M. F., & Nóvoa, X. R. (2012). The corrosion potential of stainless steel rebars in concrete: Temperature effect. *Corrosion Science*, *65*, 556–560. doi:10.1016/j.corsci.2012.09.001

Dong, S., Zhao, B., Lin, C., Du, R., Hu, R., & Zhang, G. X. (2012). Corrosion behavior of epoxy/zinc duplex coated rebar embedded in concrete in ocean environment. *Construction & Building Materials*, *28*(1), 72–78. doi:10.1016/j.conbuildmat.2011.08.026

Duari, S., Mukhopadhyay, A., Barman, T. K., & Sahoo, P. (2018). Tribological Performance Optimization of Electroless Nickel Coatings Under Lubricated Condition. In K. Kumar & J. Davim (Eds.), *Design and Optimization of Mechanical Engineering Products* (pp. 250–280). Hershey, PA: IGI Global. doi:10.4018/978-1-5225-3401-3.ch013

Duarte, R. G., Castela, A. S., Neves, R., Freire, L., & Montemor, M. F. (2014). Corrosion behavior of stainless steel rebars embedded in concrete: An electrochemical impedance spectroscopy study. *Electrochimica Acta*, *124*, 218–224. doi:10.1016/j.electacta.2013.11.154

Elsener, B., Crobu, M., Scorciapino, M. A., & Rossi, A. (2008). Electroless deposited Ni–P alloys: Corrosion resistance mechanism. *Journal of Applied Electrochemistry*, *38*(7), 1053–1060. doi:10.100710800-008-9573-8

Erdoğdu, Ş., Bremner, T. W., & Kondratova, I. L. (2001). Accelerated testing of plain and epoxy-coated reinforcement in simulated seawater and chloride solutions. *Cement and Concrete Research*, *31*(6), 861–867. doi:10.1016/S0008-8846(01)00487-2

Farina, S. B., & Duffó, G. S. (2007). Corrosion of zinc in simulated carbonated concrete pore solutions. *Electrochimica Acta*, *52*(16), 5131–5139. doi:10.1016/j.electacta.2007.01.014

Flint, G. N., & Cox, R. N. (1988). The resistance of stainless steel partly embedded in concrete to corrosion by seawater. *Magazine of Concrete Research*, *40*(142), 13–27. doi:10.1680/macr.1988.40.142.13

Gonzalez, J. A., & Andrade, C. (1982). Effect of carbonation, chlorides and relative ambient humidity on the corrosion of galvanized rebars embedded in concrete. *British Corrosion Journal*, *17*(1), 21–28. doi:10.1179/000705982798274589

Hu, X., He, C., Feng, K., Liu, S., & Walton, G. (2019). Effects of polypyrrole coated rebar on corrosion behavior of tunnel lining with the combination effect of sustained loading and pre-existing cracks when exposed to chlorides. *Construction & Building Materials*, *221*, 318–331. doi:10.1016/j.conbuildmat.2019.06.092

Jalili, M. M., Moradian, S., & Hosseinpour, D. (2009). The use of inorganic conversion coatings to enhance the corrosion resistance of reinforcement and the bond strength at the rebar/concrete. *Construction & Building Materials*, *23*(1), 233–238. doi:10.1016/j.conbuildmat.2007.12.011

Kobayashi, K., & Takewaka, K. (1984). Experimental studies on epoxy coated reinforcing steel for corrosion protection. *International Journal of Cement Composites and Lightweight Concrete*, *6*(2), 99–116. doi:10.1016/0262-5075(84)90039-3

Koleva, D. A., Hu, J., Fraaij, A. L. A., Stroeven, P., Boshkov, N., & De Wit, J. H. W. (2006). Quantitative characterisation of steel/cement paste interface microstructure and corrosion phenomena in mortars suffering from chloride attack. *Corrosion Science*, *48*(12), 4001–4019. doi:10.1016/j.corsci.2006.03.003

Liu, M., Cheng, X., Li, X., Jin, Z., & Liu, H. (2015). Corrosion behavior of Cr modified HRB400 steel rebar in simulated concrete pore solution. *Construction & Building Materials*, *93*, 884–890. doi:10.1016/j.conbuildmat.2015.05.073

Liu, M., Cheng, X., Li, X., Zhou, C., & Tan, H. (2017). Effect of carbonation on the electrochemical behavior of corrosion resistance low alloy steel rebars in cement extract solution. *Construction & Building Materials*, *130*, 193–201. doi:10.1016/j.conbuildmat.2016.10.003

Loto, C. A. (2016). Electroless nickel plating–a review. *Silicon*, *8*(2), 177–186. doi:10.100712633-015-9367-7

Luo, H., Su, H., Dong, C., & Li, X. (2017). Passivation and electrochemical behavior of 316L stainless steel in chlorinated simulated concrete pore solution. *Applied Surface Science*, *400*, 38–48. doi:10.1016/j.apsusc.2016.12.180

Macias, A., & Andrade, E. C. (1983). Corrosion rate of galvanized steel immersed in saturated solutions of $Ca(OH)_2$ in the pH range 12–13· 8. *British Corrosion Journal*, *18*(2), 82–87. doi:10.1179/bcj.1983.18.2.82

Macias, A., & Andrade, E. C. (1987a). Corrosion of galvanized steel in dilute $Ca(OH)_2$ solutions (pH 11·1–12· 6). *British Corrosion Journal*, *22*(3), 162–171. doi:10.1179/000705987798271505

Macias, A., & Andrade, E. C. (1987b). Corrosion of galvanized steel reinforcements in alkaline solutions: Part 1: Electrochemical results. *British Corrosion Journal*, *22*(2), 113–118. doi:10.1179/000705987798271631

Manning, D. G. (1996). Corrosion performance of epoxy-coated reinforcing steel: North American experience. *Construction & Building Materials*, *10*(5), 349–365. doi:10.1016/0950-0618(95)00028-3

Mukhopadhyay, A., Barman, T. K., & Sahoo, P. (2018). Electroless Nickel Coatings for High Temperature Applications. In K. Kumar & J. Davim (Eds.), *Composites and Advanced Materials for Industrial Applications* (pp. 297–331). Hershey, PA: IGI Global. doi:10.4018/978-1-5225-5216-1.ch013

Mukhopadhyay, A., & Sahoo, S. (2019). Corrosion protection of reinforcement steel rebars by the application of electroless nickel coatings. *Engineering Research Express*, *1*(1), 015021. doi:10.1088/2631-8695/ab35f0

Munot, H., Deshpande, P., & Modhera, C. (2018). Conducting Polypyrrole Coated Rebar in a Carbonated Concrete Pore Solution: Electrochemical Investigations. *Portugaliae Electrochimica Acta*, *36*(5), 365–375. doi:10.4152/pea.201805365

Omar, R., Aboraia, M. S., Oraby, E. A., Gubner, R., & Rizk, A. E. (2018). The effect of sodium citrate as a complex agent on the corrosion properties of the electroless Ni-P coating. *Materials Research Express*, *5*(12), 126511. doi:10.1088/2053-1591/aae0b9

Ožbolt, J., Balabanić, G., & Kušter, M. (2011). 3D Numerical modelling of steel corrosion in concrete structures. *Corrosion Science*, *53*(12), 4166–4177. doi:10.1016/j.corsci.2011.08.026

Ping, G., Elliott, S., Beaudoin, J. J., & Arsenault, B. (1996). Corrosion resistance of stainless steel in chloride contaminated concrete. *Cement and Concrete Research*, *26*(8), 1151–1156. doi:10.1016/0008-8846(96)00110-X

Pul, R. E. W. W., & Sprinkel, M. M. (1998). Estimating the Service Life of Epoxy Coated Reinforcing Steel. *Materials Journal*, *95*(5), 546–557.

Rahman, M. M., Islam, M. M., Khan, M. M. R., Ong, H. R., Uddin, M. T., & Islam, M. A. (2019). IBA-modified gypsum-containing epoxy resin coating for rebar: Corrosion performance and bonding characteristics. *International Journal of Plastics Technology*, *23*(1), 20–28. doi:10.100712588-019-09238-3

Roventi, G., Bellezze, T., Barbaresi, E., & Fratesi, R. (2013). Effect of carbonation process on the passivating products of zinc in $Ca(OH)_2$ saturated solution. *Materials and Corrosion*, *64*(11), 1007–1014. doi:10.1002/maco.201206868

Sagüés, A. A., Pech-Canul, M. A., & Al-Mansur, A. S. (2003). Corrosion macrocell behavior of reinforcing steel in partially submerged concrete columns. *Corrosion Science*, *45*(1), 7–32. doi:10.1016/S0010-938X(02)00087-2

Sahoo, P., & Das, S. K. (2011). Tribology of electroless nickel coatings–a review. *Materials & Design*, *32*(4), 1760–1775. doi:10.1016/j.matdes.2010.11.013

Saravanan, K., Sathiyanarayanan, S., Muralidharan, S., Azim, S. S., & Venkatachari, G. (2007). Performance evaluation of polyaniline pigmented epoxy coating for corrosion protection of steel in concrete environment. *Progress in Organic Coatings*, *59*(2), 160–167. doi:10.1016/j.porgcoat.2007.03.002

Serdar, M., Žulj, L. V., & Bjegović, D. (2013). Long-term corrosion behaviour of stainless reinforcing steel in mortar exposed to chloride environment. *Corrosion Science*, *69*, 149–157. doi:10.1016/j.corsci.2012.11.035

Shi, J., & Ming, J. (2017). Influence of defects at the steel-mortar interface on the corrosion behavior of steel. *Construction & Building Materials*, *136*, 118–125. doi:10.1016/j.conbuildmat.2017.01.007

Shi, J., Ming, J., Sun, W., & Zhang, Y. (2017). Corrosion performance of reinforcing steel in concrete under simultaneous flexural load and chlorides attack. *Construction & Building Materials*, *149*, 315–326. doi:10.1016/j.conbuildmat.2017.05.092

Singh, D. D. N., & Ghosh, R. (2006). Electroless nickel–phosphorus coatings to protect steel reinforcement bars from chloride induced corrosion. *Surface and Coatings Technology*, *201*(1), 90–101. doi:10.1016/j.surfcoat.2005.10.045

Smith, J. L., & Virmani, Y. P. (1996). Performance of epoxy coated rebars in bridge decks.

Söylev, T. A., & Richardson, M. G. (2008). Corrosion inhibitors for steel in concrete: State-of-the-art report. *Construction & Building Materials*, *22*(4), 609–622. doi:10.1016/j.conbuildmat.2006.10.013

Sudagar, J., Lian, J., & Sha, W. (2013). Electroless nickel, alloy, composite and nano coatings–A critical review. *Journal of Alloys and Compounds*, *571*, 183–204. doi:10.1016/j.jallcom.2013.03.107

Sun, C., Li, J., Shuang, S., Zeng, H., & Luo, J. L. (2018). Effect of defect on corrosion behavior of electroless Ni-P coating in CO_2-saturated NaCl solution. *Corrosion Science*, *134*, 23–37. doi:10.1016/j.corsci.2018.01.041

Tan, Z. Q., & Hansson, C. M. (2008). Effect of surface condition on the initial corrosion of galvanized reinforcing steel embedded in concrete. *Corrosion Science*, *50*(9), 2512–2522. doi:10.1016/j.corsci.2008.06.035

Tang, F., & Chen, G. (2015). Chemically reactive enamel coating of steel rebar for enhanced durability of reinforced concrete structures. *International Journal of Structural Engineering*, *6*(1), 56–72. doi:10.1504/IJSTRUCTE.2015.067992

Tang, F., Chen, G., Brow, R. K., Volz, J. S., & Koenigstein, M. L. (2012b). Corrosion resistance and mechanism of steel rebar coated with three types of enamel. *Corrosion Science*, *59*, 157–168. doi:10.1016/j.corsci.2012.02.024

Tang, F., Chen, G., Volz, J. S., Brow, R. K., & Koenigstein, M. L. (2012a). Microstructure and corrosion resistance of enamel coatings applied to smooth reinforcing steel. *Construction & Building Materials*, *35*, 376–384. doi:10.1016/j.conbuildmat.2012.04.059

Tang, F., Chen, G., Volz, J. S., Brow, R. K., & Koenigstein, M. L. (2013). Cement-modified enamel coating for enhanced corrosion resistance of steel reinforcing bars. *Cement and Concrete Composites*, *35*(1), 171–180. doi:10.1016/j.cemconcomp.2012.08.009

Tian, Y., Liu, M., Cheng, X., Dong, C., Wang, G., & Li, X. (2019). Cr-modified low alloy steel reinforcement embedded in mortar for two years: Corrosion result of marine field test. *Cement and Concrete Composites*, *97*, 190–201. doi:10.1016/j.cemconcomp.2018.12.019

Torbati-Sarraf, H., & Poursaee, A. (2018). Corrosion of coupled steels with different microstructures in concrete environment. *Construction & Building Materials*, *167*, 680–687. doi:10.1016/j.conbuildmat.2018.02.083

Treadaway, K. W. J., Cox, R. N., & Brown, B. L. (1989). Durability of corrosion resisting steels in concrete. *Proceedings - Institution of Civil Engineers*, *86*(2), 305–331. doi:10.1680/iicep.1989.1628

Verbruggen, H., Baert, K., Terryn, H., & De Graeve, I. (2019). Molybdate-phosphate conversion coatings to protect steel in a simulated concrete pore solution. *Surface and Coatings Technology*, *361*, 280–291. doi:10.1016/j.surfcoat.2018.09.056

Wang, Y. Q., Kong, G., Che, C. S., & Zhang, B. (2018). Inhibitive effect of sodium molybdate on the corrosion behavior of galvanized steel in simulated concrete pore solution. *Construction & Building Materials*, *162*, 383–392. doi:10.1016/j.conbuildmat.2017.12.035

Weishaar, A., Carpenter, M., Loucks, R., Sakulich, A., & Peterson, A. M. (2018). Evaluation of self-healing epoxy coatings for steel reinforcement. *Construction & Building Materials*, *191*, 125–135. doi:10.1016/j.conbuildmat.2018.09.197

Wronkowska, A. A. (1993). In Situ and Ex Situ Characterization of Passive Layers on $Ni_{1-x}P_x$ in Alkaline Solution. *Journal of the Electrochemical Society*, *140*(4), 995–1005. doi:10.1149/1.2056242

Wu, C., Chen, G., Volz, J. S., Brow, R. K., & Koenigstein, M. L. (2012). Local bond strength of vitreous enamel coated rebar to concrete. *Construction & Building Materials*, *35*, 428–439. doi:10.1016/j.conbuildmat.2012.04.067

Xiong, C., Li, W., Jin, Z., Gao, X., Wang, W., Tian, H., & Jiang, L. (2018). Preparation of phytic acid conversion coating and corrosion protection performances for steel in chlorinated simulated concrete pore solution. *Corrosion Science*, *139*, 275–288. doi:10.1016/j.corsci.2018.05.018

Yan, D., Reis, S., Tao, X., Chen, G., Brow, R. K., & Koenigstein, M. (2012). Effect of chemically reactive enamel coating on bonding strength at steel/mortar interface. *Construction & Building Materials*, *28*(1), 512–518. doi:10.1016/j.conbuildmat.2011.08.075

Yeomans, S. R. (1994). Performance of black, galvanized, and epoxy-coated reinforcing steels in chloride-contaminated concrete. *Corrosion*, *50*(1), 72–81. doi:10.5006/1.3293496

Yeomans, S. R. (2004). Galvanized steel in concrete: an overview. In S. R. Yeomans (Ed.), *Galvanized Steel Reinforcement in Concrete* (pp. 1–30). Oxford, UK: Elsevier Science Ltd.

Zheng, H., Dai, J. G., Poon, C. S., & Li, W. (2018). Influence of calcium ion in concrete pore solution on the passivation of galvanized steel bars. *Cement and Concrete Research*, *108*, 46–58. doi:10.1016/j.cemconres.2018.03.001

Chapter 15
Non-Destructive Testing of Carbon Fibre Reinforced Polymer (CFRP) Composite Using Thermosonic Technique

Tanmoy Bose
National Institute of Technology, Meghalaya, India

N. S. V. N. Hanuman
National Institute of Technology, Meghalaya, India

Subhankar Roy
National Institute of Technology, Meghalaya, India

ABSTRACT

Composite materials are often subjected to low velocity impacts which leads to delamination in subsequent layers. Linear ultrasound-based approaches are not accurate enough to detect it properly. The local defect resonance (LDR) based thermosonic is proved to be an efficient candidate for detection of such defects. LDR frequency excitation leads to high amplitude vibration which raises defect temperature drastically, detectable by an infrared camera. In this chapter, a numerical investigation of LDR frequency excited ultrasound thermography is carried out on delaminated carbon fibre reinforced polymer (CFRP) plate. The location and size of the delamination can be easily understood from thermal signature. The temperature gradient variation is found to be high at first and then it decreases due to higher heat conduction rate. The delamination in CFRP plate is detected by standard phased array ultrasound testing (PAUT) using flat bottom hole in aluminium plate as a case study. Delamination detection by PAUT is found to be very time consuming process compared with thermosonic technique.

DOI: 10.4018/978-1-7998-1831-1.ch015

INTRODUCTION

The composite material is a combination of two or more materials which give a unique property. The classification of the composite materials is polymer matrix composites (PMC), metal matrix composites (MMC), ceramic matrix composites (CMC). Recent days PMC material utilization is more in the field of military, aerospace, automotive, architecture, marine and energy due to its good fatigue properties, corrosion resistance, and high strength to weight ratio. The large class of PMC material are carbon fiber reinforced plastic (CFRP) and Glass fiber reinforced plastic (GFRP) which are used based on their requirements (Satishkumar, Satheeshkumar, & Naveen, 2014; Chavan & Gaikwad, 2016; Vinay, Govindaraju, & Banakar, 2015). CFRP finds more usage in automobile and aeronautical industries due to increased rigidity compared to GFRP. These type of structures should be inspected by non-destructive testing procedures to avoid further damage or its complete failure. Some defects like voids, delamination and cut fibers exist during manufacturing. This type of light weight structures are prone impact damge due to less plastic deformation, which can be observed in case of metals. Impact severely decreases the lamina strength due to formation of micro-cracks and delamination in that portion. Impact damages are of two types viz, high velocity and low velocity impact. In case of high velocity impact, the composite structures are severly damages which can visualized also. But in case of low velocity impact, the damages can not be properly visualized and the damages are also known as barely visible impact damage (BVID). Ultrasonic non-destructive testing (NDT) is very popular to find defects in the composite materials. It uses linear wave spectroscopy in the form of either contact or non-contact method. The number of industries has implemented contact-based NDT method to identify flaws in mechanical components. In wave spectroscopy, wave is propagated in anticipated direction in the material with a specific velocity and get reflected from discontinuities within the material. Ultrasonic phased array technique is developed by microelectronic parts, which is processing the wave depends velocity of the wave, attenuation, and acoustic impedance.It has a multiple-element ultrasonic transducer which steering the beam means pulsing and receiving the signals with time corrected gain (TCG). It can show low-velocity impact damages non-visible inclusions by A-scan, S-scan, and C-scan (Caminero, García-Moreno, Rodríguez, & Chacónb, 2019). But detection of low velocity impact damage is still a challenging task using this method due to its occurrence within few layers from surface. Many kinds of literature prove that Thermography technique is very efficient for localizing the defect. Moreover contactless thermography technique is very suitable for the barely visible impact damages and micro flaws in a structure. But this is permissible to the finding the cracks near to the surface (Meola et al., 2015; Katunin, Dragan, & Dziendzikowski, 2015). Moreover ultrasonic testing (UT) gives very high precise images compare to the thermography technique (Ibrahim, Smith, Wang, 2017). The non-linear wave spectroscopy is the very efficient technique to describe the non-linear behavior of the defect which is specially applied on investigation rocks. One catergory of this is nonlinear ultrasound resonant spectroscopy (NRUS) which mainly detects the deviation structural resonance to predict fault within the structure. So, it is a volumetric method but unable to detect the location of flaws present in the structure. Ultrasound excited thermography or, thermosonic or sonic IR is a non-contact defect inspection technique to visualize three dimensional defect but its usage were severely restricted due to use of high power ultrasound welding horns and high non-repeatability of experiments. The concept of local defect resonance (LDR) has made this method highly repeatable and usage with low acoustic power which will not damage the investigating structure (Solodov, Bai, Bekgulyan, & Busse, 2011).

In this book chapter, numerical simulation for ultrasound thermography is carried out on CFRP plate for localizing the delamination size. The ABAQUS software is used to develop a model of CFRP plate with delamination. Numerical simulation is carried out using explicit temperature displacement analysis by providing the LDR frequency. This LDR frequency helps to activate temperature evolution at the delamination position only. Subsequently, experiments performed on CFRP plate using phased array ultrasound using aluminium flat bottom hole (FBH) as a reference. Corresponding A-scan, B-scan, and C-scan images show the length, width, and depth of the delamination. The chapter is organized as follows: first, the complete literature review pertaining to thermosonic is depicted. Then the concept of LDR, numerical modelling and phased array ultrasound testing (PAUT) procedure is mentioned in methodology section. After that, results pertaining to thermosonic technique and PAUT is mentioned in results and discussion section followed by, the future research directions and conclusions.

BACKGROUND

Ultrasound thermography or thermosonics (or, Sonic IR) based defect detection was introduced in 1980s (Mignogna, Green, Duke, Henneke, & Reifsnider, 1981) during an investigation on heat developed due to ultrasound fatigue process. Two universities viz. Wayne state university, USA (Prof. R. L. Thomas and his group) and University of Stuttgurt, Germany (Prof. G. Busse and his group) are mainly involved in its development. Several modification of ultrasound excitation method is reported by using amplitude modulated (Rantala, Wu, & Busse, 1996), sonic pulse (Favro, Han, Ouyang, Sun, Sui, & Thomas, 2000), frequency modulated (Zweschper, Dillenz, Riegert, Scherling, & Busse, 2003), burst excitation (Dillenz, Zweschper, Riegert, & Busse, 2003) and acoustic chaos (Han et al., 2004). Introduction of lock-in thermography (Busse, Wu, & Karpen, 1992) was found to provide better quality of image and also depth determination by varying lock-in frequency. However, use of high power horn (can increase delamination in composite material) and non-repeatability of experiment is the main hindrance for its wide applicability. Clapping and rubbing at the crack or delamination face is the main source of heat generation but the process is highly frequency selective. Ultrasound wideband sweep (Gleiter, Spießberger, & Busse, 2007) was found to provide good results at low power but the reason was unknown at that time. Prof. P. Cawley from Imperial College, UK (Morbidini & Cawley, 2009) has tried to explain the heat generation process by crack damping. Later, one of the researchers (Solodov et al., 2011) from Prof. Busse's group has given its reason using the concept of local defect resonance (LDR) and open the widespread applicability of this method due to high repeatability along with low power excitation (which wouldn't damage the composite). Although the similar observation is also reported in 1990s during experimental investigation on delaminated composite plate at high frequency vibration (Tenek, Henneke, & Gunzburger, 1993) but that article has not received any attention that time. Solodov et al. (2011) was the first to show that LDR frequency excitation results in high vibrational amplitudes at the defect location only using laser vibrometry. Mainly, the clapping phenomenon is found to be maximum at that frequency which raises the local temperature also due to increases frictional heat generation. An IR camera is used to detect the temperature evolution at defect location and leads to efficient ultrasound thermography with low excitation power. The concept of LDR frequency in case of delamination can be easily explained using the concept of flat bottom hole (Solodov, Bai, & Busse, 2013). The trapped modes at flat bottom hole (FBH) is also found to be same as the circular clamped plate for the ratio of FBH radius to its thickness greater than 10 (Solodov, Derusova, & Rahammer, 2015). LDR based ultrasonic thermography is found to be effective one which reduce the excitation power by channelizing maximum energy to the defect (Solodov, 2014). The excitation area can be further increased by channelizing the ultrasound energy in a particular direction using guided mode (Rahammer, Solodov, & Kreutzbruck, 2016). For cm size defects, the LDR frequencies are found lie in acoustic zone. So, wide area non-contact excitation though speakers along with thermosonic inspection is found to be possible (Solodov, Rahammer, Gulnizkij, & Kreutzbruck, 2016). Rahammer and Kreutzbruck (2017) used frequency sweep vibrothermography based on LDR concept and it is shown to be one of the fastest method of delamination inspection in composite material. Vacuum attached piezoelectric device was used for excitation and the phase plot of thermal image was used to identify the defect. Hettler et al. (2017) used the laser vibrometry and ostu's algoritm to detect delamination in sandwich plate using piezoelectric excitation. LDR phenomena within a composite plate is also analytically modelled and the behaviour is reported like that of non-classical nonlinear oscillator (Ciampa, Scarselli, Meo, 2017). Later, the concept of LDR has been applied to detect barely visible impact damage (BVID) in composite material (Fierro, Ginzburg, Ciampa, & Meo, 2016, 2017; Dionysopoulos, Fierro, Meo, & Ciampa, 2018), crack in metallic component (Dyrwal, Meo, Ciampa, 2018) using thermal camera. In both cases, nonlinear narrow sweep excitation (NNSE) technique was implemented to detect LDR frequencies from first and second harmonic in fourier spectrum. Fierro et al. (2016, 2017) obtained the thermosonic image of impact damage in aircraft wing and good agreement was reported with laser vibrometry results. A set of piezoelectric transducers, i.e., one transmitter and receiver, were mounted between stiffners. Dionysopoulos et al. (2018) used dual excitation to calculate LDR frequencies as well as thermosonic inspection. Dyrwal et al. (2018) used the air couple ultrasound sensors for non-contact evaluation of LDR frequencies due to crack in a turbine blade. Thermosonic inspection was carried out to obtain the crack parameters in non-contact way. The previous all researches starting from Solodov et al. (2011) mainly concentrated on the LDR frequency which generate out-of-plane motion of defect. Later, it was found that in-plane LDR modes, which appears at higher frequency compared to out-of-plane LDR, are more effective for thermosonic inspection (Segers, Kersemans, Hedayatrasa, Calderon, & Paepegem, 2018). The inspection time is found to be less for in-plane LDR modes. This concept was later applied to detect BVID using thermosonic technique (Segers, Hedayatrasa, Verboven, Poelman, Paepegem, & Kersemans, 2019). Recently, the Bicoherence analysis based detection of LDR frequencies for both single and multiple circular artificial delaminations (using Teflon tape) in thin GFRP plates are reported (Roy, Bose, & Debnath, 2019). The same approach is also extended for dual excitation comprising of sweep/noise excitation at one transmitter while a single periodic frequency (SPF) excitation at the other (Roy & Bose, 2019). The second harmonic of LDR frequency as well as SPF peaks are identified easily from the bicoherence plots which can be used further for automated LDR detection. Defect depth determination in thermography or thermosonic testing is the toughest job due to lack of understanding in the three dimensional heat conduction in composite material. Generally, the defect depth is evaluated from the thermal diffusion length corresponding to phase plot of thermal image. Inception of LDR concept has drastically reduced this effort and now, the defect depth can be calculated directly from LDR equation if the area of the defect is known (Solodov, Rahammer, & Kreutzbruck, 2019).

It can be understood from the complete literature review that a vast amount of work is already done based on ultrasound thermography. The concept of LDR has healed the previous limitations of ultrasound thermography and the revealed it as defect selective imaging procedure with low excitation level and high repeatability. Now, thermosonic is widely applied on BVID and micro crack detection which is one of the challenging task for NDT society.

METHODOLOGY

Concept of Local Defect Resonance (LDR)

Any flaws present in the structure reduce stiffness and mass of the structure. The localized stiffness and mass at the flaw location generates a oscillator in that location. The natural frequency of that oscillator is known as defect resonance frequency which was attributed as "local defect resonance frequency (LDR)" by Solodov et al. (2011). The expression of LDR frequency is follows:

$$f_0 = \frac{1}{2\pi}\sqrt{\frac{K_{eff}}{M_{eff}}} \tag{1}$$

where K_{eff} = effective stiffness and M_{eff} = effective mass. When the flexural wave frequency matches with the defect natural frequency it leads to high amplitude vibration at the defect location only. In case of delamination in a composite plate, the LDR frequency can be understood by a flat bottom hole (FBH) in a plate (Figure 1). The radius of delamination is equal to the radius of FBH (r) and the thickness above the delamiantion is represented as FBH residual thickness (t). Now, the vibration for residual portion of FBH can be assumed like vibrating circular plate with elastic boundary condition. The clamped boundary condition can be assumed if r/t is greater than 10. Then the potential energy of the circular plate can be written as (Solodov et al., 2013)

$$P.E. = \frac{K_{eff}.U_{eff}^2}{2} = \frac{32\pi D u_0^2}{3r^2} \tag{2}$$

where U_{eff} is the effective displacement of the flat bottom hole, D is the bending stiffness of the plate, and u_0 is the amplitude of vibration at the centre of the plate. From Equation (2), substituting the term for effective displacement of vibration, the effective stiffness is calculated by the following relation as (Solodov et al., 2013)

$$K_{eff} = \frac{192\pi D}{r^2} \tag{3}$$

Similarly, the effective mass of the defect is determined from the relation of kinetic energy which is as follows (Solodov et al., 2013):

$$K.E. = \frac{M_{eff}.U_{eff}^2}{2} = \frac{m u_0^2}{10} \tag{4}$$

where m is the FBH mass. From Equation (4), the value of effective mass for FBH is found to be around $1.8m$.

Figure 1. Schematic of a defect in the form of a flat bottom hole

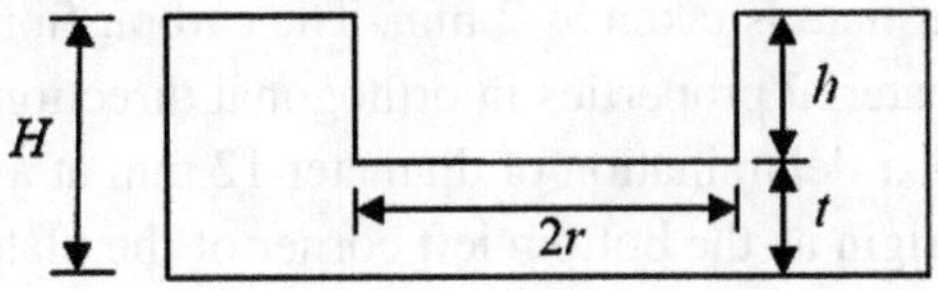

The effective stiffness, K_{eff} and the effective mass, M_{eff} are then combined to give the general relation for determining the local defect resonance (LDR) frequency. The relation for LDR frequency is written as

$$f_{LDR} = \frac{1.6\,t}{r^2}\sqrt{\frac{E}{12\rho(1-\upsilon^2)}} \qquad (5)$$

where E = modulus of elasticity, ρ = mass density of the material, and υ = Poisson's ratio.

Modeling of Thermosonic or, Ultrasound Thermography for CFRP Plate With Delamination

Thermosonic or ultrasound thermography is used for detection of size, shape and location of any defects present in a structure by means of temperature rise in the damage location. The rise in temperature is generated mainly due to resonant excitation of the damaged structure at its LDR frequency. The analytical calculation of LDR frequency is already discussed in previous section, which helps in determining the input excitation frequency to be provided on the structure. In case of unknown realistic defects, the excitation signal can be applied in form of broadband sweep. This helps in getting an idea of LDR frequency of the damage, as the temperature tends to rise whenever the sweep frequency matches with the defect frequency. During such broadband sweep excitation, the sweep rate with respect to time must be kept lower. This gives more time for generation of a temperature gradient at the damage location when sweep frequency matches defect frequency. Further, a narrow sweep excitation can be applied according to the estimated LDR frequency from the broadband excitation case. The narrow sweep excitation will further increase the temperature at the defect location which can be used for high contrast defect imaging. However, this technique is tedious, time consuming and may not be able to determine the exact LDR frequency of the defect using broadband excitation. Hence, it is recommended that the LDR frequency of the defect must be determined before performing the thermosonic defect imaging technique. Several signal processing tool like Fast Fourier Transform (FFT) and bicoherence analysis can be performed for detection of LDR frequency both numerically and experimentally (Roy, Bose, & Debnath, 2019).

The thermosonic defect imaging technique can be simulated for any structure by using explicit temperature displacement analysis in ABAQUS 6.14. When any structure having delamination is excited with its LDR frequency, high vibration amplitudes are obtained at the delamination area. These vibration leads to clapping and rubbing action between delamination surfaces which raises the temperature locally. In this chapter, the thermosonic analysis is performed in case of a CFRP plate with circular delamination at the central layer.

A CFRP plate with four layers is modelled in ABAQUS 6.14. Each layer is having a thickness of 0.5 mm and a total thickness of the plate is taken as 2 mm. The carbon fibre is considered to be as plain-woven fabric type with same material properties in orthogonal directions. The size of the CFRP plate is 180 mm×180 mm with circular delamination of diameter 12 mm at a location of (x = 120 mm, y = 60 mm and z = 1mm) taking origin as the bottom left corner of the plate. The schematic of the CFRP plate is shown in Figure 2.

The material properties for CFRP used for performing the explicit temperature displacement analysis are presented in Table 1. During the analysis, the CFRP laminates are assembled such that the delamination region is between the second and third layer from the top surface. The interaction properties of explicit surface contact type are used for providing sliding contact at the defect location. The sliding contact is kept as frictionless and hard contact in case of tangential behaviour and normal behaviour, respectively. The rest of the laminate surfaces for all the layers are tied using tie constraints. Surface to surface type tie constraints are provided with upper layer as the master surface and the lower layer as the slave surface. All the nodes of the CFRP plate is selected as a set for providing the initial ambient temperature of the

Figure 2. Schematic diagram of the CFRP plate having single delamination

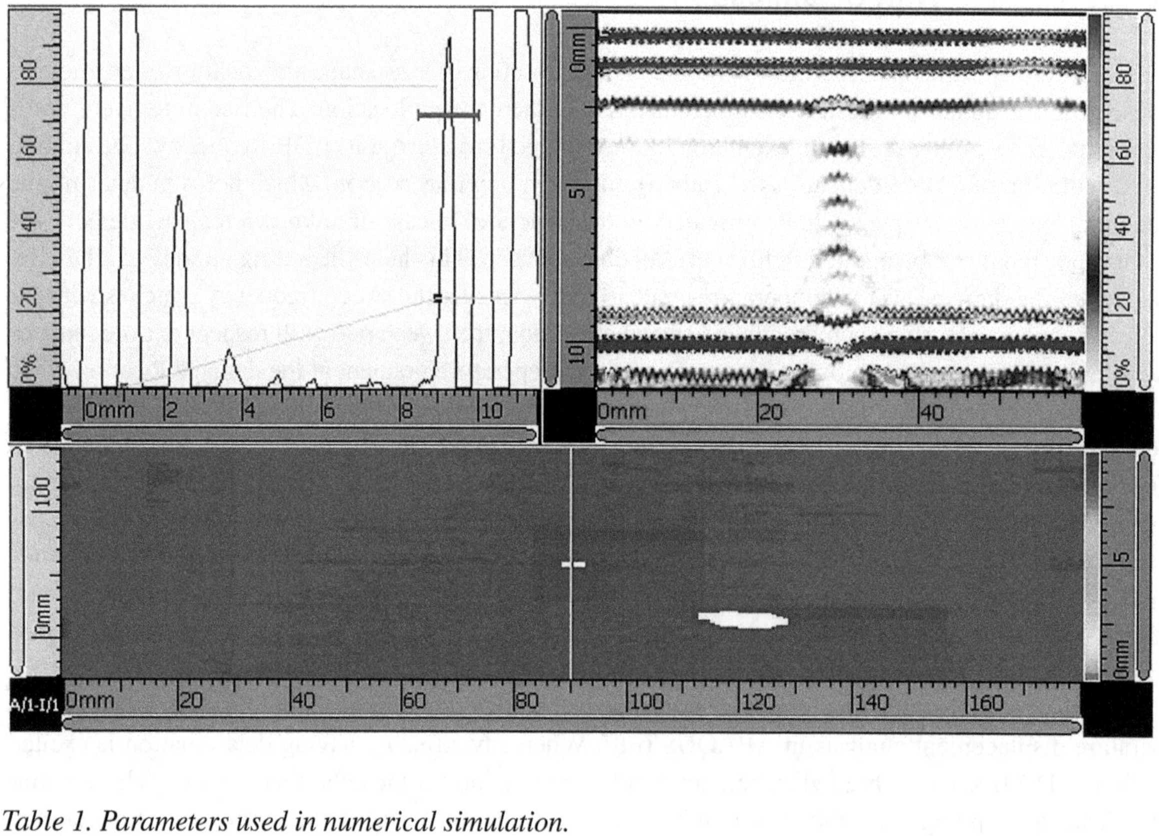

Table 1. Parameters used in numerical simulation.

Density (in kg/m³)	Young's Modulus (in GPa)	Poisson's Ratio	Specific Heat (J/kg-K)	Thermal Conductivity (W/m-K)
1546	136	0.29	2020	0.6

plate as 20K. The CFRP plate is meshed with tetrahedral elements having minimum edge length of 0.5 mm and a maximum edge length of 2 mm. The element type for the analysis is selected as C3D10MT which is a 10-noded modified thermally coupled second-order tetrahedron with hourglass control. The geometric order of the mesh element is considered as quadratic. Subsequently, a transmitter node is selected as a point of excitation on the top surface of CFRP plate. The LDR frequency for delamination is calculated from the relation expressed in Equation 5. The analytical LDR frequency of delamination is calculated to be 28.6 kHz. Hence, a single periodic continuous excitation of 28.6 kHz is applied to the transmitter node in tranverse direction to CFRP plate. The amplitude of transmitted force is taken as 50 N towards the negative z-direction. The total time period of simulation is taken as 10 ms with each time increment of 7×10^{-9} s. The time increment is calculated as per the CFL law for explicit dynamics analysis. The boundary condition of the CFRP plate is kept as simply supported at all the four sides. The field output request is kept for the whole model with parameters such as vibration displacement, nodal temperature and elemental temperature. This helps in obtaining the contours for each parameter at different time increments of the simulation. In case of history output request, the delamination surface is selected with nodal temperature in order to obtain the average temperature profile of the delamination region with respect to time. Therefore, the contour of displacement, nodal temperature, elemental temperature as well as the temperature versus time profiles can be plotted from the thermosonic analysis.

Phased Array Ultrasound Testing (PAUT) Procedure

In PAUT, sound energy propagates in several modes according to the direction of the wave and material vibration. Most generally shear and longitudinal modes are used to detect the flaws in a material. The Shear wave andlongitudinal wave described by three important factors such as wave velocity, attenuation, and acoustic impedance (Caminero, García-Moreno, Rodríguez, & Chacónb, 2019). The shear wave and longitudinal wave velocity are designated by C_s and C_l, respectively. These velocity equations are as follows (Kumar, 2009):

$$C_s = \sqrt{\frac{G}{\rho}} \tag{6}$$

$$C_l = \sqrt{\frac{E}{\rho}} \tag{7}$$

Where ρ is the density, G the shear modulus and E is Young's modulus. Based on the velocity and wave frequency, the sound field is divided into two zones i.e., near field zone and far-field zone. The near field zone formed near to the transducer there a minimum and maximum sound pressure developed. According to the focal law, following equation is used for calculating distance of the near field.

$$N = D^2 / 4\lambda \text{ where } \lambda = c / f \tag{8}$$

Where N represents near field length, D is element diameter, f is frequency, c is material sound velocity, and λ is the wavelength. The near field distance is function of beam diameter, wave velocity in material and its frequency. Next zone is described as far-field zone where more energy is dissipated due to expansion of the beam diameter.

Omniscan machine and phased array (contact type probe) are shown in Figure 3. One transducer (probe- 3.5L6A-NW1) is used for steering the beam in transmitting and receiving mode. The probe is connected to the machine, and it has a rectangular wedge (SW1-OL-IHC-C). The liquid medium (i.e. Oil, grease or water, etc.) is used to make coupling between wedge and surface of the plate. The ultrasound beam is passed via phased array wedge while the probe is sliding on the surface of the plate (Figure 3). The virtual probe aperture (VPA) is defined as

$$\text{VPA} = (\text{total elements selected} - \text{aperture size}) + 1 \tag{9}$$

The transmitting and receiving wave angle adjusted at zero degrees linear sector. Now, the wedge can be discretized into 61 ultrasound beams by Equation 9. Corresponding wedge delay is generated by software. Each ultrasound beam is created by four elements aperture (aperture size), which generates total 61 elements (Equation 9) VPA according to focal law. The pulse input voltages are 40 and 115 for aluminium plate and CFRP plate, respectively. The time corrected technique (TCG) activated for increasing amplitude of the back wall echos. Corresponding results are shown in A-scan, S-scan, and C-scan. A- scan show the percentage of amplitude with respect to time in the thickness direction, which is called time of flight. S-scan (i.e., sector sacn) represents width of the plate vs amplitude. The width depends on the probe size and orientation. Most of the cases, width considered as default for performing the C-scan. The C-scan represents the area of the plate with variation in amplitude of echos (A%). These can be performed on specific requirement of thickness of the plate. In C-scan, the probe

Figure 3. Experimental setups of ultrasound omniscan

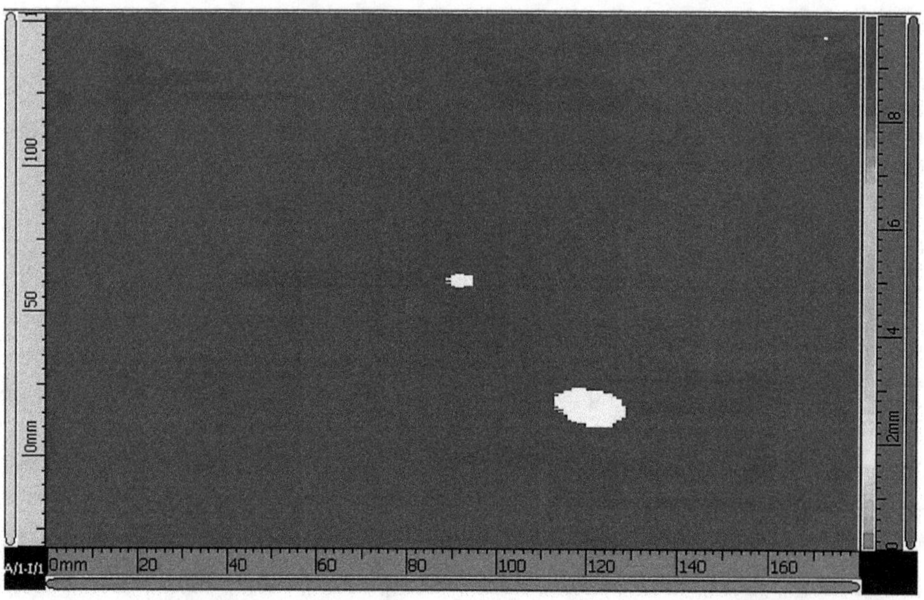

is attached to an encoder and then raster scanning is performed to obtain all the discontinuities in the plate. It is to be noted that these all three gates are operated mutually by synchronization of A- gate and I-gate (corresponding to TCG). Subsequently, adjusting the A-gate in a different position based on our requirement precise size and location of the flaws in a material can be obtained by post-processing. Some experiments performed on aluminium plate having an FBH and CFRP plate having single delamination is explained in next section.

RESULTS AND DISCUSSION

Detection of Delamination in CFRP Plate Using Thermosonic Technique

This section discusses the numerical results obtained from the explicit temperature displacement analysis performed in ABAQUS 6.14 in case of a CFRP plate with delamination. The displacement contour plot of CFRP plate at final time increment (0.01 s) is shown in Figure 4. As observed from the contour plot, the incident wave is propagated throughout the plate surface which is essential to obtain the results from the ultrasound wave spectroscopy technique. The maximum amplitude of vibration observed in the CFRP plate is 9.03×10^{-7} m when it is excited at LDR frequency.

Subsequently, the contour for nodal temperature is presented in Figure 5 which shows a significant rise in the nodal temperature particularly at the delamination region. It is observed that rest of the CFRP plate does not have any temperature difference when the plate is excited with LDR frequency. This is a characteristic of the LDR frequency which leads to temperature rise at the defect location due to rubbing and clapping action. The nodal temperature at the delamination location rises to 20.6 K from an initial ambient temperature of 20 K. Thus, a nodal temperature rise of 600mK is generated by exciting the

Figure 4. Displacement contour of CFRP plate with delamination at time increment of 0.01 s

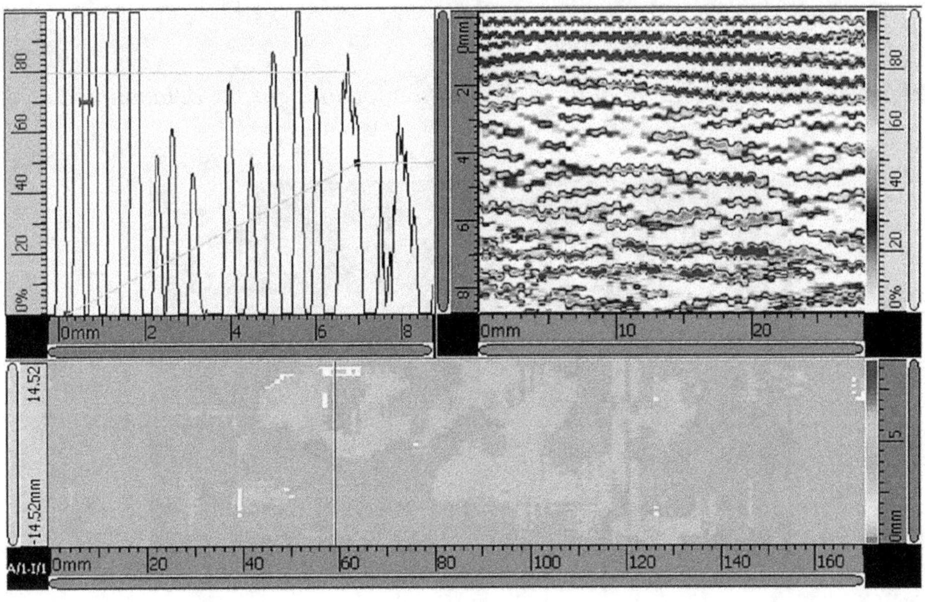

plate for 10 ms. The temperature difference will be higher if the plate is excited for a longer period of time. This will lead to a better thermal imaging of CFRP plate for detecting defects and their location. Moreover, the elemental temperature rise is shown in Figure 6. In this figure, the elemental temperature rises to 20.26 K from an ambient temperature of 20 K. The shape of the defect (i.e. circular delamination) can be easily detected in the element temperature contour. Hence, the thermosonic technique can be used for detecting the size and shape of defects in CFRP structures.

Finally, a temperature profile showing the average temperature rise at the delamination location is plotted in Figure 7. The plot shows the variation of the temperature at the delamination region with respect to the time increment through which the plate is being excited with LDR frequency. As observed

Figure 5. Nodal temperature profile over of the delaminated CFRP plate after 0.01 s

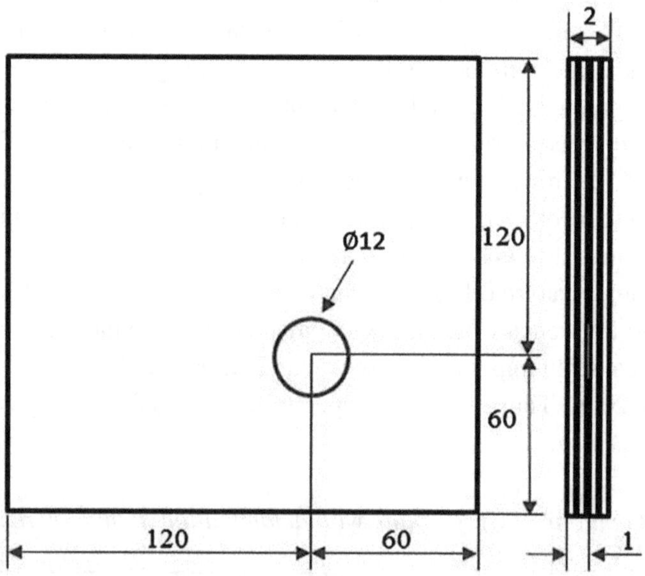

Figure 6. Rise in element temperature at the delamination location for the delaminated CFRP plate at 0.01 s

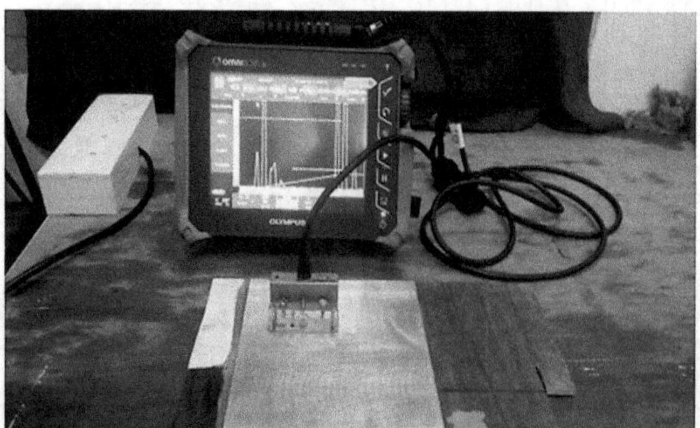

from the plot, the temperature at the delamination is constant (20 K) at ambient temperature till 1 ms after which it starts rising to around 20.14 K in 8 ms. The rise in temperature for rest of the 2 ms is comparatively slower with respect to initial rise. Hence, it can be understood that the temperature gradient is very high at inception due to low thermal conduction rate at first. As the heat conduction rate increases, the temperature gradient decreases. Therefore, the thermosonic technique of damage detection by exciting a structure at its LDR frequency is very effective and can be used for highly efficient defect selective imaging of structures.

Delamination Detection Using Phased Array Ultrasound Testing

In this section, phased array ultrasound testing (PAUT) carried out in aluminium plate having two FBHs and delaminated CFRP plate are shown. The dimensions of alumium plate is 180 mm × 180 mm × 10 mm (Figure 8a). Two FBHs of depth 9 mm are fabricated using milling process. The diameters of two FBHs are 12 and 4 mm, respectively. The dimension of CFRP plate is 190 mm × 170 mm × 7 mm. It was fabricated using hand lay-up process with 20 laminas. The delamination is created by a Teflon tape of required diameter.

The A, S and C-scan image of first FBH (bigger one) is shown in Figure 9. All procedure for scanning was mentioned in section 2.3. Here, TCG is applied 1 mm below the surface and the A-gate is placed on the back wall echo. Entire C-scan of the plate is performed and the discontinuity depth from surface is observed from S-scan. Residual thickness of the FBH is found to be 1mm (firm red lines from top surface) and its position along with diameter can be verified from C-scan. Similarly, Figure 10 shows the A-scan, S-scan, and C-scan for smaller FBH. The first echo in A-scan is obtained at 0 mm and back wall echo at 10 mm, as in the previous case. In case of S-scan, the echo generated due to presence of

Figure 7. Temperature profile showing the trend of average temperature rise at the delamination location with respect to time in case of the delaminated CFRP plate

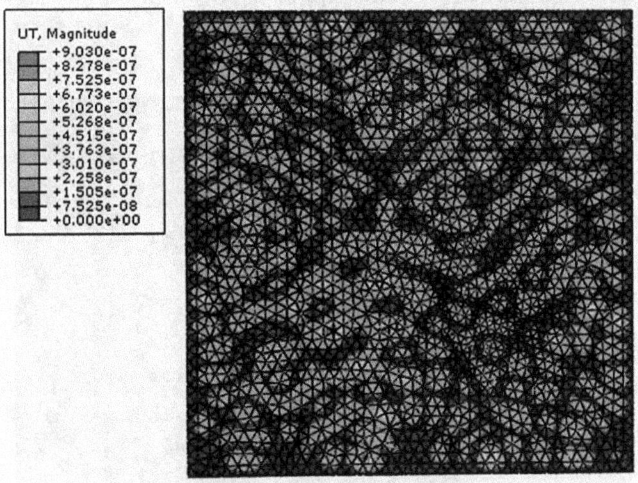

Figure 8. a) The aluminium plate having FBH; b) The delaminated CFRP plate. Note: All dimensions are in mm.

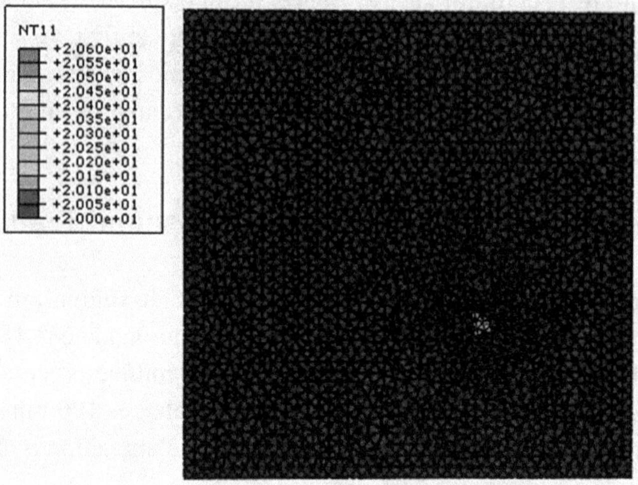

FBH is obtained at 1 mm below the surface which is the residual thickness of FBH. Here, narrower discontinuity is obtained due to less diameter than the previous case. The top view of the FBH can be observed from the C-scan of the plate. The size and dimension of both the FBHs can be easily detected from the C-scan as shown in Figure 11.

Figure 12 represents the A-scan, S-scan,and C-scan of CFRP plate having the single delamination. In A-scan, the first echo and back wall echo is represented at 0 and 7 mm, respectively. The length of the scan is 170mm (Figure 2), where scan width is fixed which depends on the size of the probe (explained in section 2.3) i.e. 61mm. Finally, we can localize the delamination from A-scan and S-scan images. In

Figure 9. Ultrasound scan image showing A-scan (upper left), S-scan (upper right) and C-scan (bottom) for the bigger FBH in aluminium plate

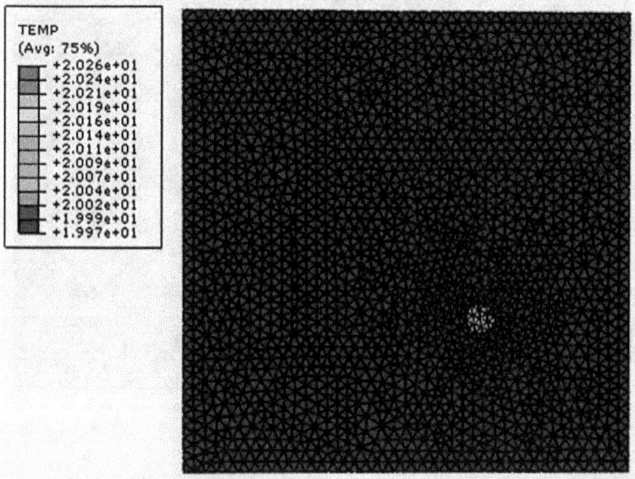

Figure 10. Ultrasound scan image showing A-scan (upper left), B-scan (upper right) and C-scan (bottom) for the smaller FBH in aluminium plate.

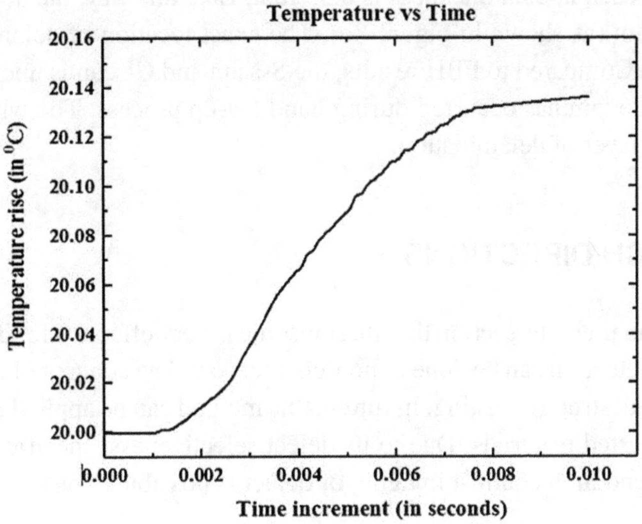

Figure 11. Ultrasound scan image showing enlarged C-scan for both the FBHs in aluminium plate depicting the location and shape of the FBH

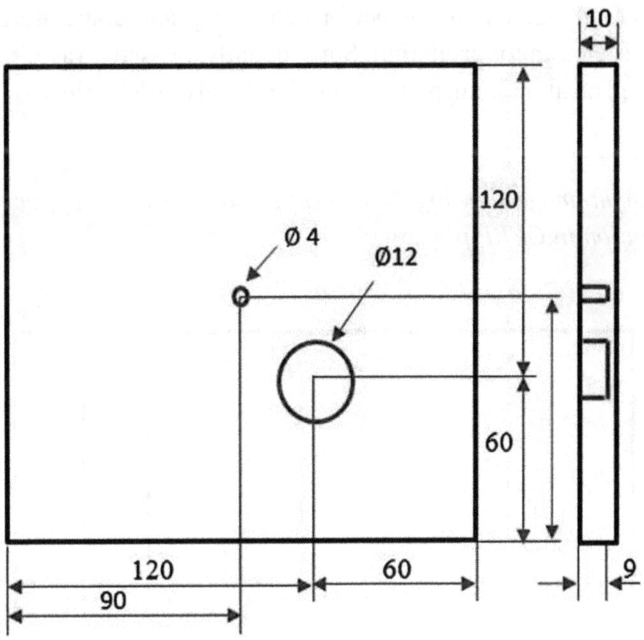

S-scan, the delamination is found to be between 0.6 to 0.8 mm. It is not possible to obtain the exact depth of the delamination due to multiple scattering and subsequent absorbtion effect. As there are 20 laminas within 7 mm thickness, each lamina thickness is 0.35 mm. Like this way, the depth of delmination can be said exactly at 0.7 mm (as shown in Figure 8b). The exact location of delamination can be identified from C-scan result. Compared to FBH results, the S-scan and C-scan is unclear due to presence of porosity in the composite laminas occurred during hand lay-up process. The white dots represents are porosity presents in the layer of delamination.

FUTURE RESEARCH DIRECTIONS

It can be understood from previous section that thermosonic is very effective for delamination detection. The method is fast and detection can be done in non-contact way. The concept of defect resonance opens a new possibility in non-destructive testing. In future, this method can be applied for porosity determination in composite and casted materials. Due to its defect selectiveness, specific defects of interest can be visualized correctly and high contrast imaging of defect is possible.

CONCLUSION

The numerical simulation for ultrasound thermography is carried out in CFRP plate having single delamination. The numerical simulation was performed by explicit temperature displacement analysis and the LDR frequency is chosen for excitation. Subsequently, phased array ultrasound testing (PAUT) experiment was performed on aluminium plate having FBH and the delaminated CFRP plate. Ultrasound

Figure 12. Ultrasound scan image showing A-scan (upper left), S-scan (upper right) and C-scan (bottom) for single delamination in CFRP plate

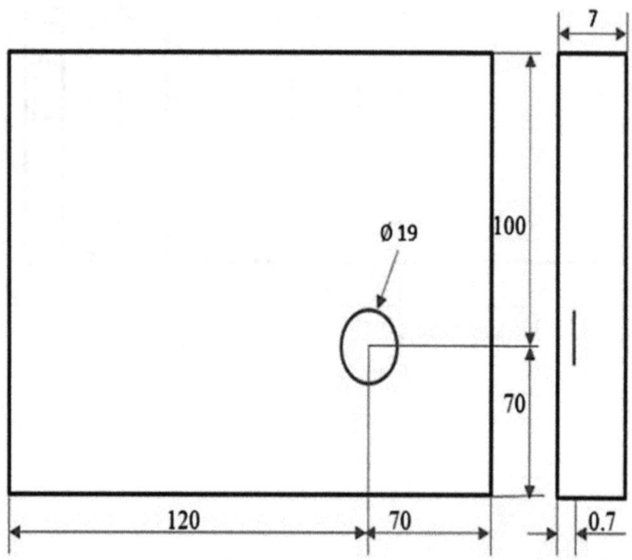

thermography is found to much quicker than other methods due to its defect selectiveness. Temperature gradient at delamination location is found to be very high at first and then it decreases due to increased heat conduction rate. PAUT method is found a time consuming method but it detects the delamination and FBH correctly. Ultrasound thermography is found to an effective method but more research is required to be carried out to increase its reliability.

REFERENCES

Busse, G., Wu, D., & Karpen, W. (1992). Thermal wave imaging with phase sensitive modulated thermography. *Journal of Applied Physics*, *71*(8), 3962–3965. doi:10.1063/1.351366

Caminero, M. A., García-Moreno, I., Rodríguez, G. P., & Chacónb, J. M. (2019). Internal damage evaluation of composite structures using phased array ultrasonic technique: Impact damage assessment in CFRP and 3D printed reinforced composites. *Composite Part B*, *165*, 131–142. doi:10.1016/j.compositesb.2018.11.091

Chavan, V. B., & Gaikwad, M. U. (2016). Review on development of glass fiber/epoxy composite material and its characterizations. *International Journal of Science, Engineering and Technology Research*, *5*, 2224–2228.

Ciampa, F., Scarselli, G., & Meo, M. (2017). On the generation of nonlinear damage resonance intermodulation for elastic wave spectroscopy. *The Journal of the Acoustical Society of America*, *141*(4), 2364–2374. doi:10.1121/1.4979256 PMID:28464640

Dillenz, A., Zweschper, T., Riegert, G., & Busse, G. (2003). Progress in phase angle thermography. *The Review of Scientific Instruments*, *74*(1), 417–419. doi:10.1063/1.1524010

Dionysopoulos, D., Fierro, G. M., Meo, M., & Ciampa, F. (2018). Imaging of barely visible impact damage on a composite panel using nonlinear wave modulation thermography. *NDT & E International*, *95*, 9–16. doi:10.1016/j.ndteint.2018.01.005

Dyrwal, A., Meo, M., & Ciampa, F. (2018). Nonlinear air-coupled thermosonics for fatigue micro-damage detection and localization. *NDT & E International*, *97*, 59–67. doi:10.1016/j.ndteint.2018.03.012

Favro, L. D., Han, X., Ouyang, Z., Sun, G., Sui, H., & Thomas, R. L. (2000). Infrared imaging of defects heated by a sonic pulse. *The Review of Scientific Instruments*, *71*(6), 2418–2421. doi:10.1063/1.1150630

Fierro, G. P. M., Ginzburg, D., Ciampa, F., & Meo, M. (2016). Nonlinear thermosonics and laser vibrometry for barely visible impact damage of a composite stiffener panel. *Proceedings of the Society for Photo-Instrumentation Engineers*, *980*, 419.

Fierro, G. P. M., Ginzburg, D., Ciampa, F., & Meo, M. (2017). Imaging of Barely Visible Impact Damage on a Complex Composite Stiffened Panel Using a Nonlinear Ultrasound Stimulated Thermography Approach. *Journal of Nondestructive Evaluation*, *36*(4), 21. doi:10.100710921-017-0449-x

Gleiter, A., Spießberger, C., & Busse, G. (2007). Improved ultrasound activated thermography using frequency analysis. *Quantitative Infrared Thermography Journal*, *4*(2), 155–164. doi:10.3166/qirt.4.155-164

Han, X., Zeng, Z., Li, W., Islam, M. S., Lu, J., Loggins, V., ... Thomas, R. L. (2004). Acoustic chaos for enhanced detectability of cracks by sonic infrared imaging. *Journal of Applied Physics*, *95*(7), 3792–3797. doi:10.1063/1.1652243

Hettler, J., Tabatabaeipour, M., Delrue, S., & Abeele, K. V. D. (2017). Detection and Characterization of Local Defect Resonances Arising from Delaminations and Flat Bottom Holes. *Journal of Nondestructive Evaluation*, *36*(1), 2. doi:10.100710921-016-0380-6

Ibrahim, M. E., Smith, R. A., & Wang, C. H. (2017). Ultrasonic detection and sizing of compressed cracks in glass and carbon-fibre reinforced plastic composites. *NDT & E International*, *92*, 111–121. doi:10.1016/j.ndteint.2017.08.004

Katunin, A., Dragan, K., & Dziendzikowski, M. (2015). Damage identification in aircraft composite structures: A case study using various non-destructive testing techniques. *Composite Structures*, *127*, 1–9. doi:10.1016/j.compstruct.2015.02.080

Kumar, P. (2009). *Elements of Fracture Mechanics. Chennai, India: McGraw Hill Education.*

Meola, C., Boccardi, S., Carlomagno, G. M., Boffa, N. D., Monaco, E., & Ricci, F. (2015). Non-destructive evaluation of carbon fibre reinforced composites with infrared thermography and ultrasonic. *Composite Structures*, *134*, 845–853. doi:10.1016/j.compstruct.2015.08.119

Mignogna, R. B., Green, R. E. Jr, Duke, J. C. Jr, Henneke, E. G. II, & Reifsnider, K. L. (1981). Thermographic investigation of high-power ultrasonic heating in materials. *Ultrasonics*, *19*(4), 159–163. doi:10.1016/0041-624X(81)90095-0

Morbidini, M., & Cawley, P. (2009). The detectability of cracks using sonic IR. *Journal of Applied Physics*, *105*(9), 1–9. doi:10.1063/1.3125444

Rahammer, M., & Kreutzbruck, M. (2017). Fourier-transform vibrothermography with frequency sweep excitation utilizing local defect resonances. *NDT & E International*, *86*, 83–88. doi:10.1016/j.ndteint.2016.11.012

Rahammer, M., Solodov, I., Bisle, W., Scherling, D., & Kreutzbruck, M. (2016). Thermosonic Testing with Phase Matched Guided Wave Excitation. *Journal of Nondestructive Evaluation*, *36*, 1–7.

Rantala, J., Wu, D., & Busse, G. (1996). Amplitude Modulated Lock-In Vibrothermography for NDE of Polymers and Composites. *Research in Nondestructive Evaluation*, *7*(4), 215–218. doi:10.1080/09349849609409580

Roy, S., & Bose, T. (2019). Efficient determination of local defect resonance frequencies from bicoherence plots using double excitations. *Mechanical Systems and Signal Processing*, *127*, 595–609. doi:10.1016/j.ymssp.2019.03.017

Roy, S., Bose, T., & Debnath, K. (2019). Detection of Local Defect Resonance Frequencies usingBicoherence Analysis. *Journal of Sound and Vibration*, *443*, 703–716. doi:10.1016/j.jsv.2018.12.006

Satishkumar, T. P., Satheeshkumar, S., & Naveen, J. (2014). Glass fiber-reinforced polymer composites – a review. *Journal of Reinforced Plastics and Composites*, *33*(13), 1258–1275. doi:10.1177/0731684414530790

Segers, J., Hedayatrasa, S., Verboven, E., Poelman, G., Paepegem, W. V., & Kersemans, M. (2019). In-plane local defect resonances for efficient vibrothermography of impacted carbon fiber-reinforced polymers (CFRP). *NDT & E International*, *102*, 218–225. doi:10.1016/j.ndteint.2018.12.005

Segers, J., Kersemans, M., Hedayatrasa, S., Calderon, J., & Paepegem, W. V. (2018). Towards in-plane local defect resonance for non-destructive testing of polymers and composites. *NDT & E International*, *98*, 130–133. doi:10.1016/j.ndteint.2018.05.007

Solodov, I. (2014). Resonant Acoustic Nonlinearity of Defects for Highly-Efficient Nonlinear NDE. *Journal of Nondestructive Evaluation*, *33*(2), 252–262. doi:10.100710921-014-0229-9

Solodov, I., Bai, J., Bekgulyan, S., & Busse, G. (2011). A local defect resonance to enhance acoustic wave-defect interaction in ultrasonic nondestructive evaluation. *Applied Physics Letters*, *99*(21), 211911. doi:10.1063/1.3663872

Solodov, I., Bai, J., & Busse, G. (2013). Resonant ultrasound spectroscopy of defects: A Case study of flat-bottomed holes. *Journal of Applied Physics*, *113*(22), 223512. doi:10.1063/1.4810926

Solodov, I., Derusova, D., & Rahammer, M. (2015). Thermosonic Chladni figures for defect-selective imaging. *Ultrasonics*, *60*, 1–5. doi:10.1016/j.ultras.2015.02.007 PMID:25773969

Solodov, I., Rahammer, M., Gulnizkij, N., & Kreutzbruck, M. (2016). Noncontact Sonic NDE and Defect Imaging Via Local Defect Resonance. *Journal of Nondestructive Evaluation*, *35*(3), 48. doi:10.100710921-016-0364-6

Solodov, I., Rahammer, M., & Kreutzbruck, M. (2019). Analytical evaluation of resonance frequencies for planar defects: Effect of a defect shape. *NDT & E International*, *102*, 274–280. doi:10.1016/j.ndteint.2018.12.008

Tenek, L. H., Henneke, E. G. II, & Gunzburger, M. D. (1993). Vibration of delaminated composite plates and some applications to non-destructive testing. *Composite Structures*, *23*(3), 253–262. doi:10.1016/0263-8223(93)90226-G

Vinay, H. B., Govindaraju, H. K., & Banakar, P. (2015). Processing and characterization of glass fiber and carbon fiber reinforced vinyl ester based composites. *International Journal of Research in Engineering and Technology*, *4*(5), 401–406. doi:10.15623/ijret.2015.0405074

Zweschper, T., Dillenz, A., Riegert, G., Scherling, D., & Busse, G. (2003). Ultrasound excited thermography using frequency modulated elastic waves. *Insight (American Society of Ophthalmic Registered Nurses)*, *45*, 178–182.

Chapter 16
Basics and Applications of Ferrofluids

Sumanta Banerjee
Heritage Institute of Technology, Kolkata, India

ABSTRACT

Engineered 'smart' fluids, called ferrofluids, which permit static and dynamic control by external magnetic fields of low or moderate strengths present a novel and challenging domain for scientists and technologists. The introduction of a 'controllable' or 'tunable' external force into the momentum transport equations opens up new fields of physical phenomena. In situations where the external field influence becomes strong enough to compete with gravitational forces or acts as a sole external agency to drive flows in the hypo-gravity environment, a new class of hydrodynamic phenomena becomes accessible. Ferrofluids exhibit special rheological properties that make them suited for a number of technological and biomedical applications. This chapter outlines the underlying physics of field-fluid interactions on one hand; on the other, it cites novel techniques in ferrofluid-actuated electronics cooling, hydrostatic & hydrodynamic bearings, extreme-boundary lubrication, damping systems, biomimetic locomotion, and medical diagnostics (e.g. magnetic drug targeting & magnetic cell sorting).

INTRODUCTION

Design of applications that employ synthetic fluids (as active or passive components) gains new possibilities, if the fluids can be conveniently positioned or moved by electromagnetic forces of low or moderate field intensities (Gast & Zukoski, 1989). As magnetic forces can be generated and controlled with minimum technical effort, e.g. by varying the electric current intensity through a coil, new ideas using an additional control parameter can be realized to suit a broad range of macro-scaled and μ-scaled applications (Liu, 1993). At small length scales (e.g. in the domain of μ-fluidics), the transport of momentum and energy is largely diffusion-limited due to very low values of flow Reynolds numbers (Das, Choi, Yu, & Pradeep, 2008). A suitable magnetically-susceptible fluid proves as a viable working medium to enhance energy and species transport in these devices (Rosensweig, 1997), as its flow pattern can be influenced by external actuations. In hypo-gravity and extra-terrestrial environments, for

DOI: 10.4018/978-1-7998-1831-1.ch016

instance, convective transport of mass, momentum and energy can only be achieved through external field actuation (without the aid of moving parts).

Effective field-fluid interaction, over a broad spectrum of technical applications, is realizable only for a magneto-rheological material with magnetic susceptibility proportional to density (Rosensweig, 1997; Liu, 1995a). The induced body force, called the Kelvin force, depends dually on the fluid magnetization and the field gradient within the medium (Rosensweig, 1997; Liu, 1995b). In paramagnetic salt solutions, for example, feeble magnetization renders the Kelvin force negligible (as compared with the g-force) even for strong magnetic fields. Liquid metals are also susceptible to Lorentz forces in strong magnetic fields; however, noticeable changes in flow-patterns are affected only by extremely high magnetic field strengths (in the order of several Tesla). In addition, liquid metals usually require a high-temperature environment. On the other hand, under-cooled metallic melts (found to exhibit magnetic ordering even in the liquid state) does not allow any handling of the liquid. Therefore, their magnetic properties cannot be used in practical designs. The viable option, then, is to produce stable suspensions of magnetic nanoparticles in appropriate carrier liquids. These suspensions (called ferrofluids) exhibit high initial susceptibility (consequently, high magnetization) even for (easily achievable) magnetic field strengths ~50mT (Liu, 1995a; Odenbach, 2002).

A ferrofluid is a colloidal suspension of single-domain magnetic nanoparticles (~3-15nm diameter) in a nonmagnetic liquid. By single-domain is meant that the magnetic moments of the individual atoms are oriented in a fixed direction. The bulk medium exhibits a magnetic behavior only when an external magnetic field is imposed, whereby the thermally-disoriented magnetic moments of the particles are aligned. The nanoparticles are coated with adsorbed surfactant layers to prevent agglomeration due to the van der Waals forces and dipole-dipole interactions. Such a stable colloid can be considered as a quasi-single-phase homogeneous liquid and the laws of continuum mechanics applied (Jiang & Liu, 1996).

The engineering utility of ferrofluids over other magneto-rheological (MR) liquids arises from the fact that their flow behavior and properties can be desirably altered in the presence of external magnetic fields of even low or moderate intensity. The possibility to exert an externally controllable force to a fluid medium, in absence of moving parts, offers broad avenues of research in basic fluid dynamics and transport phenomena. In addition, it holds promise for a wide spectrum of applications in areas of biomedicine, thermal management of miniaturized electronics applications, electrical drives and actuators, worm-like locomotion of target-delivery agents, positioning systems, hydrostatic and hydrodynamic bearings, extreme-boundary lubrication, and damping systems. Many of these concepts have been practically implemented and are currently in use (Rosensweig, 1997; Odenbach, 2002; Odenbach, 2009).

In the past 50 years since the first synthesis of commercial ferrofluid, the prospect of customized flow-control achievable by means of permanent magnets or simple coil arrangements has driven significant R&D activities globally. The applications that have been successfully implemented are all characterized by a singular trait: an external magnetic field is employed that exerts a force to move the functionalized fluid. This magnetic action, manifested as the Kelvin force, actuates transport of mass, momentum, and energy within the system. The external field that sets up the Kelvin force can therefore be engineered to control and drive flows. In other words, the external field gradients can be tailored for various pre-desired flow patterns. In addition, engineering applications exist where the external field is required simply to position a tiny blob of ferrofluid at a pre-designated site within a system. These applications, for instance, in areas of tribology and biomedicine utilize the change in rheology of ferrofluids in the presence of magnetic fields (Odenbach, 2009).

The following sections outline some of the basic theories that govern ferrofluid behavior, as well as cite some potential application areas of ferrofluids in engineering and medicine.

MAGNETIC BEHAVIOR IN SOLIDS

Ferromagnetic solids are composed of domains (Weiss, 1907) in each of which the magnetic moments of the individual atoms are oriented in a fixed direction. The domain structure for a typical ferromagnetic polycrystalline material is shown in Figure 1. Although the domain wall structures are depicted as solid lines, in reality the transition in the direction of the atomic moment vector is gradual and takes place across ~100 atoms. A ferromagnetic material breaks up into domains to minimize the field energy, as the latter would be considerable if the material were magnetized in one direction. However, the process of breaking up into domains does not continue indefinitely, as energy is required to create the domain walls that separate the domains (Rosensweig, 1997). In absence of external field, the domain magnetic moments are thermally disoriented (Figure 1a); when external field is applied, the individual moments all align up along a fixed direction (Odenbach, 2004).

According to quantum mechanics, molecular magnetism owes its origin to spin magnetic moments of atoms. When the spins on neighboring atoms change from parallel alignment to anti-parallel alignment, there is an accompanying change in the electron charge distribution in the atoms that alters the electrostatic energy of the system (Heisenberg, 1928). In the ferromagnetic state (as in nickel, cobalt, and many of their alloys), parallel alignment (of atomic magnetic moments) is energetically more favorable. The ferromagnetic ordering disappears at the Curie temperature. In contrast, anti-ferromagnetic materials (e.g. MnO, FeO, $FeCl_2$) exhibit no net moment at any temperature. This is because the spin magnetic moments alternate between adjacent atoms. The anti-ferromagnetic ordering disappears at the Néel temperature. In ferrimagnetic substances, the spin magnetic moments of unequal magnitudes alternate from one atom to the other, so that the net moment is smaller as compared to a ferromagnetic material. Ferrimagnetic compounds are depicted as $MO.Fe_2O_3$ (M:Fe, Ni, Mn, Cu, Mg) (Rosensweig, 1997; Odenbach, 2002; Odenbach, 2009).

Paramagnetism is a behavior resulting from the tendency of molecular moments to align with the applied magnetic field in the absence of long-range order. Paramagnetism is exhibited by liquid O_2, rare-earth salt solutions, and ferromagnets above the Curie temperature. Diamagnetism represents the weakest type of magnetic behavior and is prominent only in materials with closed electron shells. Inert gases, many metals (Bi, Sb, Cu, Au), and most non-metals and organic compounds are diamagnetic.

Figure 1. Ferromagnetic domain structures for polycrystalline material depicting (a) random domain orientation and (b) alignment of domain magnetic moments in presence of external magnetic field
Courtesy: https://en.wikipedia.org/wiki/Magnetic_domain

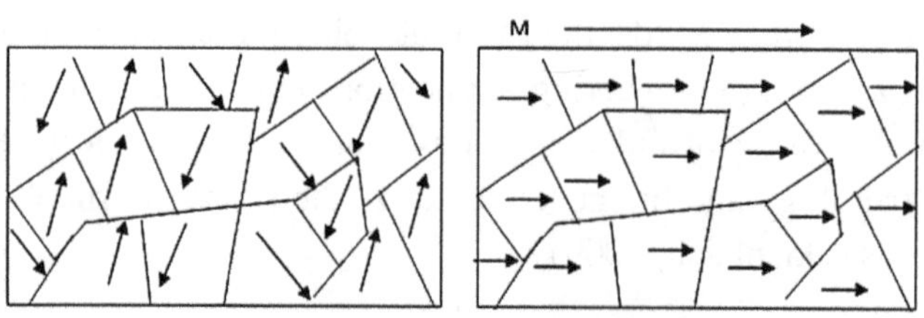

A colloidal ferrofluid consists of a collection of ferromagnetic or ferrimagnetic single-domain particles with no long-range order between them. The resulting behavior, termed superparamagnetism, is similar to paramagnetism except that the magnetization in low to moderate fields is much larger (Rosensweig, 1997). True ferrofluids do not settle out, even though a slight concentration gradient can become established after long exposure to gravitational or magnetic force field. The particles are coated with a molecular layer of dispersant and suspended in a liquid carrier. Thermal agitation keeps the particles suspended due to Brownian motion, and the surfactant coatings prevent the particles from sticking to each other. Colloidal ferrofluids are observed to retain flowability under the most intense applied magnetic fields. Also, a typical ferrofluid contains ~10^{23} particles per cubic meter, and is opaque to visible light (Odenbach, 2002).

MAGNETIC PROPERTIES OF FERROFLUIDS

The magnetic properties of a ferrofluid are of basic importance for characterizing their behavior, and they contain important information about the microscopic composition of the fluid. Basically, the magnetic behavior of ferrofluids is determined by the magnetic properties of the suspended particles. In this section, we shall distinguish between the equilibrium magnetization-which provides information about volume concentration and size of the magnetic particles-and the behavior of magnetization relaxation, which is of paramount importance in determining the use of ferrofluids in applications like hyperthermia. The latter characteristic also determines important features of the magnetically driven flow behavior of the suspension (Rosensweig, 1997; Odenbach, 2009).

EQUILIBRIUM MAGNETIZATION

As mentioned, a unique desirable feature of ferrofluids is the possibility to control their flow and physical characteristics by means of even moderate magnetic field strengths in the order of few tens of mT. Thus, all subsequent discussion concerning the field-dependent phenomena in magnetic fluids requires knowledge about their magnetic properties.

To understand the magnetic properties, one must first consider the magnetic structure of the suspended particles. For the sake of definiteness, these particles are assumed to have a mean diameter of ~10nm. Thus, for magnetite, they can be assumed to be single-domain particles (Das et al., 2008; Odenbach, 2002). Studies indicate that each unit cell of the crystal structure of magnetite (Fe_3O_4) has a volume of ~730 $\mathring{A}^3$ and contains 8 molecules of Fe_3O_4. Therefore, each of the magnetic particles in a ferrofluid contains ~ 6×10^3 molecules of Fe_3O_4, and each molecule has a magnetic moment of

$$\sim 4\mu_B \left(\mu_B = 9.27 \times 10^{-24} Am^2\right).$$

Each nanoparticle thus carries a total magnetic moment m ~ 2×10^{-19} Am^2, and can be treated as small magnetic dipoles suspended in the carrier liquid (Odenbach, 2002).

The issues related to inter-particle interactions are now considered, assuming that a ferrofluid is a system of non-interacting spherical particles. This implies that the magnetic behavior of the bulk system can be described as an ensemble of non-interacting, thermally agitated magnetic dipoles (Shliomis, 1996).

If the concentration of the magnetic particles is not too high, and the effects of magnetic inter-particle interaction neglected, the magnetic behavior of the whole liquid can be modeled as a paramagnetic system. This enables the use of the Langevin equation for the magnetization of paramagnetic systems (Rosensweig, 1997):

$$M = M_S \left(\coth \alpha - \frac{1}{\alpha} \right) \tag{1}$$

In Equation (1), the saturation magnetization $\left(M_S = \varphi M_0\right)$ of the ferrofluid is equal to the product of the volume concentration $\left(= \varphi\right)$ of suspended magnetic particles and the saturation magnetization of bulk magnetic solid $\left(= M_0\right)$. The saturation magnetization of the particle may differ from the bulk value. The relation between the energy of a particle in the magnetic field $\vec{H}$ and its thermal energy is denoted by the Langevin parameter

$$\alpha = \left(\mu_0 m H \,/\, k_B T\right); \left|\vec{H}\right| = H \,.$$

This implies that the magnetization curve of a ferrofluid shows the typical paramagnetic behavior.

The difference in magnetic behavior between a paramagnetic salt solution and a ferrofluid colloid results from the fact that the magnitude of magnetic moment of the units interacting with an external magnetic field show a difference of about four orders of magnitude. This has significant influence on the magnetization behavior, even for small magnetic fields.

Figure 2. Magnetization behavior of the ferrofluid APG513A containing 7.2 vol.% of magnetite particles with a mean diameter of 10nm
(Rosensweig, 1997)

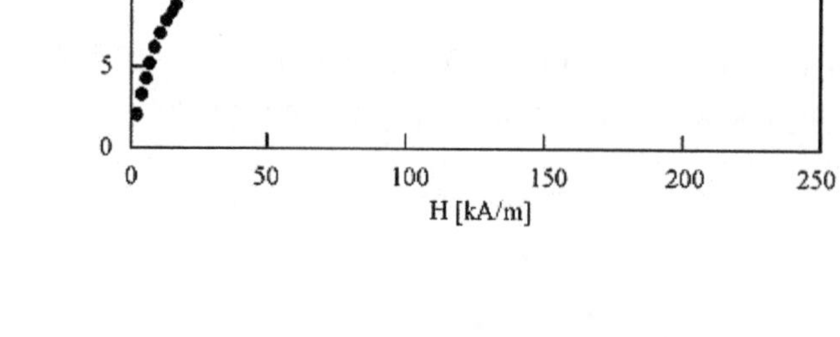

Utility of the Magnetization Curve

The magnetization curve provides important data about the composition of the fluid (see Figure 2). First of all, the saturation magnetization provides a direct measure of the volume concentration of the suspended magnetic nanoparticles. The saturation magnetization is usually obtained from the measured magnetization curve by plotting the magnetization $\left(=M\right)$ versus $1/H$. For large magnetic fields, it can be easily shown that Equation (1) can be approximated as:

$$M \approx M_S\left(1-\frac{1}{\alpha}\right) = M_S\left[1-\left(\frac{k_BT}{\mu_0 m}\right)\frac{1}{H}\right] = \varphi M_o\left(1-\frac{6}{\pi}\frac{k_BT}{\mu_0 M_o H d^3}\right) \tag{2}$$

If we extrapolate M as $1/H \to 0$ we obtain an immediate measure of M_s. In addition, for weak magnetic fields, the Langevin equation can be approximated as:

$$M \approx \frac{1}{3}\varphi M_o\left(\frac{\mu_0 m}{k_BT}\right)H = \left[\frac{1}{3}\varphi\mu_0 M_0^2\left(\frac{V}{k_BT}\right)\right]H = \chi_{in}H \tag{3}$$

It can be immediately observed that the initial susceptibility

$$\chi_{in} = \frac{\pi}{18}\varphi\mu_0 M_0^2\left(\frac{d^3}{k_BT}\right)$$

of a ferrofluid provides information about the mean particle size $\left(=d\right)$. This is because, for a given value of φ, the volume (hence the mean particle diameter) is the only unknown parameter in the initial susceptibility. This can be easily obtained from the slope of the magnetization curve at low magnetic field strengths. Moreover, appropriate techniques can additionally provide information about the size distribution of the suspended particles by analyzing the shape of the full magnetization curve (Odenbach, 2002).

It must be observed that the magnetic field mentioned here is not the external magnetic field, but the field inside the medium. The latter can differ from the external field (e.g. due to demagnetization effects). It is always the field inside the fluid which influences its behavior. Equation (3) enables to show that χ_{in} of a ferrofluid is about four orders of magnitude larger than that of a paramagnetic salt solution.

For a homogenously magnetized ferrofluid sample at low field strengths, the Kelvin force density exerted by the external field to the fluid is:

$$\vec{f}_{mag} = \mu_0\chi_{in}H\vec{\nabla}H \tag{4}$$

For the chosen standard fluid $\left(APG513A\right)$ and for a typical magnetic field at the pole of a small electromagnet

($H \approx 20kA / m; \nabla H \approx 7 \times 10^5 A / m^2$),

the magnetic force density is $\left|\vec{f}_{mag}\right| \approx 14kN / m^3$. For the same fluid (with density $\rho = 1.28 \cdot 10^3 \, kg / m^3$) the gravitational force density acting on the fluid is $\left|\vec{f}_{grav}\right| \approx 13kN / m^3$. This simple (but insightful) calculation shows that the magnetic and gravitational force intensities are in the same order of magnitude. Therefore, even low-intensity magnetic fields are able to exert forces that can significantly influence the behavior of the fluid (refer to Figure 3).

STABILITY REQUIREMENTS

It is instructive to consider the physics of certain mechanisms that are responsible for the existence of ferrofluids. By *existence* is implied the physico-chemical stability of the colloid. For this purpose, the methodology of dimensional analysis may be applied to arrive at the requisite criteria.

To start with, it is useful to write expressions for the various energy terms (per particle basis) (Odenbach, 2002).

Figure 3. Demonstration of the magnetic force acting on a ferrofluid. The fluid is attracted against gravity by the pole of a simple electromagnet
(Odenbach, 2002)

Thermal energy $k_BT; k_B = 1.38 \times 10^{-23} Nm / K$

T: Absolute temperature (K)

Magnetic energy $\mu_0 MHV; \mu_0 = 4\pi \times 10^{-7} H \cdot m$

μ_0 : Permeability of free space

Gravitational energy $\Delta\rho VgL; V = \left(\pi d^3 / 6\right) m^3$

L: Elevation in the gravitational field
d: Particle diameter

The ratios of one term to another yield dimensionless quantity that can determine the stability of the ferrofluid, as discussed below.

Stability in a Magnetic-Field Gradient

Let us consider the stability against settling of particles in a field gradient due to an external magnetic source. While the particles are attracted to the higher-intensity regions of the magnetic field, thermal motion counteracts the field force and provides statistical motions that allow the particles to sample all portions of the fluid volume. The magnetic energy $\mu_0 MHV$ represents the reversible work in removing a magnetized particle from a point (with intensity $\vec{H}$) in the field to a point in the fluid outside the field (Liu, 1995a):

$$W = -\int_H^0 \left(\mu_0 M \frac{dH}{ds} V\right) ds \approx \mu_0 MHV \tag{5}$$

Provided that some part of the fluid volume is located in a field-free region, the stability against segregation is favored by a high ratio of the thermal energy to the magnetic energy:

$$\frac{Thermal\,energy}{Magnetic\,energy} = \frac{k_B T}{\mu_0 MHV} \geq 1 \tag{6}$$

Rearranging and substituting for the volume of a sphere $\left(V = \pi d^3 / 6\right)$ gives an expression for the maximum particle size:

$$d \leq \left(\frac{6k_B T}{\pi \mu_0 MH} \right)^{0.33} \tag{7}$$

Considering the conditions existing in a beaker of magnetic fluid containing magnetite particles subject to the magnetic field gradient of a hand-held magnet

$$\left(H = 8 \cdot 10^4 A \cdot m^{-1}; M = 4.46 \cdot 10^5 A \cdot m^{-1}; T = 298K \right),$$

the particle size (computed from Equation (7)) is $d \leq 8.1 \cdot 10^{-9} m$ or $8.1nm$. Actual particle sizes of stable colloids range up to ~10nm. From the above calculation it is seen that particle size could scarcely be any larger and remain stable against segregation in a field gradient (Rosensweig, 1997; Odenbach, 2002).

Another physical phenomenon that delimits the concentration of particles in regions of more intense magnetic fields is steric hindrance. This puts an upper limit on the particle number concentration. Although concentration gradients can be established in this situation, when the (external) field is removed particles of a well-stabilized ferrofluid re-distribute spontaneously throughout the fluid volume over a period of time.

Stability Against Settling in a Gravitational field

The gravitational force pulls an individual nanoparticle downward in the container, while thermal agitation tends to keep the particle dispersed throughout the fluid matrix. Like the magnetic force, gravity furnishes a unidirectional force at any given point. The relative influence of gravity to magnetism is described by the ratio (Rosensweig, 1997):

$$\frac{Gravitational\,energy}{Magnetic\,energy} = \frac{\Delta \rho g L}{\mu_0 MH} \tag{8}$$

Considering a beaker of ferrofluid, for typical values of parameters like the height of fluid column, and the density difference

$$\left(L = 0.05m; \Delta\rho = \rho_{solid} - \rho_{fluid} = 4300kg / m^3; g = 9.8m / s^2 \right),$$

the ratio from Equation (8) is 0.047. Thus, gravity is less of a threat to the segregation of ferrofluids than is a magnetic field.

Stability Against Magnetic Agglomeration

The foregoing has considered the stability against segregating or forming large gradients of particle number concentration. The particles are assumed to be *mono-disperse*, or not attached to each other. But it is not a trivial task to achieve mono-dispersion.

A typical colloidal magnetic fluid contains ~10^{23} particles per cubic meter, and inter-particle collisions are frequent. Hence, if the particles adhere together, agglomeration will be rapid. As each nanoparticle is permanently magnetized, the maximum energy needed to separate a pair of particles (diameter d and surface-to-surface separation distance s) is when the particles are aligned. From field theory, this energy is nothing but the dipole-dipole pair energy $\left(=E_{dd}\right)$ (Odenbach, 2009):

$$E_{dd}=\frac{\pi}{9}\frac{\mu_0 M^2 d^3}{\left(l+2\right)^3};l=2s/d \tag{9}$$

Thus, when the particles are in contact, Equation (9) reduces to the following form (1=0):

$$E_{dd}=\frac{1}{12}\mu_0 M^2 V \tag{10}$$

As thermal agitation is available to disrupt the agglomerates, the effectiveness of the disruption is governed by the ratio:

$$\frac{Thermal\,energy}{Dipole\text{-}dipole\,contact\,energy}=\frac{12k_BT}{\mu_0 M^2 V}\geq 1 \tag{11}$$

For the particles to escape agglomeration, the ratio must exceed unity. This limits the particle size to:

$$d\leq\left(\frac{72k_BT}{\pi\mu_0 M}\right)^{0.33} \tag{12}$$

For the magnetic particles at room temperature, Equation (12) requires $d\leq 7.8nm$. This estimate also shows that normal magnetic fluids having particle size ~10nm are on the threshold of agglomeration but manage to escape this fate (Rosensweig, 1997).

Agglomeration Prevention Against Van Der Waals Force

In addition to assessing the stability against settling in a magnetic field or gravitational field, there is an additional consideration for the issue at hand. This is because the van der Waals forces arise spontaneously between neutral particles due to the fluctuating dipole-dipole forces that are always present. The force represents the quantum-mechanical interactions due to fluctuating orbital electrons in one particle inducing oscillating dipoles in the other. As the dipolar field varies ~ r^{-3} the field gradient varies ~ r^{-4}. If the polarization is proportional to the field (and hence ~ r^{-3}), this implies that the dipole interaction force varies ~ r^{-7} and the (interaction) energy varies ~ r^{-6}. There exist suitable analytical models (London model, Hamaker model) to predict the quantum of attraction between the nanoparticles, although a full consideration of the same is beyond the discussion of the current chapter. These models predict that, although, infinite energy is required to separate a particle pair, the magnetic energy for contacting pairs is finite. This stipulates that prevention of contact is another necessity if a stable colloid is to be synthesized. This is achieved by coating the nanoparticles by a molecular layer of dispersant (Rosensweig, 1997).

RELAXATION OF MAGNETIZATION

The field dependence of the equilibrium magnetization of a ferrofluid provides valuable insight about the fluid's make-up. For the dynamic properties of magnetic fluids, the issue of relaxation of magnetization has additional importance. Generally speaking, there are two different processes that determine the way how the magnetization of a ferrofluid can follow magnetic field changes. On the one hand, we can assume that the magnetic moment of the particle is fixed with respect to its crystal structure. These kinds of particles are called magnetically hard. The relaxation of magnetically hard particles will take place by a rotation of the whole particle. On the other hand, under certain material conditions, the magnetic moment may rotate inside the particle, i.e. relative to the crystal structure. This kind of relaxation of particles, called magnetically weak, can take place if the thermal energy is high enough to overcome the energy barrier provided by the crystallographic anisotropy of the magnetic material. Both processes can be characterized by a respective relaxation time (Odenbach, 2002).

Although magnetic nanoparticles are obtained from ferromagnetic or ferrimagnetic materials in their solid bulk form, their extremely small size gives rise to a magnetic behavior (superparamagnetism) that slightly differs from classical ferromagnetism or ferrimagnetism. As already discussed, magnetic nanoparticles usually consist of a single-domain of uniform magnetization carrying a total magnetic moment (μ) given by $\mu = M_S V$. Due to an effective (crystal) magnetic anisotropy K, the magnetic moments have preferred orientations, the so-called easy axes representing local energy minima. By thermal activation or by interaction with an external magnetic field, the magnetic moment may overcome the energy barrier $E_A = KV$ and change its orientation within the particle (refer to Figure 1). The corresponding time constant is the Néel relaxation time:

$$\tau_N = \tau_0 \exp\left(KV / k_B T\right) \tag{13}$$

The relaxation time constant $\left(= \tau_0\right)$ ranges between ~10^{-9}-10^{-11}s.

If the particles are suspended in a fluid medium (with dynamic viscosity coefficient η) the magnetic moments may also change their orientation by rotational diffusion of the particle as a whole. The time constant associated with this mechanism is the Brownian relaxation time:

$$\tau_B = 3\eta V_{hydr} / k_B T \tag{14}$$

As an example, the value of τ_B for a kerosene-based ferrofluid containing 10nm particles $\left(\eta = 0.002kg / ms; T = 25^\circ C\right)$ can be calculated as $\tau_B \approx 7.6 \cdot 10^{-7} s$. The hydrodynamic volume of the particle $\left(= V_{hydr}\right)$ includes the coating layer of the shell and may be much larger than the core volume $\left(= V\right)$. The relaxation of magnetization in a mono-disperse suspension will follow the process with the shorter relaxation time

If the expressions for τ_N and τ_B are compared, it can be observed that the response of a magnetic particle to an external field depends strongly on the particle's size. Due to the exponential relation in τ_N, the Néel relaxation time increases dramatically with the volume. This means that for larger particles, the magnetization vector is practically locked, and only the Brownian mechanism remains effective. For small particles, the magnetization vector inside the particle core is somewhat loose. This provides superparamagnetic effect even when the particles are immobilized (e.g. when a ferrofluid is solidified by freezing). When $\tau_N \ll \tau_B$ the relaxation occurs by Néel relaxation and the material is said to possess intrinsic superparamagnetism. For $\tau_B \ll \tau_N$ the Brownian mechanism is determining and the material exhibits extrinsic superparamagnetism. However, if the smaller time constant is much greater than the time scale of the experiment, the sample may be treated as ferromagnetic.

As seen from Equation (13) and Equation (14) both relaxation times grow with increasing particle size. While τ_N increases exponentially with particle size, τ_B grows linearly with the particle dimension. Thus, for small particles, τ_N will be smaller than τ_B, and the relaxation will take place by rotation of the moment inside the particle (Figure 4). From a certain critical diameter, the situation reverses and the particles become magnetically hard. This critical diameter depends on the viscosity of the carrier liquid, on the thickness of the surfactant layer, and on the magnitude of K. Close to the critical diameter, the relaxation takes place by a mixture of both processes, and this situation can be described using an effective relaxation time $\left(\tau_{eff}\right)$ combined from the Néel and Brown times for the relevant particle diameter:

$$\tau_{eff} = \tau_N \tau_B / \left(\tau_N + \tau_B\right) \tag{15}$$

In a ferrofluid, either the Brownian or the Néel mechanism leads to an apparent superparamagnetic behavior described by the Langevin law. This is because the Langevin relation describes the equilibrium response of the system, and both mechanisms permit relaxation to the equilibrium state. However, in considering dynamic aspects of magnetic flows, the rate of relaxation can become an important variable (Rosensweig, 1997; Odenbach, 2002; Berkovsky, Medvedev, & Krakov, 1993).

Figure 4. Brownian and Néel relaxation time for the magnetization of a Fe_3O_4 ferrofluid with kinematic viscosity ν = 100mm²/s as a function of particle size. The diameter d is the diameter of the magnetic core of the particle and a surfactant layer thickness of 2nm
(Odenbach, 2002)

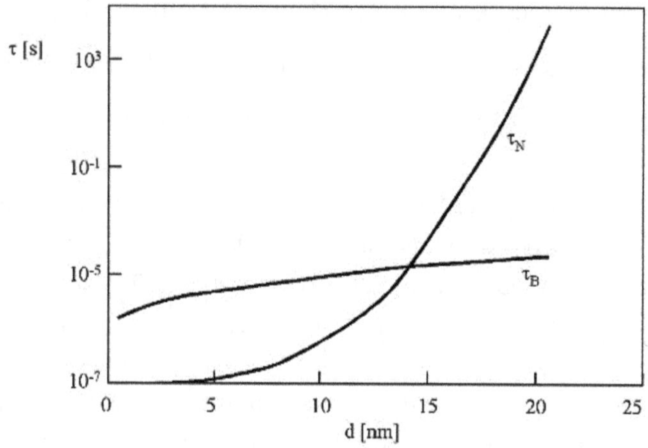

VISCOUS PROPERTIES OF FERROFLUIDS

A ferrofluid remains flowable in the presence of magnetic field, even when magnetized to saturation. However, the rheology is affected by the presence of a field. The delayed response of bulk ferromagnetic material to sudden changes of an external applied field is termed as magneto-viscous effect. In this section the magneto-viscous effect, both in the absence as well in presence of an applied field is briefly discussed.

No External Field

The viscosity of a ferromagnetic suspension is greater than that of the carrier liquid, and accompanies an increased rate of energy dissipation during viscous flow due to presence of the suspended particles. In the absence of an external field, existing theoretical models for determining the viscosity of non-magnetic colloids of solid particles (bare of coatings) suspended in liquid are available (Rosensweig, 1997; Odenbach, 2009; Berkovsky, Medvedev, & Krakov, 1993):

$$\eta / \eta_C = 1 + \frac{5}{2}\tilde{\phi} \tag{16}$$

Equation (16) relates the mixture viscosity η to the carrier fluid viscosity η_0 and solids fraction $\tilde{\phi}$ for small concentrations. For higher concentrations, Equation (16) can be suitably extended as:

$$\eta / \eta_C = \left(1 + \frac{5}{2}\tilde{\phi} + \frac{31}{5}\tilde{\phi}^2\right) \tag{17}$$

Figure 5. The dependence of the zero field viscosity of ferrofluid as a function $\tilde{\phi}$ of suspended material. The dashed line represents Equation (16), while the full line is calculated following Equation (18). The circles represent experimentally obtained data from a dilution series of a kerosene-based ferrofluid containing magnetite particles
(Odenbach, 2002)

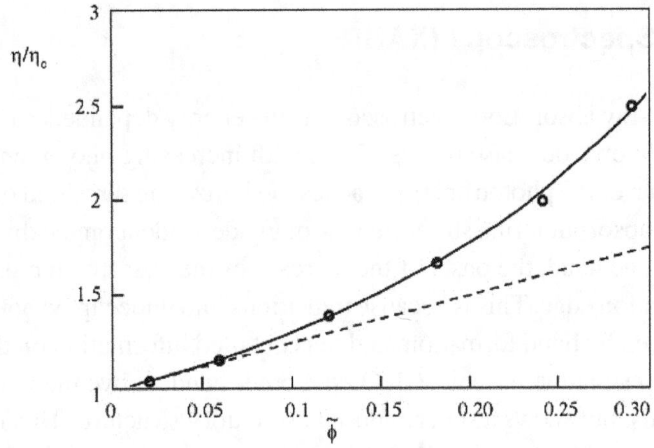

Taking into account the surface coating of the nanoparticles as against uncoated particles, a better correlation (refer to Figure 5) has been put forward ($\tilde{\phi}_C$: critical volume fraction of suspended material):

$$\eta / \eta_C = 1 / \left[1 - \frac{5}{2}\tilde{\phi} + \left(\frac{5}{2}\tilde{\phi}_C - 1 \right) \left(\frac{\tilde{\phi}}{\tilde{\phi}_C} \right)^2 \right] \tag{18}$$

External Field Present

When an external field is applied to a sample of magnetic fluid subjected to shear deformation, the magnetic particles in the fluid tend to remain rigidly aligned with the direction of the orienting field. As a result, larger gradients in the velocity field surrounding a particle are established than if the particle were not present and viscous dissipation increases in the sample as a whole.

From first principles, a theoretical model can be developed that predicts the viscosity of dilute suspensions of single-domain nano-spheres. This model accounts for Brownian rotational motion and a large value of τ_N. Neglecting Brownian motion leads to prediction of saturation even at very low field strengths (Rosensweig, 1997; Odenbach, 2009; Berkovsky, Medvedev, & Krakov, 1993). When the fluid vorticity and magnetic field are parallel, the particles can rotate freely and magnetism exerts no influence on the viscosity. The behavior then obeys Equation (16). When the directions are perpendicular to each other, the magnetic contribution to the viscosity is *maximized*.

FERROFLUID CHARACTERIZATION METHODS

In this section, some characterization methods associated with nanoparticle synthesis and preparation of MFs are introduced. The primary techniques discussed in this section are the x-ray absorption spectroscopy, the metastable impact electron spectroscopy and the x-ray diffraction method.

X-Ray Absorption Spectroscopy (XAS)

The entity measured in x-ray absorption spectroscopy is the energy dependence of the photo-absorption cross section. The well-known decrease of this entity, with increasing photon energy, is superimposed by a step structure, whenever the photon energy reaches the ionization threshold of a more tightly-bound core electron. The x-ray absorption fine structure is strongly dependent on the direct environment of the absorbing atom. On the one hand, the onset of the increase of the absorption cross section is located at energies below the ionization edge. This is because transitions into unoccupied bound states are possible. These states are influenced by bond formation, and thus detailed information on the electronic structure of the material under investigation is achieved. On the other hand, above the ionization edge the cross section is not monotonous, but shows a superimposed oscillatory structure. This is induced by coherent superposition of outgoing and back-scattered photoelectron waves with a phase factor dependent on the kinetic energy of the photoelectron. In the energy region where single scattering processes are dominant, starting typically 50-100 eV above the ionization edge, which is called Extended X-ray Absorption Fine Structure (EXAFS) region, information on number, type, degree of order, and distance of the environment of the absorber atom can be extracted analytically (Iwasawa, 1996).

There are two specific advantages of this method with respect to the analysis of nano-structured matter: First, it is a local technique, i.e. it does not rely on long-range order, and to a very good approximation it is possible to reconstruct a multi-phase system. Second, it is element-selective: By choosing the energy range of the scan, one chooses the absorption edge and thus the element on which information is gained. Consequently, it allows for development of very detailed structural models of the respective particles, including detailed understanding of the particle-matrix interactions (Ivanov & Kuznetsova, 2002).

Metastable Impact Electron Spectroscopy (MIES) and Photoelectron Spectroscopy

The characterization of the nanoparticle surface (neat, passivated, or functionalized) is of considerable relevance for technological applications such as MFs. Often a 'pre-stabilization', i.e. passivation of the particles against air and/or moisture is required. Applications, e.g. the preparation of MFs, require a peptization of the nanoparticles, i.e. their embedding into surfactant micelles, in order to form stable colloidal solutions and prevent nanoparticle aggregation. The availability of techniques (e.g. MIES, UPS) that are able to concentrate specifically onto the chemical properties and the electronic structure of the outermost layer of the nanoparticles is mandatory in order to design, characterize, and manipulate nanoparticles. Ideally, the depth resolution of the employed surface-analytical technique should be tunable in the sense that, besides the information on the outermost layer, the corresponding information on the surface-adjacent to the particles should be provided for comparison (Odenbach, 2009; Ivanov & Kuznetsova, 2002).

X-Ray Diffraction (XRD)

Chemical compounds may exist in different phases which have different crystallographic structure. Consequently, they may differ in technically-relevant properties in spite of being chemically identical. Typical examples are the modifications of iron (body-centered cubic (BCC) or face-centered cubic (FCC) and cobalt (hexagonal close packed (HCP) or face-centered cubic (FCC). For iron oxides, there are primarily two magnetic phases, the magnetite (Fe_3O_4) and the maghemite ($\gamma - Fe_2O_3$), which are important materials for synthesis of different grades of ferrofluids. Structural phase identification may be performed by lattice imaging using high-resolution transmission electron microscopy (HRTEM), together with selective-area electron diffraction (SAED) and X-ray diffraction (XRD). In particular, the XRD can be applied for phase identification of cobalt particles as well as for the dispersion of magnetite and maghemite. Methodologically, the XRD of the iron oxides is complicated by the fact that these materials exhibit very similar diffraction patterns. Although there are some typical isolated reflections for each phase, they are of minor intensity and does not allow for a sufficient signal-to-noise ratio (SINR) within reasonable time of measurement (Odenbach, 2009).

APPLICATIONS OF FERROFLUIDS

Bio-Medical Applications

Magnetic nanoparticles exhibit specific physical properties that render them viable and useful for medical diagnostics and therapy. The physical basis for all biomedical applications is the interaction of these particles with applied magnetic fields. As this interaction is confined to the particles, unwanted side-effects can be avoided or reduced. In this section, as applications for cancer therapy, the physics of magnetic drug targeting, magnetic thermo-ablation and magnetic cell sorting are briefly addressed (Trahms, 2009). Sophisticated measurement methods are also touched upon that enable the monitoring of these therapies.

Ferrofluids exhibit special rheological properties that make them particularly suited for supporting the mechanical functions of prosthetic devices like an artificial heart. For their diagnostic or therapeutic impact in medical *in vivo* and *in vitro* applications, the microscopic properties of magnetic nanoparticles are the deciding factors. These desirable properties are determined by the particular physics of the magnetic nanoparticles. To stabilize a ferrofluid colloid, the particles are prevented from aggregation and magnetic-dipolar-interaction by surrounding the metallic core with a layer of dispersant. This may increase the total particle size up to a few hundred nanometers, which is significantly smaller than cells, much larger than most organic molecules, and roughly in the size range of biological macromolecules, such as genes, proteins, and viruses (see Figure 6). Particles of such dimensions are suited to label a biologic entity and take part in metabolic processes on a nano-scale level (Berry & Curtis, 2003; Pankhurst, Connolly, Jones, & Dobson, 2003).

Magnetic nanoparticles can be made of non-toxic material. This applies to both the core and the shell. For most medical applications, the cores are made of Fe_3O_4, and the shell consists of some bio-compatible organic material. The shell can be utilized to establish a covalent or ionic binding of bio molecules, which enables a particular function of the labeled particle. One way to functionalize magnetic nanoparticles

Figure 6. The size of magnetic nanoparticles in comparison to the dimensions of biological entities (Trahms, 2009)

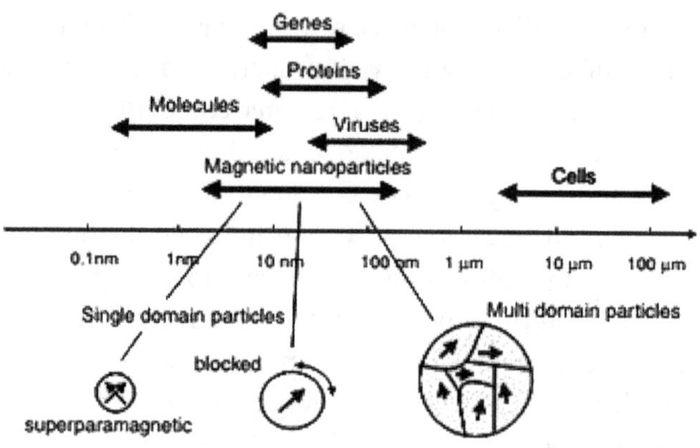

is to attach an agent that identifies the presence of a reagent by a specific binding. This is the basis of magneto-relaxometric assay, a diagnostic application method that enables the monitoring and quantification of biological binding reactions by the observation of the magnetic behavior of the nanoparticles.

Parameters that characterize binding reactions between bio molecules provide useful information for many fields of biosciences. Assessing the effectiveness of drug applications, or the selective coupling of an antibody to specific binding sites, are examples in this regard. Molecular binding can be characterized in terms of its stability, i.e. the degree of reversibility of the reaction. Further parameters that characterize the binding are the relative amount of realized bindings and the reaction kinetics. These aspects of bio-molecular binding reactions can be studied by labeling the involved reaction components with nanoparticles. The response of these magnetic labels to a sudden change of an external magnetic field reflects their mobility and a number of parameters that characterize the binding reaction. This method of measuring the relaxation of the net magnetic moment of a sample containing nanoparticles, after a magnetizing field is switched off, is referred to as magneto-relaxometry. This is an extremely sensitive method, if superconducting quantum interference devices (SQUIDs) are used for measuring the decay of the sample's magnetic moment (Matz et al., 1998). Alternatively, less-sensitive (and less expensive) magnetic sensors can be used for a magneto-relaxometric device. Another related approach is called magneto-optic relaxation. Here, the relaxing birefringence of a magnetic nanoparticle suspension in a transparent medium is measured to analyze binding reactions in terms of changes of the Brownian relaxation of the reagents. A system of magnetic nanoparticles will respond to a sudden switch-off of a magnetic field by the relaxation mechanisms described by the relations for τ_N and τ_B. The presence or absence of a Brownian relaxation path provides information on the 'mobility state' of the particle (Lange et al., 2002).

The relaxation behavior of the nanoparticles is the background for the magnetic relaxation immunoassay (MARIA), a technique that enables the specific detection of antigens by the magnetic signal of the labels. This method (refer to Figure 7) provides the quantification of the binding sites in the presence of the residual unbound molecules. Another benefit is that it is applicable to turbid media (e.g. blood, semen), where optical methods usually fail. The negligible interaction of a magnetic field with diamagnetic biological tissue makes the MARIA procedure suited for studying the binding behavior of nanoparticles *in vivo*.

Figure 7. Scheme of MARIA: The specific binding between an antibody and its antigen blocks the Brownian relaxation mechanism so that the number of antigens scales with the amount of particles that relax via the Néel relaxation mechanism
(Lange et al., 2002)

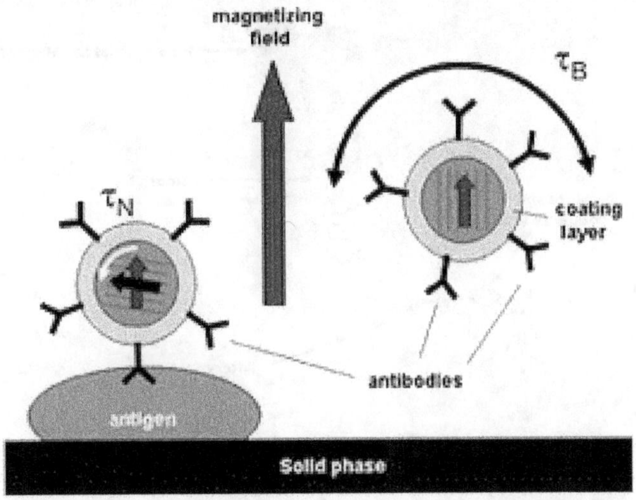

Magnetic nanoparticles, controlled non-invasively from the outside by a magnetic field, can also serve as probes in the living tissue, while the magnetic action on the diamagnetic tissue around is negligible. These particles can interact with an imposed field in various ways. In a homogeneous field the particles will align. If the field is inhomogeneous, they will be attracted towards the steepest gradient and may possibly be moved or kept from moving in a bloodstream. In this way, the particles may serve as carriers of an anti-cancer drug, which may be forced to accumulate at a particular site of the body (e.g. near a tumor). The method of magnetic drug targeting is based on this concept (Alexiou et al., 2002).

It is well-recognized that effective drug delivery to the site of tumor cells plays a crucial role in the efficacy of chemotherapy treatments of cancer. A high concentration of the agent in the tumor tissue is required to maximize its lethal impact on the cancer cells. On the other hand, the dissemination of the drug in the healthy tissue will cause systemic toxic side effects. This limits the acceptable dose of the chemotherapeutic agent. The ideal solution of this dilemma is a drug delivery system that *maximizes* the concentration of the agent in the tumor, while keeping the systemic concentration at a low and harmless level. Magnetic drug targeting is an approach to realize such a desirable drug delivery system (Figure 8).

In this concept, the drug is reversibly bound to the organic shell of the nanoparticle; it is then administered intravenously or intra-arterially to the patient. Utilizing magnetic nanoparticles as drug carriers offers the chance to focus the impact of the chemotherapeutic agent to the tumor (Alexiou et al., 2006; Jurgons et al., 2006; Rosensweig, 2002).

By applying a strong inhomogeneous magnetic field to the tumor region, the magnetic nanoparticles are captured in the tumor area for a time long enough to allow the release of the adsorbed drug. This time-exposure requires an attractive magnetic force that is stronger than the hydrodynamic drag force of the bloodstream in the vessels, in and around the tumor. According to the well-known Stokes' law, a flow-field of velocity V in a medium (viscosity η) causes a drag force (F_d) given by (Rosensweig, 1997):

Figure 8. Scheme of magnetic drug targeting
(Jurgons et al., 2006)

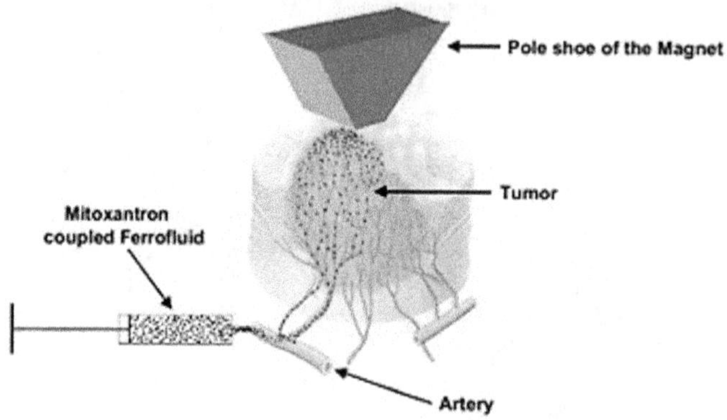

$$F_d = 6\pi\eta R V \tag{19}$$

In Equation (19) R is the hydrodynamic radius of the spherical particle (including the shell). The magnetic force $\left(\vec{F}_m\right)$ is proportional to the gradient of the magneto-static field energy $\left(\vec{B}\cdot\vec{H}\right)/2$ and the magnetic moment of the particle $\left(\mu\right)$:

$$\vec{F}_m = \frac{\mu}{2}\vec{\nabla}\left(\vec{B}\cdot\vec{H}\right) \tag{20}$$

The actual situation in a tumor blood vessel will not be as simple as modeled in Equation (19) or Equation (20). Nonetheless, the following important conclusions can be directly inferred:

- The magnetic moment μ is proportional to the volume of the nanoparticle's core and increases with the third power of its radius, while the drag force depends only linearly on the hydrodynamic radius R. So it is generally advantageous to use particles having large cores for magnetic drug targeting;
- To maximize $\vec{F}_m$, the pole shoes of an electromagnet for magnetic drug targeting should be designed to generate an inhomogeneous field that has i maximum gradient in the tumor region. Since the force scales with the square of the magnetic field strength, it is advantageous to use high field strengths;
- Since a magnetic field has a vortical structure, there is no way to establish high field gradients within the body non-invasively. This complicates the application of magnetic drug targeting to non-peripheral tumors.

If the field is alternating, static effects will average out and cancel. But what remains effective is a thermal impact. The heat deposit of an alternating magnetic field, mediated by magnetic nanoparticles, is utilized for heating up and destroying tumors. Technically, they are called magnetic hyperthermia and magnetic thermo-ablation. The mechanisms of hyperthermia and thermo-ablation are thermal-treatment strategies of tumors. Raising the temperature of a tumor by ~42-45°C is referred to as hyperthermia. Hyperthermia leads to apoptosis of the cells, i.e. a programmed malignant-cell death without the release of harmful substances into the surrounding tissue. Apoptosis is the natural way by which a multi-cellular organism discards unwanted aged cells in order to regenerate its tissue. Thermo-ablation involves generation of temperatures well above 50°C, which induces acute damage of the tumor cells and necrosis of the tissue. Although a more aggressive way, it can cause unwanted injury to healthy tissue as well. Hyperthermia alone proves insufficient for a sustainable deletion of the tumor tissue, and is usually combined with additional modalities like irradiation or chemotherapy (Hilger, Hergt, & Kaiser, 2005; Moroz, Jones, & Gray, 2002).

The common feature of both these therapeutic applications is that the action is confined locally to a restricted part of the tissue where the particles are accumulated. This makes these therapeutic strategies particularly suited for cancer therapy; while harmful side effect to the healthy tissue is minimized, the tumor is destroyed. These apply to both solid tumors as well as to distributed metastases. One can develop this concept even further and attack single tumor cells that are distributed in the blood. This is the essence of magnetically assisted cell sorting method.

Technical Applications

The possibility of magnetic control of flow and properties of ferrofluids has led to the development of a wide variety of possible applications of these liquids in various engineering fields. In this section, a concise outlook to the application possibilities shall be presented in order to shed a light on the technological potential of ferrofluids. In particular, it is the aim of this section to outline the qualitative difference between those applications already used in everyday life, and those that may result by means of magnetic fields. For a fuller discussion on the wide spectrum of applications of ferrofluids, the reader is referred to (Rosensweig, 1997; Odenbach, 2002; Odenbach, 2009; Berkovsky, Medvedev, & Krakov, 1993).

Mechanical Applications

A salient application of magnetic fluids is the sealing of rotating shafts. The situation is depicted in Figure 9. A rotating shaft made of a material with high permeability is surrounded by a permanent magnet. If the gap between the shaft and the magnet is reasonably small, i.e. in the order of a few tenth of a millimeter, the magnetic field in the gap can easily reach values ~1T. Outside the gap, the field is more or less negligible, leading to an extremely high field gradient in the region of the gap. If a ferrofluid is placed into the gap, the magnetic forces acting on the fluid can be easily high enough to hold the fluid in this position, even if a pressure difference is applied to both sides of this liquid seal.

Using the ferro-hydrodynamic Bernoulli equation as given in (Rosensweig, 1997):

$$\rho g h + p + \frac{1}{2}\rho V^2 - \mu_0 \bar{M} H = const \tag{21}$$

In Equation (21) the velocity of the fluid flow, the pressure and the mean magnetization are denoted as V, p, $\bar{M}$. The mean magnetization of the ferrofluid is defined as:

$$\bar{M} = \frac{1}{H}\int_0^H MdH \tag{22}$$

Using Equation (21) and Equation (22), one can calculate the maximum differential pressure allowed. Assuming that the shaft and, thus, the fluid is at rest (V = 0), it is true that the potential energy for horizontal points (say, (1) and (2)) in Figure 9 is identical. Therefore, the pressure difference can be expressed as:

$$\Delta p = p_1 - p_2 = \mu_0 \bar{M} \Delta H; \bar{M} = \frac{1}{\Delta H}\int_{H_1}^{H_2} MdH \tag{23}$$

For realistic values of the saturation magnetization of the fluid (M_s = 56 kA/m) and the field intensity in the gap (H = 1.5 × 104 lA/m), Equation (25) shows that the fluid can be held easily against a pressure difference of ~1 bar. Since the fluid keeps its liquid properties even under the influence of such strong magnetic fields, the friction exerted to the shaft by this magnetic fluid seal is negligibly small as compared to conventional mechanical sealing techniques (e.g. oil seals). Figure 10 depicts the principle of magnetic fluid seal for transversal and longitudinal imposed field (Stierstadt & Liu, 2015).

This advantage is commonly harnessed in various technical applications like the sealing of hard disc drive, rotating x-ray tube or rotary vacuum feed-through, where reliable sealing at low friction is required. Pressure seals of rotational feed-through are presently employed in vacuum installations in satellites, or in pumps to prevent leakage of corrosive or poisonous fluids (see Figure 11). A combination of this sealing technique with the levitation of non-magnetic materials in a ferrofluid under the influence of a magnetic field gradient enables the design of a magnetic fluid bearing (Figure 12), while the levitation

Figure 9. Principle sketch of the sealing of a rotating axis with a magnetic fluid fixed in the small gap between the axis and a surrounding permanent magnet
(Odenbach, 2002).

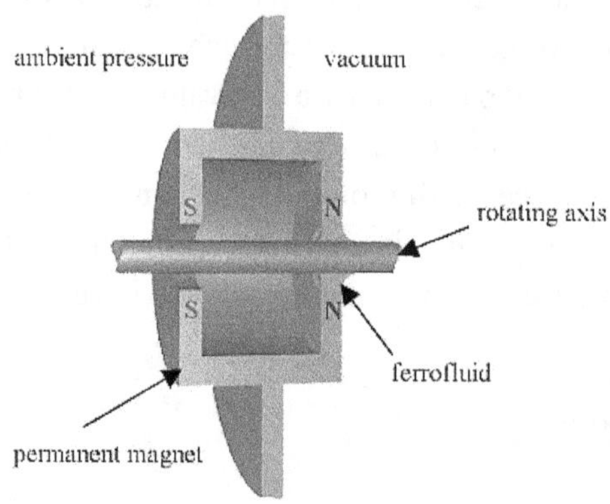

Figure 10. Principle of a magnetic fluid seal for (a) transversal and (b) longitudinal field configuration
(Stierstadt & Liu, 2015)

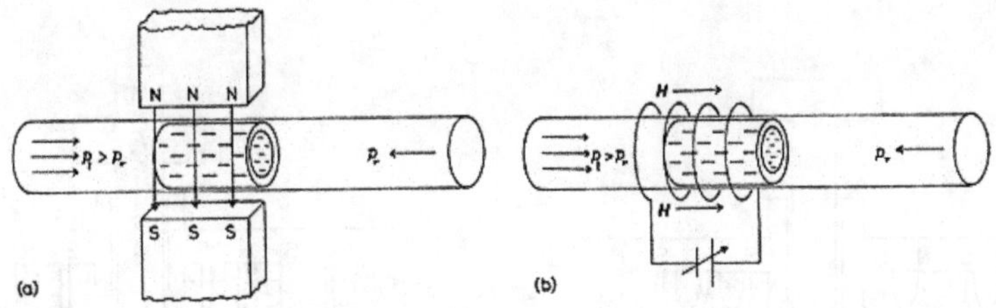

phenomenon is used in magneto-hydrostatic separation of ores (Figure 13) (Odenbach, 2009). For more details, the reader is referred to the following references (Nagaya, Ohnuma, & Sato, 1992).

Thermal Applications

The use of magnetic fluids as a heat transfer medium is nowadays commercially popular. The major application in this field is cooling of loudspeakers, which enable a significant increase of the maximum acoustical power without any geometrical changes of the speaker system. This may be achieved through magnetic positioning of the ferrofluid with the aid of permanent magnets.

The thermal management in high-power loudspeakers is the transfer of the ohmic heat from the voice coil to the speaker's structure. Since the coil has to move freely, no rigid connection is allowed. Thus, the heat transfer in conventional loudspeakers is limited by the thermal conductivity of air gap between coil and structure. Conventional non-magnetic liquids cannot be used to enhance heat transfer in the speaker gap, since they cannot be reliably located. In contrast, a magnetic fluid is spatially fixed by the strong field in the gap. The fluid is held in position, even if the coil is subjected to high amplitudes. The high thermal conductivity of the ferrofluid (~8 times higher than that of air) enables a significant increase of the maximum acoustic power of the speaker (refer to Figure 14).

Figure 11. Construction of a rotary feed-through with 12 stages of ferrofluid sealing (the magnetic yoke is obliquely hatched)
(Stierstadt & Liu, 2015)

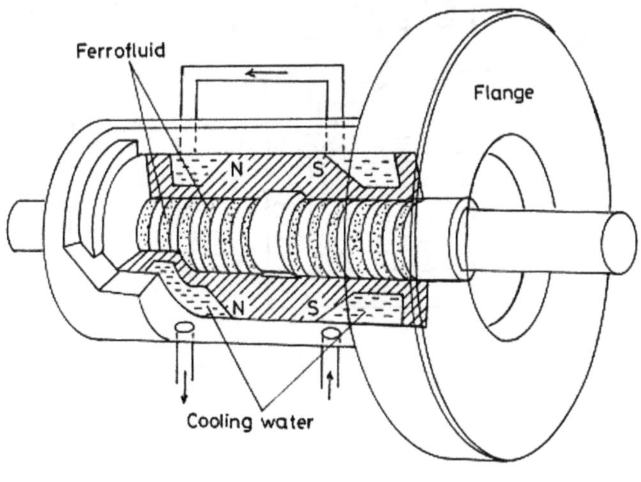

Figure 12. Magnetic bearings with supporting bodies (B) floating in a ferrofluid (F); (a) and (b) magnetic poles placed in the surrounding container (C), (c) magnetic poles inside the floating body
(Stierstadt & Liu, 2015)

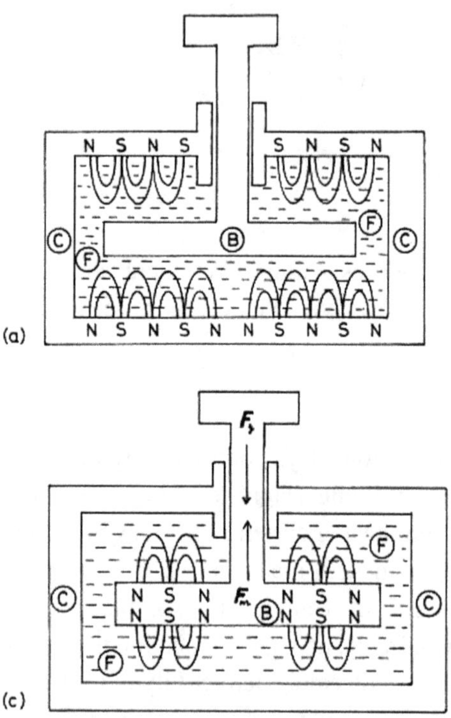

Figure 13. Sorting of small particles by levitation in a ferrofluid (F) in a strong field gradient
(Stierstadt & Liu, 2015)

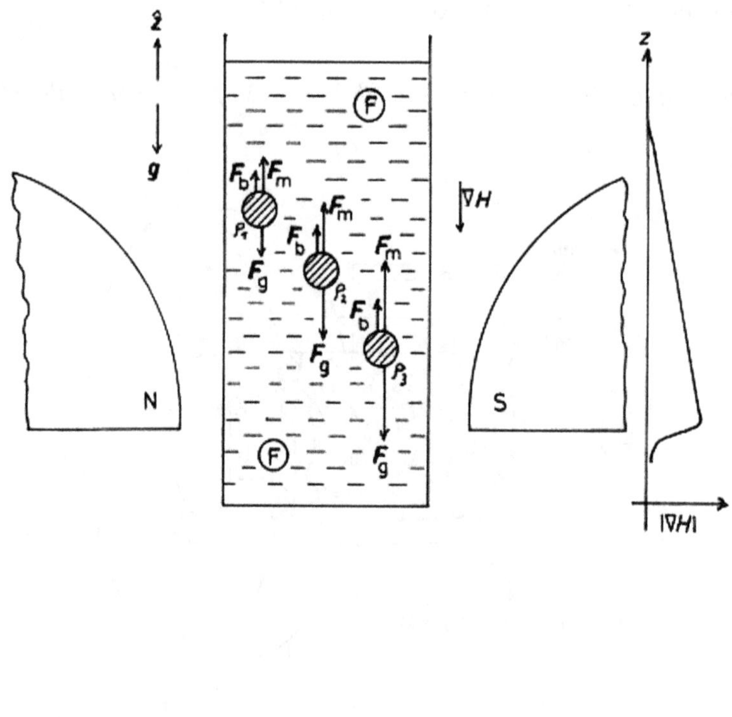

Figure 14. Sketch of a loudspeaker cooled by a magnetic fluid kept in the gap around the voice coil. The right side figure depicts the temperature of the speaker as a function of its power with and without the use of ferrofluid as cooling agent
(Odenbach, 2002)

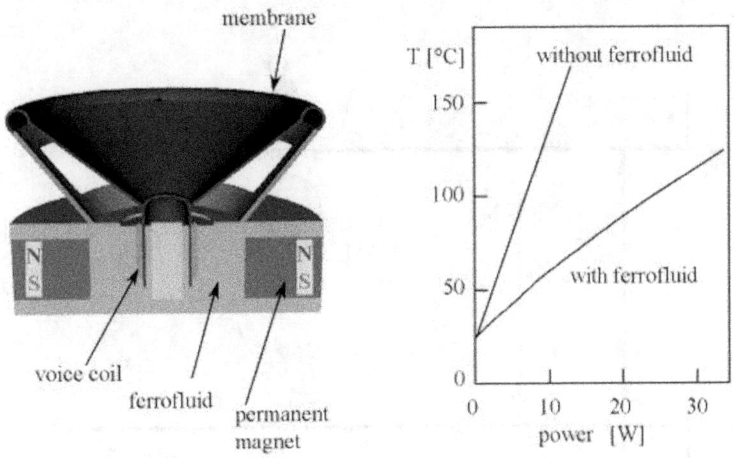

Thermomagnetic Convection

Potential application domains for ferrofluids, in enhancing heat dissipation from power-dissipating devices, rely on one or more of the following possibilities (in presence of an external magnetic field):

- Fixing the fluid at a certain spatial position
- Exerting an active control to the flow of ferrofluids by imposing steep magnetic field gradients
- Altering the rheological properties of the fluid and setting up susceptibility gradients

As already discussed, ferrofluids are widely used as magnetically positioned heat transfer medium in loudspeaker cooling. As another example from the field of heat transfer applications, field-assisted advection of ferrofluids, called thermomagnetic convection, may be induced. This phenomenon of magnetically-controlled thermal flow is based on the change of magnetization of the fluid with temperature. As seen from Equation (3) the magnetization of a magnetized ferrofluid decreases with increasing temperature. Thus, in a situation as shown in Figure 15, a temperature gradient applied to a magnetized liquid will give rise to a magnetization gradient anti-parallel to the temperature gradient. If, in addition, a magnetic field gradient parallel to the temperature gradient is applied, a destabilizing magnetic force will appear in the fluid. This force can drive a convective flow that is purely controlled by the strength and direction of the applied magnetic field (Banerjee, 2012).

At the formulation stage, the magnetic Kelvin force enters the basic hydrodynamic equations for magnetic fluids. For example, the form of the celebrated Navier-Stokes equation that governs transport processes for a ferrofluid can be written (in vector form) as (Rosensweig, 1997):

$$\frac{d\vec{V}}{dt} = -\nabla p + \nu\nabla^2\vec{V} + \underbrace{\mu_0 M\vec{\nabla}H}_{Kelvin\ force} \tag{24}$$

Figure 15. On the driving force of thermomagnetic convection. If a volume element of the fluid is displaced adiabatically in the direction of the field gradient it will be surrounded by hotter fluid in the shown arrangement. Thus, the volume element will have higher magnetization than its surrounding and will thus experience a force in the direction of the initial displacement
(Das et al., 2008).

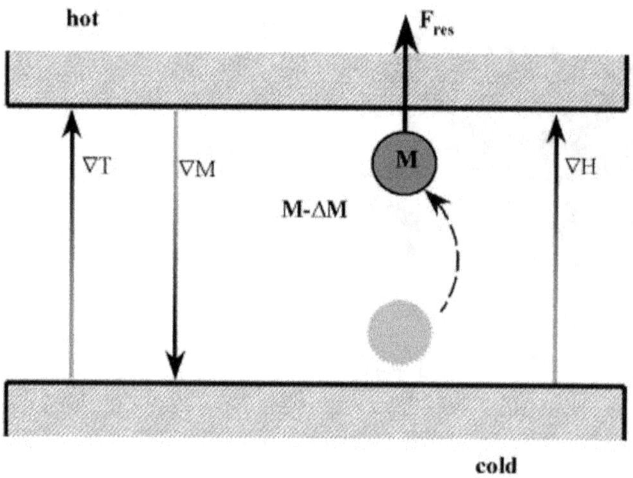

Extreme Boundary Lubrication

Almost all technical equipment and products like machine tools, robots, automobiles or consumer goods carry out a *relative* motion to guarantee the desired operating processes.

Examples for the mechanical components (or particular devices) in technical systems are gears, plain bearings, roller bearings and guides, with simultaneously active loads or stresses. Such technical systems, with relative motion of surfaces, define tribo-technical systems. They are characterized by a large number of functional parameters, and consist of a basic solid, a counter solid, an intermediate medium and an ambient medium. The basic solid and the counter solid are also described as the friction pair, which are exposed to wear in form of material removal (see Figure 16).

Therefore, in tribo-technical systems, the motion, force as well as energy are useful quantities. The kinematics, normal force, velocity, temperature and load period, as physico-technical parameters, are a part of the load spectra and act on the components of the tribo-technical system, with corresponding tribological stresses and processes acting between them. The changes at the system-structural level, as well as the dissipation quantities, are represented with the help of tribo-metrical quantities, by means of friction quantities (like friction coefficient), as well as by means of deviations of dimension at the system-component level. High wear is caused in the cases of high surface loads moving at relative velocities and solid-state contact of friction pairs. The boundary lubrication conditions occur if a friction pair carries out a relative motion at high surface load and low relative velocity. This results in a solid-state contact between the friction pair, and causes wear of the surfaces. In contrast, tribological conditions with a solid-state contact and super-adherent lubricating grease characterize the extreme boundary lubrication condition. Thus, lubricants are used to improve the tribological behavior by decreasing friction and wear under operating conditions, with high surface loads (Uhlmann, Spur, Patzwald, & Bayat, 2002).

Figure 16. Principle of a tribological system
(Nagaya et al., 1993)

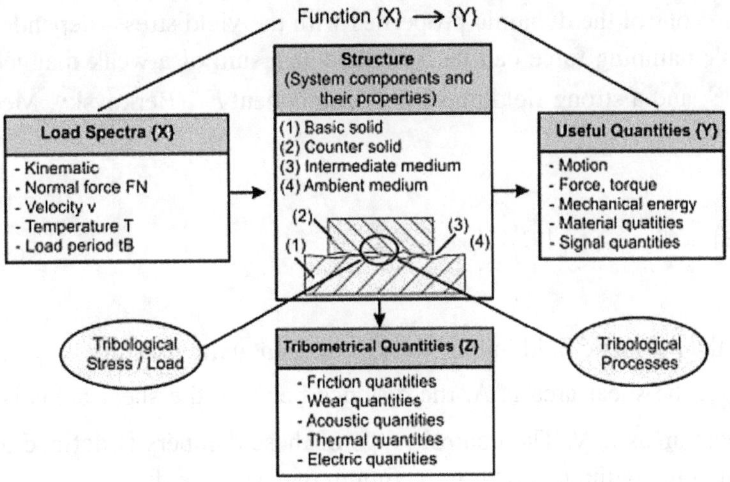

Ferrofluids hold significant promise as potential lubricants to enhance tribological characteristics. The use as a lubricant is achieved by the formation of a super-adherent grease, lubricating the surfaces of the friction pair. This means that supplying devices like pumps, pipes, or seals for lubrication are not needed anymore. Furthermore, the formation of chains of the magnetic particles in a magnetic field holds promise for the use of ferrofluids as lubricants under extreme boundary conditions. Based on experimental investigations, it can be concluded that ferrofluids can decrease the wear under boundary conditions; the variation of the applied magnetic field only slightly influences the tribological behavior. The magnetic field is suitable for the fixation of the lubricant in the contact zone of the friction pair. This enables the supply of lubricant for the friction pairs without the use of pumps or pipes. To warrant better lubrication with minimal abrasion, customized grades of ferrofluids (with carrier liquids providing better lubrication characteristics and less abrasive colloids) must be synthesized (Nagaya et al., 1993).

Magneto-Fluidic Dampers

An important commercial application of magnetic fluids is dampers. The operating modes (shear, flow and squeeze mode) of these dampers are in accordance with the three flow types (Figure 17).

Figure 17. Flow types: shear mode (left), flow mode (middle) and squeeze mode (right)
(Odenbach, 2009)

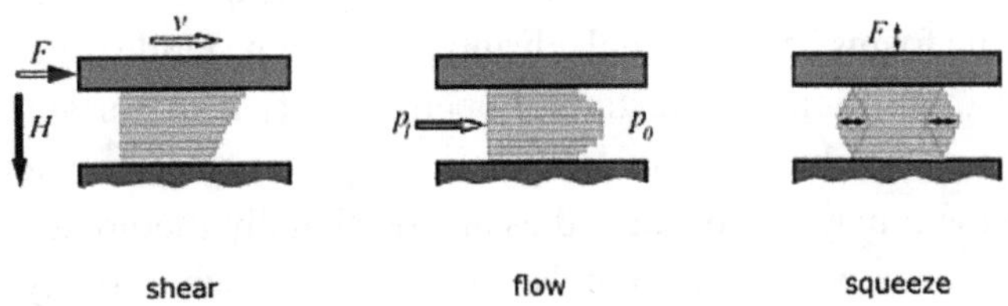

The most common damping principle uses the *shear mode*, where a volume of ferrofluid entrapped inside a magnetic coil is sheared by the movement of the damper. Applying a magnetic field to this fluid, it is possible to control the dynamic properties with the yield stress (dependent on the magnetic field). The achievable damping force can be described as a sum of a weak magnetic field-dependent viscous component F_η and a strong field dependent component F_τ (Berkovsky, Medvedev, & Krakov, 1993):

$$F\left(B\right) = F_\eta + F_\tau\left(B\right) = \frac{\eta_b VA}{h} + \tau_0\left(B\right) \cdot A \tag{25}$$

In Equation (25) the magnetic field induction is B, the Bingham viscosity is η_b, the field-dependent shear stress is $\tau_0\left(B\right)$, the shear area is A, the distance between the shear areas is h and the relative velocity between shear areas is V. The control range of these dampers is defined as the ratio (= CR) between the damping force without and with maximum magnetic field:

$$CR = \frac{F\left(B_{\max}\right)}{F_0} = \left[1 + \frac{h\tau_{0,\max}}{V\eta_b}\right] \tag{26}$$

In Equation (26) the maximum value of shear stress (at B = B_{max}) is denoted as Ä$_{0,max}$. A wide control range in conventionally-designed dampers can be realized with a large gap (h) between piston and cylinder. However, this implies the following disadvantages:

- More fluid is needed (higher costs and weight);
- Higher volume must be pervaded by the magnetic field (more energy needed to generate the field);
- Dimensions of the damper are larger;
- Achievable damping forces are lower.

Another possibility to obtain a wider control range of the damper is the use of a fluid with maximum magneto-rheological effect (ideally this implies $\tau_{0,\max} = 0$). Such an ideal fluid consists of large particles of highly permeable material. But the disadvantage of this ideal fluid arises from sedimentation processes of the particles (discussed above), which lead to inhomogeneous properties of the fluid.

In the following, a new damping concept is presented, which allows a highly-active damping volume in combination with a highly-achievable damping force as well as a small design space (Jolly & Nakano, 1998).

The basic idea of this new kind of damper, called the *foam damper*, is based on open porous foams (Figure 18). These metallic foams have the dual advantages of high stability and low weight. Especially for magneto-rheological applications, the feature of open porosity is important. The metallic foam can be directly placed in the active volume of the damper, in the presence of an external magnetic fluid. As first approximation, the foam may be considered as an additionally rheological resistance. Hence, the achievable damping forces can be shifted to higher values. The open porosity of the foam is another independent design parameter, which can be used to influence the achievable damping forces.

Figure 18. Principle of the foam damper
(Odenbach, 2009)

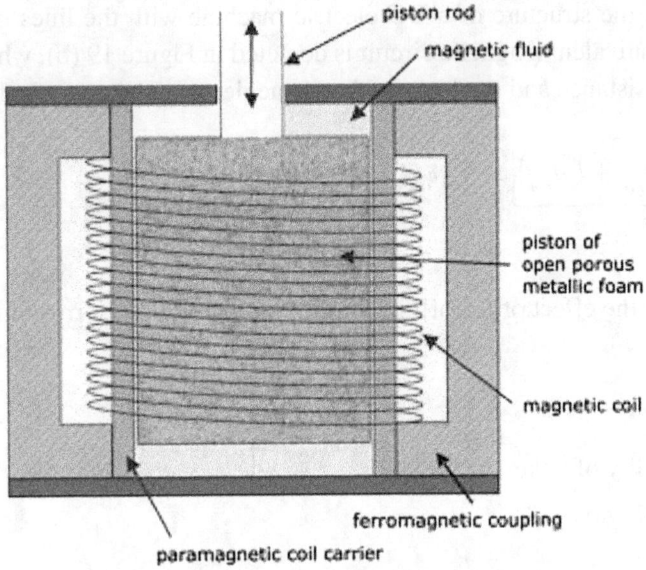

The piston consists of open porous metallic foam, located inside a magnetic coil. Moving this foam piston presses the magnetic fluid through the pores of the foam. The flow in the pores can be controlled by changing the viscosity of the fluid by applying a magnetic field. Additionally, the damping force can be adjusted by the use of different open porous metallic foams.

The advantages of the magneto-fluidic damper (based on open porous metallic foams) over conventional MRF dampers are (Sproton, Yanyo, Carlson, & El Wahed, 2002):

- Higher active volume in combination with more compact design space;
- Wider control range of the damping force;
- Higher concentration of the field flux inside the core;
- Range of the controllable damping force can be altered by using different open porous metallic foams.

Ferrofluids in Actuators and Electrical Machines

Different grades of ferrofluids are used in the gap between active magnets in electric machines and linear electric motors for the amplification of forces. Therefore, the magnetic characteristics of ferrofluids in electromagnetic fields are of substantial research interest. This justifies experimental investigations to look for ways of increasing their (response) efficiency, thermal conductivity and their dynamic behavior.

The idea of improving performance of electric machines with ferrofluids can be evaluated through a first rough estimation. For this, an outline of the theory of magnetic circuits is well suited.

Magnetic systems with small losses (closed lines of flux) can be described with the aid of magnetic circuits, which work analogous to the electric circuits. The magnetic voltage is $V_m = NI$ with N as the number of windings and I the current of the coil surrounding the path of the magnetic flux. The magnetic resistance $R_{m,i}$ is expressed as $R_{m,i} = l_i / A_i\mu_i$ with l_i, A_i, μ_i the cross-section, the length, and the permeability of segment (Odenbach, 2009).

Figure 19 (a) shows the structure of a DC electric machine with the lines of flux. To carry out a rough estimation, the equivalent magnetic circuit is depicted in Figure 19 (b), where some elements are integrated in a single resistance and the leakage flux is neglected. The analytical expression

$$B = \theta_N \mu_0 \underbrace{\left[\mu_{r,g}\mu_{r,y} \Big/ \left(l_g \mu_{r,y} + l_y \mu_{r,g}\right)\right]}_{1/R_m} \equiv \theta_N \mu_0 G_m$$

can be used to calculate the effect of ferrofluids in the gap; the various terms of the expression are:

B : Flux density
θ_N : Magnetic voltage
$\mu_{r,y}$: Relative permeability of yoke
$\mu_{r,g}$: Relative permeability of gap
l_y : length of yoke
l_g : length of air gap
R_m : Magnetic resistance
G_m : Magnetic conductance

The gain of magnetic conductivity, when the gap between the rotor and the stator is filled with ferrofluids, is now demonstrated. To compare the magnetic circuit with and without ferrofluid, three different cases are mentioned. The permeability values are typical numbers for the materials. The length

Figure 19. (a) Sketch of a DC electric machine and the lines of magnetic flux and (b) Simplified equivalent magnetic circuit of an DC electric machine
(Odenbach, 2009)

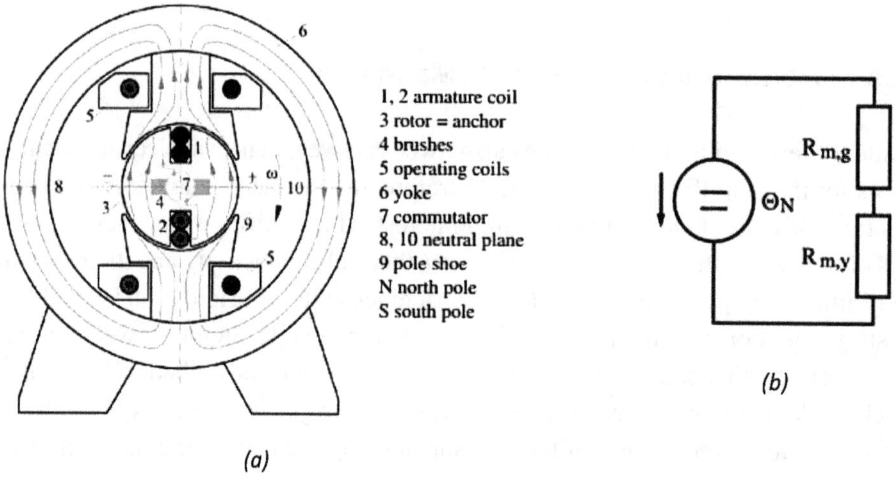

scales are typical for small electric machines: for all three cases: $1_g = 1$ mm; $1_y = 30$ mm The sample calculations are shown as under:

Gap is filled with air:

$$\mu_{r,g} = 1, \mu_{r,y} = 20000; \Rightarrow G_m = 998 Vs / A$$

Gap is filled with ferrofluid, with yoke permeability unchanged:

$$\mu_{r,g} = 2, \mu_{r,y} = 20000; \Rightarrow G_m = 1994 Vs / A$$

Gap is filled with ferrofluid, with decreased yoke permeability owing to increased field strength:

$$\mu_{r,g} = 2, \mu_{r,y} = 1000; \Rightarrow G_m = 1887 Vs / A$$

Comparing the first two calculations, an increase of the magnetic conductivity by the factor 2 occurs by introducing a ferrofluid into the gap. Taking into account the magnetic saturation of the yoke, the intensification sinks by ~5%, as can be observed in the third calculation. Raising up the magnetic conductivity in a magnetic circuit means higher flux and field strength, in accordance with classical electromagnetic theory. This implies that higher acting forces can be derived (Nethe & Stahlmann, 1998).

Theoretical and experimental works on force amplification in electric machines (using ferrofluids) between the acting magnets have been amply reported. The calculations presented here give a quantitative assessment of the (conductance-enhancement) effect of ferrofluids inserted between the stator and the rotor. Published research works also indicate that, in the case of rotating machines, the combination of the segmentation of volume filled with the ferrofluid on one hand, and the recursive determination of the field-dependent permeability on the other hand, allows to treat a non-linear problem with a linear tool as the orthogonal expansion. All these results present significant improvements of effectiveness for electric drives.

The increased force is especially interesting for the holding moment, which is a crucial performance criterion in stepping motors. Moreover, for traction engines the starting torque is an important parameter. Here, the ferrofluid offers an advantageous option since rotational friction (which lowers the force amplification) is totally absent. Taken together, the use of magnetic fluids in electric machines offers a wide variety of applications (Nethe, Scholtz, & Stahlmann, 2002).

The next step is to apply these satisfying results on rotating machines. But, here, additional difficulties are encountered. For example, the technological requirement of safe enclosure of the fluid has to be solved. A ferrofluid-based seal is a viable option for this task.

Biologically-Inspired Locomotion Systems

Biologically-inspired robots are premier examples of design in collaboration with nature. Significant progress has been made in the field of humanoid-robots, realizing pedal locomotion. However, in many situations, undulatory (non-pedal) locomotion is a vital requirement. The ability of non-pedal locomotion is an advantage of worm-like or snake-like robots. Observing the locomotion of worms, one recognizes a conversion of (mostly periodic) internal deformations into change of external position (Cebers, 2005; Nikitin, Stepanov, Mironova, & Gorbunov, 2004; Zimmermann, Zeidis, Naletova, & Turkov, 2004a; Norihiko & Nakamura, 2002).

The realization of such deformations, using ferrofluids in an applied magnetic field, is an area of active R&D. Bio-mimetic application with magnetic fluids deals with varying the controllable traveling magnetic field to enable wormlike locomotion of objects (Figure 20).

The research results have shown that there are two future-oriented trends in the domain of ferrofluid-based locomotion systems:

- The realization of locomotion based on the deformation of finite volumes of magnetic tissues (a magnetic fluid in an elastic capsule, or bodies with a magnetizable polymer). In particular, bending deformations of slender bodies (rods) and deformations due to an axial load have been reported in literature (Cebers, 2005);
- The realization of locomotion based on the deformation of a free surface of a magnetic fluid layer in a traveling magnetic field. It is well-established that traveling magnetic field can create a flux in the fluid layers, which then drives the flow of layers of magnetizable fluids (Zimmermann, Zeidis, Naletova, & Turkov, 2004b).

The investigated problems are important for many practical applications. If the magnetic field is developed from an external source, and the deformable bodies consist of magnetic polymers or elastic capsules (filled with MF), such compact locomotion devices do not require motors and other movable parts that can come into contact with a surrounding medium. This fact allows the usage of them in medicine and biology (Odenbach, 2009; Saga & Nakamura, 2002).

FUTURE RESEARCH DIRECTIONS

The above discussion showcases some of the application areas where magnetic fluids have gained prominence. It is beyond the scope of this chapter to discuss all of them here. In general, the number of different types of technical applications should lead to a variation of requirements of the ferrofluid-grades used.

Figure 20. Magnetizable elastic body in traveling magnetic field
(Odenbach, 2009; Zimmermann et al., 2004a)

These requirements can be specified by qualitative and quantitative characteristics. A primary motivation for the synthesis of ferrofluids (for existing and proposed applications) is to increase the resistance against sedimentation (i.e. ensure flowability). As mentioned above, this could be ensured by particle size of less than ~10nm. As per present global standards, most consumer products have to guarantee its functional capability for a period of at least 2 years. These periods can even extend to more than 10 years, e.g. for automotive components, or industrial goods like engines or machine tools. Therefore, the resistance against agglomeration as well as chemical consistency has to be guaranteed for the service life of the specific technical products.

First and foremost, any ferrofluid grade must be non-toxic in human exposure. Any undesired chemical reaction of a ferrofluid with the ambient media, materials or mechanical components should be avoided. In addition, the ambient temperature must not pose any negative influence on performance. As an example, ferrofluid-dampers that are used in automobiles must retain their functionality under ambient temperature ranges from ϑ = -40°C to ϑ = 80°C (depending upon different geographic regions and seasons). In future, if field-actuated ferrofluid apparatus and devices are employed in household appliances, ground-based stationary or mobile power plants, or airborne systems like in aircrafts, the range of temperature exposure is liable to broaden further. Temperature sensitivity is not a major cause for concern in high-precision ferrofluid-systems with intended use in laboratories (under controlled ambient conditions).

In addition, the use of ferrofluids is closely connected with external field actuations. Therefore, the adjustable range and absolute values of magneto-rheological response to field gradients should be as high as possible at an adequate viscosity. This increases the adaptive characteristics of the ferrofluid and enables the use of the same ferrofluid for different applications and stresses.

The future research directions can therefore be summarized as follows:

- Resistance against sedimentation up to several years
- Chemical consistency
- Thermal stability within service conditions
- High magnetization and adequately low viscosity
- High reproducibility of effects
- Non-toxic and non-corrosive
- Easy availability and ease-of-handling
- Low investment costs

Due to the conflict of objectives, trade-offs between the different requirements have to be met. An ideal long-term stable ferrofluid with high magnetization, low viscosity, thermal stability and high reproducibility (for technical applications) is bound to be associated with challenges in the manufacturing processes. Additional standard requirements like easy handling, affordability even in large amounts, and low investment costs must also be met. Finally, advances in colloidal and surface chemistry should ensure that the colloids, surfactants or carrier liquids meet the dual requirements of desired functionality and non-toxicity. The improvement of the cost-performance ratio of existing applications and the development of new applications with relatively high performance efficiency or accuracy shall be the decisive factors to consider the use of ferrofluids in the foreseeable future.

CONCLUSION

The present chapter provides a concise outline of the basics of magnetic fluids with regard to their structure, rheology, properties, characterization and some potential applications, and sheds light on field-fluid interactions that influence their magnetization properties and flow behavior.

ACKNOWLEDGMENT

The author acknowledges the Senior Research Fellowship Grant of the Council for Scientific & Industrial Research, Govt. of India, during his Doctoral tenure at Jadavpur University, which facilitated his exposure in the fascinating world of ferrohydrodynamics. In addition, the author acknowledges the infrastructural support (for literature review) extended by Heritage Institute of Technology Kolkata in penning this chapter.

REFERENCES

Alexiou, C., Jurgons, R., Seliger, C., Brunke, O., Iro, H., & Odenbach, S. (2006). Delivery of superparamagnetic nanoparticles for local chemotherapy after intra-arterial infusion and magnetic targeting. *Anticancer Research*, *27*, 2019–2022. PMID:17649815

Alexiou, C., Schmidt, A., Hulin, P., Klein, R., Bergemann, C. H., & Arnold, W. (2002). Magnetic drug targeting: Bio distribution and dependency on magnetic field strength *in vitro*. *Journal of Magnetism and Magnetic Materials*, 252, 363–366. doi:10.1016/S0304-8853(02)00605-4

Banerjee, S. (2012). *Augmentation of Natural Convection using Thermomagnetic Convection.* Unpublished doctoral dissertation, Jadavpur University, Kolkata, India.

Berkovsky, B. M., Medvedev, V., & Krakov, M. S. (1993). *Magnetic Fluids: Engineering Applications*. New York: Oxford University Press.

Berry, C. C., & Curtis, A. S. G. (2003). Functionalization of magnetic nanoparticles for applications in biomedicine. *Journal of Physics. D, Applied Physics*, *36*(13), 198–206. doi:10.1088/0022-3727/36/13/203

Cebers, A. (2005). Flexible magnetic swimmer. *Magnetohydrodynamics*, *41*, 63–72.

Das, S. K., Choi, S., Yu, W., & Pradeep, T. (2008). *Nanofluids: Science and Technology*. Hoboken, NJ: John Wiley & Sons, Inc.

Gast, A. P., & Zukoski, C. F. (1989). Electro-rheological fluids as colloidal suspensions. *Advances in Colloid and Interface Science*, *30*, 153–202. doi:10.1016/0001-8686(89)80006-5

Hilger, I., Hergt, R., & Kaiser, W. A. (2005). Towards breast cancer treatment by magnetic heating. *Journal of Magnetism and Magnetic Materials*, *293*(1), 314–319. doi:10.1016/j.jmmm.2005.02.026

Ivanov, A. O., & Kuznetsova, O. B. (2002). Magnetic properties of dense ferrofluids. *Journal of Magnetism and Magnetic Materials*, *252*, 135–137. doi:10.1016/S0304-8853(02)00639-X

Iwasawa, Y. (Ed.). (1996). *X-ray absorption fine structure for catalysts and surfaces*. Singapore: World Scientific. doi:10.1142/2807

Jiang, Y., & Liu, M. (1996). Dynamics of Dispersive and Nonlinear Media. *Physical Review Letters*, *77*(6), 1043–1046. doi:10.1103/PhysRevLett.77.1043 PMID:10062976

Jolly, M., & Nakano, M. (1998): Properties and Applications of Commercial Controllable Fluids. *Proceedings of 6th International Conference on New Actuators* (pp. 414–419). Academic Press.

Jurgons, R., Seliger, C., Hilpert, A., Trahms, L., Odenbach, S., & Alexiou, C. (2006). Drug loaded magnetic nanoparticles for cancer therapy. *Journal of Physics Condensed Matter*, *18*(38), 2893–2902. doi:10.1088/0953-8984/18/38/S24

Lange, J., Kotitz, R., Haller, A., Trahms, L., Semmler, W., & Weitschies, W. (2002). Magnetorelaxometry-a new binding specific detection method based on magnetic nanoparticles. *Journal of Magnetism and Magnetic Materials*, *252*, 381–383. doi:10.1016/S0304-8853(02)00657-1

Liu, M. (1993). Hydrodynamic Theory of Electromagnetic Fields in Continuous Media. *Physical Review Letters*, *70*(23), 3580–3584. doi:10.1103/PhysRevLett.70.3580 PMID:10053910

Liu, M. (1995a). Fluid Dynamics of colloidal magnetic and electric liquid. *Physical Review Letters*, *74*(22), 4535–4538. doi:10.1103/PhysRevLett.74.4535 PMID:10058531

Liu, M. (1995b). Electromagnetic fields in ferrofluids. *Physical Review*, *59*(3), 3669–3675. doi:10.1103/PhysRevE.59.3669

Matz, H., Drung, D., Hartwig, S., Gross, H., Kotitz, R., Muller, W., ... Trahms, L. (1998). A squid measurement system for immunoassays. *Applied Superconductivity*, *6*(10-12), 577–583. doi:10.1016/S0964-1807(99)00014-9

Moroz, P., Jones, S. K., & Gray, B. N. (2002). Magnetically mediated hyperthermia: Current status and future directions. *International Journal of Hyperthermia*, *18*(4), 267–284. doi:10.1080/02656730110108785 PMID:12079583

Nagaya, K., Ohnuma, H., & Sato, A. (1992). Analysis of a radial magnetic fluid seal and a radial bearing lubricated with a magnetic fluid. *Tribology International*, *25*(2), 107–113. doi:10.1016/0301-679X(92)90087-4

Nagaya, K., Takeda, S., Sato, A., Ikai, S., Sekiguchi, H., & Saito, N. (1993). Thrust bearing using a magnetic fluid lubricant under magnetic fields. *Tribology International*, *26*(1), 11–15. doi:10.1016/0301-679X(93)90033-W

Nethe, A., Scholz, T., Stahlmann, H.-D., & Filtz, M. (2002). Ferrofluids-A Numerical Process Model. *Transactions in Magnetism*, *38*(2), 1177–1180. doi:10.1109/20.996301

Nethe, A., & Stahlmann, H.-D. (1998). Remarks on the interaction between magnetic core coils in the presence of magnetic fluids. *Journal of Magnetism and Magnetic Materials*, *189*(2), 255–262. doi:10.1016/S0304-8853(98)00210-8

Nikitin, L. V., Stepanov, G. V., Mironova, L. S., & Gorbunov, A. I. (2004). Magneto-deformational effect and effect of shape memory in magnetoelastics. *Journal of Magnetism and Magnetic Materials*, 272-276(Part 3), 2072–2073. doi:10.1016/j.jmmm.2003.12.838

Norihiko, S., & Nakamura, T. (2002). Elucidation of propulsive force of micro-robot using magnetic fluid. *Journal of Applied Physics*, *91*(10), 7003–7005. doi:10.1063/1.1452197

Odenbach, S. (2002). *Magnetoviscous Effects in Ferrofluids*. Berlin: Springer-Verlag.

Odenbach, S. (2004). Recent progress in magnetic fluid research. *Journal of Physics Condensed Matter*, *16*(32), 1135–1150. doi:10.1088/0953-8984/16/32/R02

Odenbach, S. (Ed.). (2009). *Colloidal Magnetic Fluids: Basics, Development and Application of Ferrofluids*. Berlin: Springer. doi:10.1007/978-3-540-85387-9

Pankhurst, Q. A., Connolly, J., Jones, S. K., & Dobson, J. (2003). Applications of magnetic Nanoparticles in biomedicine. *Journal of Physics. D, Applied Physics*, *36*(13), 167–181. doi:10.1088/0022-3727/36/13/201

Rosensweig, R. E. (1997). *Ferrohydrodynamics*. Mineola, NY: Dover Publications, Inc.

Rosensweig, R. E. (2002). Heating magnetic fluid with alternating magnetic field. *Journal of Magnetism and Magnetic Materials*, *252*, 370–374. doi:10.1016/S0304-8853(02)00706-0

Saga, N., & Nakamura, T. (2002). Elucidation of propulsive force of micro robot using magnetic fluid. *Journal of Applied Physics*, *380*(10), 7003–7005. doi:10.1063/1.1452197

Shliomis, M. I. (1996). On "Maxwell's equations and vorticity: A note on the viscosity of magnetic fluids". *Journal of Magnetism and Magnetic Materials*, *159*(1-2), 236–237. doi:10.1016/0304-8853(96)00260-0

Sproston, J. L., Yanyo, L. C., Carlson, J. D., & El Wahed, A. K. (2002). Controllable Fluids in 2002-Status of ER and MR Fluid Technology. *Proceedings of 8th International Conference on New Actuators* (pp. 333–338). Academic Press.

Stierstadt, K., & Liu, M. (2015). Maxwell's stress tensor and the forces in magnetic liquids. *Zeitschrift für Angewandte Mathematik und Mechanik*, *95*(1), 4–37. doi:10.1002/zamm.201300192

Trahms, L. (2009). Biomedical applications of magnetic nanoparticles. In Colloidal Magnetic Fluids (pp. 327–358). Springer.

Uhlmann, E., Spur, G., Patzwald, R., & Bayat, N. (2002). Application of magnetic fluids in tribotechnical systems. *Journal of Magnetism and Magnetic Materials*, *252*, 337–340. doi:10.1016/S0304-8853(02)00724-2

Zimmermann, K., Zeidis, I., Naletova, V. A., & Turkov, V. A. (2004a). Waves on the surface of a magnetic fluid layer in a traveling magnetic field. *Journal of Magnetism and Magnetic Materials*, *268*(1-2), 227–231. doi:10.1016/S0304-8853(03)00503-1

Zimmermann, K., Zeidis, I., Naletova, V. A., & Turkov, V. A. (2004b). Traveling waves on a free surface of a magnetic fluid layer. *Journal of Magnetism and Magnetic Materials*, *272–276*, 2343–2344. doi:10.1016/j.jmmm.2003.12.966

Chapter 17
A Future Trend on Research Scope of Numerical Simulation on Conical Fluidized Bed

Hirakh Jyoti Das
Indian Institute of Technology, Guwahati, India

Pinakeswar Mahanta
Indian Institute of Technology, Guwahati, India & National Institute of Technology, Arunachal Pradesh, India

Rituraj Saikia
Indian Institute of Technology, Guwahati, India

ABSTRACT

Fluidized bed technology is a well-established environment friendly technology, by which energy can be generated through combustion and gasification techniques. It is widely prevalent today owing to its excellent heat transfer, mixing characteristics and compactness. The design and scale-up of the fluidized beds are vital to the enhancement of heat transfer and mixing characteristics. However, heat transfer characteristics play a key role in determining the combustion and gasification characteristics. CFD is a technique which helps to optimize the design and operation of fluidized bed combustor and gasifiers. Enhancement of computing speed and numerical techniques has led to CFD being used as a widely implemented tool to provide a bridge between laboratory scale and industrial study. In this chapter, a comprehensive review of CFD modelling and experimental study on the conical fluidized bed has been carried out. Primarily this chapter demonstrates probable future accomplishments and identifies trends and regions where further research is required.

DOI: 10.4018/978-1-7998-1831-1.ch017

INTRODUCTION

In recent times, more emphasize is given on clean coal technologies. Because of the depletion of fossil fuels, environmental security and increased pressure worldwide, there is a demand for efficient utilization of fossil fuels without emission of unburnt products. Fluidized bed combustion and gasification are such technologies, through which emissions of harmful products can be controlled (Singh, Brink, & Hupa, 2013).

The heat transfer characteristics determine the performance of fluidized bed through combustion and gasification techniques. However, the heat transfer characteristics are inevitably influenced by the hydrodynamic behaviour of a fluidized bed (Patil, Pandey, & Mahanta, 2011). The hydrodynamic behaviour of the conical fluidized bed is different from the conventional fluidized bed, which is characterized by different fluidization regimes. Hence, it is obvious to investigate the hydrodynamic behaviour and heat transfer characteristics of the conical fluidized bed riser, since, it anticipates the distribution of the species and the phases involved (Philippsen, Vilela, & Zen, 2015; Wu, 1993). The study of the hydrodynamic behaviour and heat transfer characteristics of the conical fluidized bed has received much attention from the scientific community leading to various numerical studies on it besides experimental work.

Nature of two-phase mixing plays a vital role in enhancing the heat transfer and better facilitation of the chemical reaction, which determines the performance of a fluidized bed reactor (Wu, 1993). A proper insight into complex multi-phase flow is required to advance the fluidized bed performance, leading to the improvement of multidimensional models. These multi-dimensional models form a bridge between research on a laboratory scale and activity of the fluidized bed in industries by providing the information on processes of combustion and gasification that cannot be provided by experimental data alone (Singh et al., 2013). Hence, there is a scope for the research of complex processes involved in the fluidized bed system, which are characterized by multidimensional models through numerical simulation. A proper framework of numerical simulation is required to understand the flow processes. The framework for numerical simulation is provided in Fig. 1.

The performance of the fluidized bed systems is substantially influenced by the design and scale-up. It requires a long duration to explore the impact of scale-up on fluidized bed riser performance experimentally, which leads to the high operating cost. Besides, it is challenging to attain steady-state condition fully in fluidization problems. With the changing of load or other operating parameters, its behaviour changes immensely. Therefore, the operation of the fluidized bed system in an experimental condition is very complex. Apart from that, the equations used in fluidization problems are non-linear and to obtain the exact solution of these equations is a strenuous exercise. Due to the complex non-linear equation, the accurate prediction of the numerical solution is difficult. With the rise in the computing technology and numerical tools development, computational fluid dynamics (CFD) has become a key tool for predicting the precise two-phase flow, since, the nature of the two-phase flow is complex and transient. Besides, characteristics of the fluidized bed such as design, optimization and scale-up are essential from the modelling point of view, which can easily be predicted by using CFD process (Askaripour & Dehkordi, 2016; Sau & Biswal, 2011). Hence, the above discrepancies allocated with experimental study can be shrunk with the help of numerical study so that the performance of the fluidized bed is optimized.

CFD is a numerical solution tool that includes fluid flow to solve algorithms and evaluate the issues by numerical methods. Various processes, such as physical and chemical processes are involved in the aspect of fluidized bed simulation. The models involved with physical and chemical processes along

Figure 1. A framework of numerical simulation modelling
(Myöhänen & Hyppänen, 2011; Singh et al., 2013)

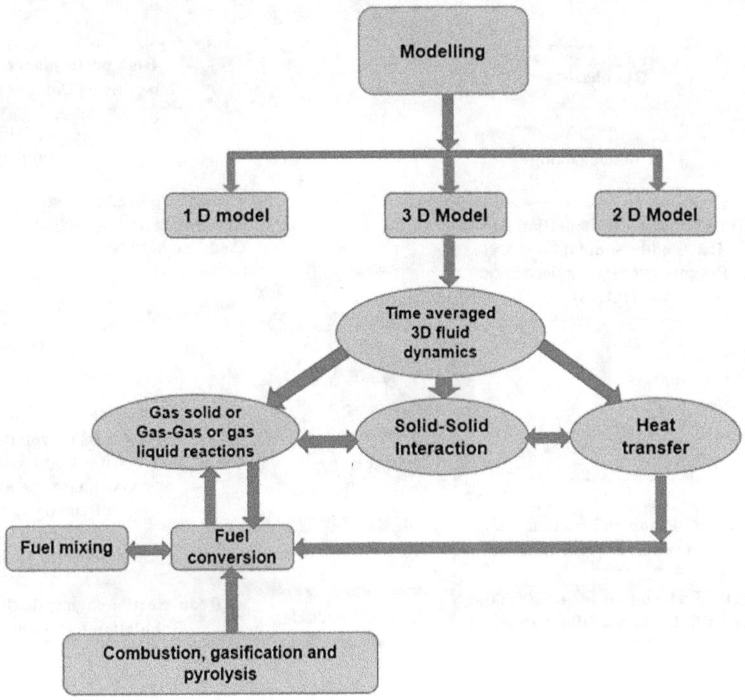

with their interrelation with other models have been shown in Fig. 2 (Pepiot, Dibble, & Foust, 2010; Singh et al., 2013).

Numerical and experimental investigations of hydrodynamic behaviour and heat transfer characteristics related to conventional fluidized bed have been carried out by many researchers. Review paper related to computational modelling in conventional fluidized bed has been published (Khan et al., 2014; Philippsen et al., 2015; Singh et al., 2013). There is a considerable amount of literature related to numerical study available on the hydrodynamic behaviour and heat transfer characteristics of fluid flow in a conical fluidized bed. However, none of the review papers studied the hydrodynamic behaviour and heat transfer characteristics on a conical fluidized bed. Given the limited information available regarding review paper for conical fluidized bed, the present chapter presents a review related to the numerical study on hydrodynamic behaviour and heat transfer characteristics in the conical fluidized bed. This chapter also aims to investigate the further requirements of probable trends of numerical simulation in the conical fluidized bed. Besides, the use of numerical models in the process of combustion and gasification is also discussed.

BACKGROUND

Fluidization involves the conversion of solid particles into a liquid-like state by streaming current of liquid or gas through the bed of particles. Based on its geometry, the fluidized bed riser can be divided into two groups viz conventional fluidized bed (cylindrical, square, rectangular in cross-section) and conical

Figure 2. Modelling of physical and chemical processes interactions in the thermochemical conversion of fuel
(Pepiot et al., 2010; Singh et al., 2013)

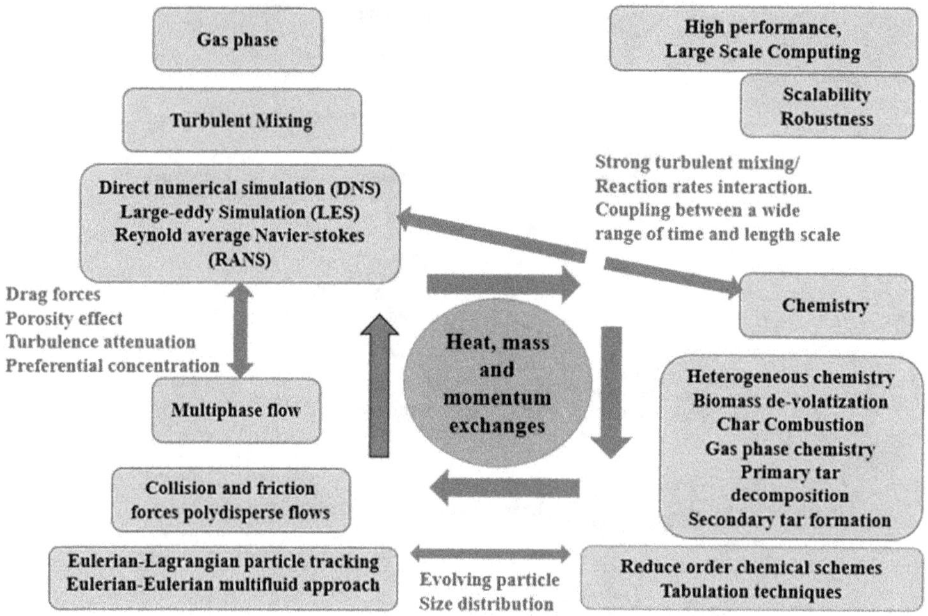

fluidized bed (Kunii & Levenspiel, 1991). In the conical fluidized bed, the area of the cross-section of the boiler is of varying dimension along the height. Due to the varying cross-section, this type of bed offers several benefits over conventional fluidized bed such as: (i) In conical bed, the fluidization can occur at a small amount of solid materials, (ii) starting time is less (iii) less pressure drop along the riser (for same height of bed material), which results in decreased operating costs (Abdelmotalib, Youssef, Hassan, Youn, & Im, 2015). Further, the conical fluidized bed shows certain advantages such as (i) a conical fluidized bed, can be operated with a broad selection of velocities, (ii) different particle sizes or densities can also be fluidized and (iii) more intensive mixing of two-phases (Padhi, Mohanty, Roy, & Sarangi, 2014). Moreover, with the use of the conical fluidized bed, higher combustion efficiency can be obtained with reduced gaseous emissions when burning various fossil fuels (Kouprianov & Permchart, 2003; Kuprianov & Arromdee, 2013; Ninduangdee & Kuprianov, 2013). Thus, conical fluidized bed is extensively used in many industries like coal gasification and combustion, liquefaction and microbial growth in which the fluidization particles change frequently (Peng & Fan, 1997; Shi, Yu, & Fan, 1984), roasting of sulphide ores (Levey, Garga, Jacobs, Heidt, & Trent, 1960; Peng & Fan, 1997), biological waste-water treatment (Peng & Fan, 1997; Scott & Hancher, 1976), destructive distillation of waste-materials, immobilization biofilm reaction, particles coating (Peng & Fan, 1997), catalytic polymerization (Bilbao, Olazar, Arandes, & Romero, 1989), fluidization of cohesive powder (Venkatesh, Chaouki, & Klvana, 1996), food processing (Depypere, Pieters, & Dewettinck, 2005), removal of organic pollutants from industrial sewages (McKay & Sze, 2008), drying of drug tablets and granules (Ridgeway, 1965) etc. Therefore, many researchers have shown their research interests towards the experimental and numerical simulation in the conical fluidized bed.

The hydrodynamic behaviour of conical fluidized bed risers is distinguished by the different flow regimes. In conventional fluidized bed; fixed bed, minimum fluidized bed, bubbling, slugging, turbulent and fast fluidized bed regimes are found to exist as shown in Fig. 3.

But in gas-solid conical bed, three types of fluidization regimes such as fixed bed, partially fluidized bed and fully fluidized bed with their specific characteristics are observed with the different superficial velocity (Jing, Hu, Wang, & Jin, 2000; Katiyar, 2017; Rasteh, Farhadi, & Bahramian, 2015).

Figure 3. Regimes of fluidization in fluidized bed riser
(Kunii & Levenspiel, 1991)

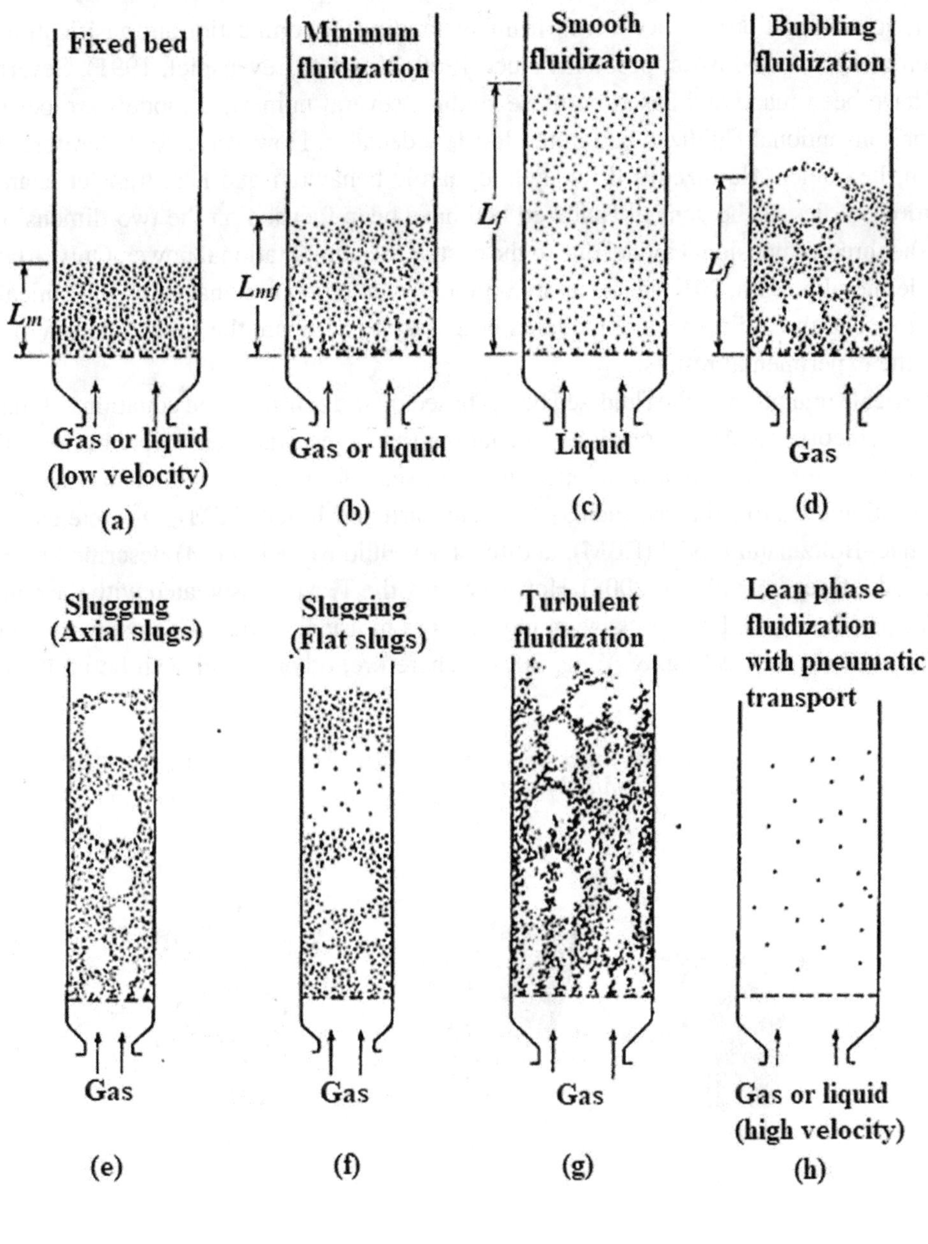

However, in a liquid-solid conical fluidized bed, five different fluidization regimes such as fixed bed, partially fluidized bed, fully fluidized bed regime, transition regime and the turbulent regime are observed (Peng & Fan, 1995, 1997).

As evident from Fig. 3, 4 and 5 that the hydrodynamic behaviour in the conical fluidized bed is different from the conventional bed. The behaviour of a conical fluidized bed is due to the development of additional stresses accompanied by the reduced pressure gradient. Thus, some additional assumptions are taken to study the behaviour of the conical fluidized bed. These assumptions are discussed later.

The gas-solid conical fluidized bed was first used by Levey et al. (Levey et al., 1960). Their application of the bed was for hydro fluorination of uranium trioxide to uranium tetrafluoride. Since then various researchers studied hydrodynamic behaviour experimentally and numerically (Askaripour & Dehkordi, 2016; Dora, Panda, Mohanty, & Roy, 2013; Padhi, Dora, Mohanty, Roy, & Sarangi, 2016; Sau & Biswal, 2011; Sirisomboon & Charernporn, 2017).

The numerical simulation plays a vital role in understanding the combustion and gasification processes and has been utilized in industrial problems since 1960 (Kunii & Levenspiel, 1991). Several numerical studies have been made and based on these studies, several numerical models are developed and improved for conventional fluidized bed in the last few decades. However, a few literatures have been published on the conical fluidized bed for hydrodynamic behaviour and heat transfer characteristics. The simulation studies of the conical fluidized bed have been focused on the two-dimensional model rather than the three-dimensional model due to the cost of the computational power. Only, Abdelmotalib and Im (Abdelmotalib & Im, 2017) made a three-dimensional computational study on conical fluidized bed combustor. The three-dimensional model is necessary to compare the exact solution of numerical results with the experimental results.

The numerical simulation of the fluidized bed is based on the conservation equations of mass, energy and momentum, incorporated with constitutive equations that depicts the interactions between the phases. There are several numerous models developed in recent years for describing the behaviour of gas-solid flow in conventional fluidized beds, such as Discrete Particle Model (DPM), Discrete element model (DEM), Lattice–Boltzmann model (LBM), and the Two-Fluid Model (TFM) described by Hoef et al. (Hoef, Annaland, Deen, & Kuipers, 2008). However, only the TFM is associated with the conical fluidized bed. The TFM is in need of a package of physical or empirical models for closure to the governing equations (Sun, Zhou, Ren, Wang, & Yang, 2011). Therefore, other models can be implemented in a

Figure 4. Flow regimes of gas-solid conical fluidized bed
(Jing et al., 2000)

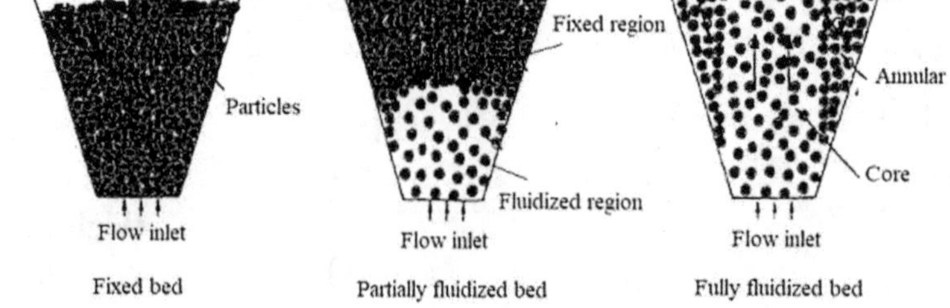

Figure 5. Flow regimes of liquid-solid conical fluidized bed
(Peng & Fan, 1997)

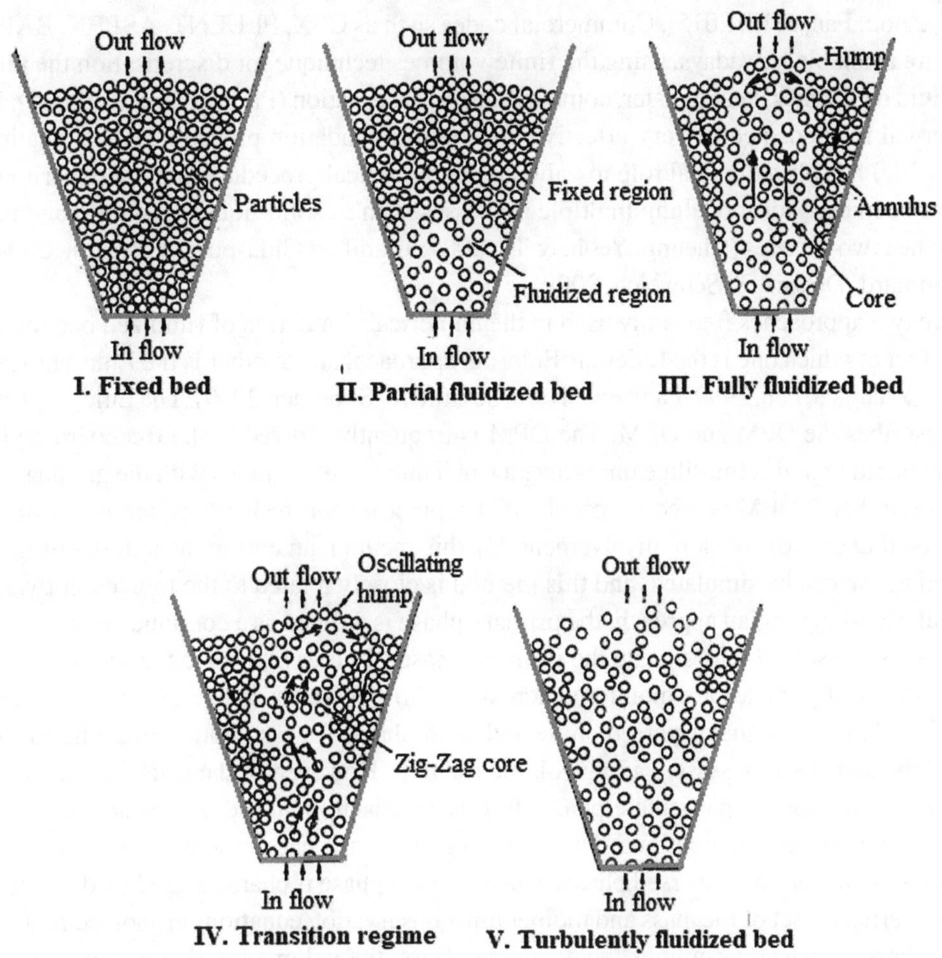

conical bed for better understanding of flow behaviour. A two-fluid Eulerian-Eulerian model incorporating the Kinetic theory of granular flow (KTGF) approach is broadly implemented to simulate the flow behaviour in a fluidized bed reactor (Busciglio, Vella, Micale, & Rizzuti, 2009; Gera & Gautam, 1995; Kuipers, Duin, Beckum, & Swaaij, 1993; Pain, Mansoorzadeh, & Oliveira, 2001; Peirano, Delloume, & Leckner, 2001; Shen, Johnsson, & Leckner, 2004). It has become the most important analogy for the constitutive law over constant viscosity model (CVM) and turbulent particle model since it offers apparent advantages in the perfect prediction of gas-solid flow and provides better results compared to experimental outcomes. This analogy considers the conservation of solid fluctuation energy to explain dispersed phase physics, such as solid shear viscosity, solid bulk viscosity, solid pressure, solid shear stress and collision frequencies between different particles along with the two-fluid model (Arastoopour, 2001; Askaripour & Dehkordi, 2016; Bi, Ellis, Abba, & Grace, 2000; Hosseini, Ahmadi, Razavi, & Zhong, 2010; Johansson, Wachem, & Almstedt, 2006). In the gas-solid flow simulation, both the hydrodynamic behaviour and heat transfer characteristics are characterized by granular temperature. Because of that, the hydrodynamic behaviour and heat transfer characteristics can, therefore, be explicitly investigated in a conical fluidized bed (Abdelmotalib et al., 2015). The constitutive equations derived from the KTGF are characterized by chemical kinetics as well as experimental correlations and models (Patankar, 1980; Shu, Wang, Zhou, Fan, & Li, 2015). Commercial codes such as CFX, FLUENT, ASPEN, BARACUDA and MFIX are available nowadays, using the finite volumes technique for discretisation the phenomena involving fluid mechanics, heat transfer, combustion and gasification (Philippsen et al., 2015). However, these numerical methods need a very effective and careful validation procedure. Computational fluid dynamics (CFD) has played a vital role to validate the numerical procedure and the experimental condition. It has been used for modelling multiple processes such as solid-liquid, gas-solid and particulate fluid for the last two decades (Cheung, Yeoh, & Tu, 2007; Jiradilok, Gidaspow, & Breault, 2007; Boutet, Larachi, Dromard, Delsart, & Schweich, 2009).

There are two approaches frequently used in the numerical simulation of fluidized bed for the multiphase flow. Out of which one is the Eulerian-Eulerian approach, and another is the Eulerian-Lagrangian approach (Li, Zhang, & Fan, 1999; Panneerselvam, Savithri, & Surender, 2009). The Eulerian-Lagrangian approach describes the DPM and DEM. The DPM is frequently utilized in the freeboard region of the bed. The freeboard region is the dilute one where a small number of particles with the gas phase are present. On the other hand, DEM is used in complicated geometries for the heterogeneous reaction because of the rotational degree of freedom involvement. By this method, an enormous number of micron-size particles and above can be simulated, and this method is closely related to the molecular dynamics.

In the Eulerian-Lagrangian approach, the primary phase is defined as a continuum where the Navier-stokes equations are solved to designate the primary phase. Whereas, to solve the secondary phase, an enormous number of particles are monitored across the flow field. Exchange of momentum, mass and energy is possible between the dispersed phase and fluid phase. An assumption made in this technique is that the dispersed phase possesses a low volume fraction. However, in the Eulerian-Eulerian model, the KTGF is utilised for the particulate phase. In this approach, both the phases are described mathematically as inter-penetrating continua. The primary phase is the gas phase while the particle phase is regarded as the secondary or dispersed phase. The dispersed phase is characterized by diameter, density and other properties. A set of the mass and momentum conservation equations are solved for each phase, and an interphase exchange term links the equations. Here, the volume fraction of both phases cannot occupy the same location at the same time. Hence, the idea of phase volume fraction is presented. These volume fractions are treated to be continuous functions of space and time, and their entirety is equal to one. The computational cost is higher in the Eulerian-Lagrangian approach than in the TFM because the former approach describes the behaviour of individual particles. Therefore, the latter approach is more conveniently used in the numerical simulation of conical fluidized beds (Abdelmotalib, Ko, & Im, 2016).

Drag models are used to obtain precise simulation results for the gas-solid momentum exchange coefficient. Sometimes, drag models are used with the KTGF based TFM to simulate the fluidized bed hydrodynamics (Hamzehei & Rahimzadeh, 2009; Li et al., 2009). There are several drag models employed in the simulation of conventional fluidized bed such as EMMS (Energy minimum multiscale method) model (Chalermsinsuwan, Piumsomboon, & Gidaspow, 2009; Wang, Ge, & Li, 2008; Wang & Liu, 2010; Wang & Li, 2007; Yang, Wang, Ge, & Li, 2003; Yang, Wang, Ge, Wang, & Li, 2004), Syamlal-O'Brien model, Gidaspow model, symmetric model, Schiller and Naumann and Wen and Yu model (Gidaspow, 1994; Syamlal & O'Brien, 1988). EMMS is analogous to the heterogeneity of particles. In conical fluidized bed riser, due to the existence of velocity gradient from bottom to top, the behaviour of gas-solid flow is uniform throughout the riser leading to homogeneous mixing of particles and gas. Therefore, EMMS model can be excluded from the conical fluidized bed simulation. However,

Chalermsinsuwan et al. (Chalermsinsuwan, Kuchonthara, & Piumsomboon, 2010) used this model to compare the different riser geometries for heterogeneous chemical reaction since they used one conventional riser and the other two conical risers. The Gidaspow model is a combination of the Wen and Yu (Yu & Wen, 1966) model and the Ergun equation (Askaripour & Dehkordi, 2016). For the value of $\varepsilon_g \geq 0.8$, Wen and Yu model is used, which is for dilute flows and in case of dense flows, the Ergun equation is used in which $\varepsilon_g < 0.8$ (Askaripour & Dehkordi, 2016). The Gidaspow model is quite useful for homogeneous bubbling action than Syamlal O'Brien model (Bakshi, Altantzis, Bates, & Ghoniem, 2015). Except for other drag models, Syamlal-O'Brien and Gidaspow models are widely used in conical gas-solid multiphase flow with coarse solid particles.

The virtual mass and lift effects are also taken into consideration in fluidized bed simulation. The lift effects are considered for particles of large diameters, while the effects of virtual forces are predominant if the density of dispersed phase is significantly greater than the fluid phase density. The equation of Lun et al. (Lun, Savage, Jeffrey, & Chepurniy, 1984) is widely used in the conical fluidized bed for calculation of the solid pressure and solid bulk viscosity that accounts for particle resistance in expansion and contraction (Abdelmotalib, Youssef, Hassan, Youn, & Im, 2016; Askaripour & Dehkordi, 2016; Sau & Biswal, 2011). However, for the solid pressure, an equation of state similar to the Van der Walls equation of state of the gas is also used sometimes (Abdelmotalib et al., 2016). In case of very dense flows, the frictional viscosity is implemented because the particles volume fraction closely approaches to the maximum packing limit. Friction produces a substantial amount of stress among solid particles. Schaeffer's expression (Schaeffer, 1987) is generally employed in dense flows to model the frictional viscosity. Several researchers used the laminar viscosity model to study the hydrodynamics of multiphase flow, especially in bubbling condition since turbulence models need high computational facilities with high speed and are difficult to converge the solution. However, the complexities encountered with the typical turbulence model can be compensated by ANSYS FLUENT (ANSYS Inc. ANSYS Fluent Theory Guide, 2011), since it offers a high precision option with an acceptable speed of convergence and can provide a better solution to keep away from these difficulties. Besides that, the knowledge of turbulence phenomenon is also necessary in some context since it can provide better mixing of multiphase flow and enhance the heat and mass transfer. In the simulation of conical bed riser, only a few works of literature are found to use the turbulence model (Askaripour & Dehkordi, 2016; Khodabandehlou, Askaripour, & Dehkordi, 2018). Most of the papers have been found to study the bubbling behaviour in conical bed, which does not require turbulence model but rather the laminar model since the Reynold number is less. The governing equations and constitutive equations which are generally used in the conical fluidized bed explained as follows:

Governing Equation

Continuity equation:

For gas

$$\frac{\partial}{\partial t}\left(\varepsilon_g \rho_g\right) + \nabla.\left(\varepsilon_g \rho_g \vec{v}_g\right) = 0 \tag{1}$$

For solid

$$\frac{\partial}{\partial t}\left(\varepsilon_s \rho_s\right)+\nabla.\left(\varepsilon_s \rho_s \vec{v}_s\right)=0 \tag{2}$$

Volume fraction equation: $\varepsilon_g + \varepsilon_s = 1$ (3)

In fluidized bed simulation, by modifying the Navier – Stokes equation, the momentum equations are balanced for both phases by introducing an interphase momentum transfer term and a solid phase source term.

Momentum equation:

For gas

$$\frac{\partial}{\partial t}\left(\varepsilon_g \rho_g \vec{v}_g\right)+\nabla.\left(\varepsilon_g \rho_g \vec{v}_g \vec{v}_g\right)=-\varepsilon_g \nabla P_g+\nabla \overline{\overline{\tau}}_g+\varepsilon_g \rho_g \vec{g}+\vec{F}_{d.g}+\vec{F}_{l.g}+\vec{F}_{vm.g} \tag{4}$$

For solid

$$\frac{\partial}{\partial t}\left(\varepsilon_s \rho_s \vec{v}_s\right)+\nabla.\left(\varepsilon_s \rho_s \vec{v}_s \vec{v}_s\right)=-\varepsilon_s \nabla P_g-\nabla P_s+\nabla \overline{\overline{\tau}}_s+\varepsilon_s \rho_s \vec{g}+\vec{F}_{d.s}+\vec{F}_{1.s}+\vec{F}_{vm.s}+\overline{S}_s \tag{5}$$

In the above expressions, the left terms indicate the temporal and spatial terms, while the various interacting forces are on the right-hand side.

Constitutive Equations for the Closure of the Governing Equations

There is an exchange of momentum between two phases through drag, lift and virtual mass forces. The various form of all these three forces acting on the two phases can be described as follows (Askaripour & Dehkordi, 2016):

$$\vec{F}_{d.g}=\beta_{gs}(\vec{v}_s-\vec{v}_g),\ \vec{F}_{d.s}=-\vec{F}_{d.g} \tag{6}$$

$$\vec{F}_{l.g}=0.5\rho_g \varepsilon_s\left(\vec{v}_s-\vec{v}_g\right).\left(\nabla\times\vec{v}_g\right),\ \vec{F}_{1.s}=-\vec{F}_{1.g} \tag{7}$$

$$\vec{F}_{vm.g}=0.5\rho_g \varepsilon_s\left(\frac{D\vec{v}_s}{Dt}-\frac{D\vec{v}_g}{Dt}\right),\ \vec{F}_{vm.s}=-\vec{F}_{vm.g} \tag{8}$$

As discussed in the earlier section that there are various drag models found in conical fluidized bed simulation. These models are used to find out the drag coefficient between two phases. The equations of these models are described below as EMMS drag model (Chalermsinsuwan et al., 2009):

$$\beta_{gs} = \frac{3}{4} C_D \frac{\varepsilon_g \varepsilon_s \rho_g \left| \vec{v}_s - \vec{v}_g \right|}{d_p} \varepsilon_g^{-2.65} \text{ for } \varepsilon_g > 0.7 \tag{9}$$

$$\beta_{gs} = 150 \frac{\varepsilon_s^2 \mu_g}{\varepsilon_g d_p^2} + 1.75 \frac{\varepsilon_s \rho_g \left| \vec{v}_g - \vec{v}_s \right|}{d_p} \text{ for } \varepsilon_g \leq 0.74 \tag{10}$$

$$v_{r,s} = 0.5 \left[A - 0.06 \operatorname{Re}_s + \sqrt{\left(0.06 \operatorname{Re}_s\right)^2 + 0.12 \operatorname{Re}_s \left(2B - A\right) + A^2} \right]$$

Syamlal- O'Brien drag model (Syamlal & O'Brien, 1988):

$$\beta_{gs} = \frac{3}{4} \frac{\varepsilon_g \varepsilon_s \rho_g}{v_{r,s}^2 d_s} C_D \left(\frac{\operatorname{Re}_s}{v_{r,s}} \right) \left| \vec{v}_s - \vec{v}_g \right| \tag{11}$$

$$C_D = \left(0.63 + \frac{4.8}{\sqrt{\operatorname{Re}_s / v_{r,s}}} \right)^2 \tag{12}$$

$$v_{r,s} = 0.5 \left[A - 0.06 \operatorname{Re}_s + \sqrt{\left(0.06 \operatorname{Re}_s\right)^2 + 0.12 \operatorname{Re}_s \left(2B - A\right) + A^2} \right] \tag{13}$$

$$A = \varepsilon_g^{4.14} \; and\; for\; \varepsilon_g \leq 0.85 \; and\; for\; \varepsilon_g > 0.85$$

Gidaspow drag model (Gidaspow, 1994):

$$\beta_{gs} = \frac{3}{4} C_D \frac{\varepsilon_g \varepsilon_s \rho_g \left| \vec{v}_s - \vec{v}_g \right|}{d_s} \varepsilon_g^{-2.65} \text{ for } \varepsilon_g > 0.8 \tag{14}$$

$$\beta_{gs} = 150 \frac{\varepsilon_s^2 \mu_g}{\varepsilon_g d_s^2} + 1.75 \frac{\varepsilon_s \rho_g \left| \vec{v}_g - \vec{v}_s \right|}{d_s} \text{ for } \varepsilon_g \leq 0.8 \tag{15}$$

Where

$$C_D = \frac{24}{\varepsilon_g \mathrm{Re}_s}\left[1 + 0.15\left(\varepsilon_g \mathrm{Re}_s\right)^{0.687}\right] \quad (16)$$

$$\mathrm{Re}_s = \frac{\rho_g d_s \left|\vec{v}_s - \vec{v}_g\right|}{\mu_g} \quad (17)$$

The granular temperature Θ_s of solid particles is similar to the thermodynamic temperature of gas-phase and can be incorporated as a measure of the fluctuation in particle velocity described as the one-third of the mean square velocity of particles, i.e.,

$$\Theta_s = \frac{1}{3} v_s'^2 \quad (18)$$

The granular temperature conservation equation:

In the laminar model, the granular temperature is calculated in terms of random oscillation of individual particles, whereas in the turbulent model it is determined by taking all particles together and measuring the oscillation of cluster of particles.

Kinetic fluctuation energy:

$$\frac{3}{2}\left[\frac{\partial}{\partial t}\left(\rho_s \varepsilon_s \Theta_s\right) + \nabla.\left(\rho_s \varepsilon_s \vec{v}_s \Theta_s\right)\right] = \left(-P_s \bar{I} + \bar{\tau}_s\right) : \nabla \vec{v}_s + \nabla.\left(k\Theta_s \nabla \Theta_s\right) - \gamma \Theta_s + \phi_{gs} \quad (19)$$

Collisional dissipation of energy (Askaripour & Dehkordi, 2016; Hamzehei, Rahimzadeh, & Ahmadi, 2010):

$$\gamma \Theta_s = \frac{12\left(1 - e_{ss}^2\right) g_{o,s}}{d_s \sqrt{\pi}} \rho_s \varepsilon_s^2 \Theta_s^{3/2} \quad (20)$$

Diffusion coefficient of energy:

$$k\Theta_s = \frac{150 \rho_s d_s \sqrt{\Theta_s \pi}}{384\left(1 + e_{ss}\right) g_{o,ss}}\left[1 + \frac{6}{5} g_{o,ss} \varepsilon_s \left(1 + e_{ss}\right)\right]^2 + 2 \rho_s \varepsilon_s^2 d_s \left(1 + e_{ss}\right) g_{o,ss} \sqrt{\frac{\Theta_s}{\pi}} \quad (21)$$

The last term on the RHS is the exchange of kinetic energy between the solid and the gas phases due to the random fluctuation in particle velocity and can be given as,

$$\phi_{gs} = -3 k_{gs} \Theta_s \quad (22)$$

In the turbulence model, however, the solid velocity fluctuation in three distinct directions of the axis produces additional stresses. The sum of laminar and turbulence granular model temperature is known as the total granular temperature. A solid pressure is calculated in the compressible regime (i.e. the regime where the solids volume fraction cannot be exceeded to maximum allowed value) for granular flows. This solid pressure is used in the dispersed phase momentum equation for the purpose of pressure gradient term. This equation introduces the granular temperature and because of which the expression becomes valid for solid pressure and viscosities. The pressure gradient term represents the solid phase force in the normal direction and is comprised of a kinetic term and a particles collision term. The equation of solid pressure as given by Lun et al. (Lun et al., 1984) is

Solid pressure:

$$P_s = \rho_s \varepsilon_s \Theta_s + 2\rho_s \left(1 + e_{ss}\right) \varepsilon_s^2 g_{o,ss} \Theta_s \tag{23}$$

However, particle to particle collision probability can be modified by a correction factor named radial distribution function and the expression for the same is given in the following form (Ma & Ahmadi, 1986)

Radial distribution function:

$$g_{o,ss} = \frac{1 + 2.5\varepsilon_s + 4.59\varepsilon_s^2 + 4.52\varepsilon_s^3}{\left[1 - \left(\frac{\varepsilon_s}{\varepsilon_{s,\max}}\right)^3\right]^{0.678}} \tag{24}$$

The solid stress tensor in the momentum equation is composed of solid shear and bulk viscosities responsible for transient momentum transfer of solid particles owing to collision and translation, respectively.

The solid shear viscosity is composed of kinetic and collisional terms. A frictional viscosity term is also incorporated to represent the change that happens when the proportions of solid quantity exceeds a critical value. Since the state of fluidized bed simulation is transient, therefore the solid shear viscosity model is important to determine the time average hydrodynamic behaviour. The solid shear viscosity is evaluated as follows (Chalermsinsuwan, Gidaspow, & Piumsomboon, 2011)

Solid shear viscosity:

$$\mu_s = \mu_{s.col} + \mu_{s.kin} + \mu_{s.fr} \tag{25}$$

Collision viscosity:

$$\mu_{s.col} = \frac{4}{5} \varepsilon_s d_s \rho_s \left(1 + e_{ss}\right) g_{o,ss} \sqrt{\frac{\Theta_s}{\pi}} \tag{26}$$

Syamlal kinematic viscosity (Syamlal & O'Brien, 1988):

$$\mu_{s.kin} = \frac{\varepsilon_s d_s \rho_s \sqrt{\Theta_s \pi}}{6\left(3-e_{ss}\right)}\left[1+\frac{2}{5}\varepsilon_s g_{o,ss}\left(1+e_{ss}\right)\left(3e_{ss}-1\right)\right] \tag{27}$$

Gidaspow kinematic viscosity (Sau & Biswal, 2011):

$$\mu_{s.kin} = \frac{10\rho_s d_s \sqrt{\Theta_s \pi}}{96\varepsilon_s\left(1+e_{ss}\right)g_{o,ss}}\left[1+\frac{4}{5}g_{o,ss}\varepsilon_s\left(1+e_{ss}\right)\right]^2 \tag{28}$$

The solid phase bulk viscosity is employed to account the resistance of solid particles to compression and expansion. However, in fluidized bed simulation, the gas phase bulk viscosity is thought to be a consistent estimation of zero. The expression of solid bulk viscosity is as follows.

Solids bulk viscosity (Lun et al., 1984):

$$k_s = \frac{4}{3}\varepsilon_s d_s \rho_s\left(1+e_{ss}\right)g_{o,ss}\sqrt{\frac{\Theta_s}{\pi}} \tag{29}$$

The frictional viscosity as defined by Schaeffer (Schaeffer, 1987) is

$$\mu_{s.\mathrm{fr}} = \frac{P_{fr}\sin\left(\phi\right)}{\sqrt{I_{2D}}} \tag{30}$$

where P_{fr} is the frictional pressure and is estimated from Johnson and Jackson (Johnson & Jackson, 1987) model,

$$P_{fr} = Fr\frac{\left(\varepsilon_s-\varepsilon_{s,cr}\right)^n}{\left(\varepsilon_{s,\max}-\varepsilon_s\right)^p},\ Fr=0.05, n=2, p=5 \tag{31}$$

where, $\varepsilon_{s,cr}$ is the critical solid volume fraction, and Fr, n, p are empirical constants.

$$I_{2D} = \frac{1}{6}\left[\left(D_{xx}-D_{yy}\right)^2+\left(D_{yy}-D_{zz}\right)^2+\left(D_{zz}-D_{xx}\right)^2\right]+D_{xy}^2+D_{yz}^2+D_{zx}^2 \tag{32}$$

$$D_{ij} = \frac{1}{2}\left[\nabla\vec{v}_s+\left(\nabla\vec{v}_s\right)^T\right] \tag{33}$$

Sometimes, the specularity coefficient is used as a wall boundary condition in numerical simulation of multiphase flows. It is employed to measure the amount of lateral momentum transferred to the wall by the fraction of particle collisions since it has a significant effect on the heat transfer and hydrodynamic behaviour on the modelling of the small-scale fluidized bed. It gives the shear to the wall. The levels of shear to the wall varies with the specularity coefficient. However, it is difficult to measure the specularity coefficient experimentally. The effect of specularity coefficient was first used by (Hui, Haff, Ungar, & Jackson, 1984) in a conventional fluidized bed. After that, it was used by (Abdelmotalib et al., 2016) to study the effect on hydrodynamic behaviour and heat transfer characteristics in a conical fluidized bed combustor. In order to have the effect of specularity coefficient, the equation of Johnson and Jackson (Johnson & Jackson, 1987) is used for the dispersed phase tangential velocity. Following are the boundary equations of the Johnson and Jackson model (Johnson & Jackson, 1987):

$$v_{s,w} = -B\frac{\partial v_{s,w}}{\partial n} \tag{34}$$

Where *B* is the slip coefficient and is inversely proportional to the specularity coefficient, i.e., the slip coefficient increases with a small reduction of specularity coefficient value. The following expression is used for the slip coefficient

$$B = \frac{6\varepsilon_s\mu_s}{\sqrt{3}\sqrt{\Theta}\pi\phi\rho_s\varepsilon_s g_o} \tag{35}$$

where, ϕ is the specularity coefficient and distinct specularity coefficient values represent the different tangential boundary conditions for wall corresponding to the free slip, partial slip and no-slip. The value of specularity coefficient refers to zero means free slip (smooth wall) whereas, for the no-slip condition the value of specularity refers to as one at which the particles are assumed to have diffused collision perfectly. The value between zero and one corresponds to the partial-slip. On the other hand, there is a loss of fluctuating kinetic energy of solids due to the inelastic collision, characterized by the restitution coefficient, which can affect the granular temperature, granular viscosity and conductivity. Its value also varies from 0 to 1. The restitution coefficient characterizes the interaction between solid particles with the wall due to which a remarkable number of solids fluctuating energy is dissipated. A zero value of the restitution coefficient means the dissipation of fluctuating kinetic energy whereas, no energy being dissipated means that the restitution coefficient is one. In general, a higher value of the restitution coefficient, 0.95 to 1 is selected. However, the loss of momentum due to collision among particles and particles and wall improves for a lower value of the restitution coefficient, which is impractical. In most of the cases, the Johnson and Jackson (Johnson & Jackson, 1987) model is applied for the restitution coefficient and is given as (Abdelmotalib, Youssef, Hassan, Youn, & Im, 2016)

$$v_{s,w} = -B\frac{\partial v_{s,w}}{\partial n} \tag{36}$$

$$B = \frac{6\varepsilon_s \mu_s}{\sqrt{3}\sqrt{\Theta}\pi\phi\rho_s\varepsilon_s g_o} \quad (37)$$

$$\Theta_w = -\frac{k_s\Theta_s}{\gamma_w}\frac{\partial\Theta_w}{\partial n} + \frac{\sqrt{3}\pi\phi\rho_s\varepsilon_s v_{s,slip}^2 g_o \Theta_s^{3/2}}{6\varepsilon_{s,\max}\gamma_w} \quad (38)$$

$$\gamma_w = \frac{\sqrt{3}\pi\left(1-e_w^2\right)\varepsilon_s\rho_s g_o \Theta_s^{3/2}}{4\varepsilon_{s,\max}} \quad (39)$$

where γ_w is the collisional dissipation at the wall and e_w is the particle-to-wall restitution co-efficient.

The conical fluidized bed is divided into two main regimes, namely, a dilute region in the core and hump (top surface) and a dense region near the wall. In this type of bed, additional stresses are developed by the sidewall, which is accompanied by a reduced pressure gradient. However, in the conventional fluidized bed, the development of stresses is not seen. In the numerical simulation, these additional stresses are incorporated with the solid phase momentum equation and expressed as the solid source term (S_s). The expression solid source term for both these two regimes is expressed as (Askaripour & Dehkordi, 2016; Khodabandehlou et al., 2018; Sau & Biswal, 2011; Wang, 2006)

$$S_s = \left(K-1\right)\varepsilon_s\rho_s g \quad (40)$$

$$K = \begin{Bmatrix} K_a & \varepsilon_g < 0.8 \\ K_c & \varepsilon_g \geq 0.8 \end{Bmatrix} \quad (41)$$

where K_a is defined as the ratio of the pressure drop in a conical fluidized-bed to that in a columnar bed and this parameter K_c is a function of some operating parameters. For solving problems, the value K_c is predetermined to be a constant value of 1.

$$K_a = \frac{\Delta P_{tapered}}{\Delta P_{columnar}} \quad (42)$$

$$K_c = f\left(\varepsilon_g \rho_s d_s \rho_g \mu_g \vec{v}_g\right) \quad (43)$$

Most of the time, RNG approach *k-ε* two-equation model is used for turbulence viscous flow modelling. This two-equation model has found wide application in industrial flow and heat transfer simulation because of its high accuracy and robustness of the turbulence. The turbulence is an additional term of the momentum equation and determined by substituting the two components, such as the mean and fluctuating components of flow variables in Navier-Stokes equations. A statistical technique is used to derive this type of model and is derived from the renormalization group (ANSYS Inc. ANSYS Fluent Theory Guide, 2011; Askaripour & Dehkordi, 2016; Orszag et al., 1993). The equation of the form is given as

$$\frac{\partial}{\partial t}\left(\rho_m k\right)+\nabla.\left(\rho_m \vec{v}_m k\right)=\nabla.\left(\alpha_k \mu_t \nabla k\right)+G_{k,m}-\rho_m \varepsilon \tag{44}$$

$$\frac{\partial}{\partial t}\left(\rho_m \varepsilon\right)+\nabla.\left(\rho_m \vec{v}_m \varepsilon\right)=\nabla.\left(\alpha_\varepsilon \mu_t \nabla \varepsilon\right)+\frac{\varepsilon}{k}\left(C_{1\varepsilon} G_{k,m}-C_{2\varepsilon}\rho_m \varepsilon\right)-R_\varepsilon \tag{45}$$

In order to solve the heat transfer problems, energy equations have to be solved, and in the CFD package, ANSYS FLUENT offers the energy equation in terms of enthalpy balance. The thermal energy equations are described in the following (*ANSYS Inc. ANSYS Fluent Theory Guide*, 2011; Askaripour & Dehkordi, 2016)

For gas,

$$\frac{\partial}{\partial t}\left(\varepsilon_g \rho_g h_g\right)+\nabla.\left(\varepsilon_g \rho_g \vec{v}_g h_g\right)=-\varepsilon_g \frac{\partial P_g}{\partial t}+\bar{\bar{\tau}}_g : \nabla \vec{v}_g+\nabla.\left(\varepsilon_g k_g \nabla T_g\right)+Q_{sg}+\Delta H_{rxn} \tag{46}$$

For solid,

$$\frac{\partial}{\partial t}\left(\varepsilon_s \rho_s h_s\right)+\nabla.\left(\varepsilon_s \rho_s \vec{v}_s h_s\right)=-\varepsilon_s \frac{\partial P_s}{\partial t}+\bar{\bar{\tau}}_s : \nabla \vec{v}_s+\nabla.\left(\varepsilon_s k_s \nabla T_s\right)-Q_{sg} \tag{47}$$

The heat transfer Q_{sg} between the two-phase is dependent upon the temperature difference of the two phases and can be evaluated by the following equation

$$Q_{sg}=h_{sg}\left(T_s-T_g\right) \tag{48}$$

where, h_{sg} is the interphase heat transfer coefficient, i.e. heat transfer coefficient between gas and solid phases. Interphase heat transfer coefficient is a function of Nusselt no, volume fraction, average particle diameter and thermal conductivity of gas and solid respectively and can be expressed as

$$h_{sg}=\frac{6k_g \varepsilon_s \varepsilon_g Nu_s}{d_s^2} \tag{49}$$

where Nu_s is the solid phase Nusselt number and to determine this non-dimensional number, Gunn (Gunn, 1978) equation is used, which is associated with the particle Reynold number and Prandtl number. The expression for Nusselt number is as follows (Gunn, 1978)

$$Nu_s = \left(7 - 10\varepsilon_g + \varepsilon_g^2\right)\left(1 + 0.7\,\mathrm{Re}_s^{0.02}\,\mathrm{Pr}^{1/3}\right) + \left(1.33 - 2.4\varepsilon_g + 1.2\varepsilon_g^2\right)\mathrm{Re}_s^{0.7}\,\mathrm{Pr}^{1/3} \tag{50}$$

where Pr is the Prandtl number and is calculated by the following way

$$\mathrm{Pr} = \frac{\mu_g C_{p,g}}{k_g} \tag{51}$$

The local wall to bed heat transfer coefficient at various fixed points located along the height of a fluidized bed can be calculated by the following relationship (Abdelmotalib et al., 2016)

$$h = \frac{\varepsilon_g k_g^{eff}\left|\frac{\partial T_g}{\partial x}\right| + \varepsilon_s k_s^{eff}\left|\frac{\partial T_s}{\partial x}\right|}{T_w - T_b} \tag{52}$$

where, x is the point, where the local heat transfer coefficient is calculated. and k_s^{eff} are the effective thermal conductivities of both gas and solid phases, respectively. Here, the effective thermal conductivity of both the phases are taken into account because the thermal conductivity of solid particles depends upon the contact of particles in a fluidized bed and is not always same as that of the particulate phase. Based on the spherical packing arrangement, Zehner and Schlunder (Zehner & Schlünder, 1970) were the first to develop the equations for effective thermal conductivity of gas and solid phases to determine effective radial thermal conductivity in a packed bed by considering the solid-phase thermal conductivity equal to the effective thermal conductivity and fluid thermal conductivity equal to microscopic thermal conductivity (Abdelmotalib et al., 2015). Their development of correlations for effective thermal conductivity of both the gas and dispersed phase has been shown as a function of the thermal properties of gas and solid material and volume fraction of solid in bulk. Further, those have been utilized for estimating the thermal conductivity of dense phase or dispersed phase in fluidized beds (Biyikli, Tuzla, & Chen, 1989; Kuipers, Prins, & Swaaij, 1992). Therefore, it is quite indispensable to have such correlations to estimate the overall bulk thermal conductivity for both the phases (Abdelmotalib et al., 2015). Syamlal and Gidaspow (Syamlal & Gidaspow, 1985) has used the correlations of Zehner and Schlunder (Zehner & Schlünder, 1970) model in a hot gas fluidized bed model and observed that the effective solid thermal conductivity is a multiple of one factor (1-ε) of the effective thermal conductivity used in Zehner and Schlunder (Zehner & Schlünder, 1970) correlations. The modified equations of effective thermal conductivity for both gas and solid phases suggested by Syamlal and Gidaspow (Syamlal & Gidaspow, 1985) and later adopted by (Kuipers et al., 1992) are given as follows:

$$k_g^{eff} = \left(1 - \sqrt{\varepsilon_s} \Big/ \varepsilon_g\right) k_g \tag{53}$$

$$k_s^{eff} = \frac{\left\{\beta X + \left(1-\beta\right)\gamma\right\} k_g}{\sqrt{\varepsilon_s}} \tag{54}$$

where,

$$\gamma = \frac{2}{1-\frac{Y}{X}}\left[\frac{X-1}{\left(1-\frac{Y}{X}\right)^2}\frac{Y}{X}\ln\frac{X}{Y} - \frac{Y-1}{1-\frac{Y}{X}} - 0.5\left(Y+1\right)\right] \tag{55}$$

$$X = \frac{k_s}{k_g},\ Y = 1.25\left(1-\frac{\varepsilon_g}{\varepsilon_s}\right)^{10/9} \text{ and } \beta = 7.26\times10^{-3} \tag{56}$$

Some of the researchers used the classical penetration theory of local heat transfer coefficient and average heat transfer coefficient described by Kuiper et al. (Kuipers et al., 1992) to validate the results of their simulation (Abdelmotalib et al., 2015). The penetration theory was found to be predominant when the heat transfers from wall to bed is through conduction only. Except that, the theory is not valid. The equations of local heat transfer coefficient and the average heat transfer coefficient given by Kuiper et al. (Kuipers et al., 1992) are as follows

$$h_{l,pen} = \sqrt{\frac{k_m\left(\rho C_p\right)_m}{\pi t}} \tag{57}$$

$$h_{\text{avg},pen} = 2\sqrt{\frac{k_m\left(\rho C_p\right)_m}{\pi t}} \tag{58}$$

where,

$$k_m = \varepsilon_g k_g + \left(1-\varepsilon_g\right)k_s \tag{59}$$

$$\left(\rho C_p\right)_m = \varepsilon_g \rho_g C_{p,g} + \left(1-\varepsilon_g\right)\rho_s C_{p,s} \tag{60}$$

The classical penetration theory is a function of time. To compare the simulation results with penetration theory, the simulation results should consider the time-specific, i.e., the results should be obtained for a certain period and then compared with the classical penetration theory. From this theory, it is seen that local and average heat transfer coefficient decrease with increasing the time.

Several numerical studies have been done in the context of conventional fluidized bed regarding the hydrodynamic, heat transfer, combustion and gasification characteristics (Fattahi, Hosseini, Ahmadi, & Parvareh, 2019; Loha, Chatterjee, & Chattopadhyay, 2011; Loha, Chattopadhyay, & Chatterjee, 2012; Sirisomboon & Laowthong, 2019; Wang, Sun, Duan, & Xu, 2019; Yang, Wang, Luo, Fan, & Chew, 2019; Zhang, Ghamari, & Ratner, 2013). However, a few literatures have been found to investigate the same numerically. A brief description of those works of literature are articulated below:

Huilin et al. (Huilin, Yunhua, Zhiheng, Ding, & Jiying, 2006) investigated the gas-solid flow in conical tapper out circulating fluidized bed riser and results were compared with that of a columnar circulating fluidized bed riser. They used the finite difference method with staggered grids for discretizing the elements. Studies of the gas-solid flow have also been numerically carried out by Chalermsinsuwan et al. (Chalermsinsuwan et al., 2010) by using the EMMS model with Gidaspow drag model. Kalo et al. (Kalo, Pant, & Upadhyay, 2014) described the flow behaviour by incorporating the two-fluid model with KTGF. Askaripour and Dehkordi (Askaripour & Dehkordi, 2016) performed 2-D simulation by using a TFM coupled with Eulerian-Eulerian model closed by the kinetic theory of Granular flow. Sau and Biswal (Sau & Biswal, 2011) performed a numerical study in a conical fluidized bed riser to investigate the hydrodynamic behaviour such as pressure drop across the bed and bed expansion ratio. The numerical simulation was performed with the use of CFD commercial software Ansys Fluent 6.1.22 (Fluent 6.1.22, User's Guide, 20.4, Eulerian Model, 2005). The agreement between the experimental and numerical result of solid volume fraction for non-spherical particles is less than for spherical particles. Similar studies have also been carried out for bi-disperse TiO_2 particles by Bahramian and Olazar (Bahramian & Olazar, 2012). They have considered similar conditions as described by the previous authors. Besides that, results were obtained for three different wall boundaries for solid materials and compared for all these three boundary conditions. No-slip boundary condition was used as one of the boundary condition based on the Schaeffer approach, while Johnson and Jackson's model (Johnson & Jackson, 1987) was applied for partial slip and the third one was used as free-slip boundary condition which does not account for the effect of friction between particles. Three different drag models as Syamlal-O'Brien (Syamlal & O'Brien, 1988), Gidaspow (Gidaspow, 1994) and Arastoopour (Arastoopour, 2001) were used in this numerical study and results were compared among all these three different drag models. Both the numerical and experimental results present similar results. Among all three different drag models, Gidaspow drag model was found to be best fitted for experimental results. Sutar and Das (Sutar & Das, 2012) numerically studied segregation of particles in a conical fluidized bed and results were compared with the experimental results. The simulation was performed by FLUENT 6.2.16 (Fluent 6.1.22, User's Guide, 20.4, Eulerian Model, 2005). Abdelmotalib et al. (Abdelmotalib et al., 2015) investigated the hydrodynamics behaviour and wall to bed heat transfer characteristics along a conical fluidized bed with 30° cone angles. Two different drag models, namely the Syamlal-O'Brien and Gidaspow (Syamlal & Gidaspow, 1985) models were used, and the results of hydrodynamic and heat transfer were compared for both these two drag models. Similar results were found for these two drag models. The predicted heat transfer results were compared with the penetration theory given by Kuiper et al. (Kuipers et al., 1992) to validate these two models. Gunn (Gunn, 1978) equation was employed to investigate the wall to bed heat transfer coefficient. The results of hydrodynamic behaviour were found to be in good agreement between these two drag models. However, the local heat transfer calculated by penetration theory was found to be different from the two drag models, which is due to the fact that the effective thermal conductivity was not considered for calculating the heat transfer coefficient by the penetration theory. Kalo et al. (Kalo, Pant, Cassanello, & Upadhyay, 2018) reported the gas-solid interaction in conical fluidized bed and compared the results with a columnar bed by using the radioactive particle tracking technique. By using a similar technique Kalo et al. (Kalo, Kamalanathan, Pant, Cassanello, & Upadhyay, 2019) also investigated the regime transition in liquid-solid conical fluidized bed.

Abdelmotalib et al. (Abdelmotalib et al., 2016) experimentally and numerically investigated the hydrodynamic and heat transfer characteristics in a conical fluidized bed combustor of height 0.4 m with 30° cone angles. Their study of hydrodynamic and wall to bed heat transfer considers the impact of both gas velocity and granular temperature models. Gunn (Gunn, 1978) equation was used to evaluate the interphase heat transfer coefficient and the results obtained by simulations were found to have good agreement with the experimental results. They also used partial differential equation (PDE) for the numerical solution of granular temperature. The use of the PDE model in the investigation of hydrodynamics was found to be excellent agreement with experimental values. Whereas, no significant impact on the heat transfer characteristics was observed for PDE model. Jia et al. (Jia, Bi, Lim, Sokhansanj, & Tsutsumi, 2019) reported the bed to wall heat transfer of biomass particles in a taper fluidized bed by considering the pulsed gas flow. Their results indicated that with the use of smaller size of biomass particles, the heat transfer increases.

Abdelmotalib et al. (Abdelmotalib et al., 2016) numerically studied the hydrodynamics and heat transfer in a conical fluidized bed combustor characteristic by considering the impact of specularity coefficient. The Johnson and Jackson equation were used for the implementation of specularity coefficient corresponding to different boundary conditions on the wall. They used four different specularity coefficients and reported the effect of those specularity coefficients on hydrodynamic and heat transfer characteristics. The results of the bed height obtained from the simulation results by using the commercial software FLUENT ("Fluent Inc, Fluent 6.3 User's Guide," 2006) was found to increase with decreasing specularity coefficient. Abdelmotalib et al. (Abdelmotalib et al., 2016) also numerically investigated the hydrodynamic behaviour and heat transfer characteristics in a gas-solid conical fluidized bed combustor. The effect of collision elasticity, parameterized by a restitution coefficient of particle-to-particle and particle-to-wall on the hydrodynamic behaviour and heat transfer characteristic have been studied. Eulerian-Eulerian model was employed to simulate the fluid flow behaviour as well as the heat transfer, and for the properties of the solid phase, particles were described by the KTGF. However, the authors ignored the effect of virtual mass force and lift forces. They used the Johnson and Jackson model for the restitution coefficient. The pressure drops across the entire particle decreases due to wall coefficient restitution. This could be the consequence of a reduction in the fluctuation energy dissipation because of collisions between particles and the wall. Increase in coefficient of restitution results in reduction of energy dissipation during particle collisions which in turn improves the elasticity of these collisions. The increase in elasticity causes the particle to separate from each other and increases the void fraction in the bed. Hence, pressure drop reduces. Abdelmotalib and Im. (Abdelmotalib & Im, 2017) carried out 3-D simulation in a conical fluidized bed combustor to study the hydrodynamic behaviour, as well as heat transfer characteristics and results were compared with 2-D combustor as well as experiments. Both Gidaspow and Symalal O'Brien (Syamlal & Gidaspow, 1985) models were used for the drag coefficient, and no significant difference in effect was found on both heat transfer and hydrodynamic characteristics. They also used the Gunn equation to estimate the heat transfer characteristics. Furthermore, the findings also showed that the three-dimensional simulation results were in excellent agreement with experimental outcomes than two dimensional numerical results at higher superficial gas velocity.

Khodabandehlou et al. (Khodabandehlou et al., 2018) numerically reported that the cone angle, particle size and density affect the size distribution of bubble fraction and rise of velocity along with the height. They proposed an equation to determine the bubble diameter and showed that the bubble diameter depends on the total area of the bubble. Their simulations were found to have excellent concurrence with the experimental results of Hulme et al. (Hulme, Clavelle, Lee, & Kantzas, 2005).

Issues, Controversies, Problems

It is observed that conventional fluidized bed boilers have various industrial applications (Arjunwadkar, Basu, & Acharya, 2016; Wang, Yang, Shen, Mao, & Long, 2016; Yang et al., 2019; Ye et al., 2017; Ye, Wang, Yu, Luo, & Cen, 2018). Despite their wide applicability; channelling, bubbling and slugging

Table 1. CFD approach in the conical fluidized bed

Author	CFD Package	Computational Model	Drag Models	Solid Boundary Condition at Wall
Hulin et al. (Huilin et al., 2006)	CFD code K-FIX		Gidaspow drag model	Johnson and Jackson
Chalermsinsuwan et al. (Chalermsinsuwan et al., 2010)	Fluent 6.2.16	Eulerian-Eulerian multifluid model with Energy Minimization Multi-Scale approach	EMMS and Gidaspow drag model	Johnson and Jackson
Sau and Biswal (Sau & Biswal, 2011)	Fluent 6.1.22	Eulerian-Eulerian multifluid model with kinetic theory of granular flow	Gidaspow drag model	Zero shear stress
Bahramian and Olazar (Bahramian & Olazar, 2012)	CFD model	Eulerian-Eulerian multifluid model with kinetic theory of granular flow	Arastoopour, Gidaspow and Syamlal-O'Brien models	Schaeffer for no-slip, Johnson and Jackson for partial slip
Sutar and Das (Sutar & Das, 2012)	Fluent 6.2.16	Eulerian-Eulerian multifluid model with kinetic theory of granular flow		No-slip
Kalo et al. (Kalo et al., 2014)	Fluent 14	Eulerian-Eulerian multifluid model with kinetic theory of granular flow	Gidaspow drag model	No-slip
Abdelmotalib et al. (Abdelmotalib et al., 2015)	Fluent 6.3	Eulerian-Eulerian multifluid model with kinetic theory of granular flow	Gidaspow and Syamlal-O'Brien models	
Askaripour and Dehkordi (Askaripour & Dehkordi, 2016)	Fluent 6.3	Eulerian-Eulerian multifluid model with kinetic theory of granular flow	Gidaspow drag model	Johnson and Jackson for Partial slip
Abdelmotalib et al. (Abdelmotalib et al., 2016)	Fluent 6.3	Eulerian-Eulerian multifluid model with kinetic theory of granular flow	Gidaspow drag model	Johnson and Jackson
Abdelmotalib et al. (Abdelmotalib et al., 2016)	Fluent 6.3	Eulerian-Eulerian multifluid model with kinetic theory of granular flow	Gidaspow drag model	
Abdelmotalib et al. (Abdelmotalib et al., 2016)	Fluent 6.3	Eulerian-Eulerian multifluid model with kinetic theory of granular flow	Syamlal–O'Brien drag model	
Abdelmotalib et al. (Abdelmotalib & Im, 2017)	Fluent 6.3	Eulerian-Eulerian multifluid model with kinetic theory of granular flow	Syamlal–O'Brien drag model	

behaviour are certain inherent drawbacks associated with this type of boiler which occurs due to the high velocity beyond the minimum velocity of fluidization. Because of that, the efficiency and quality of fluidization in large and deep beds deteriorate resulting from the inefficient gas-solid mixing and inhomogeneity in fluidization (Padhi et al., 2014; Sau, Mohanty, & Biswal, 2010). Moreover, due to the frequent changes of particle size, de-fluidization of coarser particles, agglomeration, segregation in size, back mixing of particles, elutriation loss of finer particles etc. are some hindrance that always comrade with the particles in conventional fluidized bed since the axial velocity is uniform throughout the bed (Kim, Lee, Chung, & Kim, 2000; Shi et al., 1984). For prevention of de-fluidization of coarser particles, the superficial velocity of the fluid should be increased to some extent beyond the minimum fluidization velocity. However, due to the increase in velocity, there may be high elutriation loss of finer particles, since, the transported velocity may be much higher than the terminal velocity of finer particles, which causes a high mechanical degradation (Kim et al., 2000) again. Hence, investigators need to find ways to overcome those drawbacks for improving the quality of fluidization as well as the efficiency of the fluidized bed reactor. Various techniques like vibration and rotation of bed, turbulence promoter, use of baffles, improved distributor plate and operation in multistage are widely employed to get rid of the problems associated with the conventional fluidized bed (Padhi et al., 2014; Sau et al., 2010; Shi et al., 1984). However, the pressure drop increases with the use of the above techniques. Also, additional accessories are required to be operated with some difficulties, resulting in high maintenance and operating costs (Padhi et al., 2014). Therefore, many researchers have paid attention to a new design of fluidized bed riser to find out a suitable way to overcome the problems associated with the conventional fluidized bed. This gave rise to the design of conical fluidized riser. In the conical fluidized bed, an increase in cross-sectional area along the height makes the velocity gradient occur. Because of this, the velocity at the lower region is much higher, which causes uniform fluidization particles of large size and prevention of elutriation loss of finer particles (Kim et al., 2000; Shi et al., 1984). Due to the angled wall in the conical fluidized bed, the movement of particles is not only in a restricted region but also homogeneous with reduced back mixing. (Singh, Suryanarayana, & Roy, 1992). Apart from that, in the conventional fluidized bed, the major portion of the heat produced by an exothermic reaction is liberated through the distributor plate, making a high-temperature region and thus damaging the distributor plate and sintering the particles (Shi et al., 1984). Nevertheless, the problem of the high-temperature region at the distributor plate can be compensated by the use of a conical fluidized bed. In conical fluidized bed, the air velocity near the distributor plate is higher causing the particles to move upward which results in low particle concentration at the lower region, thus reducing the reaction rate and ultimately the generation of high-temperature zone near the distributor plate is prevented (Padhi et al., 2014; Shi et al., 1984).

Although various advantages prevail in using conical fluidized bed over conventional fluidized bed, there are various issues associated with the conical fluidized bed. One of the issues related to the conical fluidized bed is the development of mechanistic and empirical models. The mechanistic models are constrained by some assumptions (Khani, 2011; Murthy et al., 2009; Padhi et al., 2016; Peng & Fan, 1997; Rasteh et al., 2015; Sau et al., 2010). Also, experimental investigations are required to develop the empirical models, which leads to the operating and maintenance cost. Another issue incorporated with the conical fluidized bed is the research gap between laboratory-scale study and industrial unit. Scale-up studies are necessary to meet the gap between these two. However, scale-up studies for conical fluidized bed is accompanied by the cone angle. Researchers studied the effect of cone angle on hydrodynamic behaviour experimentally by using different sizes of particles. The different hydrodynamic characteristics such as pressure drop, solid volume fraction, bed fluctuation ratio, expansion ratio and hysteresis are studied by the inclusion of cone angles. However, the critical angle required for scale-up study has not been mentioned in the conical fluidized bed. Since the minimum fluidization velocity is dependent upon the fluidized bed cone angle, therefore, it is required to have a critical cone angle for optimizing the fluidized bed systems. But, it would be difficult to find out the optimum cone angle experimentally since experimental scale-up study requires a prolonged duration of time. The recent development of computing speed and numerical techniques has led to CFD being used as a widely implemented tool to provide a bridge between laboratory scale and industrial study. By implementing the CFD tool, the plant can be optimized by obtaining the desired cone angle.

In the numerical study of conical fluidized bed, the Eulerian-Eulerian TFM was utilized to investigate the hydrodynamic behaviour and heat transfer characteristics. However, in the Two-Fluid model, it is necessary to solve several equations with semi-empirical parameters simultaneously to get the properties of the solid phase (Loha, Chattopadhyay, & Chatterjee, 2014b). These solid-phase properties are a function of the restitution coefficient (Loha, Chattopadhyay, & Chatterjee, 2014a). Numerical study of effect of restitution coefficient on hydrodynamic behaviour, heat transfer and gasification characteristics in conventional fluidized bed was studied by many researchers (Armstrong, Gu, & Luo, 2010; Hosseini et al., 2010; Loha et al., 2014a). No experimental study was found to investigate the effect of restitution coefficient on these characteristics, and it is very difficult to measure the restitution coefficient experimentally. Again, more studies are required to validate the numerical results with the experimental investigations. Because, if the lower value of restitution coefficient is selected then the momentum loss, due to the collision among particles and between particles and wall, increases which has no practical application. Similarly, the numerical study by considering the specularity coefficient is also in need of vigorous experimental validation, and it is also difficult to measure the specularity coefficient by experimentally. Therefore, care should be taken into consideration while studying these operating parameters in the fluidized bed systems (Loha, Chattopadhyay, & Chatterjee, 2013). Another issue related to the using of Eulerian-Lagrangian model is the higher computational cost. Although the time-dependent solid particles and variables are easily handled in a natural way, but due to the tracking of numerous particles, the cost becomes high.

Present Trends and Challenges on the Simulation of Conical Bed

The hydrodynamics of conical fluidized bed has been given much attention since the last few decades. The investigation of hydrodynamic behaviour of various conical fluidized bed riser had been studied numerically, hypothetically and experimentally across the globe by various groups of researchers in academia, industry, explore research centres and associations.

The present trend of investigation in conical fluidized bed is the use of commercial ANSYS FLUENT software and CFD with K-FIX code. Most of the studies have been found to study the investigation by using ANSYS FLUENT software. Only, Hulin et al. (Huilin et al., 2006) have considered CFD with K-FIX code. However, other commercial software's like CFX, ASPEN, MFIX and CFD with code cannot be excluded from fluidized bed simulation. The fluent software uses the finite volume method for discretization the scheme. The Eulerian-Eulerian two-fluid model incorporated with KTGF is frequently implemented to investigate the hydrodynamic behaviour and heat transfer characteristics in their numerical simulation. As per the trend, the laminar viscosity model is used for bubbling mode since it operates under moderate velocity. However, for turbulence, RNG approach k-ε model is used for the modelling.

Now a day, a new form of commercial software BARACUDA is used in dense fluidized bed simulation. The BARACUDA software is based on the concept of the Eulerian-Lagrangian approach. This software utilizes the finite volume method to make the equations discrete. The BARACUDA software can be used for both dense to dilute phase. The main advantage of this software is the validation against the experimental results of large-scale experiments and industrial units (Loha et al., 2014b). Therefore, the conical fluidized bed system can be simulated by using this BARACUDA software.

The present study on the conical fluidized bed is hydrodynamic behaviour and heat transfer characteristics with bed material and air. Simulating of combustion and gasification characteristics with chemical and physical reactions is a challenging task. The real plant (combustor and gasifier) is a dense bed, and it comprises of different sizes of particles. In real gasification process, the high gas yields fuel particles remain unburnt rendering the particles to burn again. However, it is difficult to simulate the industrial gasification processes by the CFD model, and it is challenging. The scale-up studies of conical fluidized bed by introducing the cone angle is carried out experimentally. Numerical simulation of scale-up studies can be carried out to meet the energy demands of the industrial fluidized bed.

The current trend is the assessment of heat transfer and thermodynamic study through the gasification process in a fluidized bed by integrating the energy and exergy analysis. Chemical equilibrium model is used to study the thermodynamic analysis of the gasification process (Loha, Chattopadhyay, & Chatterjee, 2011). CFD can be the most promising tool to optimize the thermodynamic processes involved in the fluidized bed.

In the present scenario, the Gunn equation is used to calculate the local heat transfer coefficient in a conical fluidized bed. The local heat transfer coefficient depends on particle Nusselt number, which is further dependent upon particle Reynold number, Prandtl number and local void fraction at which local heat transfer coefficient is determined. However, in practice, the average heat transfer coefficient is very essential to understand the bulk mean temperature of the bed rather than the local heat transfer coefficient. Hence, it is quite necessary to implement the CFD with one's code to get better heat transfer results.

Future of CFD In Conical Fluidized Bed

Although hydrodynamic behaviour and heat transfer characteristics in the conical fluidized bed is studied by ANSYS Fluent, which provides the benefit of an understanding of physics involved with the fluidized bed, the development of information technology and growth of CFD will also help us in the development of combustion and gasification models. The future application of CFD in conical fluidized bed needs to be effective and qualitative other than quantitative.

The continuous advancement of information technology and the development of CFD software in forthcoming years will turn the complex design and optimization with chemical reactions possible. All these advancements will lead to the transformation of CFD into a discipline and an incredible simulation tool of design and scale-up in the context of the conical fluidized bed. As a result, there will be progressively far-reaching and rapid acceptance of the utilization of the CFD in a conical fluidized bed in the near future.

CONCLUSION

This chapter describes the hydrodynamic behaviour and heat transfer characteristics in conical fluidized bed numerically. The problems associated with numerical simulation has been incorporated. The governing equations, constitutive equations, boundary conditions and other different models are also described. Numerical simulation acts as a significant role in predicting the hydrodynamic behaviour and heat transfer characteristics, which will be helpful for investigations of combustion and gasification characteristics.

Majority of studies used the Two-Fluid Eulerian-Eulerian model for their numerical simulation. However, for freeboard condition, the DPM within the Eulerian-Lagrangian model is still required to be explored. Apart from that, the Two-Fluid Eulerian-Eulerian model is also needed to be explored in the conical fluidized bed for combustion and gasification processes.

Various mechanistic and empirical models of hydrodynamic behaviour are developed for conical fluidized bed. However, none of them satisfy the fluidized bed's industrial aspects. There are various critical issues incorporated with the conical fluidized bed regarding the development of CFD models since operating conditions between the commercial fluidized bed, and laboratory-scale fluidized bed are different. After a comprehensive review of literature, the authors make remarks about the improvement of fluidized bed system that in order to optimize the performance of conical fluidized bed, interdependence of CFD model on hydrodynamic behaviour and heat transfer characteristics are required to be understood since it affects the design of a fluidized bed. Finally, the issues, problems and discrepancies are thoroughly discussed and the probable techniques to overcome those problems are also recommended. Based on the review, the possible future studies to be done are also discussed below:

- 3-D simulations can be taken up for better agreement with the experimental results. CFD with own codes by using finite difference method can be implemented to compare the results with experimental results.
- Effect of scale-up on heat transfer characteristics of conical fluidized bed can be studied numerically by considering the desired optimum cone angle for better combustion and gasification models.
- Based on the simulation, models can be developed to realize the gap between laboratory-scale of study and industrial units.

Nomenclature

B_s: Slip coefficient

C_D: Drag coefficient

$C_{1\varepsilon}$: Model constant

$C_{2\varepsilon}$: Model constant

D_{ij} : Rate of strain tensor for solid phase

$\frac{D\vec{v}_g}{Dt}$: Substantial time derivative for gas velocity

$\frac{D\vec{v}_s}{Dt}$: Substantial time derivative for solids velocity

e_w : Particle-to-wall restitution coefficient
e_{ss} : Particle–particle restitution coefficient
$\overrightarrow{F_{dg}}$: Drag forces for gas phase
$\overrightarrow{F_{d,s}}$: Drag forces for solid phase
$\overrightarrow{F_{l,g}}$: Lift force for gas phase
$\overrightarrow{F_{l,s}}$: Lift force for solid phase
$\overrightarrow{F_{vm,g}}$: Virtual masses for gas phase
$\overrightarrow{F_{vm,s}}$: Virtual masses for solid phase
g : Acceleration due to gravity
$G_{k,m}$: Generation of turbulence kinetic energy due to velocity gradient
$g_{o,ss}$: Solid radial distribution function
$h_{avg,pen}$: Average heat transfer coefficient according to penetration theory
H_g : Enthalpy of gas
h_s : Static bed height
H_s : Enthalpy of solids
h_{sg} : Interphase heat transfer coefficient
ΔH_{rxn} : Rate of heat reaction
I_{2D} : Second invariant of the deviatory stress tensor
k_a : Ratio of the pressure drop in a tapered fluidized bed to that in a columnar bed
k_c : Function of some operating parameters = 1
k_g : Thermal conductivity of gas
k_g^{eff} : Effective thermal conductivity of gas
k_m : Mean thermal conductivity
k_s : Thermal conductivity of solids
k_s^{eff} : Effective thermal conductivity of solids
n : Constant
Nu_s : Solid phase Nusselt number
Pr: Prandtl number
P_{fr} : Frictional pressure
P_g : Gas phase pressure
∇P_g : Gas phase momentum equation
P_s : Solid phase pressure
ΔP : Pressure drop
ΔP_{mff} : Pressure drop at full fluidization condition
∇P_s : Solids phase momentum equation

Q_{sg} : Heat transfer between gas and solid phases
Re_s : Solid Reynold number
$\overline{S}_s$: Solid phase source term
T_b : Bed temperature
T_g : Temperature of gas
T_s : Temperature of solids
T_w : Wall temperature
U : Superficial velocity of fluid
v : Velocity vector
$\vec{v}_g$: Local velocity of the gas phase
$\vec{v}_m$: Mixture velocity
$\vec{v}_s$: Local velocity of the solid phase
$v_s'^2$: Mean square velocity of particles
x : Distance of the point from the wall where the local heat transfer coefficient is calculated

- *Greek symbol*

$\left(\rho C_p\right)_m$: Mean heat capacity
ρ_s : Density of solid
ρ_g : Density of gas
ε_s : Volume fraction of the solid phase
$\varepsilon_{s,cr}$: Critical solid volume fraction
$\varepsilon_{s,\max}$: Maximum solid volume fraction
ϕ : Angle of internal friction
ϕ_s : Sphericity of solid particle
ϕ_{sp} : Specularity coefficient
α_k : Inverse turbulent Prandtl number for k
α_ε : Inverse turbulent Prandtl number for ε
β_{gs} : Drag coefficient between the gas and solid phase
μ_g : Viscosity of gas
μ_s : Solid viscosity
$\mu_{s.col}$: Collisional viscosity
$\mu_{s.fr}$: Frictional viscosity
$\mu_{s.kin}$: Kinetic viscosity
μ_t : Turbulent viscosity
$\overline{\overline{\tau}}_g$: Gas phase stress tensor

$\overline{\overline{\tau}}_s$: Solid phase stress tensor

k_{Θ_s} : Diffusion coefficient

$k_{\Theta_s} \nabla \Theta_s$: The transport of energy due to diffusion

Θ_s : Granular temperature

γ_{Θ_s} : Dissipation of energy due to collision

γ_w : Collisional dissipation at the wall

$\left(-p_s \overline{I} + \overline{\tau}_s\right) : \nabla \overrightarrow{\nu_s}$: Generation of energy by the solid stress tensor

REFERENCES

Abdelmotalib, H. M., & Im, I.-T. (2017). Three dimensional modeling of heat transfer and bed flow in a conical fluidized bed reactor. *International Journal of Heat and Mass Transfer*, *106*, 1335–1344. doi:10.1016/j.ijheatmasstransfer.2016.10.117

Abdelmotalib, H. M., Ko, D. G., & Im, I.-T. (2016). A study on wall-to-bed heat transfer in a conical fluidized bed combustor. *Applied Thermal Engineering*, *99*, 928–937. doi:10.1016/j.applthermaleng.2016.01.054

Abdelmotalib, H. M., Youssef, M. A., Hassan, A. A., Youn, S. B., & Im, I.-T. (2015). Numerical study on the wall to bed heat transfer in a conical fluidized bed combustor. *International Journal of Precision Engineering and Manufacturing*, *16*(7), 1551–1559. doi:10.100712541-015-0205-z

Abdelmotalib, H. M., Youssef, M. A. M., Hassan, A. A., Youn, S. B., & Im, I.-T. (2016). Influence of the specularity coefficient on hydrodynamics and heat transfer in a conical fluidized bed combustor. *International Communications in Heat and Mass Transfer*, *75*, 169–176. doi:10.1016/j.icheatmasstransfer.2016.04.018

Abdelmotalib, H. M., Youssef, M. A. M., Hassan, A. A., Youn, S. B., & Im, I.-T. (2016). Numerical study on heat transfer in a conical fluidized bed combustor considering particle elasticity. *International Journal of Heat and Mass Transfer*, *92*, 236–243. doi:10.1016/j.ijheatmasstransfer.2015.08.069

Ansys Inc. (2011). ANSYS Fluent Theory Guide.

Arastoopour, H. (2001). Numerical simulation and experimental analysis of gas/solid flow systems: 1999 Fluor-Daniel Plenary lecture. *Powder Technology*, *119*(2-3), 59–67. doi:10.1016/S0032-5910(00)00417-4

Arjunwadkar, A., Basu, P., & Acharya, B. (2016). A review of some operation and maintenance issues of CFBC boilers. *Applied Thermal Engineering*, *102*, 672–694. doi:10.1016/j.applthermaleng.2016.04.008

Armstrong, L. M., Gu, S., & Luo, K. H. (2010). Study of wall-to-bed heat transfer in a bubbling fluidised bed using the kinetic theory of granular flow. *International Journal of Heat and Mass Transfer*, *53*(21-22), 4949–4959. doi:10.1016/j.ijheatmasstransfer.2010.05.047

Askaripour, H., & Molaei Dehkordi, A. (2016). Effects of initial static bed height on fractional conversion and bed pressure drop in tapered-in and tapered-out fluidized bed reactors. *International Journal of Multiphase Flow*, *79*, 50–61. doi:10.1016/j.ijmultiphaseflow.2015.08.006

Bahramian, A., & Olazar, M. (2012). Fluidization of micronic particles in a conical fluidized bed: Experimental and numerical study of static bed height effect. *AIChE Journal. American Institute of Chemical Engineers*, *58*(3), 730–744. doi:10.1002/aic.12621

Bakshi, A., Altantzis, C., Bates, R. B., & Ghoniem, A. F. (2015). Eulerian–Eulerian simulation of dense solid–gas cylindrical fluidized beds: Impact of wall boundary condition and drag model on fluidization. *Powder Technology*, *277*, 47–62. doi:10.1016/j.powtec.2015.02.056

Bi, H. T., Ellis, N., Abba, I. A., & Grace, J. R. (2000). A state-of-the-art review of gas-solid turbulent fluidization. *Chemical Engineering Science*, *55*(21), 4789–4825. doi:10.1016/S0009-2509(00)00107-X

Bilbao, J., Olazar, M., Arandes, J. M., & Romero, A. (1989). Optimization of the operation in a reactor with continuous catalyst circulation in the gaseous benzyl alcohol polymerization. *Chemical Engineering Communications*, *75*(1), 121–134. doi:10.1080/00986448908940672

Biyikli, S., Tuzla, K., & Chen, J. C. (1989). A phenomenological model for heat transfer in freeboard of fluidized beds. *Canadian Journal of Chemical Engineering*, *67*(2), 230–236. doi:10.1002/cjce.5450670209

Busciglio, A., Vella, G., Micale, G., & Rizzuti, L. (2009). Analysis of the bubbling behaviour of 2D gas solid fluidized beds. *Chemical Engineering Journal*, *148*(1), 145–163. doi:10.1016/j.cej.2008.11.010

Chalermsinsuwan, B., Gidaspow, D., & Piumsomboon, P. (2011). Two- and three-dimensional CFD modeling of Geldart A particles in a thin bubbling fluidized bed: Comparison of turbulence and dispersion coefficients. *Chemical Engineering Journal*. doi:10.1016/j.cej.2011.04.007

Chalermsinsuwan, B., Kuchonthara, P., & Piumsomboon, P. (2010). CFD modeling of tapered circulating fluidized bed reactor risers: Hydrodynamic descriptions and chemical reaction responses. *Chemical Engineering and Processing: Process Intensification, 49*(11), 1144-1160. doi:10.1016/j.cep.2010.08.016

Chalermsinsuwan, B., Piumsomboon, P., & Gidaspow, D. (2009). Kinetic theory based computation of PSRI riser: Part II-Computation of mass transfer coefficient with chemical reaction. *Chemical Engineering Science, 64(6), 1212-1222.* doi:10.1016/j.ces.2008.11.006

Cheung, S. C. P., Yeoh, G. H., & Tu, J. Y. (2007). On the numerical study of isothermal vertical bubbly flow using two population balance approaches. *Chemical Engineering Science*, *62*(17), 4659–4674. doi:10.1016/j.ces.2007.05.030

Deiva Venkatesh, R., Chaouki, J., & Klvana, D. (1996). Fluidization of cryogels in a conical column. *Powder Technology*, *89*(3), 179–186. doi:10.1016/S0032-5910(96)03181-6

Depypere, F., Pieters, J. G., & Dewettinck, K. (2005). Expanded bed height determination in a tapered fluidised bed reactor. *Journal of Food Engineering*, *67*(3), 353–359. doi:10.1016/j.jfoodeng.2004.04.039

Dora, D. T. K., Panda, S. R., Mohanty, Y. K., & Roy, G. K. (2013). Hydrodynamics of gas–solid fluidization of a homogeneous ternary mixture in a conical bed: Prediction of bed expansion and bed fluctuation ratios. *Particuology*, *11*(6), 681–688. doi:10.1016/j.partic.2013.02.009

Fattahi, M., Hosseini, S. H., Ahmadi, G., & Parvareh, A. (2019). Numerical simulation of heat transfer coefficient around different immersed bodies in a fluidized bed containing Geldart B particles. *International Journal of Heat and Mass Transfer*, *141*, 353–366. doi:10.1016/j.ijheatmasstransfer.2019.06.058

Fluent 6.1.22, User's Guide, 20.4, Eulerian Model. (2005). Fluent Inc.

Fluent Inc. (2006). Fluent 6.3 User's Guide.

Gera, D., & Gautam, M. (1995). Bubble rise velocity in two-dimensional fluidized beds. *Powder Technology*, *84*(3), 283–285. doi:10.1016/0032-5910(95)02993-C

Gidaspow, D. (1994). *Multiphase Flow and Fluidization: Continuum and Kinetic Theory Descriptions*. New York: Academic Press.

Gunn, D. J. (1978). Transfer of heat or mass to particles in fixed and fluidised beds. *International Journal of Heat and Mass Transfer*, *21*(4), 467–476. doi:10.1016/0017-9310(78)90080-7

Hamzehei, M., & Rahimzadeh, H. (2009). Experimental and Numerical Study of Hydrodynamics with Heat Transfer in a Gas-Solid Fluidized-Bed Reactor at Different Particle Sizes. *Industrial & Engineering Chemistry Research*, *48*(6), 3177–3186. doi:10.1021/ie801413q

Hamzehei, M., Rahimzadeh, H., & Ahmadi, G. (2010). Computational and Experimental Study of Heat Transfer and Hydrodynamics in a 2D Gas-Solid Fluidized Bed Reactor. *Industrial & Engineering Chemistry Research*, *49*(11), 5110–5121. doi:10.1021/ie900510a

Hosseini, S. H., Ahmadi, G., Saeedi Razavi, B., & Zhong, W. (2010). Computational fluid dynamic simulation of hydrodynamic behavior in a two-dimensional conical spouted bed. *Energy & Fuels*, *24*(11), 6086–6098. doi:10.1021/ef100612r

Hui, K., Haff, P. K., Ungar, J. E., & Jackson, R. (1984). Boundary conditions for high-shear grain flows. *Journal of Fluid Mechanics*, *145*(-1), 223. doi:10.1017/S0022112084002883

Huilin, L., Yunhua, Z., Zhiheng, S., Ding, J., & Jiying, J. (2006). Numerical simulations of gas-solid flow in tapered risers. *Powder Technology*, *169*(2), 89–98. doi:10.1016/j.powtec.2006.07.023

Hulme, I., Clavelle, E., van der Lee, L., & Kantzas, A. (2005). CFD Modeling and Validation of Bubble Properties for a Bubbling Fluidized Bed. *Industrial & Engineering Chemistry Research*, *44*(12), 4254–4266. doi:10.1021/ie049837j

Jia, D., Bi, X., Lim, C. J., Sokhansanj, S., & Tsutsumi, A. (2019). Heat transfer in a tapered fluidized bed of biomass particles with pulsed gas flow. *Particuology*, *42*, 2–14. doi:10.1016/j.partic.2018.01.007

Jing, S., Hu, Q., Wang, J., & Jin, Y. (2000). Fluidization of coarse particles in gas – solid conical beds. *Chemical Engineering and Processing*, *39*(4), 379–387. doi:10.1016/S0255-2701(99)00103-8

Jiradilok, V., Gidaspow, D., & Breault, R. W. (2007). Computation of gas and solid dispersion coefficients in turbulent risers and bubbling beds. *Chemical Engineering Science*, *62*(13), 3397–3409. doi:10.1016/j.ces.2007.01.084

Johansson, K., van Wachem, B. G. M., & Almstedt, A. E. (2006). Experimental validation of CFD models for fluidized beds: Influence of particle stress models, gas phase compressibility and air inflow models. *Chemical Engineering Science*, *61*(5), 1705–1717. doi:10.1016/j.ces.2005.09.014

Johnson, P. C., & Jackson, R. (1987). Frictional–collisional constitutive relations for granular materials, with application to plane shearing. *Journal of Fluid Mechanics*, *176*(-1), 67. doi:10.1017/S0022112087000570

Kalo, L. Kamalanathan, P., Pant, H. J., Cassanello, M. C., & Upadhyay, R. K. (2019). Mixing and regime transition analysis of liquid-solid conical fluidized bed through RPT technique. *Chemical Engineering Science.*

Kalo, L. Pant, H. J., Cassanello, M. C., & Upadhyay, R. K. (2018). Time series analysis of a binary gas-solid conical fluidized bed using radioactive particle tracking (RPT) technique data. *Chemical Engineering Journal.*

Kalo, L., Pant, H. J., & Upadhyay, R. K. (2014). Numerical and Experimental Investigation of Conical and Cylindrical Gas-Solids Fluidized Bed with Head to Head Comparison. Proceedings of ISCRE 23 & APCRE 7, Bangkok, Thailand. Academic Press.

Katiyar, R. (2017). Conical fluidized beds - Case Study for a gas-solid system involving coarse particles.

Khan, M. J. H., Hussain, M. A., Mansourpour, Z., Mostoufi, N., Ghasem, N. M., & Abdullah, E. C. (2014). CFD simulation of fluidized bed reactors for polyolefin production – A review. *Journal of Industrial and Engineering Chemistry*, *20*(6), 3919–3946. doi:10.1016/j.jiec.2014.01.044

Khani, M. H. (2011). Models for prediction of hydrodynamic characteristics of gas-solid tapered and mini-tapered fluidized beds. *Powder Technology*, *205*(1-3), 224–230. doi:10.1016/j.powtec.2010.09.018

Khodabandehlou, R., Askaripour, H., & Molaei Dehkordi, A. (2018). Numerical investigation of gas bubble behavior in tapered fluidized beds. *Particuology*, *38*, 152–164. doi:10.1016/j.partic.2017.05.013

Kim, H. G., Lee, I. O., Chung, U. C., & Kim, Y. H. (2000). Fluidization Characteristics of Iron Ore Fines of Wide Size Distribution in a Cold Tapered Gas-Solid Fluidized Bed. *ISIJ International*, *40*(1), 16–22. doi:10.2355/isijinternational.40.16

Kouprianov, V. I., & Permchart, W. (2003). Emissions from a conical FBC fired with a biomass fuel. *Applied Energy*, *74*(3-4), 383–392. doi:10.1016/S0306-2619(02)00194-0

Kuipers, J. A. M., Prins, W., & Van Swaaij, W. P. M. (1992). Numerical calculation of wall-to-bed heat-transfer coefficients in gas-fluidized beds. *AIChE Journal. American Institute of Chemical Engineers*, *38*(7), 1079–1091. doi:10.1002/aic.690380711

Kuipers, J. A. M., van Duin, K. J., van Beckum, F. P. H., & van Swaaij, W. P. M. (1993). Computer simulation of the hydrodynamics of a two-dimensional gas-fluidized bed. *Computers & Chemical Engineering*, *17*(8), 839–858. doi:10.1016/0098-1354(93)80067-W

Kunii, D., & Levenspiel, O. (1991). *Fluidization Engineering*. Butterworth-Heinemann.

Kuprianov, V. I., & Arromdee, P. (2013). Combustion of peanut and tamarind shells in a conical fluidized-bed combustor: A comparative study. *Bioresource Technology*, *140*, 199–210. doi:10.1016/j.biortech.2013.04.086 PMID:23693147

Laborde-Boutet, C., Larachi, F., Dromard, N., Delsart, O., & Schweich, D. (2009). CFD simulation of bubble column flows: Investigations on turbulence models in RANS approach. *Chemical Engineering Science*, *64*(21), 4399–4413. doi:10.1016/j.ces.2009.07.009

Levey, R. P., Garga, A. D. L., Jacobs, S. C., Heidt, H. M., & Trent, P. E. (1960). Fluid Bed Conversion of UO3 to UF4. *Chemical Engineering Progress*, *56*, 43–48.

Li, P., Lan, X., Xu, C., Wang, G., Lu, C., & Gao, J. (2009). Drag models for simulating gas–solid flow in the turbulent fluidization of FCC particles. *Particuology*, *7*(4), 269–277. doi:10.1016/j.partic.2009.03.010

Li, Y., Zhang, J., & Fan, L.-S. (1999). Numerical simulation of gas–liquid–solid fluidization systems using a combined CFD-VOF-DPM method: Bubble wake behavior. *Chemical Engineering Science*, *54*(21), 5101–5107. doi:10.1016/S0009-2509(99)00263-8

Loha, C., Chatterjee, P. K., & Chattopadhyay, H. (2011). Performance of fluidized bed steam gasification of biomass - Modeling and experiment. *Energy Conversion and Management*, *52*(3), 1583–1588. doi:10.1016/j.enconman.2010.11.003

Loha, C., Chattopadhyay, H., & Chatterjee, P. K. (2011). Thermodynamic analysis of hydrogen rich synthetic gas generation from fluidized bed gasification of rice husk. *Energy*, *36*(7), 4063–4071. doi:10.1016/j.energy.2011.04.042

Loha, C., Chattopadhyay, H., & Chatterjee, P. K. (2012). Assessment of drag models in simulating bubbling fluidized bed hydrodynamics. *Chemical Engineering Science*, *75*, 400–407. doi:10.1016/j.ces.2012.03.044

Loha, C., Chattopadhyay, H., & Chatterjee, P. K. (2013). Euler-Euler CFD modeling of fluidized bed: Influence of specularity coefficient on hydrodynamic behavior. *Particuology*, *11*(6), 673–680. doi:10.1016/j.partic.2012.08.007

Loha, C., Chattopadhyay, H., & Chatterjee, P. K. (2014a). Effect of coefficient of restitution in Euler-Euler CFD simulation of fluidized-bed hydrodynamics. *Particuology*, *15*, 170–177. doi:10.1016/j.partic.2013.07.001

Loha, C., Chattopadhyay, H., & Chatterjee, P. K. (2014b). Three dimensional kinetic modeling of fluidized bed biomass gasification. *Chemical Engineering Science*, *109*, 53–64. doi:10.1016/j.ces.2014.01.017

Lun, C. K. K., Savage, S. B., Jeffrey, D. J., & Chepurniy, N. (1984). Kinetic theories for granular flow: Inelastic particles in Couette flow and slightly inelastic particles in a general flowfield. *Journal of Fluid Mechanics*, *140*, 223–256. doi:10.1017/S0022112084000586

Ma, D., & Ahmadi, G. (1986). An equation of state for dense rigid sphere gases. *The Journal of Chemical Physics*, *84*(6), 3449–3450. doi:10.1063/1.450229

McKay, G., & Sze, M. F. F. (2008). The removal of organic pollutants from industrial effluents via tapered bed adsorption columns. *International Journal of Environmental Technology and Management*, *9*(1), 20. doi:10.1504/IJETM.2008.017857

Murthy, J. S. N., Reddy, V. S., Tasleem, S., Kumar, D. V., Krishna, C. S., & Sankarshana, T. (2009). Hydrodynamics of gas-solid fluidization in tapered beds. *Canadian Journal of Chemical Engineering*, *87*(3), 359–374. doi:10.1002/cjce.20180

Myöhänen, K., & Hyppänen, T. (2011). A Three-Dimensional Model Frame for Modelling Combustion and Gasification in Circulating Fluidized Bed Furnaces. *International Journal of Chemical Reactor Engineering*, *9*(1). doi:10.1515/1542-6580.2571

Ninduangdee, P., & Kuprianov, V. I. (2013). Study on burning oil palm kernel shell in a conical fluidized-bed combustor using alumina as the bed material. *Journal of the Taiwan Institute of Chemical Engineers*. doi:10.1016/j.jtice.2013.06.011

Orszag, S. A., Yakhot, V., Flannery, W. S., Boysan, F., Choudhury, D., Maruzewski, J., & Patel, B. (1993). Renormalization group modeling and turbulence simulations. *Proceedings of the International Conference on Near-Wall Turbulent Flows*. Academic Press.

Padhi, R. K., Dora, D. T. K., Mohanty, Y. K., Roy, G. K., & Sarangi, B. (2016). Hydrodynamics of three-phase fluidization of homogeneous ternary mixture in a conical conduit— Experimental and statistical analysis. *Chinese Journal of Chemical Engineering*, *24*(10), 1335–1343. doi:10.1016/j.cjche.2016.05.032

Padhi, R. K., Mohanty, Y. K., Roy, G. K., & Sarangi, B. (2014). Hydrodynamics studies of gas-solid fluidization in non cylindrical conduits for spherical and non-spherical particles- A review. International Journal of Scientific and Research Publications, 4.

Pain, C. C., Mansoorzadeh, S., & de Oliveira, C. R. E. (2001). A study of bubbling and slugging fluidised beds using the two-fluid granular temperature model. *International Journal of Multiphase Flow*, *27*(3), 527–551. doi:10.1016/S0301-9322(00)00035-5

Panneerselvam, R., Savithri, S., & Surender, G. D. (2009). CFD simulation of hydrodynamics of gas–liquid–solid fluidised bed reactor. *Chemical Engineering Science*, *64*(6), 1119–1135. doi:10.1016/j.ces.2008.10.052

Patankar, S. (1980). *Numerical heat transfer and fluid flow*. CRC press.

Patil, R. S., Pandey, M., & Mahanta, P. (2011). Parametric studies and effect of scale-up on wall-to-bed heat transfer characteristics of circulating fluidized bed risers. *Experimental Thermal and Fluid Science*, *35*(3), 485–494. doi:10.1016/j.expthermflusci.2010.11.012

Peirano, E., Delloume, V., & Leckner, B. (2001). Two- or three-dimensional simulations of turbulent gas–solid flows applied to fluidization. *Chemical Engineering Science*, *56*(16), 4787–4799. doi:10.1016/S0009-2509(01)00141-5

Peng, Y., & Fan, L. T. (1995). Hysteresis in liquid-solid tapered fluidized beds. *Chemical Engineering Science*, *50*(16), 2669–2671. doi:10.1016/0009-2509(95)00105-E

Peng, Y., & Fan, L. T. (1997). Hydrodynamic characteristics of fluidization in liquid-solid tapered beds. *Chemical Engineering Science*, *52*(14), 2277–2290. doi:10.1016/S0009-2509(97)00061-4

Pepiot, P., Dibble, C. J., & Foust, T. D. (2010). Computational fluid dynamics modeling of biomass gasification and pyrolysis. In *Computational modeling in lignocellulosic biofuel production* (pp. 273–298). American Chemical Society.

Philippsen, C. G., Vilela, A. C. F., & Zen, L. D. (2015). Fluidized bed modeling applied to the analysis of processes: Review and state of the art. *Journal of Materials Research and Technology*, *4*(2), 208–216. doi:10.1016/j.jmrt.2014.10.018

Rasteh, M., Farhadi, F., & Bahramian, A. (2015). Hydrodynamic characteristics of gas-solid tapered fluidized beds: Experimental studies and empirical models. *Powder Technology*, *283*, 355–367. doi:10.1016/j.powtec.2015.06.002

Ridgeway, K. (1965). The Tapering Fluidized Bed—A New Processing Tool. *Chemical & Process Engineering*, *6*, 317–321.

Sau, D. C., & Biswal, K. C. (2011). Computational fluid dynamics and experimental study of the hydrodynamics of a gas–solid tapered fluidized bed. *Applied Mathematical Modelling*, *35*(5), 2265–2278. doi:10.1016/j.apm.2010.11.037

Sau, D. C., Mohanty, S., & Biswal, K. C. (2010). Experimental studies and empirical models for the prediction of bed expansion in gas-solid tapered fluidized beds. *Chemical Engineering and Processing: Process Intensification*. doi:10.1016/j.cep.2010.02.010

Schaeffer, D. G. (1987). Instability in the evolution equations describing incompressible granular flow. *Journal of Differential Equations*, *66*(1), 19–50. doi:10.1016/0022-0396(87)90038-6

Scott, C. D., & Hancher, C. W. (1976). Use of a tapered fluidized bed as a Continuous bioreactor. *Biotechnology and Bioengineering*, *18*(10), 1393–1403. doi:10.1002/bit.260181006 PMID:963280

Shen, L., Johnsson, F., & Leckner, B. (2004). Digital image analysis of hydrodynamics two-dimensional bubbling fluidized beds. *Chemical Engineering Science*, *59*(13), 2607–2617. doi:10.1016/j.ces.2004.01.063

Shi, Y. F., Yu, Y. S., & Fan, L. T. (1984). Incipient Fluidization Condition for a Tapered Fluidized Bed. *Industrial & Engineering Chemistry Fundamentals*, *23*(4), 484–489. doi:10.1021/i100016a018

Shu, Z., Wang, J., Zhou, Q., Fan, C., & Li, S. (2015). Evaluation of multifluid model for heat transfer behavior of binary gas-solid flow in a downer reactor. *Powder Technology*, *281*, 34–48. doi:10.1016/j.powtec.2015.04.055

Singh, R. I., Brink, A., & Hupa, M. (2013). CFD modeling to study fluidized bed combustion and gasification. *Applied Thermal Engineering*, *52*(2), 585–614. doi:10.1016/j.applthermaleng.2012.12.017

Singh, R. K., Suryanarayana, A., & Roy, G. K. (1992). Prediction of minimum velocity and minimum bed pressure drop for gas-solid fluidization in conical conduits. *Canadian Journal of Chemical Engineering*, *70*(1), 185–189. doi:10.1002/cjce.5450700127

Sirisomboon, K., & Charernporn, P. (2017). Effects of air staging on emission characteristics in a conical fluidized-bed combustor firing with sunflower shells. *Journal of the Energy Institute*. doi:10.1016/j.joei.2015.12.001

Sirisomboon, K., & Laowthong, P. (2019). Experimental investigation and prediction of heat transfer in a swirling fluidized-bed combustor. *Applied Thermal Engineering*, *147*, 718–727. doi:10.1016/j.applthermaleng.2018.10.097

Sun, J., Zhou, Y., Ren, C., Wang, J., & Yang, Y. (2011). CFD simulation and experiments of dynamic parameters in gas–solid fluidized bed. *Chemical Engineering Science*, *66*(21), 4972–4982. doi:10.1016/j.ces.2011.06.035

Sutar, H., & Das, C. K. (2012). Mixing and Segregation Characteristics of Binary Granular Material in Tapered Fluidized Bed: A CFD Study. *Engineering*, *4*(4), 215–227. doi:10.4236/eng.2012.44029

Syamlal, M., & Gidaspow, D. (1985). Hydrodynamics of fluidization: Prediction of wall to bed heat transfer coefficients. *AIChE Journal. American Institute of Chemical Engineers*, *31*(1), 127–135. doi:10.1002/aic.690310115

Syamlal, M., & O'Brien, T. J. (1988). Simulation of granular layer inversion in liquid fluidized beds. *International Journal of Multiphase Flow*, *14*(4), 473–481. doi:10.1016/0301-9322(88)90023-7

van der Hoef, M. A., van Sint Annaland, M., Deen, N. G., & Kuipers, J. A. M. (2008). Numerical Simulation of Dense Gas-Solid Fluidized Beds: A Multiscale Modeling Strategy. *Annual Review of Fluid Mechanics*, *40*(1), 47–70. doi:10.1146/annurev.fluid.40.111406.102130

Wang, J., Ge, W., & Li, J. (2008). Eulerian simulation of heterogeneous gas-solid flows in CFB risers: EMMS-based sub-grid scale model with a revised cluster description. *Chemical Engineering Science*, *63*(6), 1553–1571. doi:10.1016/j.ces.2007.11.023

Wang, J., & Liu, Y. (2010). EMMS-based Eulerian simulation on the hydrodynamics of a bubbling fluidized bed with FCC particles. *Powder Technology*. doi:10.1016/j.powtec.2009.09.022

Wang, L. Sun, J., Duan, S., & Xu, L. (2019). Simulation of cylindrical fluidized bed between granular flow and annulus surface. *International Journal of Thermal Sciences*. doi:10.1016/j.ijthermalsci.2018.12.026

Wang, L. Yang, D., Shen, Z., Mao, K., & Long, J. (2016). Thermal-hydraulic calculation and analysis of a 600 MW supercritical circulating fluidized bed boiler with annular furnace. *Applied Thermal Engineering*. doi:10.1016/j.applthermaleng.2015.11.014

Wang, W., & Li, J. (2007). Simulation of gas–solid two-phase flow by a multi-scale CFD approach—Of the EMMS model to the sub-grid level. *Chemical Engineering Science*, *62*(1-2), 208–231. doi:10.1016/j.ces.2006.08.017

Wang, Z. (2006). *Experimental studies and CFD simulations of conical spouted bed hydrodynamics*. University of British Columbia.

Wu, Z. (1993). *Coal, Lignite, and Peat; Fluidized-bed combustion technology assessment, chemical Reactions; Fluidization*, Pollutants.

Yang, N., Wang, W., Ge, W., & Li, J. (2003). CFD simulation of concurrent-up gas-solid flow in circulating fluidized beds with structure-dependent drag coefficient. *Chemical Engineering Journal*, *96*(1-3), 71–80. doi:10.1016/j.cej.2003.08.006

Yang, N., Wang, W., Ge, W., Wang, L., & Li, J. (2004). Simulation of heterogeneous structure in a circulating fluidized-bed riser by combining the two-fluid model with the EMMS approach. *Industrial & Engineering Chemistry Research*, *43*(18), 5548–5561. doi:10.1021/ie049773c

Yang, S., Wang, S., Luo, K., Fan, J., & Chew, J. W. (2019). Numerical investigation of the back-mixing and non-uniform characteristics in the three-dimensional full-loop circulating fluidized bed combustor with six parallel cyclones. *Applied Thermal Engineering*. doi:10.1016/j.applthermaleng.2019.03.032

Ye, C., Wang, Q., Luo, Z., Xie, G., Jin, K., Siyil, M., & Cen, K. (2017). Characteristics of Coal Partial Gasification on a Circulating Fluidized Bed Reactor. *Energy & Fuels*, *31*(3), 2557–2564. doi:10.1021/acs.energyfuels.6b02889

Ye, C., Wang, Q., Yu, L., Luo, Z., & Cen, K. (2018). Characteristics of coal partial gasification experiments on a circulating fluidized bed reactor under CO2–O2 atmosphere. *Applied Thermal Engineering*, *130*, 814–821. doi:10.1016/j.applthermaleng.2017.11.071

Yu, Y., & Wen, C. (1966). Mechanics of fluidization. *Chemical Engineering Progress*. doi:10.1016/S0032-0633(98)00014-2

Zehner, P., & Schlünder, E. U. (1970). Wärmeleitfähigkeit von Schüttungen bei mäßigen Temperaturen. *Chemieingenieurtechnik (Weinheim)*, *42*(14), 933–941. doi:10.1002/cite.330421408

Zhang, X., Ghamari, M., & Ratner, A. (2013). Numerical modeling of co-firing a light density biomass, oat (Avena sativa) hulls, and chunk coal in fluidized bed boiler. *Biomass and Bioenergy*, *56*, 239–246. doi:10.1016/j.biombioe.2013.05.006

Chapter 18
Supply Chain Management via System Dynamics in Flexible Manufacturing System

Arzu Eren Şenaras
https://orcid.org/0000-0002-3862-4551
Bursa Uludag University, Turkey

Onur Mesut Şenaras
Bursa Uludag University, Turkey

ABSTRACT

System dynamics is a method that allows analysts to separate complex social, and behavioral systems into components, to visualize them by reconstructing them as a whole again, and to develop a simulation model. In this chapter, a system dynamics model is developed for a flexible manufacturing system. The case study is developed using Vensim package program. To manage a flexible manufacturing system it is important to determine production schedule rules and examine results. System dynamics is a tool to test these rules without any production loss.

INTRODUCTION

System dynamics approach was developed by Jay Forrester from MIT during the 1950's to analyze especially the complex behavior in administration with computer simulation in the social sciences. System dynamics is a form of the systems approach as a methodology to understand the dynamic behavior of complex systems. The basis of system dynamics is to understand how system structures cause system behavior and system events (Sezen, 2009, p. 298).

Jay Forrester initially constructed his first dynamic model during his meeting with the management of General Electric Corporation. Big fluctuations in production, inventory, labor force and profitability were compelling GE management. Despite the hard efforts of the management, these fluctuations were mostly associated with external factors. Especially, the fluctuations in the business were related to received

DOI: 10.4018/978-1-7998-1831-1.ch018

orders. Forrester interacted with the management to observe the system operations in other departments. In the first model he developed, he observed that simulations were necessary since the system could not be monitored analytically. He demonstrated that the corporation could experience serious fluctuations due to management policies even when the demand is considered constant with the weekly simulation he ran. Later on, he designed the computer simulation for the same problem. In his later studies, Forrester demonstrated how the feedback control theory could be adapted for complex administration and human systems. He published his initial findings in an article in Harvard Business Review. Later on, he developed this study to write his famous work "Industrial Dynamics" (Lane and Sterman, 2011; Ramage and Shipp, 2009, pp. 100-101).

SHORT LITERATURE REWIEV

Olhager and Selldin (2004) investigated supply chain management strategies and practices in a sample of 128 Swedish manufacturing firms. They specifically studied issues related to the supply chain design, integration, planning and control, and communication tools for managing supply chains. The main findings indicate the following. The extent to which suppliers and customers are involved in supply chain planning and control is expected to increase steadily over the next 2 years.

Kehoe and Boughton (2001) studied to describe current research which examines the classification of manufacturing supply chains and positions Internet-based applications in order to identify the operations management challenges for the next generation of manufacturing planning and control systems.

Lee and Kim (2007) studied a review of the development and use of multi-agent modelling techniques and simulations in the context of manufacturing systems and supply chain management (SCM). The objective of the paper is twofold. First, it presents a comprehensive literature review of current multi-agent systems (MAS) research applications in the field of manufacturing systems and SCM. Second, it aims to identify and evaluate some key issues involved in using MAS methods to model and simulate manufacturing systems.

Ghadge et.al. (2018). studied to assess the impact of additive manufacturing (AM) implementation on aircraft supply chain (SC) networks. Additive and conventional manufacturing spare part inventory control systems are studied and compared, revealing insights into SC performance. The paper offers guidance on the adaption of AM in aircraft SCs and AM's impact on spare part inventory systems. The study provides robust evidence for making critical managerial decisions on SC re-design driven by a new and disruptive technology. Next-generation SC and logistics will replace the current demand for fulfilling material products by AM machines.

Mönch et al. (2018) proposed a classification scheme for the relevant literature. The remainder of this paper then focuses on Strategic Network Design models for this industry, supply chain coordination through contracting and semiconductor supply chain simulation. Part II discusses Demand Planning, Inventory Management and Capacity Planning, while Part III addresses Master Planning, Production Planning and Demand Fulfilment. They presented an overview of the semiconductor supply chain from a decision-making and functional perspective to provide a foundation for a review of the industrial engineering and operations research literature addressing supply chain problems in this industry. They then reviewed the literature on strategic supply chain network design, the use of contracting mechanisms to coordinate the different agents in a supply chain, the very recently emerging literature on sustainability issues in these supply chains, and semiconductor supply chain simulation.

Hong et al. (2018) studied empirically investigates the impact of SSCM practices on supply chain (SC) dynamic capabilities and enterprise performance (including economic, environmental and social performance) by explicitly focusing on the mediation effect of SC dynamic capabilities on the link between SSCM practices and enterprise performance.

Dolgui et al. (2019) presented a survey on the applications of optimal control to scheduling in production, supply chain and Industry 4.0 systems with a focus on the deterministic maximum principle. The first objective was to derive major contributions, application areas, limitations, as well as research and application recommendations for the future research. The second objective was to explain control engineering models in terms of industrial engineering and production management. To achieve these objectives, optimal control models, qualitative methods of performance analysis and computational methods for optimal control were considered. They provided a brief historic overview and clarify major mathematical fundamentals whereby the control engineering terms are brought into correspondence with industrial engineering and management.

SYSTEM DYNAMICS

System dynamics deals with how things change over time. Almost all are interested in how the past formed the present moment and how today's actions determine the future. System dynamics is designed to model, analyze and improve socio-economic and administrative systems using a feedback perspective. Dynamic structured administrative problems are modeled by mathematical equations and using computer software. Dynamic constructions of model variables are obtained using computer simulations (Forrester, 1995; Forrester, 1962; Ford, 1999; Sterman, 2000).

The main principle of system dynamics is that the ongoing accumulation of the complex behavior of organizational and social systems (human, material, financial assets, information, biological and psychological states) is also the result of balancing and empowering feedback mechanisms. (Richardson, 1999).

System dynamics is an interdisciplinary problem-solving methodology that utilizes several significant thinking skills such as dynamic thinking and cause-and-effect thinking. System dynamics is a disciplined collaborative approach that could accelerate learning by combining a multifaceted perspective that provides insight into complex and interactive issues (Richmond, 2010, Soderquist and Overakker, 2010; Ferencik, 2014).

System dynamics is a method that allows analysts to separate complex social, and behavioral systems into components, to visualize them by reconstructing them as a whole again, and to develop a simulation model (Tang and Vijay, 2001, p. 3).

There is never one single correct answer in system dynamics. Since system dynamics represents the relationships within the current system, it presents multiple possible approaches that could be applied, instead of providing a single correct answer. Each approach could provide some of the desired result, as well as some unexpected ones. In fact, another significant characteristic of system thinking is to become aware of why we preferred a solution over others (Sterman, 2000, pp. 127-133). System dynamics was designed to help managers make decisions similar to other information technologies. The aim of system dynamics studies is to assist decisions by analyzing old policies and designing new ones (Wheeler, 1994: 80).

System dynamics models are not derived statistically from time series data. They are statements related to policies and system structure that guide decisions. Models include admissions made about the system. A model is only as good as the expertise that underlies its formulation. A good computer model is distinguished from a bad one by the accuracy of its reflection of the system it represents. Several other mathematical models are limited due to the nonlinear nature of the real system and the lack of their inclusion of several feedback cycles. On the other hand, system dynamics computer models could reflect real system behavior (Forrester, 1995, p. 6).

System dynamics is based on information feedback theory and provides symbols for system management. These symbols include diagrams, equations, and programming languages for computer simulation (Morecroft and Sterman, 1994, p. 15).

The objective of system dynamics is not optimization, but to examine the system behavior in the long run at a macro level against specific changes and to help determine the strategies. System dynamics aims to achieve a holistic view of the system and to determine how interactions influence the system as a whole. It is often used for large organizations and global events. However, it could also be used for micro-level and small organizations (taking into account their intermittent structure inflows and levels). The most significant function of system dynamics is to improve system behavior by explaining the structure and behavior of the system. The system dynamics has a broad approach to problems to facilitate monitoring, development, and reorganization of the system performance. As a result, the improvement of the performance of one part of the system prevents the deterioration of another part of the system (Hesan et al., 2014; Başkaya, 1997, p. 93; Özbayrak vd., 2007; Towil, 1996). In system dynamics, generally a "what if" analysis is conducted.

STRUCTURE OF THE SYSTEM DYNAMICS MODEL

In order to analyze the dynamics of the actual system, system models are investigated. Models are simplified structures designed to examine the system behavior (Karnopp et al., 1990, p. 4). Models are constructed to better understand the economic performance of alternative decisions and their impact on environmental quality. Models are important tools for obtaining new knowledge (Ruth and Hannon, 2012, p. 4).

The model must have a clear objective, and this objective should provide a solution to a specific problem. The clear objective is the most important component of a successful model. A clear model, of course, could be inaccurate, too broad and difficult to understand. However, an explicit goal allows the users of the model to ask questions that demonstrate whether solving the problem is beneficial (Sterman, 1991, p. 5).

According to Forrester (1969), four hierarchical structures must be defined to model the dynamic behavior of a system:

1. A closed boundary around the system
2. Feedback loops as basic structural elements within the closed boundary
3. Level (status) variables indicating the totals of the feedback loops
4. Rate (flow) variables that indicate the actions within the feedback loops
    a. Objective
    b. Observed situation
    c. Determination of the difference
    d. Actions based on the difference

To develop the whole systems concept, boundaries in which the interactions forming the system's characteristic behavior occur must be established (Forrester, 1969, p. 12).

In order to achieve the expected objectives in the dynamic model, the model should include the following characteristics (Erkut, 1983, pp. 42-43):

1. The ability to define the predicted cause and effect relationship
2. A simple mathematical structure
3. Incorporating concepts adequate for the economic and social structure
4. Ability to expand to several variables without exceeding computer application limits
5. Ability to produce and process continuous interactions.

SYSTEM DYNAMICS MODELING PROCESS

Forrester's system dynamics modeling steps could be listed as follows (Forrester, 1962, p. 13):

- Identification of the problem.
- Separation (isolation) of the factors that include the observed symptoms.
- Arriving at new decisions through connecting the decisions to actions and actions to information feedback by following cause and effect information feedback cycles
- Formulation of acceptable and adequate (formal) decision-making policies that define the results of these decisions using existing information flows.
- Construction of the mathematical model for the interactions between decision-making policies, sources of information and system components.
- Formulation of system behavior as defined in the model
- Comparison of the results with all adequate data in the existing system
- Review of the model until it represents the actual system
- Discovery of organizational relationships and policy amendments that could improve system behavior through model redesign
- Obtaining the improved performance by implementing the changes in the actual system as stipulated by the model

DELAYS IN SYSTEM DYNAMICS

An important characteristic of various dynamic systems is the presence of delays that control the system, which are much more difficult to comprehend than it might seem at the first glance. Delays are a characteristic of dynamic systems and affect both material and information flow. The most common ones are exponential and pipeline-type delays (Sezen and Günal, 2009, p. 326).

Delays are inherent in many administrative processes. For example, it takes time to manufacture a product, to hire or lay out workers, to build a new factory, or to set up services (Kirkwood, 1998, p. 73).

In principle, there are delays in all flow channels. On the other hand, entering a time delay for each flow results in a quite unnecessarily detailed model, since most of these affect system behavior only very little. When introducing delays into the system, the number of delays could be reduced by two types of shortcuts. First is the case where the system delay is so short that its effect could be ignored. Second, the consecutive delays, which are actually separate, are usually combined and expressed as a single delay (Forrester, 1962, p. 86).

The delays are actually conversion processes that accept the speed of the inflow and transfer it out at the outflow speed. Delays are a special case of levels. Time delays are indicated as packets in the flow channels. By changing the time relation in the inflow, they create the outflow time relation (Forrester, 1962, p. 87).

TIME PROCESSING IN SYSTEM DYNAMICS

System dynamics approach usually utilizes the time framing method. The modeler must determine the length of the time frame in advance. One important issue to note here is the accurate selection of the time frame length. If the time frame is extremely long for model behavior, the modeler will miss significant variations, hence, the simulation could provide misleading results. If the time frame is extremely short for model behavior, then model would be examined even at times when a change in the state is not possible, causing excessive collection of data and excessive computer processing time. It is recommended to select a sufficiently short time frame length generally (Sezen, 2009, p. 320).

DEVELOPED SD MODEL

In this chapter, a system dynamics model will be developed for flexible manufacturing system. The case study will be developed using Vensim package program. To manage flexible manufacturing system it is important to determine production schedule rule and examine results. System dynamics is a power tool to test these rules without any production loss.

Our system consists of one CNC (computer numerical control) machine which is capable of produce 4 different type of products whit out any setup time. After machining each product is transferred to its conveyor and then an AGV transfers this product to its conveyors behind final assembly line. In assembly line products are mounted to final product (4 final products) according to the assembly line program that is determined by the customer orders that we know only distribution.

This simple model also seems a little bit confusing in fact system developer face with much more confusing problem with changing setup time, with different machining time. So, it is necessary to find rules for production.

It is shown below a model that is capable to manage production.

CONSUMPTION

First, it can begin with consumption of the final line.

As it is said before only distribution is known about final product schedule. We use random function here as

```
Consumption = RANDOM NORMAL(0, 1, 0.5, 0.1, 1)
```

Figure 1. Production Assembly Line

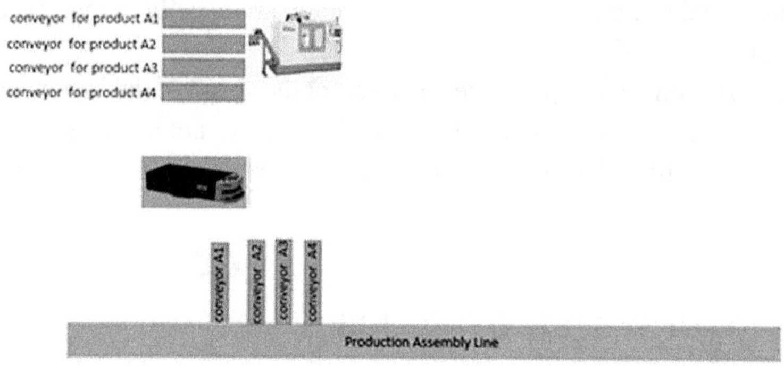

Distribution can be different but here it used random normal distribution with mean 1 and standard deviation 0,5. Here, like the Monte Carlo simulation, we create normal distribution and then according to the value we assign product type as below

```
"a1-band" = IF THEN ELSE(Consumption<=0.1, 1, 0)
"a2-band" = IF THEN ELSE(Consumption <=0.3:AND: Consumption >0.1, 1, 0)
"a3-band" = IF THEN ELSE(Consumption <=0.7:AND: Consumption >0.3, 1, 0)
"a4-band" = IF THEN ELSE(Consumption >0.7, 1, 0)
if Consumption is lower than 0.1 that means a1 product is in the queue
if Consumption is between 0.1 -0.3 that means a2 product is in the queue
if Consumption is between 0.3 -0.7 that means a3 product is in the queue
if Consumption is bigger than 0.7 that means a4 product is in the queue
```

Transfer With AGV

As said before products are transferred with AGV between conveyor near CNC and behind assembly line. It is assumed that we have unlimited AGV and transfer is done in one minute but we must create rule about transfer. It is done as below

Figure 2. SD model using Vensim

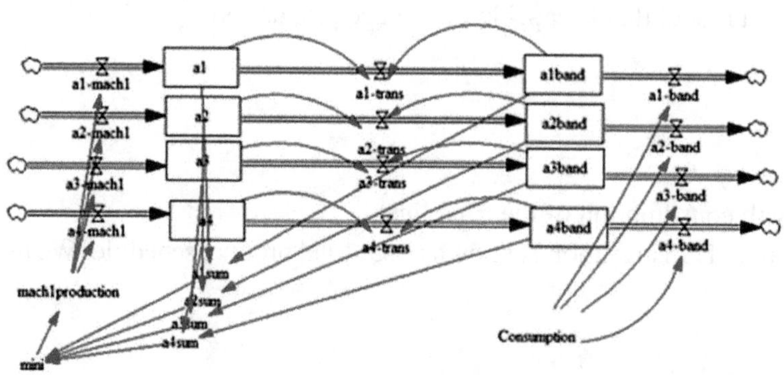

```
"a1-trans" =IF THEN ELSE(a1band<2:AND:a1>1, 1, 0)
"a2- trans" =IF THEN ELSE(a2band<2:AND:a2>1, 1, 0)
"a3- trans" =IF THEN ELSE(a3band<2:AND:a3>1, 1, 0)
"a4- trans" =IF THEN ELSE(a4band<2:AND:a4>1, 1, 0)
```

If product on conveyor near CNC is greater than one and product on conveyor behind assembly line less than 2 transfer between conveyor is realized. Further study AGV constraint can be added to the model.

Production Schedule

For production Schedule it is assumed that we increase the minimum number of products.

First minimum number of product type is calculated and then minimum number of product type is manufactured so we increase minimum number of product type.

```
mini=IF THEN ELSE(a1sum<a2sum:AND:a1sum<a3sum:AND:a1sum<a4sum, 1, IF THEN ELSE
(a2sum<a1sum:AND:a2sum<a3sum:AND:a2sum<a4sum, 2, IF THEN ELSE(a3sum<a1sum: AND
:a3sum<a2sum:AND:a3sum<a4sum, 3, 4)))
```

The equations in SD model are shown as below:

Figure 3. Total number of A1 products in system

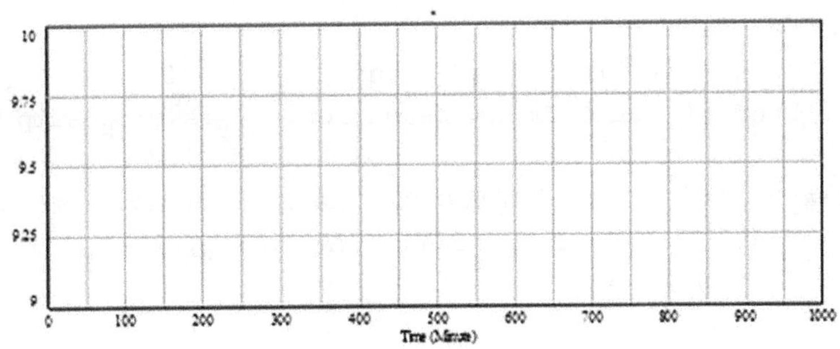

Figure 4. Total number of A2 products in system

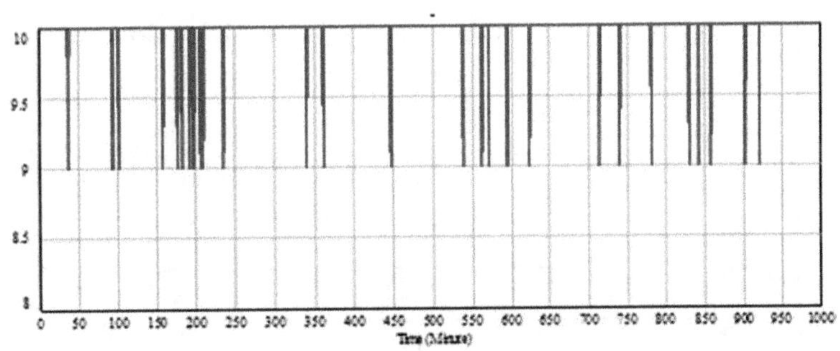

Figure 5. Total number of A3 products in system

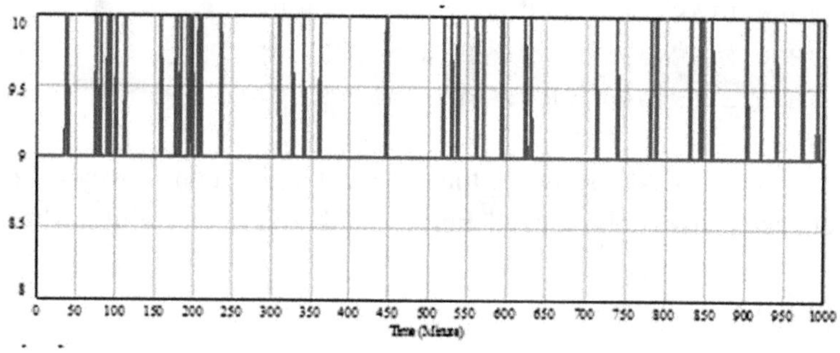


```
a1= INTEG (          "a1-mach1"-"a1-trans",5)
"a1-band"=IF THEN ELSE(Consumption<=0.1, 1, 0)
"a1-mach1"=IF THEN ELSE(mach1production=1, 1, 0)
"a1-trans"=IF THEN ELSE(a1band<2:AND:a1>1, 1, 0)
a1band= INTEG ("a1-trans"-"a1-band",5)
a1sum=1+a1band
a2= INTEG ("a2-mach1"-"a2-trans",          5)
"a2-band"=IF THEN ELSE(Consumption<=0.3:AND:Consumption>0.1, 1,
0)
"a2-mach1"=IF THEN ELSE(mach1production=2, 1, 0)
"a2-trans"=IF THEN ELSE(a2band<2:AND:a2>1, 1, 0)
a2band= INTEG ("a2-trans"-"a2-band",5)
a2sum=a2+a2band
a3= INTEG ("a3-mach1"-"a3-trans",          5)
"a3-band"=IF THEN ELSE(Consumption<=0.7:AND:Consumption>0.3, 1,
0)
"a3-mach1"=          IF THEN ELSE(mach1production=3, 1, 0)
"a3-trans"=IF THEN ELSE(a3band<2:AND:a3>1, 1, 0)
a3band= INTEG ("a3-trans"-"a3-band",5)
a3sum=a3+a3band
```

Figure 6. Total number of A4 products in system

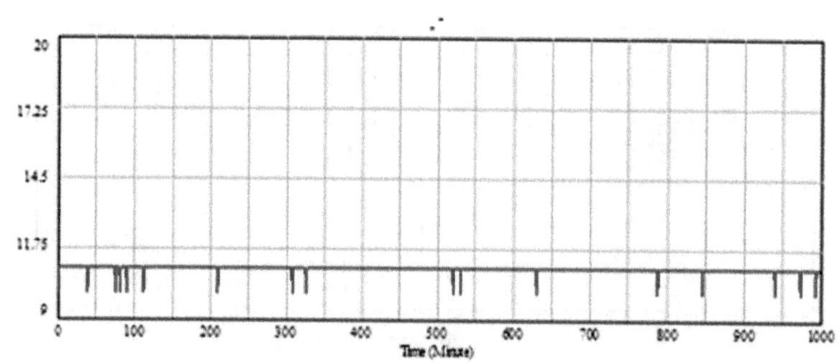

```
(19)        a4= INTEG ("a4-mach1"-"a4-trans",        5)
(20)        "a4-band"=IF THEN ELSE(Consumption>0.7, 1, 0)
(21)        "a4-mach1"=IF THEN ELSE(mach1production=4, 1, 0)
(22)        "a4-trans"=IF THEN ELSE(a4band<2:AND:a4>1, 1, 0)
(23)        a4band= INTEG ("a4-trans"-"a4-band",5)
(24)        a4sum=a4+a4band
(25)        Consumption=RANDOM NORMAL(0, 1, 0.5, 0.1, 1)
(26)        FINAL TIME  = 1000
       Units: Minute
(27)        INITIAL TIME  = 0
       Units: Minute
       The initial time for the simulation.
(28)        mach1production=mini
(29)        mini=       IF THEN ELSE(a1sum<a2sum:AND:a1sum<a3sum:AND:a1sum<a
4sum, 1, IF THEN ELSE(a2sum<a1sum:AND:a2sum<a3sum:AND:a2sum<a4sum, 2, IF THEN
ELSE(a3sum<a1sum         :AND:a3sum<a2sum:AND:a3sum<a4sum, 3, 4)))
(30)        SAVEPER  =     TIME STEP
(31)        TIME STEP  = 1
```

CONCLUSION

There are a lot of rules that can be applied to find good production schedule, we only apply a brief method but with this method there is always stocks of product. 4 type of product in the system are shown below. As it can be seen number of stocks is always near 10 and it can be said that system is steady state.

Further constraint can be added to the model. Especially setup time can be added when CNC change product. Also number of AGV, number of CNC can be added to the model. But the main idea is to model real situation with acceptable assumption and then create rule and test on the model to discover how system behavior along to the run.

REFERENCES

Towil, D.R. (1996). Industrial Dynamics Modelling of Supply Chains. *Logistic Information Management*, *9*(4), 43-56.

Dolgui, A., Ivanov, D., Sethi, S. P., & Sokolov, B. (2019). Scheduling in production, supply chain and Industry 4.0 systems by optimal control: Fundamentals, state-of-the-art and applications. *International Journal of Production Research*, *57*(2), 411–432. doi:10.1080/00207543.2018.1442948

Erkut, H. (1983). *Sistem Dinamiğinin Temelleri*. İstanbul: İTÜ Fen Edebiyat Fakültesi Ofset Atölyesi.

Ferencik, R.C., Soderquist, K.M., & Phillips, M.A. (2014). Building Systems Thinking Capacity: An Essential Skill Set For Policymakers. *Proceedings of the 32nd International Conference of the System Dynamics Society*, Delft, Netherlands. Academic Press.

Ford, A. (1999). *Modeling the Environment: An Introduction to System Dynamics Modeling of Environmental Systems*. Washington, DC: Island Press.

Forrester, J. W. (1962). *Industrial Dynamics*. Waltham, MA: Pegasus Communications.

Forrester, J. W. (1969). *Urban Dynamics. Cambridge*, MA: The MIT Press.

Forrester, J.W. (1995). *Road Map 1: Counterintuitive Behaviour Of Social Systems*. MIT System Dynamics In Education Project.

Ghadge, A., Karantoni, G., Chaudhuri, A., & Srinivasan, A. (2018). Impact of additive manufacturing on aircraft supply chain performance. *Journal of Manufacturing Technology Management*, *29*(5), 846–865. doi:10.1108/JMTM-07-2017-0143

Hong, J., Zhang, Y., & Ding, M. (2018). Sustainable supply chain management practices, supply chain dynamic capabilities, and enterprise performance. *Journal of Cleaner Production*, *172*, 3508–3519. doi:10.1016/j.jclepro.2017.06.093

Jan, O., & Erik, S. (2000). Supply Chain Management Survey of Swedish Manufacturing Firms. *International Journal of Production Economics*, *89*, 353–361.

Karnopp, D. C., Margolis, D. L., & Rosenberg, R. C. (1990). *System Dynamics A Unified Approach* (2nd ed.). United States of America: John Wiley & Sons, Inc.

Kehoe, D., & Boughton, N. (2001). Internet based supply chain management: A classification of approaches to manufacturing planning and control. *International Journal of Operations & Production Management*, *21*(4), 516–525. doi:10.1108/01443570110381417

Kirkwood Craig, W. (1998). *System Dynamics Methods: A Quick Introduction*. College Of Business Arizona State University.

Lane, D. J., & Sterman, J. D. (2011). Profiles in Operations Research. In Profiles in Operations Research: Pioneers and Innovators (Ch. 20, pp. 363-386). New York, Springer: .

Lee, J. H., & Kim, C. O. (2008). Multi-agent systems applications in manufacturing systems and supply chain management: A review paper. *International Journal of Production Research*, *46*(1), 233–265. doi:10.1080/00207540701441921

Mönch, L., Uzsoy, R., & Fowler, J. W. (2018). A survey of semiconductor supply chain models part I: Semiconductor supply chains, strategic network design, and supply chain simulation. *International Journal of Production Research*, *56*(13), 4524–4545. doi:10.1080/00207543.2017.1401233

Morecroft, J. D. W., & Sterman, J. D. (1994). *Modelling For Learning Organizations*. Portland, OR: Productivity Press.

Mustafa, Ö., Papadopoulou, T. C., & Melek, A. (2007). System Dynamics Modelling of A Manufacturing Supply Chain System. *Simulation Modelling Practice and Theory, 15*(10), 1338–1355. doi:10.1016/j.simpat.2007.09.007

Ramage, M., & Ship, K. (2009). *Sytem Thinkers, The Open University Walton Hall.* Milton Keynes, United Kingdom: Springer London.

Reza, H., Amineh, G., & Virginia, D. (2014). Using Difference Equation to Model Discrete-time Behavior in System Dynamics Modeling. *Proceedings of the 32nd International Conference of the System Dynamics Society*, Delft, Netherlands. Academic Press.

Richardson George, P. (1999). *Feedback Thought in Social Science And Systems Theory.* Waltham: Pegasus Communications, Inc.

Richmond, B. (2010). The thinking in systems thinking: eight critical skills. *Tracing Connections: Voices of Systems Thinkers*, 3-21.

Ruth, M., & Hannon, B. (2012). *Modeling Dynamic Economic Systems* (2nd ed.). Springer. doi:10.1007/978-1-4614-2209-9

Sezen, H. K., & Günal, M. (2009). *Yöneylem Araştırmasında Benzetim.* Bursa: Ekin Yayınevi.

Soderquıst, C., & Overakkers, S. (2010). Education for Sustainable Development: A Systems Thinking Approach. *Global Environmental Research, 14*, 193–202.

Sterman, J. D. (1991). *A Skeptic's Guide to Computer Models.* Cambridge, USA: MIT System Dynamics In Education Project.

Sterman, J. D. (2000). Business Dynamics Systems Thinking And Modelling. In *A Complex World.* New York: McGraw-Hill.

Victor, T., & Samutra, V. (2001). System Dynamics Origins, Development And Future Prospects of a Method. In Research Seminar in Engineering Systems. Academic Press.

Wheeler Frederick, P. (1994). How to Get Managers To Use System Dynamics. *Proceedings of the International System Dynamics Conference*, Stirling, Scotland. Academic Press.

Zehra, B. (1997). *Sistem Dinamiğinin İşletmelerde Uygulanması.* Bursa: Ekin Kitabevi Yayınları.

KEY TERMS AND DEFINITIONS

Decision Functions: States of the policy that determines how to convert available information in stock into a decision.

Delays: A characteristic of dynamic systems and affect both material and information flow.

Flow: It is defined as the increase or decrease in the unit time interval in stocks and denoted by f (*t*).

Information Delay: They are usually caused by the transfer of information and the delay in the process of taking an action after the information is received. Information delays could occur in vertical hierarchical administrations.

Information Flow: Creation, control and distribution of information is the most significant task of business management. It might be difficult to model the information properly in the business process.

Material Flow: It includes stocks and flow rates of physical goods such as raw materials, inventories in the process or finished goods.

Material Delays: In this structure, the material moves forward from the beginning to the end of the pipeline without any change during a period of time, similar to a water running through a pipeline.

Stocks: The present values of the variables that are formed by the accumulated difference between the inflow and outflow.

System Dynamics: The basis of system dynamics is to understand how system structures cause system behavior and system events

System Dynamic Language: This language consists of four components: stocks, flows, decision functions and information flow.

Chapter 19
Total Quality Management and Quality Engineering

Shubhajit Das
https://orcid.org/0000-0002-4297-3324
National Institute of Technology, Arunachal Pradesh, India

Kakoli Roy
National Institute of Technology, Arunachal Pradesh, India

Tage Nampi
National Institute of Technology, Arunachal Pradesh, India

ABSTRACT

Total quality management (TQM) is a set of the systematic management approaches for the continuous improvement of quality standards of products, services, and business relations with employees and consumers. This chapter mainly focusses on the eight key principles of TQM, the involvement of workers, leadership, process approach, strategic approach, continuous improvement, together for a factual approach to decision-making and communication. This chapter also discusses a four-part management model that implements continuous quality improvements and process control in different stages of an organization based on the Deming cycle or the Shewhart cycle. Quality engineering encompasses a broad range of methodologies and tools, which include quality management systems, advanced product quality planning (APQP), tools like quality function development (QFD), failure modes, and effects analysis (FMEA), statistical process control (SPC), and are widely accepted methodologies used in industries.

INTRODUCTION

In the late 1970s and early 1980s, the American businesses struggle to stay competitive in the market due to their loss in the global markets. To sustain in the competitive environment, the industries started to adopt the improvement program to regain their quality and services in the market. These quality programs have been successfully implemented in Japanese industries. Quality in goods and services is one

DOI: 10.4018/978-1-7998-1831-1.ch019

of the important concerns for consumers. One of the improvement programmes was the Total Quality Management (TQM) program. In TQM, the goal was customer satisfaction. TQM is the combination of all processes and procedures within an organization to accomplish unceasing improvement in the goods and service quality (Omachonu, 2004). It is a management approach. The major objective of TQM is that everyone in the organization should involve directly or indirectly to focus on the meeting of the consumer's needs and demands. Originally, TQM was adopted in the manufacturing sector but gradually it is now implemented in every industrial sector and organization. TQM is now observed as a universal tool for quality everywhere. It not only benefits the consumers but also TQM increases the efficiency of an organization. The implementation of TQM in organizations is not an easy task. It involves time, commitment, resources and continuous support from the management. The creation of TQM can be attributed to a former Bell telephone American employee Dr. Walter Shewhart who was also a physicist and statistician. The different Quality Gurus Deming, Juran, Crosby, Feigenbaum, Ishikawa, Shewhart, Taguchi, etc., have made a significant impact to improve quality in products, businesses, and services.

HISTORY AND BACKGROUND OF TQM

History of TQM

In the 18th century, the quality did not exist. The development of quality started in the 20th century. Its significance has transformed and developed over time. During World War II, in the 1940s, quality became statistical. TQM was initially originated in the early 1950s in the USA. In the early 1930s, a former Bell telephone employee named Dr. Walter Shewhart who was also an American physicist and statistician is often talked about to as the "Father of the Quality Control." He established the modern statistical approaches with the Shewhart Cycle of learning and Improvement Cycle. His work and methods were later developed by other statisticians like William Edwards Deming and Joseph Moses Juran. Shewhart, Deming, and Juran were referred to as the three authors of the quality improvement program. Deming in the late 1920s, who was an employee at Western Electric Company based in Chicago, found that the employees have low enthusiasm leading to inefficient productivity. Using Shewhart's theory, Deming framed a management process that is statistically controlled and allows the managers to intervene when and how in an organization's process. The U.S Department sent Deming to Japan where he taught statistical analysis techniques and quality improvement. His methods were adopted and later developed by Japanese engineers. Juran taught the concepts of controlling quality and management practices. The main focus of the TQM is to provide continuous quality improvement of products, services, and business relations. It first originated in the manufacturing sector but later applied to all the organizations.

Background of TQM

Right from the times, when the quality gurus have given TQM a broader meaning, it has been adopted in many top organizations around the world. The Toyota in Japan and Maruti in India has implemented TQM at that time. Many organizations nowadays adopt TQM to develop and improve the objectives of quality. In businesses and engineering, product development started where various departments work independently. The U.S awards Malcolm Baldrige National Quality Award while Japan awards Deming prize to their companies for quality excellence in management and implementation. TQM is a decision-making process. It is an outline of principles in a systematic method. It performs as a tool to determine and develop quality and efficiency. The principles of TQM include relationships between customer/supplier prevailing in both within the organization and between the organizations. The principles enable an organization to break any traditional obstructions and to recognize if the company implements TQM.

TRADITIONAL AND ADVANCED QUALITY MANAGEMENT

Traditional Quality Management

Traditional quality management varies in beliefs, execution, and measurement. The mangers supervise their workers towards the short-term objectives of the organization and the profits earned. The organization defines its standards of quality and decides whether a particular product will be acceptable to the market. If imperfections are identified through traditional quality management, administrators hold the employee accountable who is responsible for the cause of the defect. While in TQM, employees and managers improve the quality by varying the different methods used to manufacture a product. In traditional quality management, managers sometimes use fear to threaten them and fire their workers. They solve problems based on their specific knowledge and skills. The workers are given specific roles and those who are openly involved in manufacturing a product are accountable for its quality. The products with the defect are reproduced to identify the problems and resolve them on a case by case basis. On the other hand, TQM eliminates waste and increases efficiencies so that a product is produced correctly. There is continuous process improvement.

Principles of TQM

The management plan of TQM focuses on the requirements of the customers. The strategy to upgrade the quality levels of the product is included. The involvement of the leaders and workers of an organization in the quality and product processes is also important. There are eight key principles of TQM that are shown in Figure 1.

1. **Customer-Oriented Approach:** The activities and profit of the organization mainly depend on customer needs, requirements, and satisfaction. The consumers decide whether the services and products provided to them have maintained the standards of quality. The organizations must implement TQM keeping in mind the interest of their consumers and try to live up to their expectations.
2. **Leadership:** Good and strong leadership in an organization is necessary to ensure that the objectives of the organization are achieved. The top management's commitment ensures increased motivation among its employees.
3. **Involvement of Workers:** The employees of the organization actively participate to work towards the common business objective of the organization. When the company provides a better working environment to its workers, their involvement in achieving the common vision facilitates complete employee commitment, better empowerment of the employees, high performance of the working systems and better decision making in the organization.
4. **Process Approach:** Process thinking is one of the important principles of TQM. TQM looks for better ways for developing new performance measures and better ways of working standards that

Figure 1. The 8 principles of TQM

target the satisfaction of the customers and improving the product quality standards to provide greater effectiveness, efficiency, and quality outcomes.

5. **Strategic Approach:** For the improvement of efficiency and effectiveness of the business operations, there must be strategic management to identify, understand and manage the processes to achieve the goals of the organization. The strategic plan ensures that quality becomes the main aspect of all business activities.
6. **Continuous Improvement:** TQM focuses on continuous improvement of a company's business. The company should work towards improving the performance and quality of their services and processes.
7. **Factual Approach to Decision Making:** For improved and better understanding, the decisions of the organization are entirely based on data and facts. It does not depend upon the worker's emotions either based on their personalities.
8. **Communication:** There should be engaged and active communication between the suppliers and customers, higher authorities and the workers at all levels of an organization leading to proper planning, decision making and increased motivation among the employees.

METHODOLOGIES IN TQM

Deming/Shewhart Cycle (PDSA Cycle)

The Demming Cycle, or Demming Wheel, or PDCA cycle, or PDSA cycle is a four-part management model that implements continuous quality improvement and process control in different stages of an organization. The cycle is based on the Shewhart Cycle. The PDCA cycle is sometimes changed to the PDSA cycle, where the first two principles are the same but instead of the third principle 'Check', it is 'Study' where the processes are studied. The cycle is made up of:

1. **Plan:** Planning is a major part of TQM. During this stage, the course of action is determined, the problem is identified and analyzed, the hypothesis is developed of the issues concerned and the potential solutions proposed using various problem-solving techniques are tested. In the context, the

employees find answers to some questions like, what resources are required? What is the problem need to solve? What resources are available? What are the predicted solutions?

2. **Do:** The employees develop solutions for the problem determined in the previous stage. The main motive of this stage is to execute the plan. Various procedures and strategies are established and implemented to overcome the problems faced by the employees.
3. **Check/Study:** It is one of the most important stages of the PDCA cycle, that allows the solution to be evaluated and any plan adapted is to be monitored. The issues found are to be identified quickly and the required actions are to be measured from the data collected. The results obtained and the goals achieved are considered carefully to allow any necessary adjustments to proceed further to the final stage.
4. **Act:** This stage is both the final stage of the cycle and the first stage of the next cycle. The final stage is the accumulation of the planning, testing and analyzing whether the desired improvement and actions are achieved, and the purpose to act upon what has been learned. Employees decide on their changes, report the results and address other problems. Once the organization achieves the desired goal, the PDCA cycle is complete. After the investigation of previous mistakes, the complete PDCA/PDSA cycle is repeated with new guidelines to ensure that the products, services, and processes are improved.

The continuous improvement of the PDCA/PDSA cycle makes it a standardized model. Since it is simple and easy to adapt, it can be implemented in any complex company for improving the performance and efficiency of the organization. The cycle also offers many advantages:

- It prevents wastage of time.
- The continuous improvement of the cycle ensures that a process is error-free, as the cycle goes through several repeated stages and planning.
- The reliability of the cycle enables better planning, better performance in the organization, increasing the benefits and productivity of the organization.
- When a new method is checked and analyzed successfully, the company implements the method in all its stages.

PDCA is an approach toward TQM by providing feedback for continuous improvement of quality. Quality has always been one of the major factors in determining the profit and success of an organization. The quality of products benefits the organization over its competitors. Consumers generally consider the quality of services and products. With the emergence of quality management, the importance of quality has increased in recent years.

Deming's 14 Points to Improve Quality

Deming has summarised his management philosophy in fourteen rules of management, which are as follows:

1. Construct a devotion of purpose to improve product quality and facilities.
2. Implement a new quality thinking.
3. Stop dependency on quality inspection.

4. End the exercise to award business to suppliers based on price alone.
5. Continuously develop product quality and services.
6. Impart on-the-job training to the workers.
7. Institute headship to help people and machines to do an efficient job.
8. Remove fear within the work environment so that people can work effectively for the organization.
9. Break down the obstacles between the departments and people so that they can grind as a team.
10. Ditch slogans, targets, and exhortations.
11. Remove work quotas for the administration.
12. Eradicate obstacles that prevent work teams of their pride in their work.
13. Inspire education and self-improvement for everyone.
14. Include everyone to work to achieve organizational transformation.

QUALITY ENGINEERING METHODS AND TOOLS

Quality Engineering Methods

In the race to gain new customers or increase market share, companies are continuously developing and presenting new or improved products. In today's highly competitive global marketplace, consumers want to buy high-quality products at reasonable prices. Through the active use of quality engineering approaches and tools during the product conception, development and implementation processes, industries can ensure that the voice of customers is heard and incorporated into the product design, that quality is designed and built into the product and waste is identified and eliminated in the manufacturing processes.

Quality Management Systems

A Quality Management System (QMS)is defined as an organizational structure consisting of a set of measures, methods, and resources to systematically increase the effectiveness and competence of quality objectives (Deshpande, 2016). The other key factors which are necessary for the efficacy and efficiency are the appropriate work training, motivation throughout to achieve the goals, how the workers perform the activities, how the goal is achieved and so on. The resources like the materials, people's time, and equipment investment must also be provided. The implementation of QMS plays a crucial role in the concerned organization. The widely accepted standards having different objectives and fields of applications are used for this purpose.

International Standard ISO 9001

The ISO 9000 family is a set of standards, which provide an international agreement on excellent management practices to enable an organization to provide quality products, and facilities, which meet consumer requirements. It includes four standards:

1. **ISO 9000:2005:** This standard sets the definitions of the terms used in all the standards with a description of the 8 quality principles and an introduction to QMS.

2. **ISO 9001:2008:** This standard is used when an organization provides products and services regularly to meet the various quality requirements and satisfaction of the consumers. It focuses on the effectiveness of QMS.
3. **ISO 9004:2009:** This standard guides the organizations to achieve and maintain long-term goals. It is not intended for certification usage.
4. **ISO 9011:2002:** The standard provides guidelines to achieve quality goals.

International Standard ISO 17025

This standard applies to all organizations that perform any tests or calibrations giving them technical competence recognition. Any laboratory which meets the requirements of ISO 17025 functions under a QMS for testing and calibration activities. It can be used by third parties like the Accreditation bodies.

Principles of Good Laboratory Practice

These practices are applied to test the items for safety to human health and environment in pharmaceutical, cosmetics, industrial chemicals, pesticides and, food additives.

Spanish Standards UNE 166000 Family of Standards

These standards apply to either R&D projects or the organizations performing R&D activities.

Advanced Product Quality Planning

Advanced Product Quality Planning (APQP) is an outline of methods and practices used to improve quality products in the industry. It is used mainly in the automobile industry. APQP supports product development and quality plans to meet the requirements of internal and external consumers (Thisse, 1998). APQP comprised of one planning stage and five concurrent phases. The process is often represented with the Plan, Do, Study, Cycle (PDSA cycle). The APQP and Control Plan (APQP/CP) reference manual was developed and distributed in 1994 by Chrysler, Ford, and General Motors. The manual provides guidelines to help in the development of excellence of products or services to content the consumer's needs. The manual emphasizes the five stages of quality product planning as shown in Figure 2. There are other reference manuals like Measurement Systems Analysis (MSA), Statistical Process Control (SPC) and Potential Failure Mode and Effects Analysis (FMEA). The suppliers must demonstrate that the techniques and methods from these manuals are met. Only the required techniques and methods applicable should be used. These instructions do not set any requirements. Instead, the QS 9000 manual set the supplier's quality system requirements.

APQP can be implemented in an existing organization in the following steps:

1. **Determine the Necessary Training:** The organization must know where to implement and effectively use the techniques described in the APQP/CP manual. They should also know how the roles of the employees fit in the different activities of the organization. The organizations must have an understanding of how and where the methods and techniques are beneficial to them. The

Figure 2. Stages of APQP

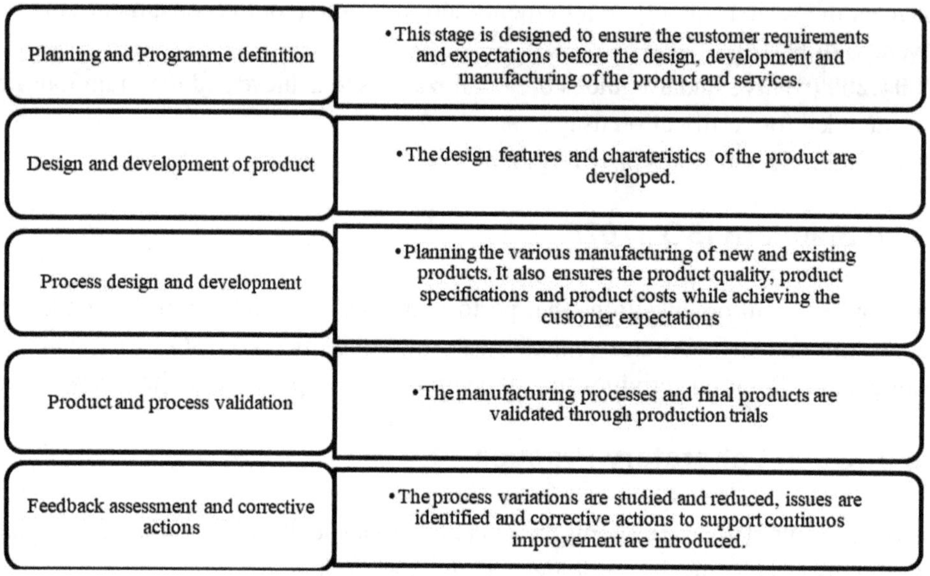

training for the employees must be provided by the organization and the supervisors and managers throughout the organization must develop a training plan.

2. **Identification of Multidisciplinary Approaches in the Organization:** The team members should comprise of who knows the organization well and analyze the organization and product fulfillment activities that practice multidisciplinary approaches. The teams develop an outline of the product's fulfillment cycle which leads to the identification of many redundant activities.
3. **Identification of what Current Activities are Advance Quality Planning activities:** The identification and provision of training are required since the team acquaintance of the content of the APQP/CP manual and activities of the organization is necessary. After studying the flowchart in Step 2, the APQP within the organization is identified.
4. **Identification of Advance Quality Planning Activities in the Organization:** The use of the APQP/CP manual helps to determine the various undertakings the organization should be doing to improve the significance to the customers. The existing activities should be expanded effectively.
5. **Development of Plan:** The organization should decide how to include the activities to the multidisciplinary activities already taking place within the association. For the APQP process to be successful, set up the system and write the processes that permit the maximum authorization and tractability possible within the team in the setting up, inaccuracy, and handling of the quality planning activities.

Quality Function Deployment

Quality Function Deployment (QFD) is a process to recognize consumer requirements and conversion of these requirements into product designs. It was created by a planning specialist of Japanese origin Yoji Akaowho worked in the Mitsubishi shipyards in 1966 to help product planners view the new and developing products through the lenses of consumer, organization, and technology. QFD was later adopted by Toyota and was introduced in the U.S in the early 1980s by automobile industries and electronic manufacturers. QFD practices few principles of Concurrent engineering where the cross-functional teams are included in all stages of developing a product. Each of the four stages of QFD uses a matrix to transform consumer necessities from early planning phases through control of production. Each matrix denotes a specific characteristic of the product's requirements. In each of the phases, relationships between the elements are assessed. From each phase, the most important aspects are placed in the next matrix. The stages are shown in Figure 3.

The organizing arrangement in which QFD uses to establish information is a planning tool known as the House of the Quality. The design engineering team and marketers first create consumer attributes for the product. These attributes are placed in rows in the central matrix of the House of Quality. The application of QFD is seen at the Chrysler Motors Corporation in 1986. Chrysler executed the QFD in four stages i.e.

1. Spreading awareness
2. Develop effective case studies and examples to inspire teams
3. Working out on QFD techniques within the company
4. Implementing QFD as its business philosophy.

The implementation of QFD in Chrysler resulted in the innovative design changes with wide acceptance in the market while the launch of the LH platform for mid-size cars was also effective. The total product design cycle took approximately 36 months while the historic cycles took about 62 to 54 months. Only 740 people were required in the QFD, while 1600 people were required in the historical environment. The possible benefits of employing the QFD process are:

- It emphasizes the needs of customer-friendly products.
- It uses techniques of concurrent engineering.
- It reduces the number of design changes.
- Effective QFD reduces the project time and wastage of resources.

Figure 3. Stages of QFD

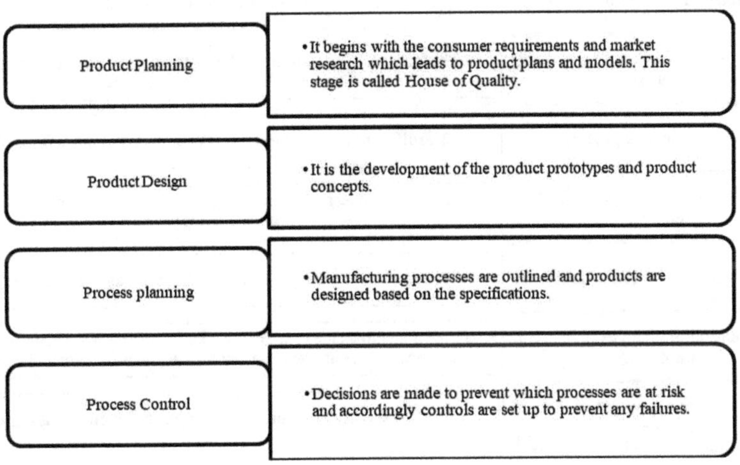

Failure Mode and Effect Analysis

Nowadays, customers have high expectations of the company and manufacturers to provide them with reliable and quality products and services. However, the faults and problems are detected in the later stages of development; it can turn out to be expensive and time-consuming to the process. The main challenge to the manufacturers is to design quality products and ensure that any problems or defects to not arise in the process. One of the tools to identify potential problems and risks is Failure Mode and Effects Analysis (FMEA). FMEA is an organized way to recognize and address possible difficulties or failures, and their effects on the system or a process before an undesirable event takes place. It is a risk assessment tool to determine the chances of failure and potential problems and impact. It allows manufacturers to mitigate failure in the early stages of development. In comparison, root cause analysis is a structured way to address problems after they occur. The different stages of FMEA are shown in Figure 4.

The advantages of FMEA are:

- To evaluate processes, products, and services
- Reduction in costs, process development time
- Decrease in waste
- Increases customer satisfaction
- Improve the quality and reliability of products.
- Accident prevention
- Standardized approach to risk management

Statistical Process Control

Manufacturing companies are facing much competition in global markets. Statistical Process Control (SPC) is a method to measure and control quality by continually improving the manufacturing processes. Data is collected from various machines in the form of process or product measurements. The collected

Figure 4. Stages of FMEA

Selection of a process to analyse	• A process that is problematic is choosen.
Choosing a team facilatator and team members	• The facilatator is appointed and team members are selected directly to analyse the chosen problem.
Description of the process	• The different steps of the process to be analysed are defined clearly.
Identification of the potential problems	• The team conduct brainstorming sessions to gnerate solutions of various problems in the process.
Choosing and elimination of problems	• The team selects the problems which are the focus of improvement.
Implementation of changes to prevent problems	• The team identifies the root cause of the action selected and determines how the changes are implemented without any loss or harm to the people.
Measurement of success of the changes	• The success evaluation of the implementation plan is done.

data used are evaluated, monitored and controlled in a process. The SPC is applied to shift an organization from detection-based quality controls to prevention-based quality controls. The objective of SPC is to continually improve by reducing variations. The possible benefits of SPC are as follows:

- Elimination for inspection in the supply chain.
- Reduction in scrap and inspection cost
- Increased motivation among the workers
- Increase in productivity
- Low investments lead to a decrease in costs
- Better communication throughout the organization

Root Cause Analysis

Root cause analysis (RCA) is performed with various techniques, methods, and principles. It is a process to recognize the root causes of a problem and define the appropriate solutions. RCA presumes that systems and events are interconnected. By tracing back the steps of an activity, we can discover where and how the actual problem started. There are three types of causes on which the RCA works, namely

- Physical causes: Failure of material items. (e.g. a bike's brakes stopped working)
- Human causes: Errors caused by humans lead to physical causes.
- Organizational causes: When the work of the people in an organization is faulty.

There are seven steps in the RCA which are summarized as follows (Serrat, 2017):

1. Selection of the organization's problem
2. Defining the problem
3. Identification of the probable causes
4. Implementation of the corrective actions
5. Analysis of the effectiveness of the actions taken, and if necessary, modify the existing corrective actions
6. Check whether the procedures can be applied to other areas within the organization or the factory
7. Check to see the new procedures are followed effectively

RCA helps to tackle the problems in the following ways:

- It helps to see the workers the real cause of the problem
- Problems are sorted with priority basis
- Effective and beneficial practices are employed in the organization
- Saves time by fixing the important issues

5-Whys

The 5-why analysis is frequently used to reduce waste in lean manufacturing. The technique was developed by Sakichii Toyoda for the Toyota Industries Corporation. Taichi Ohno defined the 5-whys technique as "the basis of Toyota's scientific method by repeating why five times, the nature of the problem as well as its solution becomes clear". The tool is used extensively beyond Toyota, and now it is used with Kaizen, lean manufacturing and Six Sigma. The best approach is to recognize the root causes of the problem by going deeper with each "why?" and act upon it. The 5-whys principle is a simple problem-solving tool to determine the cause-effect relationship of an event or a situation. If the approach is implemented correctly, it can be effective to lead an organization to develop corrective and preventive actions. It is also identified as Why Tree and is considered an unpretentious form of RCA (Gangidi, 2019). The technique involves repeatedly asking the question "Why?" five times so that the nature of the problem and solution becomes clear. Usually, the solution to the previous question is used to frame the next question. Without asking enough of 5 whys, one may not find the root cause of the problem but rather end up with the symptom. Remedial actions of a root cause can be "preventive" while that of a symptom is usually "detective". The common perception behind this method is to involve the human mind in a route to search some ideas by framing a series of questions and to encourage problem-oriented thinking (Myszewski, 2013). The following five steps are applied to implement the 5-whys method in a team:

Step 1: A team is formed and a clear and concise statement of the problem is developed. Then the decision is taken how many members are required to resolve the issue or it can be approached singlehandedly.
Step 2: The team leader asks the first "why" to its members: "Why the problem has occurred?" The various responses of the members are recorded or written on the board or index cards.
Step 3: Ask the successive "Whys" to the members and write down the answers on the board or in the cards. This exercise is repeated until no further "Whys" gives useful information and the root cause of the problem arrives.
Step 4: The systematic solution among the dozen causes collected from the previous sessions is discussed with the team and the logic behind the solution is analyzed.
Step 5: The team develops various corrective actions to eliminate the root causes and further actions to prevent the problems to reoccur again.

The various benefits of 5-whys technique are stated below:

- Identifies the root cause of a problem.
- It helps to define the connection between diverse root causes of a problem.
- It doesn't use statistical analysis.

QUALITY ENGINEERING TOOLS

Ishikawa Cause and Effect Diagrams

The Fishbone analysis is commonly referred to as "Ishikawa Diagram" or cause and effect diagrams. In the 1960's Ishikawa diagrams were popularised by Kaoru Ishikawa. He developed quality management procedures in the Kawasaki shipyards, and in the process became one of the founding fathers of modern quality management. This diagram is used in many organizations for diagnosing or taking actions in which the root cause of the problem is known (Liliana, 2016). The structure of the figure looks like the skeleton of a fish. The diagrams work from right to left, with the large bones of the fish subdividing into smaller bones. The problem is shown at the head or mouth of the fish and the entire possible multiple and sub causes are shown on the smaller bones of the skeleton as shown in Figure 5. There are four steps to use the tool:

1. Identification of the problem
2. Identification of the main aspects involved
3. Listing out the potential causes
4. Diagram analysis

The possible causes of variation are listed in several categories such as:

1. People: The manpower engaged to produce a product.
2. Methods: The need for rules, regulations, policies to ensure continuous quality.
3. Machines: It includes assembly lines and robots.
4. Materials: The required materials to produce a product.
5. Measurements: The measurements used to monitor the quality of the product.
6. Environment: This is outside the control of an organization that can affect the product.

Toyota developed 5M's to identify causes of variation in manufacturing i.e. man, machine, material, method, and measurement. The Fishbone analysis has the following advantages which are:

- It creates a visual illustration of the cause.
- The diagram shows areas of weaknesses, which can be rectified.
- As the possible solutions to solve the problem are discussed, the team members remain focussed on the goal.
- It enhances brainstorming the reasons to get a result for the causes.

Stratification

When data from a variety of sources are grouped, the meaning of the data can be difficult to understand. Stratification or flowchart is defined as the arrangement of data, people, and objects into separate groups. This technique is used in combination with other data analysis tools. The purpose of flowcharts is to distribute the data and extract the information to solve a problem. Stratification is used to determine the bottlenecks and breakdowns in the process. It is used to show the different workflow in the organization

Figure 5. Typical Ishikawa diagram

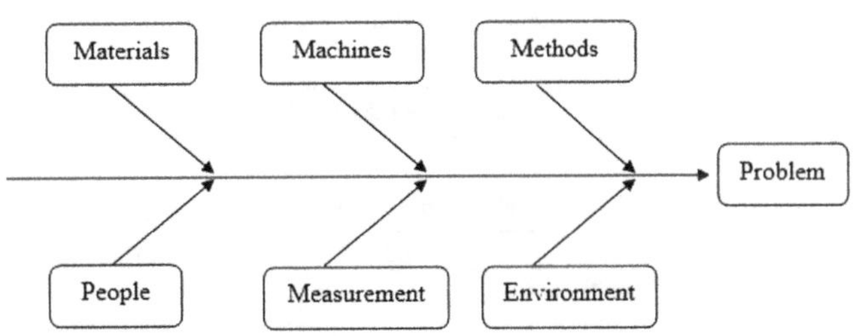

and the execution of new processes within it. It is applied before the collection of data or when data is collected from various sources or when data analysis requires the separation of different sources.

Check Sheets

A check sheet or a checklist is an organized table for accumulating data and examining these data. When the information composed is measurable, the check sheet can also be called a tally sheet. Check sheets list down the significant events in a tabular format and keep on updating or marking the status on their existence, which helps in accepting the growth, defect patterns and source of defects. Check sheets have become outdated now. They have been replaced by modern-day Business Process Management (BPM) software, which facilitates data that are more complicated to be automatically recorded.

Statistical Control Charts

Control charts or Shewhart's charts or process control charts are graphs where the data are plotted alongside a time-ordered sequence in the X-axis. These charts include a centerline, an upper control limit, and a lower control limit. The centerline represents the mean. These lines are determined from historical data. The designed control limits give the projected process variation from a period of baseline data. These charts are very popular in Quality Control Techniques and show whether a process is stable or unstable within specified conditions. It helps to identify how manufacturing activity is in line with the consumer's expectations. The different steps associated with a control chart are shown in Figure 6.

Statistical control charts are used in various industries to determine patient wait times, billing errors, healthcare, data entry errors, operating expenses in finance, customer support calls, customer support calls in software development to name a few. Implementing control charts can help in the following ways:

- Reduction in the post-production inspection
- Figure out process capability based on previous trends
- Evaluation of the impact of process changes
- Visualization of the performance of the processes over time
- Make decisions about processes which are fine and which require immediate actions

Figure 6. Steps of Control Chart

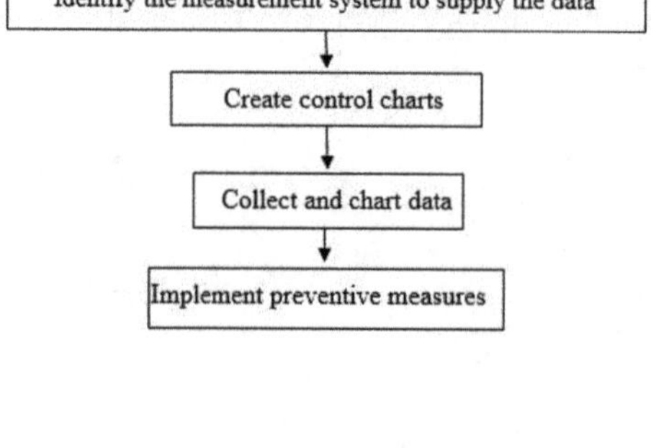

Scatter Diagrams

Scatter diagrams are tools used to show the relationship between two random variables. They are a form of a sheet of paper upon which the data points corresponding to the variables of interest are scattered. The relationship between the variables to find the "root cause" of the problems during root cause analysis. From the shape of the pattern of the data points, the association between the two variables can be found out and the best suitable correlation is applied based on it. They are also known as Scatter Plots or X-Y Diagrams. The plots can be useful to verify a cause and effect analysis. The different stages of a scatter diagram are shown in Figure 7.

Pareto Analysis

Pareto analysis is a QC tool that ranks the data classifications in the descending order from the highest frequency of occurrences to the lowest frequency of occurrences. The total frequency is equated to 100% (Karuppusami, 2006). Pareto Analysis uses the Pareto Principle which is also known as the 80/20 rule. In 1896, an Italian origin economist, Vilfredo Pareto in his book, "Cours d'economie politique", named the rule. The idea behind the principle is that when 20% of the work is done, 80% of the benefit of the entire work can be generated. That 20% of the problems are the "vital few" and the remaining problems are the "trivial many." The principle is based on the law of "significant few versus the trivial many" which is the unequal distribution of things. A Pareto diagram puts data in a hierarchical order, which allows the most important problems to be corrected first. This technique is used mostly to identify and calculate nonconformities, even if it summarizes all types of data. The bars are arranged according to their importance along the abscissa from left to right in the decreasing order and the cumulative total is shown by a line graph. The Pareto analysis allows individuals to make effective changes, which are needed. The principle is useful to determine which areas to focus to attain the greatest efficiency. The working of Pareto Analysis is shown in Figure 8.

Pareto Analysis has several benefits which are:

Figure 7. Stages of Scatter diagram

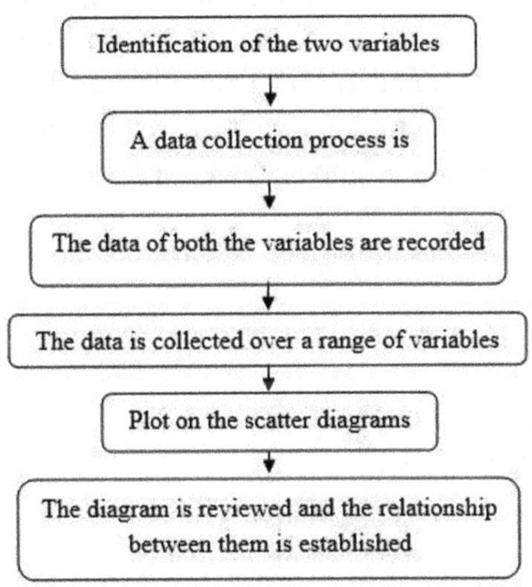

Figure 8. Working of Pareto analysis

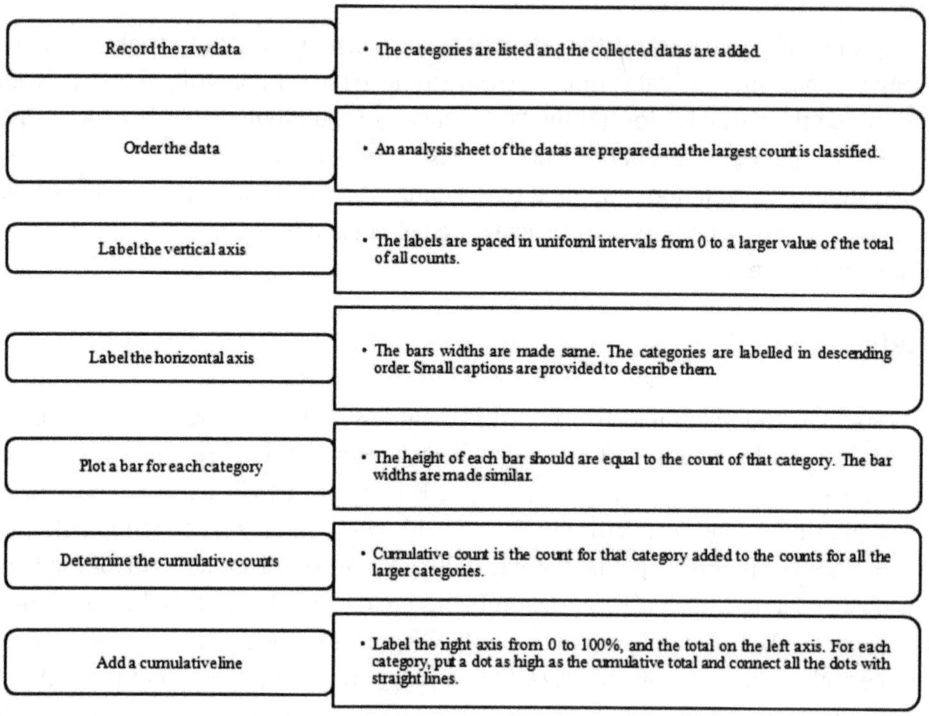

- Helps the team to focus on the inputs that have the maximum impact
- Displays the significant inputs that matter in a straightforward, visual format
- Provides an easy way to validate that any changes in the process had the desired outcome

Histogram

Histograms provide visual interpretations of the data by showing the number of data points that comes within a specified range of values called "bins" and allow a quick description of the manufacturability of the data. There are no gaps between the vertical bars of the histogram. When there is a large number of observations, a histogram helps to summarize and visualize the data easily. The histogram is useful when it is necessary to communicate useful information visually. The following is an example of how the histogram is used. Example: Ramesh is a branch manager at ABC bank. He has received customer feedback that the wait times for a client to be served by a customer service representative are too long. He writes down the time spent by each waiting customer. The findings of the times spent by 20 customers waiting is shown in Figure 9.

From the histogram, it is observed that,

- 0 customers waiting between 1 and 14 seconds
- 3 customers waiting between 1 and 23 seconds
- 6 customers waiting between 1 and 32 seconds
- 10 customers waiting between 1 and 41 seconds

Figure 9. Customer wait time

43.1	42.2
35.6	45.5
37.6	30.3
36.5	31.4
45.3	35.6
43.5	45.2
40.3	54.1
50.2	45.6
47.3	36.5
31.2	43.1

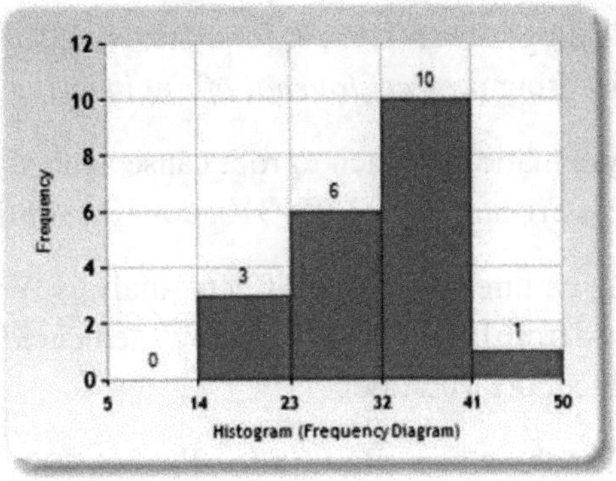

- 1 customer waiting between 1 and 50 seconds

Ramesh can conclude from the histogram that the majority of customer's waiting time is between 31 and 41 seconds.

CONCLUSION

Implementing TQM effectively in an organization gives the necessary thrust and life to the organization. TQM has a better system than traditional quality management to produce high-quality products or services to meet the consumer's requirements. It has created a better work environment for the employees. The team members focus more on the needs of the consumers. Since TQM is encouraged to apply across all organizations worldwide, the companies now compete across the global markets. The various quality tools are implemented in the companies to gain customer satisfaction, continuous quality, and process improvement and increase profit benefits. Also, constant up-gradation of skills of all the workers is achieved by continuous education and training within the organization. Through the active use of quality engineering methods and tools during the product conception, development and execution processes, industries can confirm that the voice of customers is heard and combined into the product design, that quality is designed and built into the product and waste is identified and eliminated in the manufacturing processes. Proper implementation of quality engineering tools and procedures can have a huge influence on product cost and quality and the industry as a whole.

SCOPE FOR FUTURE WORK

The chapter is aimed at addressing different quality management tools and implementation of TQM in various organizations. The future study is aimed to conduct case studies of various quality tools and evaluation and contribution of TQM in different industrial sectors.

REFERENCES

Deshpande, A. M., Siddhalingeshwar, I. G., & Ekabote, N. (2016). Implementation of advanced product quality planning in engineering project. *Journal of Engineering Education Transformations.*

Gangidi, P. (2019). A systematic approach to root cause analysis using 3× 5 why's technique. *International Journal of Lean Six Sigma, 10*(1), 295–310. doi:10.1108/IJLSS-10-2017-0114

Guidance For Performing Failure Mode And Effects Analysis With Performance Improvement Projects. (2019). *Cms.gov.* Retrieved from https://www.cms.gov/Medicare/Provider-Enrollment-and-Certification/QAPI/downloads/GuidanceForFMEA.pdf

Karuppusami, G., & Gandhinathan, R. (2006). Pareto analysis of critical success factors of total quality management: A literature review and analysis. *The TQM Magazine, 18*(4), 372–385. doi:10.1108/09544780610671048

Liliana, L. (2016). A new model of Ishikawa diagram for quality assessment. *Proceedings of the IOP Conference Series: Materials Science and Engineering.* IOP Publishing. 10.1088/1757-899X/161/1/012099

Myszewski, J.M. (2013). On improvement story by 5 whys. *The TQM journal, 25*(4), 371-383.

Omachonu, V. K., & Ross, J. E. (2004). *Principles of total quality.* CRC Press.

Pareto Analysis. (n.d.). *Improhealth.org.* Retrieved from http://www.improhealth.org/fileadmin/Documents/Improvement_Tools/Pareto_Analysis.pdf

PARETO ANALYSIS. (n.d.). *Improhealth.org.* Retrieved from .http://www.improhealth.org/fileadmin/Documents/Improvement_Tools/Pareto_Analysis.pdf

Serrat, O. (2017). The five whys technique. In *Knowledge solutions* (pp. 307–310). Singapore: Springer.

Thisse, L. C. (1998). Advanced quality planning: A guide for any organization. *Quality Progress, 31*(2), 73–77.

Compilation of References

Abdel-Basset, M., Manogaran, G., Mohamed, M., & Chilamkurti, N. (2018). Three-way decisions based on neutrosophic sets and AHP-QFD framework for supplier selection problem. *Future Generation Computer Systems*, *89*, 19–30. doi:10.1016/j.future.2018.06.024

Abdelmotalib, H. M., & Im, I.-T. (2017). Three dimensional modeling of heat transfer and bed flow in a conical fluidized bed reactor. *International Journal of Heat and Mass Transfer*, *106*, 1335–1344. doi:10.1016/j.ijheatmasstransfer.2016.10.117

Abdelmotalib, H. M., Ko, D. G., & Im, I.-T. (2016). A study on wall-to-bed heat transfer in a conical fluidized bed combustor. *Applied Thermal Engineering*, *99*, 928–937. doi:10.1016/j.applthermaleng.2016.01.054

Abdelmotalib, H. M., Youssef, M. A. M., Hassan, A. A., Youn, S. B., & Im, I.-T. (2016). Influence of the specularity coefficient on hydrodynamics and heat transfer in a conical fluidized bed combustor. *International Communications in Heat and Mass Transfer*, *75*, 169–176. doi:10.1016/j.icheatmasstransfer.2016.04.018

Abdelmotalib, H. M., Youssef, M. A. M., Hassan, A. A., Youn, S. B., & Im, I.-T. (2016). Numerical study on heat transfer in a conical fluidized bed combustor considering particle elasticity. *International Journal of Heat and Mass Transfer*, *92*, 236–243. doi:10.1016/j.ijheatmasstransfer.2015.08.069

Abdelmotalib, H. M., Youssef, M. A., Hassan, A. A., Youn, S. B., & Im, I.-T. (2015). Numerical study on the wall to bed heat transfer in a conical fluidized bed combustor. *International Journal of Precision Engineering and Manufacturing*, *16*(7), 1551–1559. doi:10.100712541-015-0205-z

Abiteboul, S., & Polyzotis, N. (2007). The Data Ring: Community Content Sharing. Proceedings of the Third Biennial Conference on Innovative Data Systems Research (CIDR 2007), Asilomar, CA, January 7-10. Academic Press.

Abiteboul, S., Benjelloun, O., & Milo, T. (2002, December). Web services and data integration. *Proceedings of the Third International Conference on Web Information Systems Engineering WISE 2002*. (pp. 3-6). IEEE.

Acosta, M., Diaz, R., & Anaya, A. (2009). Review of techniques of multiple criteria decision analysis (MCDA) as a support to complex problems: Demand predictions. *Revista Científica Guillermo de Ockham*, *7*(2), 91–110.

Alexiou, C., Jurgons, R., Seliger, C., Brunke, O., Iro, H., & Odenbach, S. (2006). Delivery of superparamagnetic nanoparticles for local chemotherapy after intra-arterial infusion and magnetic targeting. *Anticancer Research*, *27*, 2019–2022. PMID:17649815

Alexiou, C., Schmidt, A., Hulin, P., Klein, R., Bergemann, C. H., & Arnold, W. (2002). Magnetic drug targeting: Bio distribution and dependency on magnetic field strength *in vitro*. *Journal of Magnetism and Magnetic Materials*, 252, 363–366. doi:10.1016/S0304-8853(02)00605-4

Alirezaei, S., Monirvaghefi, S. M., Salehi, M., & Saatchi, A. (2007). Wear behavior of Ni–P and Ni–P–Al_2O_3 electroless coatings. *Wear, 262*(7-8), 978–985. doi:10.1016/j.wear.2006.10.013

Alonso, M. C., Luna, F. J., & Criado, M. (2019). Corrosion behavior of duplex stainless steel reinforcement in ternary binder concrete exposed to natural chloride penetration. *Construction & Building Materials, 199*, 385–395. doi:10.1016/j.conbuildmat.2018.12.036

Alonso, P., & Villegas, E. (2005). La importancia de los criterios cualitativos de los pronósticos en los agronegocios. *Revista Mexicana de Agronegocios, 9*(16), 464–477.

Alpas, A. & Zhang, J. (1992). Effect of SiC particulate reinforcement on the dry sliding wear of aluminium-silicon alloys (a356). *Wear, 155*(1), 83-104.

Alwadi, A., Gawanmeh, A., Parvin, S., & Al-Karaki, J. N. (2017). Smart Solutions for RFID Based Inventory Management Systems. *Survey, 18*(4), 347–360.

Alwadi, A., Kilby, J., & Gawanmeh, A. (2017). Tracking and Automating a Library System Using Radio Frequency Identification Technology. *International Journal on Smart Sensing and Intelligent Systems, 10*(2), 425–450. doi:10.21307/ijssis-2017-219

Aminoff, A., Kettunen, O., & Pajunen-Muhonen, H. (2002). Research on factors affecting warehousing efficiency. *International Journal of Logistics, 5*(1), 45–57.

Anaya, T. J. (2011). *Logística integral: La gestión operativa de la empresa*. Madrid: ESIC EDITORIAL.

Andrade, M. C., & Macias, E. A. (1988). Galvanized reinforcements in concrete. In A. D. Wilson, J. W. Nichols, & H. J. Prosser (Eds.), *Surface Coatings* (pp. 137–154). New York: Elsevier Applied Science. doi:10.1007/978-94-009-1351-6_5

Ansys Inc. (2011). ANSYS Fluent Theory Guide.

Antoniou, G., & Harmelen, F. V. (2004). *Semantic Web Primer*. MIT Press.

Apachitei, I., Tichelaar, F. D., Duszczyk, J., & Katgerman, L. (2002). The effect of heat treatment on the structure and abrasive wear resistance of autocatalytic Ni-P and Ni-P-SiC coatings. *Surface and Coatings Technology, 149*(2-3), 263–278. doi:10.1016/S0257-8972(01)01492-X

Araghi, A., & Paydar, M. H. (2010). Electroless deposition of Ni–P–B_4C composite coating on AZ91D magnesium alloy and investigation on its wear and corrosion resistance. *Materials & Design, 31*(6), 3095–3099. doi:10.1016/j.matdes.2009.12.042

Arango Serna, M. D., Urán Serna, C. A., & Pérez Ortega, G. (2010). Fuzzy mathematical programming applied to the material requirements planning (MRP). *Revista Técnica de la Facultad de Ingeniería. Universidad del Zulia, 33*(1), 77–86.

Arango, M., Cano, J., & Álvarez, K. (2012). Modelos de sistemas mrp cerrados integrando incertidumbre. *Revista EIA*, 61-76.

Arastoopour, H. (2001). Numerical simulation and experimental analysis of gas/solid flow systems: 1999 Fluor-Daniel Plenary lecture. *Powder Technology, 119*(2-3), 59–67. doi:10.1016/S0032-5910(00)00417-4

Arenas, M. A., Casado, C., Nobel-Pujol, V., & de Damborenea, J. (2006). Influence of the conversion coating on the corrosion of galvanized reinforcing steel. *Cement and Concrete Composites, 28*(3), 267–275. doi:10.1016/j.cemconcomp.2006.01.010

Arep, H. B., Min, G. C., Tar, A. F. B. M., & Jaafar, R. B. (2016). Material selection for automotive fender design using integrated AHP-TOPSIS. *Journal of Advanced Manufacturing Technology, 10*(2), 29–44.

Aribo, S., Fakorede, A., Ige, O., & Olubambi, P. (2017). Erosion-corrosion behaviour of aluminum alloy 6063 hybrid composite. *Wear, 376*, 608–614. doi:10.1016/j.wear.2017.01.034

Arjunwadkar, A., Basu, P., & Acharya, B. (2016). A review of some operation and maintenance issues of CFBC boilers. *Applied Thermal Engineering, 102*, 672–694. doi:10.1016/j.applthermaleng.2016.04.008

Armbrust, M., Fox, A., Griffith, R., Joseph, A. D., Katz, R. H., Konwinski, A., & Zaharia, M. (2009). Above the clouds: a Berkeley view of cloud computing. UC Berkley. Retrieved from http://www.eecs.berkeley.edu/Pubs/TechRpts/2009/EECS-2009-28.html

Armstrong, L. M., Gu, S., & Luo, K. H. (2010). Study of wall-to-bed heat transfer in a bubbling fluidised bed using the kinetic theory of granular flow. *International Journal of Heat and Mass Transfer, 53*(21-22), 4949–4959. doi:10.1016/j.ijheatmasstransfer.2010.05.047

Arrieta, J. G. (2011). Aspectos a considerar para una buena gestión en los almacenes de las empresas (Centros de Distribución, CEDIS). *Journal of Economics, Finance and Administrative Science*, 84–86.

Askaripour, H., & Molaei Dehkordi, A. (2016). Effects of initial static bed height on fractional conversion and bed pressure drop in tapered-in and tapered-out fluidized bed reactors. *International Journal of Multiphase Flow, 79*, 50–61. doi:10.1016/j.ijmultiphaseflow.2015.08.006

Aydin, K. (2007c). Vibratory characteristics of axially loaded Timoshenko beams with arbitrary number of cracks. *Journal of Vibration and Acoustics, 129*(3), 341–354. doi:10.1115/1.2731411

Babu, T. M., Sugin, M. A., & Muthukrishnan, N. (2012). Investigation on the characteristics of surface quality on machining of hybrid metal matrix composite (Al-SiC-B4C). *Procedia Engineering, 38*, 2617–2624. doi:10.1016/j.proeng.2012.06.308

Bahramian, A., & Olazar, M. (2012). Fluidization of micronic particles in a conical fluidized bed: Experimental and numerical study of static bed height effect. *AIChE Journal. American Institute of Chemical Engineers, 58*(3), 730–744. doi:10.1002/aic.12621

Baird, J. L. (1928). Improvements in or relating to apparatus for transmitting views or images to a distance. *Patent# GB292* 185.

Baker, P., & Canessa, M. (2009). Warehouse design: A structured approach. *European Journal of Operational Research, 193*(2), 425–436. doi:10.1016/j.ejor.2007.11.045

Baker, T., Aldawsari, B., & England, D. (2015). Trusted energy-efficient cloud-based services brokerage platform. *International Journal of Intelligent Computing Research, 6*(4), 630–639. doi:10.20533/ijicr.2042.4655.2015.0078

Baker, T., Al-Dawsari, B., Tawfik, H., Reid, D., & Ngoko, Y. (2015). GreeDi: An energy efficient routing algorithm for big data on cloud. *Ad Hoc Networks, 35*, 83–96. doi:10.1016/j.adhoc.2015.06.008

Baker, T., Asim, M., Tawfik, H., Aldawsari, B., & Buyya, R. (2017). An energy-aware service composition algorithm for multiple cloud-based IoT applications. *Journal of Network and Computer Applications, 89*, 96–108. doi:10.1016/j.jnca.2017.03.008

Baker, T., Ngoko, Y., Tolosana-Calasanz, R., Rana, O. F., & Randles, M. (2013, December). Energy efficient cloud computing environment via autonomic meta-director framework. *Proceedings of the 2013 Sixth International Conference on Developments in eSystems Engineering* (pp. 198-203). IEEE. 10.1109/DeSE.2013.43

Bakshi, A., Altantzis, C., Bates, R. B., & Ghoniem, A. F. (2015). Eulerian–Eulerian simulation of dense solid–gas cylindrical fluidized beds: Impact of wall boundary condition and drag model on fluidization. *Powder Technology*, *277*, 47–62. doi:10.1016/j.powtec.2015.02.056

Balaraju, J. N., & Rajam, K. S. (2005). Electroless deposition of Ni-Cu-P, Ni-W-P and Ni-W-Cu-P alloys. *Surface and Coatings Technology*, *195*(2–3), 154–161. doi:10.1016/j.surfcoat.2004.07.068

Balaraju, J. N., & Seshadri, S. K. (1999). Preparation and characterization of electroless Ni-P and Ni-P-Si_3N_4 composite coatings. *Transactions of the Institute of Metal Finishing*, *77*(2), 84–86. doi:10.1080/00202967.1999.11871253

Ballou, R. H. (2004). *Logística administración de la cadena de suministro*. Mexico: Pearson Educación.

Banerjee, S. (2012). *Augmentation of Natural Convection using Thermomagnetic Convection*. Unpublished doctoral dissertation, Jadavpur University, Kolkata, India.

Bansal, R. (2003). Coming Soon to a Wal-Mart Near You. *IEEE Antennas & Propagation Magazine*, *45*(6), 105–106. doi:10.1109/MAP.2003.1282186

Barham, P., Dragovic, B., Fraser, K., Ho, A., Neugebauer, R., Pratt, I., & Warfield, A. (2003). Xen and the art of virtualization. In Proceedings of the nineteenth acm Symposium on Operating Systems Principles(SOSP) (No.5, pp. 164-177 10.1145/945445.945462

Bartholdi, J., & Hackman, S. (2014). *Warehouse & Distribution Science*. Atlanta, GA: The Supply Chain and Logistics Institute.

Batihan, A. C., & Kadioglu, F. S. (2016a). Vibration analysis of a cracked beam on an elastic foundation. *International Journal of Structural Stability and Dynamics*, *16*(05), 1–18. doi:10.1142/S0219455415500066

Batini, C., Lenzerini, M., & Navathe, S. B. (1986). A Comparative Analysis of Methodologies for Database Schema Integration. *ACM Computing Surveys*, *18*(4), 323–364. doi:10.1145/27633.27634

Baumgaertel, H., Ehm, H., & Kasprzik, A. (2018). Collaboration in Supply Chains for Development of CPS Enabled by Semantic Web Technologies. In M. Rabe, A. A. Juan, N. Mustafee et al. (Eds.), *Proceedings of the 2018 Winter Simulation Conference* (pp. 3627-3638). Academic Press.

Bayardo, R. J., Bohrer, B., Brice, R. S., Cichocki, A., Fowler, J., Helal, A., . . . Woelk, D. (1997). InfoSleuth: Agent-Based Semantic Integration of Information in Open and Dynamic Environments. *Proceeding of the 1997 ACM SIGMOD International Conference on Management of Data (SIGMOD 1997)*, Tucson, Arizona (pp. 195-206). ACM. 10.1145/253260.253294

Begnum, K. (2012). Simplified cloud-oriented virtual machine management with MLN. *The Journal of Supercomputing*, *61*(2), 251–266. doi:10.100711227-010-0424-0

Behera, R., & Sutradhar, G. (2012). Machinability of LM6/SiCp metal matrix composites with tungsten carbide cutting tool inserts. *Journal of Engineering and Applied Sciences (Asian Research Publishing Network)*, *7*(2), 216–221.

Behzadian, M., Otaghsara, S. K., Yazdani, M., & Ignatius, J. (2012). A state-of the-art survey of TOPSIS applications. *Expert Systems with Applications*, *39*(17), 13051–13069. doi:10.1016/j.eswa.2012.05.056

Bellezze, T., Malavolta, M., Quaranta, A., Ruffini, N., & Roventi, G. (2006). Corrosion behaviour in concrete of three differently galvanized steel bars. *Cement and Concrete Composites*, *28*(3), 246–255. doi:10.1016/j.cemconcomp.2006.01.011

Bellezze, T., Timofeeva, D., Giuliani, G., & Roventi, G. (2018). Effect of soluble inhibitors on the corrosion behaviour of galvanized steel in fresh concrete. *Cement and Concrete Research*, *107*, 1–10. doi:10.1016/j.cemconres.2018.02.008

Beneventtano, D., Bergamaschi, S., Castano, S., Corni, A., Guidetti, R., Malvezzi, G., . . . Vincini, M. (2000). Information Integration: the MOMIS Project Demonstration. *Proceeding of the 26th Very Large Database Conference*, Cairo, Egypt (pp. 611-614). Academic Pres.

Benham, A., Kumar, K. A., Vettivel, S., Pandian, M. S., & Paul, C. G. (2018). The effect of matrix treatment and SiCp addition on the mechanical behaviour of glass fibre reinforced epoxy composites. *Materials Today: Proceedings*, *5*(2), 7029–7036.

Berkovsky, B. M., Medvedev, V., & Krakov, M. S. (1993). *Magnetic Fluids: Engineering Applications*. New York: Oxford University Press.

Berry, C. C., & Curtis, A. S. G. (2003). Functionalization of magnetic nanoparticles for applications in biomedicine. *Journal of Physics. D, Applied Physics*, *36*(13), 198–206. doi:10.1088/0022-3727/36/13/203

Bertolini, L. (2008). Steel corrosion and service life of reinforced concrete structures. *Structure and Infrastructure Engineering*, *4*(2), 123–137. doi:10.1080/15732470601155490

Bhimaraya, A. M. (2003). Impact of house of quality on buyer satisfaction. *Productivity*, *43*(4), 603–610.

Bhushan, R. K., Kumar, S., & Das, S. (2010). Effect of machining parameters on surface roughness and tool wear for 7075 al alloy SiC composite. *The International Journal of Advanced Manufacturing Technology*, *50*(5-8), 459-469.

Bi, H. T., Ellis, N., Abba, I. A., & Grace, J. R. (2000). A state-of-the-art review of gas-solid turbulent fluidization. *Chemical Engineering Science*, *55*(21), 4789–4825. doi:10.1016/S0009-2509(00)00107-X

Bikri, K. E., Benamar, R., & Bennouna, M. (2006b). Geometrically nonlinear free vibration of clamped-clamped beams with an edge crack. *Computers & Structures*, *84*(7), 485–502. doi:10.1016/j.compstruc.2005.09.030

Bilbao, J., Olazar, M., Arandes, J. M., & Romero, A. (1989). Optimization of the operation in a reactor with continuous catalyst circulation in the gaseous benzyl alcohol polymerization. *Chemical Engineering Communications*, *75*(1), 121–134. doi:10.1080/00986448908940672

Binetti, G., Naso, D., & Turchiano, B. (2013). Decentralized task allocation for surveillance systems with critical tasks. *Robotics and Autonomous Systems*, *61*(12), 1653–1664. doi:10.1016/j.robot.2013.06.007

Biyikli, S., Tuzla, K., & Chen, J. C. (1989). A phenomenological model for heat transfer in freeboard of fluidized beds. *Canadian Journal of Chemical Engineering*, *67*(2), 230–236. doi:10.1002/cjce.5450670209

Blanco, M. T., Andrade, C., & Macias, A. (1984). SEM Study of the Corrosion Products of Galvanized Reinforcements Immersed in Solutions in the pH Range 12·6 to 13·6. *British Corrosion Journal*, *19*(1), 41–48. doi:10.1179/000705984798273524

Bowersox, D. J., Closs, D. J., & Cooper, M. (2007). *Administración y Logística en la cadena de suministros*. México: McGraw-Hill Interamericana.

Broszeit, E. (1982). Mechanical, thermal and tribological properties of electro and chemodeposited composite coatings. *Thin Solid Films*, *95*(2), 133–142. doi:10.1016/0040-6090(82)90235-8

Bunma, M., Subarporn, P., Bobuangern, R., & Patthannkitdamrong, T., Thuan- wong, T., and Patcharawit, T. (2011). Process parameter-microstructure- mechanical property relations of SiCp-reinforced aluminum composites produced by powder-injection casting. *Journal of Metals, Materials, and Minerals*, *21*(2), 85–93.

Busciglio, A., Vella, G., Micale, G., & Rizzuti, L. (2009). Analysis of the bubbling behaviour of 2D gas solid fluidized beds. *Chemical Engineering Journal*, *148*(1), 145–163. doi:10.1016/j.cej.2008.11.010

Busse, G., Wu, D., & Karpen, W. (1992). Thermal wave imaging with phase sensitive modulated thermography. *Journal of Applied Physics*, *71*(8), 3962–3965. doi:10.1063/1.351366

Bustos-Flores, C. E., & Chacón, P. G. (2007). *El MRP En la gestión de inventarios*. Visión Gerencial.

Buyya, R., Beloglazov, A., & Abawajy, J. (2010). Energy-efficient management of data center resources for cloud computing: a vision, architectural elements, and open challenges. *Proceedings of the 2010 International Conference on Parallel and Distributed Processing Techniques and Applications, PDPTA 2010*. Academic Press.

Buyya, R., Yeo, C. S., Venugopal, S., Broberg, J., & Brandic, I. (2009). Cloud computing and emerging IT platforms: Vision, hype and reality for delivering computing as the 5th utility. *Future Generation Computer Systems*, *25*(6), 599–616. doi:10.1016/j.future.2008.12.001

Cáceres, C. D., Reyes, V. J., García, C. M., & Sanchez, R. C. (2015). Modelo de Programación Lineal para Planeación de Requerimiento de Materiales. *Revista Tecnológica ESPOL*, 24-33.

Çalışkan, H., Kurşuncu, B., Kurbanoğlu, C., & Güven, Ş. Y. (2013). Material selection for the tool holder working under hard milling conditions using different multi criteria decision making methods. *Materials & Design*, *45*, 473–479. doi:10.1016/j.matdes.2012.09.042

Caloud, P., Choi, W., Latombe, J. C., Le Pape, C., & Yim, M. (1990, July). Indoor automation with many mobile robots. *Proceedings of the IEEE International Workshop on Intelligent Robots and Systems, Towards a New Frontier of Applications* (pp. 67-72). IEEE. 10.1109/IROS.1990.262370

Caminero, M. A., García-Moreno, I., Rodríguez, G. P., & Chacónb, J. M. (2019). Internal damage evaluation of composite structures using phased array ultrasonic technique: Impact damage assessment in CFRP and 3D printed reinforced composites. *Composite Part B*, *165*, 131–142. doi:10.1016/j.compositesb.2018.11.091

Cano, A. J. (2011). *Modelo de un Sistema MRP Cerrado Integrando Incertidumbre en los Tiempos de Entrega, Disponibilidad de la Capacidad de Fabricación e Inventarios. Universidad Nacional de Colombia*. Medellín: Magister Tesis.

Capozucca, R., & Magagnini, E. (2017a). Analysis of cracked RC beams under vibration. *Journal of Physics: Conference Series*, *842*, 012076. doi:10.1088/1742-6596/842/1/012076

Cardenas, D. M., & Urquiaga, A. J. (2007). Logística de operaciones: integrando las decisiones estratégicas para la competitividad. *Ingeniería Industrial*, *28*(1), 37–41.

Carey, M., Haas, L., Schwarz, P., Arya, M., Cody, W., Fagin, R., . . . Wimmers, E. (1995). Towards Heterogeneous Multimedia Information Systems: The Garlic Approach. *Proceeding of the 5th International Workshop on Research Issues in Data Engineering-Distributed Object Management (RIDE-DOM 1995)*, Taipei, Taiwan, March 6-7 (pp. 124-131). Academic Press. 10.1109/RIDE.1995.378736

Casanovas, A., & Cuatrecasas, L. (2011). *Logistica Empresarial. Gestión Integral de la información y material en la empresa*. Barcelona: Ediciones Gestión 2000.

Castro, H., Rodríguez, C., Belzunce, F. J., & Canteli, A. F. (2003). Mechanical properties and corrosion behaviour of stainless steel reinforcing bars. *Journal of Materials Processing Technology*, *143*, 134–137. doi:10.1016/S0924-0136(03)00393-5

Cavallini, C., Giorgetti, A., Citti, P., & Nicolaie, F. (2013). Integral aided method for material selection based on quality function deployment and comprehensive VIKOR algorithm. *Materials & Design*, *47*, 27–34. doi:10.1016/j.matdes.2012.12.009

Cebers, A. (2005). Flexible magnetic swimmer. *Magnetohydrodynamics*, *41*, 63–72.

Ceri, S., & Widom, J. (1993). Managing Semantic Heterogeneity with Production Rules and Persistent Queues. *Proceedings of the 19th Very Large Data Base Conference*, Dublin, Ireland (pp. 108-119). Academic Press.

Chadawada, R., Sarfaraz, A., Jenab, K., & Pourmohammadi, H. (2015). Integration of AHP-QFD for selecting facility location. *Benchmarking: An International Journal*, *22*(3), 411–425. doi:10.1108/BIJ-06-2013-0064

Chalermsinsuwan, B., Kuchonthara, P., & Piumsomboon, P. (2010). CFD modeling of tapered circulating fluidized bed reactor risers: Hydrodynamic descriptions and chemical reaction responses. *Chemical Engineering and Processing: Process Intensification, 49*(11), 1144-1160. doi:10.1016/j.cep.2010.08.016

Chalermsinsuwan, B., Gidaspow, D., & Piumsomboon, P. (2011). Two- and three-dimensional CFD modeling of Geldart A particles in a thin bubbling fluidized bed: Comparison of turbulence and dispersion coefficients. *Chemical Engineering Journal*. doi:10.1016/j.cej.2011.04.007

Chalermsinsuwan, B., Piumsomboon, P., & Gidaspow, D. (2009). Kinetic theory based computation of PSRI riser: Part II-Computation of mass transfer coefficient with chemical reaction. *Chemical Engineering Science, 64(6), 1212-1222.* doi:10.1016/j.ces.2008.11.006

Chang, C. W., Wu, C. R., Lin, C. T., & Lin, H. L. (2007). Evaluating digital video recorder systems using analytic hierarchy and analytic network processes. *Information Sciences*, *177*(16), 3383–3396. doi:10.1016/j.ins.2007.02.010

Chang, P., & Wang, Y. (2006). Fuzzy Delphi and back-propagation model for sales forecasting in PCB industry. *Expert Systems with Applications*, *30*(4), 715–726. doi:10.1016/j.eswa.2005.07.031

Chang, Y., Zhu, X., & Haghani, A. (2019). Modeling and solution of joint storage space allocation and handling operation for outbound containers in rail-water intermodal container terminals. *IEEE Access: Practical Innovations, Open Solutions*.

Chan, L. K., & Wu, M. L. (2002). Quality function deployment: A comprehensive review of its concepts and methods. *Quality Engineering*, *15*(1), 23–35. doi:10.1081/QEN-120006708

Chao, X., & Yiqiang, Z. (1998). Analysis of multi-server queues with station and server vacations. *European Journal of Operational Research*, *110*(2), 392–406. doi:10.1016/S0377-2217(97)00253-1

Chasalevris, A. C., & Papadopoulos, C. A. (2006a). Identification of multiple cracks in beams under bending. *Mechanical Systems and Signal Processing*, *20*(7), 1631–1673. doi:10.1016/j.ymssp.2006.03.008

Chase, J. S., Anderson, D. C., Thakar, P. N., Vahdat, A. M., & Doyle, R. P. (2001). Managing energy and server resources in hosting centers. *Proceedings of the 18th ACM Symposium on Operating Systems Principles* (pp. 103-116). New York, NY: ACM. 10.1145/502034.502045

Chase, R., Jacobs, F., & Aquilano, N. (2014). *Operations and supply chain management* (13th ed.). Mexico: Mc Graw Hill.

Chati, M. R., & Mukharjee, S. (1997c). Modal analysis of a cracked beam. *Journal of Sound and Vibration*, *207*(2), 249–270. doi:10.1006/jsvi.1997.1099

Chaudhuri, S., & Dayal, U. (1997). An Overview of Data Warehousing and OLAP Technology. *SIGMOD Record*, *26*(1), 65–74. doi:10.1145/248603.248616

Chavan, V. B., & Gaikwad, M. U. (2016). Review on development of glass fiber/epoxy composite material and its characterizations. *International Journal of Science, Engineering and Technology Research*, *5*, 2224–2228.

Chawathe, S. GarciaMolina, H., Hammer, J., Ireland, K., Papakonstantinou, Y, Ullman, J., &Widom, J. (1994). The TSIMMIS Project: Integration of Heterogeneous Information Sources. *Proceeding of the 1000th Anniversary Meeting of the Information Processing Society of Japan (IPSJ)*, Tokyo, Japan (pp. 7-18). Academic Press.

Chawla, N., Deng, X., & Schnell, D. (2006). Thermal expansion anisotropy in extruded SiC particle reinforced 2080 aluminum alloy matrix composites. *Materials Science and Engineering A*, *426*(1-2), 314–322. doi:10.1016/j.msea.2006.04.054

Chebiyyam, M., Malviya, R., Bose, S. K., & Sundarrajan, S. (2009). Server consolidation: Leveraging the benefits of virtualization. *Infosys Research*, *7*(1), 65–75.

Chen, C. C. (2009). Integration of quality function deployment and process management in the semiconductor industry. *International Journal of Production Research*, *47*(6), 1469–1484. doi:10.1080/00207540701549616

Chen, L., Langevin, A., & Riopel, D. (2011). A tabu search algorithm for the relocation problem in a warehousing system. *International Journal of Production Economics*, *129*(1), 147–156. doi:10.1016/j.ijpe.2010.09.012

Chen, P., & Hoshi, T. (1992). High-performance machining of SiC whisker- reinforced aluminium composite by self-propelled rotary tools. *CIRP Annals*, *41*(1), 59–62. doi:10.1016/S0007-8506(07)61152-4

Chen, P.-S., Huang, C.-Y., Yu, C.-C., & Hung, C.-C. (2017). The examination of key performance indicators of warehouse operation systems based on detailed case studies. *Journal of Information and Optimization Sciences*, *38*(2), 367–389. doi:10.1080/02522667.2016.1224465

Chen, S. C., Yang, C. C., Lin, W. T., Yeh, T. M., & Lin, Y. S. (2007). Construction of key model for knowledge management system using AHP-QFD for semiconductor industry in Taiwan. *Journal of Manufacturing Technology Management*, *18*(5), 576–597. doi:10.1108/17410380710752671

Chen, X. H., Chen, C. S., Xiao, H. N., Liu, H. B., Zhou, L. P., Li, S. L., & Zhang, G. (2006). Dry friction and wear characteristics of nickel/carbon nanotube electroless composite deposits. *Tribology International*, *39*(1), 22–28. doi:10.1016/j.triboint.2004.11.008

Cheung, S. C. P., Yeoh, G. H., & Tu, J. Y. (2007). On the numerical study of isothermal vertical bubbly flow using two population balance approaches. *Chemical Engineering Science*, *62*(17), 4659–4674. doi:10.1016/j.ces.2007.05.030

Chiavenato, I. (2006). *Introducción a la teoría general de la administración.* México: MCGraw-Hill Interamericana.

Chiavenato, I. (1993). *Iniciación a la administración de materiales*. México: McGraw-Hill.

Choi, T., Wing-Kwan, Y., Cheng, T. C. E., & Yue, X. (2018). Optimal Scheduling, Coordination, and the Value of RFID Technology in Garment Manufacturing Supply Chains. *IEEE Transactions on Engineering Management*, *65*(1), 72–84. doi:10.1109/TEM.2017.2739799

Chondros, T. G. (2001). The continuous crack flexibility model for crack identification. *Fatigue & Fracture of Engineering Materials & Structures*, *24*(10), 643–650. doi:10.1046/j.1460-2695.2001.00442.x

Chondros, T. G., & Dimarogonas, A. D. (1980). Identification of cracks in welded joints of complex structures. *Journal of Sound and Vibration*, *69*(4), 531–538. doi:10.1016/0022-460X(80)90623-9

Chondros, T. G., Dimarogonas, A. D., & Yao, J. A. (1998). Continuous cracked beam vibration theory. *Journal of Sound and Vibration*, *215*(1), 17–34. doi:10.1006/jsvi.1998.1640

Christides, S., & Barr, A. D. S. (1984a). One-dimensional theory of cracked Bernoulli-Euler beams. *International Journal of Mechanical Sciences*, *26*(11-12), 639–648. doi:10.1016/0020-7403(84)90017-1

Ciampa, F., Scarselli, G., & Meo, M. (2017). On the generation of nonlinear damage resonance intermodulation for elastic wave spectroscopy. *The Journal of the Acoustical Society of America*, *141*(4), 2364–2374. doi:10.1121/1.4979256 PMID:28464640

Codd, E. F. (1970). *The Relational Model for Database Management*. Addison Wesley.

Coello, C. A. C., Pulido, G. T., & Lechuga, M. S. (2004). Handling multiple objectives with particle swarm optimization. *IEEE Transactions on Evolutionary Computation*, *8*(3), 256–279. doi:10.1109/TEVC.2004.826067

Collins, J. (1980). *Integrated Manufacturing- a state of art*. IET.

Corbett, C. J., & Karmarkar, U. S. (2001). Competition and Structure in Serial Supply Chains with Deterministic Demand. *Management Science*, *47*(7), 966–978. doi:10.1287/mnsc.47.7.966.9799

Criado, M., Bastidas, D. M., Fajardo, S., Fernández-Jiménez, A., & Bastidas, J. M. (2011). Corrosion behaviour of a new low-nickel stainless steel embedded in activated fly ash mortars. *Cement and Concrete Composites*, *33*(6), 644–652. doi:10.1016/j.cemconcomp.2011.03.014

Cure, O., & Blin, G. (2015). *RDF Database Systems Triples Storage and SPARQL Query Processing*. Waltham, MA: Morgan Kaufmann.

Dante, R. C. (2015). *Handbook of friction materials and their applications*. Woodhead Publishing.

Das, D., & Mukherjee, K. (2008). Development of an AHP-QFD framework for designing a tourism product. *International Journal of Services and Operations Management*, *4*(3), 321–344. doi:10.1504/IJSOM.2008.017297

Das, S. K., Choi, S., Yu, W., & Pradeep, T. (2008). *Nanofluids: Science and Technology*. Hoboken, NJ: John Wiley & Sons, Inc.

Davim, J. P. (2002). Diamond tool performance in machining metal-matrix composites. *Journal of Materials Processing Technology*, *128*(1-3), 100–105. doi:10.1016/S0924-0136(02)00431-4

Davim, J. P., & Antonio, C. C. (2001). Optimisation of cutting conditions in the machining of aluminium matrix composites using a numerical and experimental model. *Journal of Materials Processing Technology*, *112*(1), 78–82. doi:10.1016/S0924-0136(01)00551-9

Davison, A. J., & Murray, D. W. (1998, June). Mobile robot localisation using active vision. *Proceedings of the European conference on computer vision* (pp. 809-825). Springer.

De Felice, F., & Petrillo, A. (2010). A multiple choice decision analysis: An integrated QFD–AHP model for the assessment of customer needs. *International Journal of Engineering Science and Technology*, *2*(9), 25–38.

De Koster, R. B., Johnson, A. L., & Roy, D. (2017). Warehouse design and management. *International Journal of Production Research*, *55*(21), 6327–6330. doi:10.1080/00207543.2017.1371856

De Koster, R., Le-Duc, T., & Roodbergen, K. (2007). Design and control of warehouse order picking: A literature review. *European Journal of Operational Research*, *182*(2), 481–501. doi:10.1016/j.ejor.2006.07.009

Deiva Venkatesh, R., Chaouki, J., & Klvana, D. (1996). Fluidization of cryogels in a conical column. *Powder Technology*, *89*(3), 179–186. doi:10.1016/S0032-5910(96)03181-6

Demir, A., & Mermertas, V. (2008a). Natural frequencies of annular plates with circumferential cracks by means of sector element. *Engineering Fracture Mechanics*, *75*(5), 1143–1155. doi:10.1016/j.engfracmech.2007.04.016

Demir, L., Tunali, S., & Tursel, E. D. (2014). The state of the art on buffer allocation problem: A comprehensive survey. *Journal of Intelligent Manufacturing*, *25*(3), 371–392. doi:10.100710845-012-0687-9

Dentsoras, A., & Dimarogonas, A. D. (1983c). Resonance controlled fatigue crack propagation in a beam under longitudinal vibrations. *International Journal of Fracture, 23*(1), 15–22. doi:10.1007/BF00020154

Depypere, F., Pieters, J. G., & Dewettinck, K. (2005). Expanded bed height determination in a tapered fluidised bed reactor. *Journal of Food Engineering, 67*(3), 353–359. doi:10.1016/j.jfoodeng.2004.04.039

Deshpande, A. M., Siddhalingeshwar, I. G., & Ekabote, N. (2016). Implementation of advanced product quality planning in engineering project. *Journal of Engineering Education Transformations.*

Deuis, R., Subramanian, C., & Yellup, J. (1997). Dry sliding wear of aluminium composites—A review. *Composites Science and Technology, 57*(4), 415–435. doi:10.1016/S0266-3538(96)00167-4

Deus, J. M., Freire, L., Montemor, M. F., & Nóvoa, X. R. (2012). The corrosion potential of stainless steel rebars in concrete: Temperature effect. *Corrosion Science, 65*, 556–560. doi:10.1016/j.corsci.2012.09.001

Dhar, V., & Stein, R. (1997). *Seven methods for transforming corporate data into business intelligence, Nueva Jersey.* Prentice-Hall.

Dhoka, D., & Choudary, Y. (2013). ABC Classification for Inventory Optimization. *IOSR Journal of Business and Management, 15*(1), 38–41. doi:10.9790/487X-1513841

Dijkstra, E. W. (1959). A note on two problems in connexion with graphs. *Numerische Mathematik, 1*(1), 269–271. doi:10.1007/BF01386390

Dillenz, A., Zweschper, T., Riegert, G., & Busse, G. (2003). Progress in phase angle thermography. *The Review of Scientific Instruments, 74*(1), 417–419. doi:10.1063/1.1524010

Dimarogonas, A. D. (1970). *Dynamic response of cracked rotors*. Schenectady, NY: General Electric Co.

Dimarogonas, A. D. (1996). Vibration of cracked structures: The state of the art review. *Engineering Fracture Mechanics, 55*(5), 831–857. doi:10.1016/0013-7944(94)00175-8

Dimarogonas, A. D., & Papadopoulos, C. A. (1983a). Vibration of cracked shafts in bending. *Journal of Sound and Vibration, 91*(4), 583–593. doi:10.1016/0022-460X(83)90834-9

Dionysopoulos, D., Fierro, G. M., Meo, M., & Ciampa, F. (2018). Imaging of barely visible impact damage on a composite panel using nonlinear wave modulation thermography. *NDT & E International, 95*, 9–16. doi:10.1016/j.ndteint.2018.01.005

Dirscherl, M., Esser, G., & Schmidt, M. (2006). Ultrashort pulse laser bending. *Journal of Laser Micro Nanoengineering, 1*(1), 54–60. doi:10.2961/jlmn.2006.01.0011

Dixit, U. S., Joshi, S. N., & Kant, R. (2015). Laser forming systems: A review. *International Journal of Mechatronics and Manufacturing Systems, 8*(3/4), 160–205. doi:10.1504/IJMMS.2015.073077

Doebling, S. W., Ferrar, C. R., & Prome, M. B. (1998a). A summary review of vibration-based damage Identification methods. *The Shock and Vibration Digest, 30*(2), 91–105. doi:10.1177/058310249803000201

Dolgui, A., Ivanov, D., Sethi, S. P., & Sokolov, B. (2019). Scheduling in production, supply chain and Industry 4.0 systems by optimal control: Fundamentals, state-of-the-art and applications. *International Journal of Production Research, 57*(2), 411–432. doi:10.1080/00207543.2018.1442948

Dong, S., Zhao, B., Lin, C., Du, R., Hu, R., & Zhang, G. X. (2012). Corrosion behavior of epoxy/zinc duplex coated rebar embedded in concrete in ocean environment. *Construction & Building Materials, 28*(1), 72–78. doi:10.1016/j.conbuildmat.2011.08.026

Dora, D. T. K., Panda, S. R., Mohanty, Y. K., & Roy, G. K. (2013). Hydrodynamics of gas–solid fluidization of a homogeneous ternary mixture in a conical bed: Prediction of bed expansion and bed fluctuation ratios. *Particuology*, *11*(6), 681–688. doi:10.1016/j.partic.2013.02.009

Doshi, B. (1986). Queueing systems with vacations-a survey. *Queueing Systems*, *1*(1), 29–66. doi:10.1007/BF01149327

Duari, S., Mukhopadhyay, A., Barman, T. K., & Sahoo, P. (2018). Tribological Performance Optimization of Electroless Nickel Coatings Under Lubricated Condition. In K. Kumar & J. Davim (Eds.), *Design and Optimization of Mechanical Engineering Products* (pp. 250–280). Hershey, PA: IGI Global. doi:10.4018/978-1-5225-3401-3.ch013

Duarte, R. G., Castela, A. S., Neves, R., Freire, L., & Montemor, M. F. (2014). Corrosion behavior of stainless steel rebars embedded in concrete: An electrochemical impedance spectroscopy study. *Electrochimica Acta*, *124*, 218–224. doi:10.1016/j.electacta.2013.11.154

Duc, D. N., Lee, H., Konidala, D. M., & Kim, K. (2009). Open issues in RFID security. *Proceedings of the International Conference on Internet Technology and Security Transactions (ICITST)*. Academic Press.

DuCharme, B. (2013). *Learning SPARQL: Querying and Updating with SPARQL 1.1*. Sebastopol, CA: O'Reilly Publishing Company.

Durao, F., Carvalho, J. F. S., Fonseka, A., & Garcia, V. C. (2014). A systematic review on cloud computing. *The Journal of Supercomputing*, *68*(3), 1321–1346. doi:10.100711227-014-1089-x

Durga Prasad, K. G., Venkata Subbaiah, K., & Narayana Rao, K. (2012). Aligning the competitive strategy with supply chain strategy through QFD. *Journal of Advances in Management Research*, *9*(2), 189–198. doi:10.1108/09727981211271931

Dursun, M., & Arslan, Ö. (2018). An integrated decision framework for material selection procedure: A case study in a detergent manufacturer. *Symmetry*, *10*(11), 657. doi:10.3390ym10110657

Dwivedi, S., Kumar, S., & Kumar, A. (2012). Effect of turning parameters on surface roughness of a356/5% SiC composite produced by electromagnetic stir casting. *Journal of Mechanical Science and Technology*, *26*(12), 3973–3979. doi:10.100712206-012-0914-5

Dyrwal, A., Meo, M., & Ciampa, F. (2018). Nonlinear air-coupled thermosonics for fatigue micro-damage detection and localization. *NDT & E International*, *97*, 59–67. doi:10.1016/j.ndteint.2018.03.012

Ebrahimian-Hosseinabadi, M., Azari-Dorcheh, K., & Vaghefi, S. M. M. (2006). Wear behavior of electroless Ni-P-B_4C composite coatings. *Wear*, *260*(1-2), 123–127. doi:10.1016/j.wear.2005.01.020

Economic Commission for Latin America and the Caribbean [CEPAL]. (2018). *Acerca de Microempresas y Pymes*. Retrieved from: https://www.cepal.org/es/temas/pymes/acerca-microempresas-pymes

Efendigil, T., Önüt, S., & Kahraman, C. (2009). A decision support system for demand forecasting with artificial neural networks and neuro-fuzzy models: A comparative analysis. *Expert Systems with Applications*, *36*(3), 6697–6707. doi:10.1016/j.eswa.2008.08.058

El-Gallab, M., & Sklad, M. (1998). Machining of Al/SiC particulate metal matrix composites: Part ii: Workpiece surface integrity. *Journal of Materials Processing Technology*, *83*(1-3), 277–285. doi:10.1016/S0924-0136(98)00072-7

Elnozahy, E., Kistler, M., & Rajamony, R. (2003). Energy-efficient server clusters. *Proceedings of the International Workshop on Power-Aware Computer Systems* (pp. 179-197). Springer. 10.1007/3-540-36612-1_12

Elsener, B., Crobu, M., Scorciapino, M. A., & Rossi, A. (2008). Electroless deposited Ni–P alloys: Corrosion resistance mechanism. *Journal of Applied Electrochemistry*, *38*(7), 1053–1060. doi:10.100710800-008-9573-8

Emmett, S. (2005). *Excellence in Warehouse Management: How to Minimise Costs and Maximise Value*. United Kingdom: John Wiley & Sons Ltd.

Emmett, S. (2005). *Excellence in Warehouse*. England: John Wiley & Sons, Ltd.

Erdoğdu, Ş., Bremner, T. W., & Kondratova, I. L. (2001). Accelerated testing of plain and epoxy-coated reinforcement in simulated seawater and chloride solutions. *Cement and Concrete Research*, *31*(6), 861–867. doi:10.1016/S0008-8846(01)00487-2

Erkut, H. (1983). *Sistem Dinamiğinin Temelleri*. İstanbul: İTÜ Fen Edebiyat Fakültesi Ofset Atölyesi.

Escoda, I., Ortega, A., Sanz, A., & Herms, A. (1997). Demand forecast by neuro-fuzzy techniques. *Proceedings of the Sixth IEEE International Conference*, 3, 1381-1386.

Escudero, J. (2014). *Logística de almacenamiento. Madrid*, Spain: Ediciones Paraninfo.

Fagin, R., Kolaitis, P. G., Miller, R. J., & Popa, L. (2005a). Data Exchange: Semantics and Query Answering. *Theoretical Computer Science*, *336*(1), 89–124. doi:10.1016/j.tcs.2004.10.033

Fagin, R., Kolaitis, P. G., Popa, L., & Tan, W. (2005b). Composing schema mappings: Second-order dependencies to the rescue. *ACM Transactions on Database Systems*, *30*(4), 994–1055. doi:10.1145/1114244.1114249

Fan, X., Weber, W. D., & Barroso, L. A. (2007). Power provisioning for a warehouse sized computer. *Proceedings of the 34th Annual International Symposium on Computer Architecture* (pp. 13-23). 10.1145/1250662.1250665

Farahnakian, F., Pahikkala, T., Liljeberg, P., Plosila, J., & Tenhunen, H. (2014). Multiagent based architecture for dynamic vm consolidation in cloud data centers. *Proceedings of the 40th IEEE Euromicro Conference on Software Engineering and Advanced Applications Software Engineering and Advanced Applications* (pp. 111-118). IEEE Press.

Farahnakian, F., Pahikkala, T., Liljeberg, P., Plosila, J., & Tenhunen, H. (2014). Multi agent based architecture for dynamic vm consolidation in cloud data centers. *Proceedings of the 40th IEEE Euromicro Conference on Software Engineering and Advanced Applications* (pp. 111–119). IEEE Press. 10.1109/SEAA.2014.56

Farina, S. B., & Duffó, G. S. (2007). Corrosion of zinc in simulated carbonated concrete pore solutions. *Electrochimica Acta*, *52*(16), 5131–5139. doi:10.1016/j.electacta.2007.01.014

Fattahi, M., Hosseini, S. H., Ahmadi, G., & Parvareh, A. (2019). Numerical simulation of heat transfer coefficient around different immersed bodies in a fluidized bed containing Geldart B particles. *International Journal of Heat and Mass Transfer*, *141*, 353–366. doi:10.1016/j.ijheatmasstransfer.2019.06.058

Favro, L. D., Han, X., Ouyang, Z., Sun, G., Sui, H., & Thomas, R. L. (2000). Infrared imaging of defects heated by a sonic pulse. *The Review of Scientific Instruments*, *71*(6), 2418–2421. doi:10.1063/1.1150630

Feller, E., Morin, C., & Esnault, A. (2012). A case for fully decentralized dynamic VM consolidation in clouds. *Proceedings of the IEEE International Conference on Cloud Computing Technology and Science* (pp. 26-33). IEEE Press. 10.1109/CloudCom.2012.6427585

Ferencik, R.C., Soderquist, K.M., & Phillips, M.A. (2014). Building Systems Thinking Capacity: An Essential Skill Set For Policymakers. *Proceedings of the 32nd International Conference of the System Dynamics Society*, Delft, Netherlands. Academic Press.

Fetene, B. N., Kumar, V., Dixit, U. S., & Echempati, R. (2018). Numerical and experimental study on multi-pass laser bending of AH36 steel strips. *Optics & Laser Technology*, *99*(1), 291–300. doi:10.1016/j.optlastec.2017.09.014

Fierro, G. P. M., Ginzburg, D., Ciampa, F., & Meo, M. (2016). Nonlinear thermosonics and laser vibrometry for barely visible impact damage of a composite stiffener panel. *Proceedings of the Society for Photo-Instrumentation Engineers, 980*, 419.

Fierro, G. P. M., Ginzburg, D., Ciampa, F., & Meo, M. (2017). Imaging of Barely Visible Impact Damage on a Complex Composite Stiffened Panel Using a Nonlinear Ultrasound Stimulated Thermography Approach. *Journal of Nondestructive Evaluation, 36*(4), 21. doi:10.100710921-017-0449-x

Flint, G. N., & Cox, R. N. (1988). The resistance of stainless steel partly embedded in concrete to corrosion by seawater. *Magazine of Concrete Research, 40*(142), 13–27. doi:10.1680/macr.1988.40.142.13

Flores, B., Stading, G., & Klassen, R. (2007). The business forecasting process: A comparison of differences between small and large Canadian manufacturing and service firms. *International Journal of Management and Enterprise Development, 4*(4), 387–402. doi:10.1504/IJMED.2007.013152

Flores, B., & Whybark, D. (1988). Know Your ABC. *Management Decision, 26*(3), 20–24. doi:10.1108/eb001494

Fluent 6.1.22, User's Guide, 20.4, Eulerian Model. (2005). Fluent Inc.

Fluent Inc. (2006). Fluent 6.3 User's Guide.

Fogliatto, F. S., Da Silveira, G. J., & Royer, R. (2003). Flexibility-driven index for measuring mass customization feasibility on industrialized products. *International Journal of Production Research, 41*(8), 1811–1829. doi:10.1080/1352816031000074991

Ford, A. (1999). *Modeling the Environment: An Introduction to System Dynamics Modeling of Environmental Systems.* Washington, DC: Island Press.

Forquin, P., Tran, L., Louvigne, P.-F., Rota, L., & Hild, F. (2003). Effect of aluminum reinforcement on the dynamic fragmentation of SiC ceramics. *International Journal of Impact Engineering, 28*(10), 1061–1076. doi:10.1016/S0734-743X(03)00034-4

Forrester, J.W. (1995). *Road Map 1: Counterintuitive Behaviour Of Social Systems.* MIT System Dynamics In Education Project.

Forrester, J. W. (1962). *Industrial Dynamics.* Waltham, MA: Pegasus Communications.

Forrester, J. W. (1969). *Urban Dynamics. Cambridge*, MA: The MIT Press.

Frank, C., Garg, A., Raheja, A., & Sztandera, L. (2003). Forecasting women's apparel sales using mathematical modeling. *International Journal of Clothing Science and Technology, 15*(2), 107–125. doi:10.1108/09556220310470097

Frazelle, E., & Sojo, R. (2007). *Logística de almacenamiento y manejo de materiales de clase mundial.* Bogotá: Norma S.A.

Friedman, M., Levy, A., & Millstein, T. (1999). Navigational plans for data integration. *Proceedings of AAAI*, Menlo Park, CA (pp. 67-73). AAAI Press/ MIT Press.

Gaither, N., & Frazier, G. (2000). *Administración de Producción y Operaciones.* México: International Thomson Editores.

Gangidi, P. (2019). A systematic approach to root cause analysis using 3× 5 why's technique. *International Journal of Lean Six Sigma, 10*(1), 295–310. doi:10.1108/IJLSS-10-2017-0114

Gangil, M., & Pradhan, M. K. (2016). Optimization the machining parameters by using VIKOR Method during the EDM process of Titanium alloy.

Gangil, M., Pradhan, M. K., & Purohit, R. (2017). Review on modelling and optimization of electrical discharge machining process using modern Techniques. *Materials Today: Proceedings*, *4*(2), 2048–2057.

Gao, H., Liu, X., Zhang, S., & Qi, J. (2018). Synergistic effect of glass fibre and al powder on the mechanical properties of glass-ceramics. *Ceramics International*, *44*(13), 15167–15175. doi:10.1016/j.ceramint.2018.05.155

García, C. A. (1999). *Almacenes: Planeación, organización y control*. México: TRILLAS.

Garcia-Solaco, M., Saltor, F., & Castellanos, M. (1996). Semantic Heterogeneity in Multidatabase Systems, In O.A. Bukhres & A.K. Elmagarmid (Eds.), Object-Oriented Multidatabase Systems, A Solution for Advanced Applications (pp. 129-202). Prentice-Hall.

Gast, A. P., & Zukoski, C. F. (1989). Electro-rheological fluids as colloidal suspensions. *Advances in Colloid and Interface Science*, *30*, 153–202. doi:10.1016/0001-8686(89)80006-5

Geiger, M., & Vollertsen, F. (1993). The mechanisms of laser forming. *CIRP Annals*, *42*(1), 301–304. doi:10.1016/S0007-8506(07)62448-2

Gera, D., & Gautam, M. (1995). Bubble rise velocity in two-dimensional fluidized beds. *Powder Technology*, *84*(3), 283–285. doi:10.1016/0032-5910(95)02993-C

Ghadge, A., Karantoni, G., Chaudhuri, A., & Srinivasan, A. (2018). Impact of additive manufacturing on aircraft supply chain performance. *Journal of Manufacturing Technology Management*, *29*(5), 846–865. doi:10.1108/JMTM-07-2017-0143

Ghiani, G., Laporte, G., & Musmanno, R. (2013). *Introduction to Logistics Systems Management*. Chichester: John Wiley & Sons, Ltd. doi:10.1002/9781118492185

Gidaspow, D. (1994). *Multiphase Flow and Fluidization: Continuum and Kinetic Theory Descriptions*. New York: Academic Press.

Gill, S. S., Chana, I., Singh, M., & Buyya, R. (2018). CHOPPER: An intelligent QoS-aware autonomic resource management approach for cloud computing. *Cluster Computing*, *21*(2), 1203–1241. doi:10.100710586-017-1040-z

Gleiter, A., Spießberger, C., & Busse, G. (2007). Improved ultrasound activated thermography using frequency analysis. *Quantitative Infrared Thermography Journal*, *4*(2), 155–164. doi:10.3166/qirt.4.155-164

Goetschalckx, M., McGinnis, L., Sharp, G., Bodner, D., Govindaraj, T., & Huang, K. (2001). Development of a design methodology for warehousing systems: hierarchical framework. *Proceedings of the Industrial Engineering Research*. Academic Press.

Gollo, M. H., Nadi, G., Mehdi, M., & Abbaszadeh, M. (2015). Experimental and numerical study of spiral scan paths on cap laser forming. *Journal of Laser Applications*, *27*(1), 0120023.

Gonzalez, C. (2013). *Control de aprovisionamiento de materias primas*. Madrid, España: Paraninfo.

Gonzalez, J. A., & Andrade, C. (1982). Effect of carbonation, chlorides and relative ambient humidity on the corrosion of galvanized rebars embedded in concrete. *British Corrosion Journal*, *17*(1), 21–28. doi:10.1179/000705982798274589

Groover, M. P. (2007). *Automation, production systems, and computer-integrated manufacturing*. Prentice Hall Press.

Gruber, T. R. (1993). A translation approach to portable ontology specifications. *Knowledge Acquisition*, *5*(2), 199–220. doi:10.1006/knac.1993.1008

Gudmundson, P. (1982). Eigen frequency changes of structures due to cracks, notches or other geometrical changes. *Journal of the Mechanics and Physics of Solids*, *30*(5), 339–353. doi:10.1016/0022-5096(82)90004-7

Gudmundson, P. (1983b). The dynamic behavior of slender structures with cross-sectional cracks. *Journal of the Mechanics and Physics of Solids*, *31*(4), 329–345. doi:10.1016/0022-5096(83)90003-0

Guidance For Performing Failure Mode And Effects Analysis With Performance Improvement Projects. (2019). *Cms.gov*. Retrieved from https://www.cms.gov/Medicare/Provider-Enrollment-and-Certification/QAPI/downloads/GuidanceForFMEA.pdf

Gu, J., Goetschalckx, M., & McGinnis, L. F. (2007). Research on warehouse operation: A comprehensive review. *European Journal of Operational Research*, *177*(1), 1–21. doi:10.1016/j.ejor.2006.02.025

Gu, J., Goetschalckx, M., & McGinnis, L. F. (2010). Warehouse performance measurement: A literature. *European Journal of Operational Research*, *203*, 539–549. doi:10.1016/j.ejor.2009.07.031

Gunasekaran, A., Goyal, S. K., Virtanen, I., & Yli-Olli, P. (1993). Production planning and control in automated manufacturing systems. *Production Planning and Control*, *4*(1), 2–16.

Gunasekaran, A., Marri, H., & Menci, F. (1999). Improving the effectiveness of warehousing operations: A case study. *Industrial Management & Data Systems*, *99*(8), 328–329. doi:10.1108/02635579910291975

Gunasekaran, A., Rai, B., & Griffin, M. (2011). Resilience and competitiveness of small and medium size enterprises: An empirical research. *International Journal of Production Research*, *49*(18), 5489–5509. doi:10.1080/00207543.2011.563831

Gunn, D. J. (1978). Transfer of heat or mass to particles in fixed and fluidised beds. *International Journal of Heat and Mass Transfer*, *21*(4), 467–476. doi:10.1016/0017-9310(78)90080-7

Halev, G. (2017). *Integrated manufacturing System. In Expectation and Disappointment of Industrial Innovation.*

Halevy, A. Y. (2003). Data Integration: A Status Report. In *Datenbanksysteme in Business, Technology and Web (BTW 2003)*, Leipzig, Germany, February 26-28. Academic Press.

Halevy, A. Y. (2001). Answering queries using views: A survey. *The VLDB Journal*, *10*(4), 270–294. doi:10.1007007780100054

Hamdy, A. S., Shoeib, M. A., Hady, H., & Abdel Salam, O. F. (2007). Corrosion behavior of electroless Ni-P alloy coatings containing tungsten or nano-scattered alumina composite in 3.5% NaCl solution. *Surface and Coatings Technology*, *202*(1), 162–171. doi:10.1016/j.surfcoat.2007.05.030

Hammer, J., & Mcleod, D. (1993). An Approach to Resolving Semantic Heterogeneity in a Federation of Autonomous, Heterogeneous Database Systems. *International Journal of Intelligent and Cooperative Information Systems*, *2*(1), 51–83. doi:10.1142/S0218215793000046

Hamzehei, M., & Rahimzadeh, H. (2009). Experimental and Numerical Study of Hydrodynamics with Heat Transfer in a Gas-Solid Fluidized-Bed Reactor at Different Particle Sizes. *Industrial & Engineering Chemistry Research*, *48*(6), 3177–3186. doi:10.1021/ie801413q

Hamzehei, M., Rahimzadeh, H., & Ahmadi, G. (2010). Computational and Experimental Study of Heat Transfer and Hydrodynamics in a 2D Gas-Solid Fluidized Bed Reactor. *Industrial & Engineering Chemistry Research*, *49*(11), 5110–5121. doi:10.1021/ie900510a

Hanke, J., & Wichern, D. (2010). *Business forecasting* (9th ed.). Mexico: Pearson Education.

Hanumaiah, N., Ravi, B., & Mukherjee, N. P. (2006). Rapid hard tooling process selection using QFD-AHP methodology. *Journal of Manufacturing Technology Management, 17*(3), 332–350. doi:10.1108/17410380610648290

Han, X., Zeng, Z., Li, W., Islam, M. S., Lu, J., Loggins, V., ... Thomas, R. L. (2004). Acoustic chaos for enhanced detectability of cracks by sonic infrared imaging. *Journal of Applied Physics, 95*(7), 3792–3797. doi:10.1063/1.1652243

Harman, P. M. (1998). *The natural philosophy of James Clerk Maxwell.* Cambridge, England: Cambridge University Press.

Hart, P. E., Nilsson, N. J., & Raphael, B. (1968). A formal basis for the heuristic determination of minimum cost paths. *IEEE transactions on Systems Science and Cybernetics, 4*(2), 100-107.

Hassan, M. (2002). A framework for the design of warehouse layout. *Facilities, 20*(13/14), 432–440. doi:10.1108/02632770210454377

Hauser, J. R., & Clausing, D. (1988). The House of Quality. *Harvard Business Review, 66*(3), 63–73.

Haykin, S. (2009). *Neural Networks and Learning Machines.* Canada: Pearson Education.

Hazan, Y., Zimmermann, D., Z'graggen, M., Roos, S., Aneziris, C., Bollier, H., ... Graule, T. (2010). Homogeneous electroless Ni–P/SiO_2 nanocomposite coatings with improved wear resistance and modified wear behavior. *Surface and Coatings Technology, 204*(21-22), 3464–3470. doi:10.1016/j.surfcoat.2010.04.007

Heizer, J., & Render, B. (2010). *Principles of operations management* (8th ed.). Mexico: Pearson Educación.

Helu, M., Vijayaraghavan, A., & Dornfeld, D. (2011). Evaluating the relationship between use phase environmental impacts and manufacturing process precision. *CIRP Annals, 60*(1), 49–52. doi:10.1016/j.cirp.2011.03.020

Henneberry, J. M. (1987). Occupiers and their use of accommodation on science parks and hightechnology developments. *Journal of Property Research, 4*(2), 109–144.

Hennige, T. (2000). Development of irradiation strategies for 3D-laser forming. *Journal of Materials Processing Technology, 103*(1), 102–108. doi:10.1016/S0924-0136(00)00392-7

Hettler, J., Tabatabaeipour, M., Delrue, S., & Abeele, K. V. D. (2017). Detection and Characterization of Local Defect Resonances Arising from Delaminations and Flat Bottom Holes. *Journal of Nondestructive Evaluation, 36*(1), 2. doi:10.100710921-016-0380-6

Hilera, J. R., & Martínez, V. J. (1995). *Redes neuronales artificiales: fundamentos, modelos y aplicaciones.* Madrid, España: RA-MA Editorial.

Hilger, I., Hergt, R., & Kaiser, W. A. (2005). Towards breast cancer treatment by magnetic heating. *Journal of Magnetism and Magnetic Materials, 293*(1), 314–319. doi:10.1016/j.jmmm.2005.02.026

Hillier, S. F., & Lieberman, J. G. (2015). Introduction to Operations Research (10th ed.). New York: McGraw-Hill Education.

Hitomi, K. (1991). Strategic integrated manufacturing systems: The concept and structures. *International Journal of Production Economics, 25*(1-3), 5–12. doi:10.1016/0925-5273(91)90125-D

Hofer, P., Eisl, C., & Mayr, A. (2015). Forecasting in Austrian companies. *Journal of Applied Accounting Research, 16*(3), 359–382. doi:10.1108/JAAR-10-2014-0113

Hong, J., Zhang, Y., & Ding, M. (2018). Sustainable supply chain management practices, supply chain dynamic capabilities, and enterprise performance. *Journal of Cleaner Production, 172*, 3508–3519. doi:10.1016/j.jclepro.2017.06.093

Hosking, F., Portillo, F. F., Wunderlin, R., & Mehrabian, R. (1982). Composites of aluminium alloys: Fabrication and wear behaviour. *Journal of Materials Science, 17*(2), 477–498. doi:10.1007/BF00591483

Hosseini, S. H., Ahmadi, G., Saeedi Razavi, B., & Zhong, W. (2010). Computational fluid dynamic simulation of hydrodynamic behavior in a two-dimensional conical spouted bed. *Energy & Fuels, 24*(11), 6086–6098. doi:10.1021/ef100612r

Ho, W. (2008). Integrated analytic hierarchy process and its applications–A literature review. *European Journal of Operational Research, 186*(1), 211–228. doi:10.1016/j.ejor.2007.01.004

Ho, W., & Prasanta, K. (2011). Strategic sourcing: A combined QFD and AHP approach in manufacturing. *Supply Chain Management, 16*(6), 446–461. doi:10.1108/13598541111171093

Hsu, M. H. (2005b). Vibration analysis of edge-cracked beam on elastic foundation with axial loading using the differential quadrature method. *Computer Methods in Applied Mechanics and Engineering, 194*(1), 1–17. doi:10.1016/j.cma.2003.08.011

Hu, Y., Yao, Z., & Xu, X. (2011). Experimental investigation on the bending deformation of thin 1060 pure aluminum sheet by laser peen forming. *Proceedings of the international conference on shot peening*, South Bend, IN (pp. 341-346). Academic Press.

Huang, L., Ding, Y., Zhou, M., Jin, Y., & Hao, K. (2018). Multiple-Solution Optimization Strategy for Multirobot Task Allocation. *IEEE Transactions on Systems, Man, and Cybernetics. Systems*, 1–12. doi:10.1109/TSMC.2018.2847608

Huang, Y. S., Zeng, X. T., Annergren, I., & Liu, F. M. (2003). Development of electroless Ni–P-PTFE–SiC composite coating. *Surface and Coatings Technology, 167*(2-3), 207–211. doi:10.1016/S0257-8972(02)00899-X

Huhns, M. N., & Singh, M. P. (1997). Agents on the Web: Ontologies for Agents. *IEEE Internet Computing, 1*(6), 81–83. doi:10.1109/4236.643942

Hui, K., Haff, P. K., Ungar, J. E., & Jackson, R. (1984). Boundary conditions for high-shear grain flows. *Journal of Fluid Mechanics, 145*(-1), 223. doi:10.1017/S0022112084002883

Huilin, L., Yunhua, Z., Zhiheng, S., Ding, J., & Jiying, J. (2006). Numerical simulations of gas-solid flow in tapered risers. *Powder Technology, 169*(2), 89–98. doi:10.1016/j.powtec.2006.07.023

Hull, R. (1997). Managing semantic heterogeneity in databases: A theoretical perspective. *Proceedings of the sixteenth ACM SIGACT-SIGMOD-SIGART symposium on Principles of database systems*, Tucson, AZ (pp. 51-61). ACM. 10.1145/263661.263668

Hulme, I., Clavelle, E., van der Lee, L., & Kantzas, A. (2005). CFD Modeling and Validation of Bubble Properties for a Bubbling Fluidized Bed. *Industrial & Engineering Chemistry Research, 44*(12), 4254–4266. doi:10.1021/ie049837j

Hurson, A. R., & Bright, M. W. (1991). Multidatabase Systems: An Advanced Concept in Handling Distributed Data. *Advances in Computers, 32*, 149–200. doi:10.1016/S0065-2458(08)60247-8

Hussein, A., Adel, M., Bakr, M., Shehata, O. M., & Khamis, A. (2014). Multi-robot task allocation for search and rescue missions. *Journal of Physics: Conference Series, 570*(5), 052006. doi:10.1088/1742-6596/570/5/052006

Hussein, A., & Khamis, A. (2013, December). Market-based approach to multi-robot task allocation. *Proceedings of the 2013 International Conference on Individual and Collective Behaviors in Robotics (ICBR)* (pp. 69-74). IEEE. 10.1109/ICBR.2013.6729278

Hu, X., He, C., Feng, K., Liu, S., & Walton, G. (2019). Effects of polypyrrole coated rebar on corrosion behavior of tunnel lining with the combination effect of sustained loading and pre-existing cracks when exposed to chlorides. *Construction & Building Materials*, *221*, 318–331. doi:10.1016/j.conbuildmat.2019.06.092

Ibrahim, M. E., Smith, R. A., & Wang, C. H. (2017). Ultrasonic detection and sizing of compressed cracks in glass and carbon-fibre reinforced plastic composites. *NDT & E International*, *92*, 111–121. doi:10.1016/j.ndteint.2017.08.004

Inter-American Development Bank [IDB]. (2010). *La logística de cargas en América Latina y el Caribe: una agenda para mejorar su desempeño*. Retrieved from: https://publications.iadb.org/bitstream/handle/11319/6269/La%20log%C3%ADstica%20de%20cargas%20en%20Am%C3%A9rica%20Latina%20y%20el%20Caribe%3A%20una%20agenda%20para%20mejorar%20su%20desempe%C3%B1o.pdf?sequence=1

Inter-American Development Bank [IDB]. (2011). *La logística como motor de la competitividad en América Latina y el Caribe. Banco Interamericano de Desarrollo*, Retrieved from: https://publications.iadb.org/bitstream/handle/11319/5091/La%20log%C3%ADstica%20como%20motor%20de%20la%20competitividad%20en%20Am%C3%A9rica%20Latina%20y%20el%20Caribe.pdf?sequence=1&isAllowed=y

Irwin, G. R. (1957). Analysis of stresses and strains near the end of a crack transversing a plate. *Journal of Applied Mechanics*, *24*, 361–364.

Islam, R., Ahmed, M., & Alias, M. H. (2007). Application of Quality Function Deployment in redesigning website: A case study on TV3. *International Journal of Business Information Systems*, *2*(2), 195–216. doi:10.1504/IJBIS.2007.011619

Ivanov, A. O., & Kuznetsova, O. B. (2002). Magnetic properties of dense ferrofluids. *Journal of Magnetism and Magnetic Materials*, *252*, 135–137. doi:10.1016/S0304-8853(02)00639-X

Iwasawa, Y. (Ed.). (1996). *X-ray absorption fine structure for catalysts and surfaces*. Singapore: World Scientific. doi:10.1142/2807

Jalham, I. S. (2006). Computer-aided quality function deployment method for material selection. *International Journal of Computer Applications in Technology*, *26*(4), 190–196. doi:10.1504/IJCAT.2006.010764

Jalili, M. M., Moradian, S., & Hosseinpour, D. (2009). The use of inorganic conversion coatings to enhance the corrosion resistance of reinforcement and the bond strength at the rebar/concrete. *Construction & Building Materials*, *23*(1), 233–238. doi:10.1016/j.conbuildmat.2007.12.011

James, S. J., Venkatesan, K., Kuppan, P., & Ramanujam, R. (2014). Hybrid aluminium metal matrix composite reinforced with SiC and TiB2. *Procedia Engineering*, *97*, 1018–1026. doi:10.1016/j.proeng.2014.12.379

Janic, M., & Reggiani, A. (2002). An application of the multiple criteria decision making (mcdm) analysis to the selection of a new hub airport. *European Journal of Transport and Infrastructure Research*, *2*(2).

Jan, O., & Erik, S. (2000). Supply Chain Management Survey of Swedish Manufacturing Firms. *International Journal of Production Economics*, *89*, 353–361.

Jarke, M., Lenzerini, M., Vassiliou, Y., & Vassiliadis, P. (2000). *Fundamentals of Data Warehouses*. Springer. doi:10.1007/978-3-662-04138-3

Jia, D., Bi, X., Lim, C. J., Sokhansanj, S., & Tsutsumi, A. (2019). Heat transfer in a tapered fluidized bed of biomass particles with pulsed gas flow. *Particuology*, *42*, 2–14. doi:10.1016/j.partic.2018.01.007

Jiang, Y., & Liu, M. (1996). Dynamics of Dispersive and Nonlinear Media. *Physical Review Letters*, *77*(6), 1043–1046. doi:10.1103/PhysRevLett.77.1043 PMID:10062976

Jianping, C., Yimin, Y., & Yunbiao, W. (2009, November). Multi-robot task allocation based on robotic utility value and genetic algorithm. *Proceedings of the 2009 IEEE International Conference on Intelligent Computing and Intelligent Systems* (Vol. 2, pp. 256-260). IEEE. 10.1109/ICICISYS.2009.5357957

Jiaqiang, G., Lei, L., Yating, W., Bin, S., & Wenbin, H. (2006). Electroless Ni–P–SiC composite coatings with superfine particles. *Surface and Coatings Technology*, *200*(20-21), 5836–5842. doi:10.1016/j.surfcoat.2005.08.134

Jing, S., Hu, Q., Wang, J., & Jin, Y. (2000). Fluidization of coarse particles in gas – solid conical beds. *Chemical Engineering and Processing*, *39*(4), 379–387. doi:10.1016/S0255-2701(99)00103-8

Jiradilok, V., Gidaspow, D., & Breault, R. W. (2007). Computation of gas and solid dispersion coefficients in turbulent risers and bubbling beds. *Chemical Engineering Science*, *62*(13), 3397–3409. doi:10.1016/j.ces.2007.01.084

Johansson, K., van Wachem, B. G. M., & Almstedt, A. E. (2006). Experimental validation of CFD models for fluidized beds: Influence of particle stress models, gas phase compressibility and air inflow models. *Chemical Engineering Science*, *61*(5), 1705–1717. doi:10.1016/j.ces.2005.09.014

Johnson, A., & McGinnis, L. (2010). Performance measurement in the warehousing industry. *IIE Transactions*, *43*(3), 220–230. doi:10.1080/0740817X.2010.491497

Johnson, D. (2002). RFID tags improve tracking, quality on Ford line in Mexico. *Control Engineering*, *49*(11), 16.

Johnson, P. C., & Jackson, R. (1987). Frictional–collisional constitutive relations for granular materials, with application to plane shearing. *Journal of Fluid Mechanics*, *176*(-1), 67. doi:10.1017/S0022112087000570

Jolly, M., & Nakano, M. (1998): Properties and Applications of Commercial Controllable Fluids. *Proceedings of 6th International Conference on New Actuators* (pp. 414–419). Academic Press.

Jose, K., & Pratihar, D. K. (2016). Task allocation and collision-free path planning of centralized multi-robots system for industrial plant inspection using heuristic methods. *Robotics and Autonomous Systems*, *80*, 34–42. doi:10.1016/j.robot.2016.02.003

Jouela, S. L. J. (2007). *Future trends in automation*. Finland: Helsinki University of Technology.

Ju, C., & Son, H. (2018). Multiple UAV systems for agricultural applications: Control, implementation, and evaluation. *Electronics (Basel)*, *7*(9), 162. doi:10.3390/electronics7090162

Jurgons, R., Seliger, C., Hilpert, A., Trahms, L., Odenbach, S., & Alexiou, C. (2006). Drug loaded magnetic nanoparticles for cancer therapy. *Journal of Physics Condensed Matter*, *18*(38), 2893–2902. doi:10.1088/0953-8984/18/38/S24

Kalloniatis, C., Kavakli, E., Gritzalis, S., Forrest, W., & Kindler, N. (2014). Addressing privacy in Traditional and Cloud-Based Systems [IJAIE]. *International Journal of Applied Industrial Engineering*, *2*(1), 14–40. doi:10.4018/ijaie.2014010102

Kalo, L., Pant, H. J., & Upadhyay, R. K. (2014). Numerical and Experimental Investigation of Conical and Cylindrical Gas-Solids Fluidized Bed with Head to Head Comparison. Proceedings of ISCRE 23 & APCRE 7, Bangkok, Thailand. Academic Press.

Kalo, L. Kamalanathan, P., Pant, H. J., Cassanello, M. C., & Upadhyay, R. K. (2019). Mixing and regime transition analysis of liquid-solid conical fluidized bed through RPT technique. *Chemical Engineering Science.*

Kalo, L. Pant, H. J., Cassanello, M. C., & Upadhyay, R. K. (2018). Time series analysis of a binary gas-solid conical fluidized bed using radioactive particle tracking (RPT) technique data. *Chemical Engineering Journal.*

Kant, R., & Joshi, & S.N. (2013). Finite element simulation of multi-pass laser bending process with forced cooling. *Proceedings of the International Conference on Precision, Meso, Micro and Nano Engineering*, Kerala, India (pp. 772-777). Academic Press.

Kaplan, J., Forrest, W., & Kindler, N. (2008). *Revolutionizing Data Center Energy Efficiency (Tech. Report)*. McKinsey and Company.

Karakis, I., Tanyas, M., & Baskak, M. (2018). Hierarchical Warehouse Design Approach for Distribution Centres. *International Journal of Business and Management Invention, 7*(2), 69–82.

Kärkkäinen, M. (2003). Increasing efficiency in the supply chain for short shelf life goods using RFID tagging. *International Journal of Retail & Distribution Management, 31*(10), 529–536. doi:10.1108/09590550310497058

Karnopp, D. C., Margolis, D. L., & Rosenberg, R. C. (1990). *System Dynamics A Unified Approach* (2nd ed.). United States of America: John Wiley & Sons, Inc.

Karuppusami, G., & Gandhinathan, R. (2006). Pareto analysis of critical success factors of total quality management: A literature review and analysis. *The TQM Magazine, 18*(4), 372–385. doi:10.1108/09544780610671048

Kasaei, A., Abedian, A., & Milani, A. S. (2014). An application of Quality Function Deployment method in engineering materials selection. *Materials & Design, 55*, 912–920. doi:10.1016/j.matdes.2013.10.061

Katiyar, R. (2017). Conical fluidized beds - Case Study for a gas-solid system involving coarse particles.

Katunin, A., Dragan, K., & Dziendzikowski, M. (2015). Damage identification in aircraft composite structures: A case study using various non-destructive testing techniques. *Composite Structures, 127*, 1–9. doi:10.1016/j.compstruct.2015.02.080

Kehoe, D., & Boughton, N. (2001). Internet based supply chain management: A classification of approaches to manufacturing planning and control. *International Journal of Operations & Production Management, 21*(4), 516–525. doi:10.1108/01443570110381417

Ke, L. L., Yang, J., & Kitipornchai, S. (2009a). Post buckling analysis of edge cracked functionally graded Timoshenko beams under end shortening. *Composite Structures, 90*(2), 152–160. doi:10.1016/j.compstruct.2009.03.003

Key, A. (2015, July 22). 5 Types of inspection equipment every manufacturer should know.

Khani, M. H. (2011). Models for prediction of hydrodynamic characteristics of gas-solid tapered and mini-tapered fluidized beds. *Powder Technology, 205*(1-3), 224–230. doi:10.1016/j.powtec.2010.09.018

Khan, M. J. H., Hussain, M. A., Mansourpour, Z., Mostoufi, N., Ghasem, N. M., & Abdullah, E. C. (2014). CFD simulation of fluidized bed reactors for polyolefin production – A review. *Journal of Industrial and Engineering Chemistry, 20*(6), 3919–3946. doi:10.1016/j.jiec.2014.01.044

Kharat, M. G., Khadke, S. S., Raut, R. D., Kamble, S. S., Kamble, S. J., & Kharat, M. G. (2016). Application of hybrid firefly algorithm-tabu search technique to minimize the makespan in job shop scheduling problem. *International Journal of Applied Industrial Engineering, 3*(2), 1–21. doi:10.4018/IJAIE.2016070101

Khazaei, H., Misic, J., & Misic, V. B. (2012). Performance Analysis of Cloud Computing Centers Using M/G/m/m+r Queuing Systems. *IEEE Transactions on Parallel and Distributed Systems, 23*(5), 936–943. doi:10.1109/TPDS.2011.199

Khiem, N. T., & Hang, P. T. (2017b). Analysis and identification of multiple-cracked beam subjected to moving harmonic load. *Journal of Vibration and Control*, 1–20.

Khodabandehlou, R., Askaripour, H., & Molaei Dehkordi, A. (2018). Numerical investigation of gas bubble behavior in tapered fluidized beds. *Particuology*, *38*, 152–164. doi:10.1016/j.partic.2017.05.013

Kikidis, M. L., & Papadopoulos, C. A. (1992c). Slenderness ratio effect on cracked beam. *Journal of Sound and Vibration*, *155*(1), 1–11. doi:10.1016/0022-460X(92)90641-A

Kilickap, E., Cakir, O., Aksoy, M., & Inan, A. (2005). Study of tool wear and surface roughness in machining of homogenised SiC-p reinforced aluminium metal matrix composite. *Journal of Materials Processing Technology*, *164*, 862–867. doi:10.1016/j.jmatprotec.2005.02.109

Kim, H. G., Lee, I. O., Chung, U. C., & Kim, Y. H. (2000). Fluidization Characteristics of Iron Ore Fines of Wide Size Distribution in a Cold Tapered Gas-Solid Fluidized Bed. *ISIJ International*, *40*(1), 16–22. doi:10.2355/isijinternational.40.16

Kim, W., Choi, I., Gala, S., & Scheevel, M. (1993). On Resolving Schematic Heterogeneity in Multidatabase Systems. *Distributed and Parallel Databases*, *1*(3), 252–279. doi:10.1007/BF01263333

Kim, W., & Seo, J. (1991). Classifying Schematic and Data Heterogeneities in Multidatabase Systems. *IEEE Computer*, *24*(12), 12–18. doi:10.1109/2.116884

Kinsella, B. (2003). The Wal-Mart factors. *Industrial Engineering (American Institute of Industrial Engineers)*, *32–36*(November).

Kirkwood Craig, W. (1998). *System Dynamics Methods: A Quick Introduction*. College Of Business Arizona State University.

Kisa, M., & Brandon, J. A. (2000). The effects of closure of cracks on the dynamic of a cracked cantilever beams. *Journal of Sound and Vibration*, *238*(1), 1–18. doi:10.1006/jsvi.2000.3099

Kisa, M., Brandon, J., & Topcu, M. (1998b). Free vibration analysis of cracked beams by a combination of finite elements and component mode synthesis methods. *Computers & Structures*, *67*(4), 215–223. doi:10.1016/S0045-7949(98)00056-X

Kisa, M., & Gurel, M. A. (2007b). Free vibration analysis of uniform and stepped cracked beams with circular cross sections. *International Journal of Engineering Science*, *45*(2-8), 364–380. doi:10.1016/j.ijengsci.2007.03.014

Kitipornchai, S., Ke, L. L., Yang, J., & Xiang, Y. (2009b). Nonlinear vibration of edge cracked functionally graded Timoshenko beams. *Journal of Sound and Vibration*, *324*(3-5), 962–982. doi:10.1016/j.jsv.2009.02.023

Knospe, H., & Pohl, H. (2004). RFID security. *Information Security Technical Report*, *9*(4), 39–50. doi:10.1016/S1363-4127(05)70039-X

Kobayashi, K., & Takewaka, K. (1984). Experimental studies on epoxy coated reinforcing steel for corrosion protection. *International Journal of Cement Composites and Lightweight Concrete*, *6*(2), 99–116. doi:10.1016/0262-5075(84)90039-3

Koleva, D. A., Hu, J., Fraaij, A. L. A., Stroeven, P., Boshkov, N., & De Wit, J. H. W. (2006). Quantitative characterisation of steel/cement paste interface microstructure and corrosion phenomena in mortars suffering from chloride attack. *Corrosion Science*, *48*(12), 4001–4019. doi:10.1016/j.corsci.2006.03.003

Kouprianov, V. I., & Permchart, W. (2003). Emissions from a conical FBC fired with a biomass fuel. *Applied Energy*, *74*(3-4), 383–392. doi:10.1016/S0306-2619(02)00194-0

Kourentzes, N. (2013). Intermittent demand forecasts with neural networks. *International Journal of Production Economics*, *143*(1), 198–206. doi:10.1016/j.ijpe.2013.01.009

Krajewski, L., Ritzman, L., & Malhotra, M. (2008). *Operation management process and values chain* (8th ed.). Mexico: Pearson Education.

Krawczuk, M. (1992b). Finite Timoshenko-type beam element with a crack. *Engineering Transactions*, *40*, 229–248.

Krawczuk, M. (1993a). Natural vibrations of rectangular plates with a through crack. *Archive of Applied Mechanics*, *63*, 491–504.

Krawczuk, M., & Ostachowicz, W. M. (1992a). Parametric Vibrations of beam with crack. *Archive of Applied Mechanics*, *62*(7), 463–473. doi:10.1007/BF00810237

Krawczuk, M., & Ostachowicz, W. M. (1993b). Transverse natural vibrations of a cracked beam loaded with a constant axial force. *Journal of Vibration and Acoustics*, *115*(4), 428–524. doi:10.1115/1.2930381

Krawczuk, M., & Ostachowicz, W. M. (1995). Modelling and vibration analysis of a cantilever composite beam with a transverse open crack. *Journal of Sound and Vibration*, *183*(1), 69–89. doi:10.1006/jsvi.1995.0239

Krebsonsecurity. (2014). Target hackers broke in via HVAC company. Retrieved from http://krebsonsecurity.com/2014/02/target-hackers-broke-in-via-hvac-company

Kuipers, J. A. M., Prins, W., & Van Swaaij, W. P. M. (1992). Numerical calculation of wall-to-bed heat-transfer coefficients in gas-fluidized beds. *AIChE Journal. American Institute of Chemical Engineers*, *38*(7), 1079–1091. doi:10.1002/aic.690380711

Kuipers, J. A. M., van Duin, K. J., van Beckum, F. P. H., & van Swaaij, W. P. M. (1993). Computer simulation of the hydrodynamics of a two-dimensional gas-fluidized bed. *Computers & Chemical Engineering*, *17*(8), 839–858. doi:10.1016/0098-1354(93)80067-W

Kumar, R., & Ray, A. (2014). Selection of material for optimal design using multi-criteria decision making. *Procedia materials science, 6*, 590-596.

Kumar, D., Loganathan, C., & Narayanasamy, R. (2011). Effect of glass in an aluminum matrix on workability and strain hardening behavior of powder metallurgy composite. *Materials & Design*, *32*(4), 2413–2422. doi:10.1016/j.matdes.2010.12.008

Kumar, P. (2009). *Elements of Fracture Mechanics. Chennai, India: McGraw Hill Education.*

Kumar, P., Herbert, M., & Rao, S. (2014). Demand forecasting Using Artificial Neural Network Based on Different Learning Methods: Comparative Analysis. *International Journal for Research in Applied Science and Engineering Technology*, *2*(4), 364–374.

Kumar, T. S., Shivashankar, G., Dhotey, K., & Singh, J. (2017). Experimental study wear rate of glass fibre reinforced epoxy polymer composites filled with aluminium powder. *Materials Today: Proceedings*, *4*(10), 10764–10768.

Kunii, D., & Levenspiel, O. (1991). *Fluidization Engineering*. Butterworth-Heinemann.

Kuprianov, V. I., & Arromdee, P. (2013). Combustion of peanut and tamarind shells in a conical fluidized-bed combustor: A comparative study. *Bioresource Technology*, *140*, 199–210. doi:10.1016/j.biortech.2013.04.086 PMID:23693147

Laborde-Boutet, C., Larachi, F., Dromard, N., Delsart, O., & Schweich, D. (2009). CFD simulation of bubble column flows: Investigations on turbulence models in RANS approach. *Chemical Engineering Science*, *64*(21), 4399–4413. doi:10.1016/j.ces.2009.07.009

LaMarco, N. & Shoenberger, E. (2018). Types of Manufacturing System.

Landers, T., & Rosenberg, R. L. (1982). An Overview of MULTIBASE. *Proceedings of the Second International Symposium on Distributed Data Bases (DDB 1982)*, Berlin, Germany, September 1-3 (pp. 153-184). Academic Press.

Lane, D. J., & Sterman, J. D. (2011). Profiles in Operations Research. In Profiles in Operations Research: Pioneers and Innovators (Ch. 20, pp. 363-386). New York, Springer: .

Lange, J., Kotitz, R., Haller, A., Trahms, L., Semmler, W., & Weitschies, W. (2002). Magnetorelaxometry-a new binding specific detection method based on magnetic nanoparticles. *Journal of Magnetism and Magnetic Materials*, *252*, 381–383. doi:10.1016/S0304-8853(02)00657-1

Larsen-Basse, J. (1990). Role of microstructure and mechanical properties in abrasion. *Scripta Metallurgica et Materialia*, *24*(5), 821–826. doi:10.1016/0956-716X(90)90119-2

Le-Duc, T., & Koster, R. (2005). Travel distance estimation and storage zone optimization in a 2-block classbased storage strategy warehouse. *International Journal of Production Research*, *43*(17), 3561–3581. doi:10.1080/00207540500142894

Lee, C. K. (2008). Corrosion and wear-corrosion resistance properties of electroless Ni-P coatings on GFRP composite in wind turbine blades. *Surface and Coatings Technology*, *202*(19), 4868–4874. doi:10.1016/j.surfcoat.2008.04.079

Lee, C. K. (2009). Structure, electrochemical and wear-corrosion properties of electroless nickel-phosphorus deposition on CFRP composites. *Materials Chemistry and Physics*, *114*(1), 125–133. doi:10.1016/j.matchemphys.2008.08.088

Lee, D. H. (2018). Resource-based task allocation for multi-robot systems. *Robotics and Autonomous Systems*, *103*, 151–161. doi:10.1016/j.robot.2018.02.016

Lee, J. H., & Kim, C. O. (2008). Multi-agent systems applications in manufacturing systems and supply chain management: A review paper. *International Journal of Production Research*, *46*(1), 233–265. doi:10.1080/00207540701441921

Lee, Y. C., & Zomaya, A. Y. (2010). Energy efficient utilization of resources in cloud computing systems. *The Journal of Supercomputing*, *60*(2), 268–280. doi:10.100711227-010-0421-3

Lenzerini, M. (2002). Data integration: A theoretical preceptive. *Proceedings of Twenty-first ACM SIGACT-SIGMOD-SIGART Symposium on Principles of Database System*, Madison, WI (pp. 1-14). ACM.

Levey, R. P., Garga, A. D. L., Jacobs, S. C., Heidt, H. M., & Trent, P. E. (1960). Fluid Bed Conversion of UO3 to UF4. *Chemical Engineering Progress*, *56*, 43–48.

Lewis, A. (2018, February). A Breakdown of manufacturing. *System*, 26.

Li, P. (2019). Introductory Chapter: New Trends in Industrial Automation. In New Trends in Industrial Automation. IntechOpen.

Lien, C. H., Bai, Y. W., Lin, M. B., Chang, C. Y., & Tsai, M. Y. (2006). Web server power estimation, modeling and management. *Proceedings of 14th IEEE International Conference on Networks* (Vol. 2, pp. 1-6). IEEE Press. 10.1109/ICON.2006.302688

Li, J., Li, Y. K., Chen, X., Lee, P. P., & Lou, W. (2014). A hybrid cloud approach for secure authorized deduplication. *IEEE Transactions on Parallel and Distributed Systems*, *26*(5), 1206–1216. doi:10.1109/TPDS.2014.2318320

Liliana, L. (2016). A new model of Ishikawa diagram for quality assessment. *Proceedings of the IOP Conference Series: Materials Science and Engineering*. IOP Publishing. 10.1088/1757-899X/161/1/012099

Lim, S., Gupta, M., Ren, L., & Kwok, J. (1999). The tribological properties of Al-Cu/SiCp metal-matrix composites fabricated using the rheocasting technique. *Journal of Materials Processing Technology*, *89*, 591–596. doi:10.1016/S0924-0136(99)00067-9

Lin, H. P., & Chang, S. C. (2006c). Forced responses of cracked cantilever beams subjected to a concentrated moving load. *International Journal of Mechanical Sciences*, *48*(12), 1456–1463. doi:10.1016/j.ijmecsci.2006.06.014

Lin, M. C., Wang, C. C., & Chen, T. C. (2006). A strategy for managing customer-oriented product design. *Concurrent Engineering*, *14*(3), 231–244. doi:10.1177/1063293X06068390

Li, P., Lan, X., Xu, C., Wang, G., Lu, C., & Gao, J. (2009). Drag models for simulating gas–solid flow in the turbulent fluidization of FCC particles. *Particuology*, *7*(4), 269–277. doi:10.1016/j.partic.2009.03.010

Liu, C., & Kroll, A. (2012, April). A centralized multi-robot task allocation for industrial plant inspection by using A* and genetic algorithms. *Proceedings of the International Conference on Artificial Intelligence and Soft Computing* (pp. 466-474). Springer. 10.1007/978-3-642-29350-4_56

Liu, G., Yang, L., Wang, L., Wang, S., Chongyang, L., & Wang, J. (2010). Corrosion behavior of electroless deposited Ni–Cu–P coating in fiue gas condensate. *Surface and Coatings Technology*, *204*(21-22), 3382–3386. doi:10.1016/j.surfcoat.2010.03.056

Liu, J., Shao, Y. M., & Zhu, W. D. (2017c). Free vibration analysis of a cantilever beam with a slant edge crack. *Journal of Mechanical Engineering Science*, *231*(5), 823–843. doi:10.1177/0954406216631006

Liu, M. (1993). Hydrodynamic Theory of Electromagnetic Fields in Continuous Media. *Physical Review Letters*, *70*(23), 3580–3584. doi:10.1103/PhysRevLett.70.3580 PMID:10053910

Liu, M. (1995a). Fluid Dynamics of colloidal magnetic and electric liquid. *Physical Review Letters*, *74*(22), 4535–4538. doi:10.1103/PhysRevLett.74.4535 PMID:10058531

Liu, M. (1995b). Electromagnetic fields in ferrofluids. *Physical Review*, *59*(3), 3669–3675. doi:10.1103/PhysRevE.59.3669

Liu, M., Cheng, X., Li, X., Jin, Z., & Liu, H. (2015). Corrosion behavior of Cr modified HRB400 steel rebar in simulated concrete pore solution. *Construction & Building Materials*, *93*, 884–890. doi:10.1016/j.conbuildmat.2015.05.073

Liu, M., Cheng, X., Li, X., Zhou, C., & Tan, H. (2017). Effect of carbonation on the electrochemical behavior of corrosion resistance low alloy steel rebars in cement extract solution. *Construction & Building Materials*, *130*, 193–201. doi:10.1016/j.conbuildmat.2016.10.003

Liu, T. (2006). Study on corrosive wear resistance of a Ni–P–Cr_2O_3 composite electroless plating with low-content phosphor in NaOH. *Run Hua Yu Mi Feng*, *4*, 90–91.

Liu, X. (2013). A Programming Model for the Cloud Platform. *International Journal of Advanced Science and Technology*, *57*, 75–82.

Liu, Y., & Bucknall, R. (2018). Efficient multi-task allocation and path planning for unmanned surface vehicle in support of ocean operations. *Neurocomputing*, *275*, 1550–1566. doi:10.1016/j.neucom.2017.09.088

Liu, Y., & Zhao, Q. (2004). Study of electroless Ni–Cu–P coatings and their anti-corrosion properties. *Applied Surface Science*, *228*(1-4), 57–62. doi:10.1016/j.apsusc.2003.12.031

Li, Y., Zhang, J., & Fan, L.-S. (1999). Numerical simulation of gas–liquid–solid fluidization systems using a combined CFD-VOF-DPM method: Bubble wake behavior. *Chemical Engineering Science*, *54*(21), 5101–5107. doi:10.1016/S0009-2509(99)00263-8

Li, Z., Chen, Z., Liu, S., Zheng, F., & Dai, A. (2008). Corrosion and wear properties of electroless Ni–P plating layer on AZ91D magnesium alloy. *Transactions of Nonferrous Metals Society of China*, *18*(4), 819–824. doi:10.1016/S1003-6326(08)60142-9

Li, Z., Xu, B., Yang, L., Chen, J., & Li, K. (2009, June). Quantum evolutionary algorithm for multi-robot coalition formation. *Proceedings of the first ACM/SIGEVO Summit on Genetic and Evolutionary Computation* (pp. 295-302). ACM. 10.1145/1543834.1543874

Loha, C., Chatterjee, P. K., & Chattopadhyay, H. (2011). Performance of fluidized bed steam gasification of biomass - Modeling and experiment. *Energy Conversion and Management*, *52*(3), 1583–1588. doi:10.1016/j.enconman.2010.11.003

Loha, C., Chattopadhyay, H., & Chatterjee, P. K. (2011). Thermodynamic analysis of hydrogen rich synthetic gas generation from fluidized bed gasification of rice husk. *Energy*, *36*(7), 4063–4071. doi:10.1016/j.energy.2011.04.042

Loha, C., Chattopadhyay, H., & Chatterjee, P. K. (2012). Assessment of drag models in simulating bubbling fluidized bed hydrodynamics. *Chemical Engineering Science*, *75*, 400–407. doi:10.1016/j.ces.2012.03.044

Loha, C., Chattopadhyay, H., & Chatterjee, P. K. (2013). Euler-Euler CFD modeling of fluidized bed: Influence of specularity coefficient on hydrodynamic behavior. *Particuology*, *11*(6), 673–680. doi:10.1016/j.partic.2012.08.007

Loha, C., Chattopadhyay, H., & Chatterjee, P. K. (2014a). Effect of coefficient of restitution in Euler-Euler CFD simulation of fluidized-bed hydrodynamics. *Particuology*, *15*, 170–177. doi:10.1016/j.partic.2013.07.001

Loha, C., Chattopadhyay, H., & Chatterjee, P. K. (2014b). Three dimensional kinetic modeling of fluidized bed biomass gasification. *Chemical Engineering Science*, *109*, 53–64. doi:10.1016/j.ces.2014.01.017

Lopez, E. Acosta, M. Espinoza, E. & Delfín, F. (2007). Globalization and its effects on the management of micro, small and medium enterprises and Latin America. *Equidad y desarrollo*, *8*, 113-125.

Loto, C. A. (2016). Electroless nickel plating–a review. *Silicon*, *8*(2), 177–186. doi:10.100712633-015-9367-7

Loya, J. A., Rubio, L., & Fernandez-Saez, J. (2006d). Natural frequencies for bending vibration of Timoshenko cracked beams. *Journal of Sound and Vibration*, *290*(3-5), 640–653. doi:10.1016/j.jsv.2005.04.005

Lozowski, D. (2016). Advancing with automation. *Chemical Engineering (Albany, N.Y.)*, *123*(3), 5.

Luckett, D. (2004). The supply chain. *BT Technology Journal*, *22*(3), 50–55. doi:10.1023/B:BTTJ.0000047119.22852.38

Lu, G., Evans, P., & Zangari, G. (2004). Investigations of the effect of chromate conversion coatings on the corrosion resistance of Ni-based alloys. *Journal of the Electrochemical Society*, *49*, 1461–1473.

Lun, C. K. K., Savage, S. B., Jeffrey, D. J., & Chepurniy, N. (1984). Kinetic theories for granular flow: Inelastic particles in Couette flow and slightly inelastic particles in a general flowfield. *Journal of Fluid Mechanics*, *140*, 223–256. doi:10.1017/S0022112084000586

Luo, H., Su, H., Dong, C., & Li, X. (2017). Passivation and electrochemical behavior of 316L stainless steel in chlorinated simulated concrete pore solution. *Applied Surface Science*, *400*, 38–48. doi:10.1016/j.apsusc.2016.12.180

Macias, A., & Andrade, E. C. (1983). Corrosion rate of galvanized steel immersed in saturated solutions of $Ca(OH)_2$ in the pH range 12–13· 8. *British Corrosion Journal*, *18*(2), 82–87. doi:10.1179/bcj.1983.18.2.82

Macias, A., & Andrade, E. C. (1987a). Corrosion of galvanized steel in dilute $Ca(OH)_2$ solutions (pH 11·1–12· 6). *British Corrosion Journal*, *22*(3), 162–171. doi:10.1179/000705987798271505

Macias, A., & Andrade, E. C. (1987b). Corrosion of galvanized steel reinforcements in alkaline solutions: Part 1: Electrochemical results. *British Corrosion Journal*, *22*(2), 113–118. doi:10.1179/000705987798271631

Ma, D., & Ahmadi, G. (1986). An equation of state for dense rigid sphere gases. *The Journal of Chemical Physics*, *84*(6), 3449–3450. doi:10.1063/1.450229

Madhukumar, U. (2018). Characterization of glass particulate reinforced aluminium alloy6061 metal matrix composites. *Materials Today: Proceedings*, *5*(8), 7604–7608.

Mafi, I. R., & Dehghanian, C. (2011). Comparison of the coating properties and corrosion rates in electroless Ni–P/PTFE composites prepared by different types of surfactants. *Applied Surface Science*, *257*(20), 8653–8658. doi:10.1016/j.apsusc.2011.05.043

Mahalik, N. P. (2014). Advances in packaging methods, processes and systems. *Challenges*, *5*(2), 374–389.

Manikonda, R. D., Kosaraju, S., Raj, K. A., & Sateesh, N. (2018). Wear behavior analysis of silica carbide-based aluminum metal matrix composites. *Materials Today: Proceedings*, *5*(9), 20104–20109.

Manning, D. G. (1996). Corrosion performance of epoxy-coated reinforcing steel: North American experience. *Construction & Building Materials*, *10*(5), 349–365. doi:10.1016/0950-0618(95)00028-3

Marín, R. (2014). *Almacén de clase mundial*. Medellín: Esumer.

Martínez, S. I., Vértiz, C. G., López, P. J., Jiménez, L. G., & Moncayo, M. L. (2014). *Investigación de Operaciones*. México: GRUPO EDITORIAL PATRIA.

Marur, S. R., & Prathap, G. (2005d). Non-linear beam vibration problems and simplifications in finite element models. *Computational Mechanics*, *35*(5), 352–360. doi:10.100700466-004-0622-9

Marya, M., & Edwards, G. R. (2000). Factors affecting the laser bending of Ti-6Al-2Sn-4Zr-2Mo. *Journal of Laser Applications*, *12*(4), 149–159. doi:10.2351/1.521926

Marzolla, M., Babaoglu, O., & Panzieri, F. (2011). Server consolidation in clouds through gossiping. *Proceedings of the IEEE International Symposium on World of Wireless, Mobile and Multimedia Networks (WoWMoM)* (pp. 1-6). IEEE Press. 10.1109/WoWMoM.2011.5986483

Mastura, M. T., Sapuan, S. M., Mansor, M. R., & Nuraini, A. A. (2018). Comparative Study on Weighting Customer Requirements using Fuzzy Analysis Hierarchy Process with Extent Analysis and Analytic Hierarchy Process. *Journal of Advanced Manufacturing Technology*, *12*(1), 477–490.

Mathew, V., Sitaraman, R. K., & Shenoy, P. (2012). Energy-aware load balancing in content delivery networks. *Proceedings of the 31st Annual IEEE International Conference on Computer Communications (IEEE INFOCOM)* (pp. 954-962). IEEE Press. 10.1109/INFCOM.2012.6195846

Matz, H., Drung, D., Hartwig, S., Gross, H., Kotitz, R., Muller, W., ... Trahms, L. (1998). A squid measurement system for immunoassays. *Applied Superconductivity*, *6*(10-12), 577–583. doi:10.1016/S0964-1807(99)00014-9

Mayyas, A., Shen, Q., Mayyas, A., Shan, D., Qattawi, A., & Omar, M. (2011). Using quality function deployment and analytical hierarchy process for material selection of body-in-white. *Materials & Design*, *32*(5), 2771–2782. doi:10.1016/j.matdes.2011.01.001

Mazanoglu, K. (2015a). A novel methodology using simplified approaches for identification of cracks in beams. *Latin American Journal of Solids and Structures*, *12*(13), 2460–2479. doi:10.1590/1679-78251723

McKay, G., & Sze, M. F. F. (2008). The removal of organic pollutants from industrial effluents via tapered bed adsorption columns. *International Journal of Environmental Technology and Management*, *9*(1), 20. doi:10.1504/IJETM.2008.017857

Mecalux, S. A. (2010). *Manual técnico de almacenaje. Conceptos básicos sobre sistemas, equipos de manutención y procesos logísticos*. Barcelona: Mecalux S A.

Medina, L. A., Nogueira, R. D., & Sosa, N. E. (2002). *EL SISTEMA MRP*. Retrieved from https://www.academia.edu/1104284/EL_SISTEMA_MRP

Mena, E., Kashyap, V., Sheth, A. P., & Illarramendi, A. (1996). OBSERVER: An Approach for Query Processing in Global Information Systems based on Interoperation across Pre-existing Ontologies. *Proceedings of the First IFCIS International Conference on Cooperative Information Systems (CoopIS 1996)*, Brussels, Belgium, June 19-21 (pp. 14-25). IEE Computer Society. 10.1109/COOPIS.1996.554955

Mendez, G., & Lopez, E. (2014). Methodology to demand forecasting under multiproduct environments and high variability. *Tecnura*, *18*(40), 89–102.

Meola, C., Boccardi, S., Carlomagno, G. M., Boffa, N. D., Monaco, E., & Ricci, F. (2015). Non-destructive evaluation of carbon fibre reinforced composites with infrared thermography and ultrasonic. *Composite Structures*, *134*, 845–853. doi:10.1016/j.compstruct.2015.08.119

Miclo, R., Fontalini, F., Lauras, M., Lamothe, J., & Milian, B. (2016). An empirical study of Demand-Driven MRP. *Proceedings of the 6th International Conference on Information Systems, Logistics and Supply Chain*, Bordeaux. Academic Press.

Mignogna, R. B., Green, R. E. Jr, Duke, J. C. Jr, Henneke, E. G. II, & Reifsnider, K. L. (1981). Thermographic investigation of high-power ultrasonic heating in materials. *Ultrasonics*, *19*(4), 159–163. doi:10.1016/0041-624X(81)90095-0

Ministry of Economic. (2012). *Pequeñas empresas*. Retrieved from http://www.2006-2012.economia.gob.mx/mexico-emprende/empresas/mediana-empresa

Ministry of Economic. (2012). *Pequeñas empresas*. Retrieved from http://www.2006-2012.economia.gob.mx/mexico-emprende/empresas/pequena-empresa

Ministry of Economics. (2011). *Evolución del desempeño logístico de las cadenas de suministro en México*. Retrieved from http://www.elogistica.economia.gob.mx/swb/work/models/elogistica/Resource/3/1/images/evolucion2011.pdf

Ministry of Economy. (2010). Trade and Industry, Establishment of the Japan data center council [Press Release].

Miño-Cascante, G., Saumell-Fonseca, E., Toledo-Borrego, A., & Moreno, R. (2015). *Planeación de requerimientos de materiales por el sistema MRP. Caso Laboratorio Farmacéutico Oriente* (pp. 208–2019). Cuba: SCIELO.

Mishra, S. K., Puthal, D., Sahoo, B., Jayaraman, P. P., Jun, S., Zomaya, A. Y., & Ranjan, R. (2018). Energy-efficient VM-placement in cloud data center. *Sustainable computing: informatics and systems*, *20*, 48-55.

Mittal, K., Chandra Tewari, P., Khanduja, D., & Kaushik, P. (2016). Application of Fuzzy TOPSIS MADM approach in ranking & underlining the problems of plywood industry in India. *Cogent Engineering*, *3*(1), 1155839. doi:10.1080/23311916.2016.1155839

Mohammadi, M., & Ghorbani, M. (2011). Wear and corrosion properties of electroless nickel composite coatings with PTFE and/or MoS_2 particles. *Journal of Coatings Technology and Research*, *8*(4), 527–533. doi:10.100711998-011-9329-y

Mönch, L., Uzsoy, R., & Fowler, J. W. (2018). A survey of semiconductor supply chain models part I: Semiconductor supply chains, strategic network design, and supply chain simulation. *International Journal of Production Research, 56*(13), 4524–4545. doi:10.1080/00207543.2017.1401233

Morbidini, M., & Cawley, P. (2009). The detectability of cracks using sonic IR. *Journal of Applied Physics, 105*(9), 1–9. doi:10.1063/1.3125444

Morecroft, J. D. W., & Sterman, J. D. (1994). *Modelling For Learning Organizations*. Portland, OR: Productivity Press.

Morlidge, S., & Player, S. (2010). *Future ready: How to master business forecasting*. Chichester, England: John Wiley & Sons.

Moroz, P., Jones, S. K., & Gray, B. N. (2002). Magnetically mediated hyperthermia: Current status and future directions. *International Journal of Hyperthermia, 18*(4), 267–284. doi:10.1080/02656730110108785 PMID:12079583

Muhuri, P. K., & Rauniyar, A. (2017). Immigrants based adaptive genetic algorithms for task allocation in multi-robot systems. *International Journal of Computational Intelligence and Applications, 16*(04), 1750025. doi:10.1142/S1469026817500250

Mukherjee, D., & Rajagopal, D. (1992). Electrodeposition of amorphous nickel alloys. *Metal Finishing, 90*, 15–19.

Mukhopadhyay, A., Barman, T. K., & Sahoo, P. (2018). Electroless Nickel Coatings for High Temperature Applications. In K. Kumar & J. Davim (Eds.), *Composites and Advanced Materials for Industrial Applications* (pp. 297–331). Hershey, PA: IGI Global. doi:10.4018/978-1-5225-5216-1.ch013

Mukhopadhyay, A., & Sahoo, S. (2019). Corrosion protection of reinforcement steel rebars by the application of electroless nickel coatings. *Engineering Research Express, 1*(1), 015021. doi:10.1088/2631-8695/ab35f0

Munot, H., Deshpande, P., & Modhera, C. (2018). Conducting Polypyrrole Coated Rebar in a Carbonated Concrete Pore Solution: Electrochemical Investigations. *Portugaliae Electrochimica Acta, 36*(5), 365–375. doi:10.4152/pea.201805365

Murtazaev, A., & Oh, S. (2011). Sercon: Server consolidation algorithm using live migration of virtual machines for green computing. *IETE Technical Review, 28*(3), 212–231. doi:10.4103/0256-4602.81230

Murthy, J. S. N., Reddy, V. S., Tasleem, S., Kumar, D. V., Krishna, C. S., & Sankarshana, T. (2009). Hydrodynamics of gas-solid fluidization in tapered beds. *Canadian Journal of Chemical Engineering, 87*(3), 359–374. doi:10.1002/cjce.20180

Mustafa, Ö., Papadopoulou, T. C., & Melek, A. (2007). System Dynamics Modelling of A Manufacturing Supply Chain System. *Simulation Modelling Practice and Theory, 15*(10), 1338–1355. doi:10.1016/j.simpat.2007.09.007

Myöhänen, K., & Hyppänen, T. (2011). A Three-Dimensional Model Frame for Modelling Combustion and Gasification in Circulating Fluidized Bed Furnaces. *International Journal of Chemical Reactor Engineering, 9*(1). doi:10.1515/1542-6580.2571

Myszewski, J.M. (2013). On improvement story by 5 whys. *The TQM journal, 25*(4), 371-383.

Nagaya, K., Ohnuma, H., & Sato, A. (1992). Analysis of a radial magnetic fluid seal and a radial bearing lubricated with a magnetic fluid. *Tribology International, 25*(2), 107–113. doi:10.1016/0301-679X(92)90087-4

Nagaya, K., Takeda, S., Sato, A., Ikai, S., Sekiguchi, H., & Saito, N. (1993). Thrust bearing using a magnetic fluid lubricant under magnetic fields. *Tribology International, 26*(1), 11–15. doi:10.1016/0301-679X(93)90033-W

Nahvi, H., & Jabbari, M. (2005a). Crack detection in beams using experimental modal data and finite element model. *International Journal of Mechanical Sciences, 47*(10), 1477–1497. doi:10.1016/j.ijmecsci.2005.06.008

Nandwana, B. P., & Maiti, S. K. (1997a). Modelling of vibration of beam in presence of inclined edge or international crack for its possible detection based on frequency measurements. *Engineering Fracture Mechanics*, *58*(3), 193–205. doi:10.1016/S0013-7944(97)00078-7

Nandwana, B. P., & Maiti, S. K. (1997b). Detection of the location and size of a crack in stepped cantilever beams based on measurements of natural frequencies. *Journal of Sound and Vibration*, *203*(3), 435–446. doi:10.1006/jsvi.1996.0856

Narayanan, T. S. N. S., & Seshadri, S. K. (2004). Formation and characterization of borohydride reduced electroless nickel deposits. *Journal of Alloys and Compounds*, *365*(1-2), 197–205. doi:10.1016/S0925-8388(03)00680-7

Nathuji, R., & Schwan, K. (2007). Virtual Power: coordinated power management in virtualized enterprise systems. *Proceedings of Twenty-First ACM SIGOPS Symposium on Operating Systems Principles* (pp. 265- 278). ACM. 10.1145/1294261.1294287

National Institute of Statistics and Geography [INEGI]. (2014). *Censos económicos 2014*. Retrieved from http://internet.contenidos.inegi.org.mx/contenidos/Productos/prod_serv/contenidos/espanol/bvinegi/productos/nueva_estruc/702825077952.pdf

National Institute of Statistics and Geography [INEGI]. (2018). *Sistema de Cuentas Nacionales de México*. Retrieved from http://www.inegi.org.mx/est/contenidos/Proyectos/scN/

National Research Council. (1998). *Manufacturing Process Controls for the Industries of the Future*. National Academies Press.

Nethe, A., Scholz, T., Stahlmann, H.-D., & Filtz, M. (2002). Ferrofluids-A Numerical Process Model. *Transactions in Magnetism*, *38*(2), 1177–1180. doi:10.1109/20.996301

Nethe, A., & Stahlmann, H.-D. (1998). Remarks on the interaction between magnetic core coils in the presence of magnetic fluids. *Journal of Magnetism and Magnetic Materials*, *189*(2), 255–262. doi:10.1016/S0304-8853(98)00210-8

Nikitin, L. V., Stepanov, G. V., Mironova, L. S., & Gorbunov, A. I. (2004). Magneto-deformational effect and effect of shape memory in magnetoelastics. *Journal of Magnetism and Magnetic Materials*, 272-276(Part 3), 2072–2073. doi:10.1016/j.jmmm.2003.12.838

Ninduangdee, P., & Kuprianov, V. I. (2013). Study on burning oil palm kernel shell in a conical fluidized-bed combustor using alumina as the bed material. *Journal of the Taiwan Institute of Chemical Engineers*. doi:10.1016/j.jtice.2013.06.011

Norihiko, S., & Nakamura, T. (2002). Elucidation of propulsive force of micro-robot using magnetic fluid. *Journal of Applied Physics*, *91*(10), 7003–7005. doi:10.1063/1.1452197

Novak, M., Vojtech, D., & Vitu, T. (2010). Influence of heat treatment on tribological properties of electroless Ni–P and Ni–P–Al_2O_3 coatings on Al–Si casting alloy. *Applied Surface Science*, *256*(9), 2956–2960. doi:10.1016/j.apsusc.2009.11.057

Novakovic, J., & Vassiliou, P. (2009). Vacuum thermal treated electroless Ni-P-TiO_2 composite coatings. *Electrochimica Acta*, *54*(9), 2499–2503. doi:10.1016/j.electacta.2008.12.015

O'Grady, P. J. (Ed.). (2012). *Controlling automated manufacturing systems*. Springer Science & Business Media.

O'Sullivan, D. (2009). *Industrial Automation*. National University of Ireland Galway.

Odenbach, S. (2002). *Magnetoviscous Effects in Ferrofluids*. Berlin: Springer-Verlag.

Odenbach, S. (2004). Recent progress in magnetic fluid research. *Journal of Physics Condensed Matter*, *16*(32), 1135–1150. doi:10.1088/0953-8984/16/32/R02

Odenbach, S. (Ed.). (2009). *Colloidal Magnetic Fluids: Basics, Development and Application of Ferrofluids*. Berlin: Springer. doi:10.1007/978-3-540-85387-9

Oghazi, P., Rad, F. F., Karlsson, K., & Haftor, H. (2018). RFID and ERP Systems in Supply Chain Management. *European Journal of Management and Business Economics*, 171-182

Omachonu, V. K., & Ross, J. E. (2004). *Principles of total quality*. CRC Press.

Omar, R., Aboraia, M. S., Oraby, E. A., Gubner, R., & Rizk, A. E. (2018). The effect of sodium citrate as a complex agent on the corrosion properties of the electroless Ni-P coating. *Materials Research Express*, *5*(12), 126511. doi:10.1088/2053-1591/aae0b9

Opricovic, S., & Tzeng, G. H. (2004). Compromise solution by MCDM methods: A comparative analysis of VIKOR and TOPSIS. *European Journal of Operational Research*, *156*(2), 445–455. doi:10.1016/S0377-2217(03)00020-1

Organization for Economic Co-operation and Development [OECD]. (2013). *Mexico: Key Issues and Polices*. Paris: OECD Publishing.

Organization for Economic Co-operation and Development [OECD]. (2017). *Enhancing the contributions of SMEs in a global and digitalised economy*. Retrieved from: https://www.oecd.org/mcm/documents/C-MIN-2017-8-EN.pdf

Orszag, S. A., Yakhot, V., Flannery, W. S., Boysan, F., Choudhury, D., Maruzewski, J., & Patel, B. (1993). Renormalization group modeling and turbulence simulations. *Proceedings of the International Conference on Near-Wall Turbulent Flows*. Academic Press.

Ostachowicz, W. M., & Krawczuk, M. (1993c). Coupled torsional and bending vibrations of a rotor with an open crack. *Archive of Applied Mechanics*, *62*, 191–201.

Ostachowitz, W. M., & Krawczuk, M. (1991). Analysis of the effect of cracks on the natural frequencies of a cantilever beams. *Journal of Sound and Vibration*, *150*(2), 191–201. doi:10.1016/0022-460X(91)90615-Q

Ožbolt, J., Balabanić, G., & Kušter, M. (2011). 3D Numerical modelling of steel corrosion in concrete structures. *Corrosion Science*, *53*(12), 4166–4177. doi:10.1016/j.corsci.2011.08.026

Padhi, R. K., Mohanty, Y. K., Roy, G. K., & Sarangi, B. (2014). Hydrodynamics studies of gas-solid fluidization in non cylindrical conduits for spherical and non-spherical particles- A review. International Journal of Scientific and Research Publications, 4.

Padhi, R. K., Dora, D. T. K., Mohanty, Y. K., Roy, G. K., & Sarangi, B. (2016). Hydrodynamics of three-phase fluidization of homogeneous ternary mixture in a conical conduit— Experimental and statistical analysis. *Chinese Journal of Chemical Engineering*, *24*(10), 1335–1343. doi:10.1016/j.cjche.2016.05.032

Pafelias, T. (1974). Dynamic behavior of a cracked rotor. General Electric Co.

Pain, C. C., Mansoorzadeh, S., & de Oliveira, C. R. E. (2001). A study of bubbling and slugging fluidised beds using the two-fluid granular temperature model. *International Journal of Multiphase Flow*, *27*(3), 527–551. doi:10.1016/S0301-9322(00)00035-5

Pal, K. (2018). A Big Data Framework for Decision Making in Supply Chain. In M.V. Kumar, G.D. Putnik, K. Jayakrishna et al. (Eds.), Emerging Applications in Supply Chains for Sustainable Business Development (Ch. 1, pp. 1-22). Hershey PA: IGI Global Publication.

Pal, K. (2019). Quality Assurance Issues for Big Data Applications in Supply Chain Management. In P.K. Gupta, T. Oren, & M. Singh (Eds.), Predictive Intelligence Using Big Data and the Internet of Things (Ch. 3, pp. 51-76). Hershey PA: IGI Global.

Palaniappa, M., & Seshadri, S. K. (2008). Friction and wear behavior of electroless Ni-P and Ni-W-P alloy coatings. *Wear*, *265*(5–6), 735–740. doi:10.1016/j.wear.2008.01.002

Pal, K. (2017). Supply Chain Coordination Based on Web Services. In H. K. Chan, N. Subraanian, & M. D. Abdulrahman (Eds.), *Supply Chain Management in the Big Data Era* (pp. 137–171). Hershey, PA: IGI Global. doi:10.4018/978-1-5225-0956-1.ch009

Pal, K. (2018). Building High Quality Big Data-Based Applications in Supply Chains. In A. Kumar & S. Saurav (Eds.), *Edited Book – 'Supply Chain Management Strategies and Risk Assessment in Retail Environments* (pp. 1–24). Hershey, PA: IGI Global. doi:10.4018/978-1-5225-3056-5.ch001

Pal, K. (2019). *Quality Assurance Issues for Big Data Applications in Supply Chain Management, Predictive Intelligence Using Big Data and The Internet of Things* (P. K. Gupta, T. Oren, & M. Singh, Eds.). Hershey, PA: IGI Global.

Panigrahi, B., & Pohit, G. (2016b). Nonlinear modelling and dynamic analysis of cracked Timoshenko functionally graded beams based on neutral surface approach. *Journal of Mechanical Engineering Sciences, 230*(9), 1486-1497.

Pankhurst, Q. A., Connolly, J., Jones, S. K., & Dobson, J. (2003). Applications of magnetic Nanoparticles in biomedicine. *Journal of Physics. D, Applied Physics*, *36*(13), 167–181. doi:10.1088/0022-3727/36/13/201

Panneerselvam, R., Savithri, S., & Surender, G. D. (2009). CFD simulation of hydrodynamics of gas–liquid–solid fluidised bed reactor. *Chemical Engineering Science*, *64*(6), 1119–1135. doi:10.1016/j.ces.2008.10.052

Papadopoulos, C. A., & Dimarogonas, A. D. (1987). Coupled longitudinal and bending vibrations of a rotating shaft with an open crack. *Journal of Sound and Vibration*, *117*(1), 81–93. doi:10.1016/0022-460X(87)90437-8

Papaeconomou, N., & Dimarogonas, A. (1989). Vibration of cracked beams. *Computational Mechanics*, *4*, 130–137.

Paramasivan, K. (2019). *Study of laser forming of sheet metal with and without cutout* [Doctoral dissertation]. Jadavpur University, Kolkata, India.

Paramasivan, K., Das, S., Marimuthu, S., & Misra, D. (2015). Thermo-mechanical Simulation of the Laser Forming of AISI 304 Stainless Steel Square Plate with a Central Cutout. *Lasers in Engineering*, *32*(5-6), 393–410.

Paramasivan, K., Das, S., Marimuthu, S., & Misra, D. (2019). Numerical Study of the Mechanism of Laser Forming Process. *International Journal of Laser Science*, *1*(3-4), 347–382.

Paramasivan, K., Das, S., & Misra, D. (2014). A study on the effect of rectangular cutout on laser forming of AISI 304 plates. *International Journal of Advanced Manufacturing Technology*, *72*(9), 1513–1525. doi:10.100700170-014-5695-9

PARETO ANALYSIS. (n.d.). *Improhealth.org*. Retrieved from .http://www.improhealth.org/fileadmin/Documents/Improvement_Tools/Pareto_Analysis.pdf

Pareto Analysis. (n.d.). *Improhealth.org*. Retrieved from http://www.improhealth.org/fileadmin/Documents/Improvement_Tools/Pareto_Analysis.pdf

Park, C. M., & Wang, G. N. (2009). Job allocation and scheduling in multi robotic tasks considering collision free operation. *Intelligent Automation & Soft Computing*, *15*(2), 249–261. doi:10.1080/10798587.2009.10643029

Parker, L. E. (1994, February). An experiment in mobile robotic cooperation. *Proceedings of the ASCE Specialty Conference on Robotics for Challenging Environments*, Albuquerque, NM. Academic Press.

Patankar, S. (1980). *Numerical heat transfer and fluid flow*. CRC press.

Patil, N. A., Mulik, S. S., Wangikar, K. S., & Kulkarni, A. P. (2018). Characterization of glass laminate aluminium reinforced epoxy-a review. *Procedia Manufacturing*, *20*, 554–562. doi:10.1016/j.promfg.2018.02.083

Patil, R. S., Pandey, M., & Mahanta, P. (2011). Parametric studies and effect of scale-up on wall-to-bed heat transfer characteristics of circulating fluidized bed risers. *Experimental Thermal and Fluid Science*, *35*(3), 485–494. doi:10.1016/j.expthermflusci.2010.11.012

Patra, S. S. (2018). Energy-efficient task consolidation for cloud data center. *International Journal of Cloud Applications and Computing*, *8*(1), 117–142. doi:10.4018/IJCAC.2018010106

Pedram, M. (2012). Energy-efficient datacenters. *IEEE Transactions on Computer-Aided Design of Integrated Circuits and Systems*, *31*(10), 164–177. doi:10.1109/TCAD.2012.2212898

Peirano, E., Delloume, V., & Leckner, B. (2001). Two- or three-dimensional simulations of turbulent gas–solid flows applied to fluidization. *Chemical Engineering Science*, *56*(16), 4787–4799. doi:10.1016/S0009-2509(01)00141-5

Peng, Y., & Fan, L. T. (1995). Hysteresis in liquid-solid tapered fluidized beds. *Chemical Engineering Science*, *50*(16), 2669–2671. doi:10.1016/0009-2509(95)00105-E

Peng, Y., & Fan, L. T. (1997). Hydrodynamic characteristics of fluidization in liquid-solid tapered beds. *Chemical Engineering Science*, *52*(14), 2277–2290. doi:10.1016/S0009-2509(97)00061-4

Pepiot, P., Dibble, C. J., & Foust, T. D. (2010). Computational fluid dynamics modeling of biomass gasification and pyrolysis. In *Computational modeling in lignocellulosic biofuel production* (pp. 273–298). American Chemical Society.

Petroski, H. J. (1981). Simple static and dynamic models for the cracked elastic beam. *International Journal of Fracture*, *17*(4), 71–76. doi:10.1007/BF00036201

Petroski, H. J. (1984b). Stability of a crack in a cantilever beam undergoing large plastic deformation after impact. *International Journal of Pressure Vessels and Piping*, *16*(4), 285–295. doi:10.1016/0308-0161(84)90019-X

Philippsen, C. G., Vilela, A. C. F., & Zen, L. D. (2015). Fluidized bed modeling applied to the analysis of processes: Review and state of the art. *Journal of Materials Research and Technology*, *4*(2), 208–216. doi:10.1016/j.jmrt.2014.10.018

Ping, G., Elliott, S., Beaudoin, J. J., & Arsenault, B. (1996). Corrosion resistance of stainless steel in chloride contaminated concrete. *Cement and Concrete Research*, *26*(8), 1151–1156. doi:10.1016/0008-8846(96)00110-X

Pochet, Y., & Wolsey, L. (2006). *Production Planning by Mixed Integer Programming*. New York: Springer.

Popovic, D. (n.d.). Automation and control in production processes. *Automation and Control, 3*, 2.

Pottinger, R., & Halevy, A. (2001). MiniCon: A Scalable Algorithm for Answering Queries Using Views. *The Very Large Database Journal*, *10*, 182–198.

Pradhan, M. K. (2010). *Experimental investigation and modelling of surface integrity, accuracy and productivity aspects in EDM of AISI D2 steel* [Doctoral dissertation].

Pradhan, M. K., Gangil, M., & Kumar, P. R. (2016, December). Application of MCDM for Optimization of EDM of LM6 Silicon Carbide Boron Carbide Hybrid Composite. *Proceedings of the 6th International and 27th All India Manufacturing, Technology, Design, and Research conference (AIMTDR 2016)* (pp. 557-562). Academic Press.

Pradhan, M. K. (2012). *Process simulation, modelling, and estimation of the temperature and residual stresses electrical discharge machining of AISI D2 steel.* ISCI.

Pradhan, M. K. (2013). Estimating the effect of process parameters on MRR, TWR and radial overcut of EDMed AISI D2 tool steel by RSM and GRA coupled with PCA. *International Journal of Advanced Manufacturing Technology, 68*(1-4), 591–605. doi:10.100700170-013-4780-9

Pradhan, M. K. (2018). Optimization of EDM process for MRR, TWR and radial overcut of D2 steel: A hybrid RSM-GRA and entropy weight-based TOPSIS approach. *International Journal of Industrial and Systems Engineering, 29*(3), 273–302. doi:10.1504/IJISE.2018.093047

Pradhan, M. K., & Singh, B. (2019). *Machinability and Multi-response Optimization of EDM of Al7075/SIC/WS2 Hybrid Composite Using the PROMETHEE Method. In Optimization for Engineering Problems* (pp. 39–75). John Wiley and Sons.

Pradhan, M. K., & Tiwari, A. (2018). Modeling and Optimization of End Milling Parameters on Ti-6Al-4V Titanium Alloy using TLBO Algorithm and TOPSIS Algorithm. *International Journal of Machining and Machinability of Materials, 20*(6), 513–535. doi:10.1504/IJMMM.2018.096379

Prasad, K. D., Sravani, K. D. S., & Manasa, B. L. (2019). Kano-HOQ-GRA Hybrid Methodology for Customer-Driven Product Development. In Optimizing Current Strategies and Applications in Industrial Engineering (pp. 120–139). Hershey, PA: IGI Global. doi:10.4018/978-1-5225-8223-6.ch005

Prasad, B. (1998). Review of QFD and related deployment techniques. *Journal of Manufacturing Systems, 17*(3), 221–234. doi:10.1016/S0278-6125(98)80063-0

Prasad, K. D., Subbaiah, K. V., & Rao, K. N. (2014). Multi-objective optimization approach for cost management during product design at the conceptual phase. *Journal of Industrial Engineering International, 10*(1), 48. doi:10.100740092-014-0048-8

Prasad, K. G. D., Subbaiah, K., & Prasad, M. (2017). Supplier evaluation and selection through DEA-AHP-GRA integrated approach-A case study. *Uncertain Supply Chain Management, 5*(4), 369–382. doi:10.5267/j.uscm.2017.4.001

Prasad, K., & Chakraborty, S. (2013). A quality function deployment-based model for materials selection. *Materials & Design, 49*, 525–535. doi:10.1016/j.matdes.2013.01.035

Prasad, K., Maity, S. R., & Chakraborty, S. (2016). Quality function deployment-based expert system for materials selection. *International Journal of Materials Science and Engineering, 4*(2), 80–93.

Prud'hommeaux, E., & Seaborne, A. (2013). SPARQL Query Language for RDF. W3. Retrieved from http://www.w3.org/TR/rdf-sparql-query

Pul, R. E. W. W., & Sprinkel, M. M. (1998). Estimating the Service Life of Epoxy Coated Reinforcing Steel. *Materials Journal, 95*(5), 546–557.

Qian, B., & Cheng, H. H. (2018). Bio-Inspired Coalition Formation Algorithms for Multirobot Systems. *Journal of Computing and Information Science in Engineering, 18*(2), 021010. doi:10.1115/1.4039638

Qian, G. L., Gu, S. N., & Jiang, J. S. (1990c). The dynamic behavior and crack detection of a beam with a crack. *Journal of Sound and Vibration, 138*(2), 233–243. doi:10.1016/0022-460X(90)90540-G

Rabinowicz, E. (1995). *Friction and Wear of Materials*. New York: John Wiley and Sons.

Rabizadeh, T., & Allahkaram, S. R. (2011). Corrosion resistance enhancement of Ni–P electroless coating by incorporation of nano-SiO_2 particles. *Materials & Design, 32*(1), 133–138. doi:10.1016/j.matdes.2010.06.021

Rabizadeh, T., Allahkaram, S. R., & Zarebidaki, A. (2010). An investigation on effects of heat treatment on corrosion properties of Ni-P electroless nano-coatings. *Materials & Design*, *31*(7), 3174–3179. doi:10.1016/j.matdes.2010.02.027

Rahammer, M., & Kreutzbruck, M. (2017). Fourier-transform vibrothermography with frequency sweep excitation utilizing local defect resonances. *NDT & E International*, *86*, 83–88. doi:10.1016/j.ndteint.2016.11.012

Rahammer, M., Solodov, I., Bisle, W., Scherling, D., & Kreutzbruck, M. (2016). Thermosonic Testing with Phase Matched Guided Wave Excitation. *Journal of Nondestructive Evaluation*, *36*, 1–7.

Raharjo, H., Xie, M., & Brombacher, A. C. (2006). Prioritizing quality characteristics in dynamic quality function deployment. *International Journal of Production Research*, *44*(23), 5005–5018. doi:10.1080/00207540600547414

Rahman, M. M., Islam, M. M., Khan, M. M. R., Ong, H. R., Uddin, M. T., & Islam, M. A. (2019). IBA-modified gypsum-containing epoxy resin coating for rebar: Corrosion performance and bonding characteristics. *International Journal of Plastics Technology*, *23*(1), 20–28. doi:10.100712588-019-09238-3

Rajadell, M., & Sánchez, J. (2010). *Lean manufacturing- La evidencia de una necesidad*. Madrid: Ediciones Díaz de Santos.

Rajawat, S.S. (2017). *What is an Algorithm and what are the types of algorithm with examples.*

Raju, Y., Kang, P., Moroz, A., & Clement, R., Hopwell, A., & Duffy A. (2015). Investigating the Demand for Short-shelf Life Food Products for SME Wholesalers. *International Journal of Social, Behavioral, Educational, Economic, Business and Industrial Engineering*, *9*(6), 2051–2055.

Ramage, M., & Ship, K. (2009). *Sytem Thinkers, The Open University Walton Hall.* Milton Keynes, United Kingdom: Springer London.

Rantala, J., Wu, D., & Busse, G. (1996). Amplitude Modulated Lock-In Vibrothermography for NDE of Polymers and Composites. *Research in Nondestructive Evaluation*, *7*(4), 215–218. doi:10.1080/09349849609409580

Rasteh, M., Farhadi, F., & Bahramian, A. (2015). Hydrodynamic characteristics of gas-solid tapered fluidized beds: Experimental studies and empirical models. *Powder Technology*, *283*, 355–367. doi:10.1016/j.powtec.2015.06.002

Ravinder, H., & Misra, R. (2014). ABC Analysis For Inventory Management: Bridging The Gap Between Research And Classroom. *American Journal of Business Education*, *7*(3), 257–263. doi:10.19030/ajbe.v7i3.8635

Ravindran, P., Manisekar, K., Rathika, P., & Narayanasamy, P. (2013). Tribological properties of powder metallurgy-processed aluminium self-lubricating hybrid composites with SiC additions. *Materials & Design*, *45*, 561570.

Rehman, F., & Yan, X. T. (2003). Product design elements as means to realise functions in mechanical conceptual design. *Proceedings of ICED 03, the 14th International Conference on Engineering Design DS 31*, Stockholm (pp. 213-214). Academic Press.

Reza, H., Amineh, G., & Virginia, D. (2014). Using Difference Equation to Model Discrete-time Behavior in System Dynamics Modeling. *Proceedings of the 32nd International Conference of the System Dynamics Society*, Delft, Netherlands. Academic Press.

Rezaee, M., & Hassannejad, R. (2011a). A new approach to free vibration analysis of a beam with a breathing crack based on mechanical energy balance method. *Guti Lixue Xuebao*, *24*, 185–194.

Richards, G. (2014). *Warehouse Management: A Complete Guide to Improving Efficiency and Minimizing Costs in the Modern Warehouse* (2nd ed.). Londres: Kogan Page.

Richardson George, P. (1999). *Feedback Thought in Social Science And Systems Theory*. Waltham: Pegasus Communications, Inc.

Richmond, B. (2010). The thinking in systems thinking: eight critical skills. *Tracing Connections: Voices of Systems Thinkers*, 3-21.

Ridgeway, K. (1965). The Tapering Fluidized Bed—A New Processing Tool. *Chemical & Process Engineering*, *6*, 317–321.

Rieback, M. R., Crispo, B., & Tanenbaum, A. S. (2006). Is your cat infected with a computer virus? Proceedings of PerCon (pp. 169-179). Academic Press. doi:10.1109/PERCOM.2006.32

RightScale. (2012). Designing Private and Hybrid Clouds: Architectural Best Practices [white paper]. Retrieved from http://www.rightscale.com/lp/private-hybrid-cloud-white-paper.php

Rimal, B. P., Choi, E., & Lumb, I. (2009). A taxonomy and survey of cloud computing systems. *Proceedings of the Fifth International Joint Conference on INC, IMS and IDC* (pp. 44-51). Academic Press. 10.1109/NCM.2009.218

Rivera, P. J., Ortega, P. E., & Pereyra, Q. J. (2014). Diseño e Implementacipon del Sistema MRP en PyMes. *Revista de la Faculta de Ingenieria Industrial*, 48-55.

Rivoire, S., Ranganathan, P., & Kozyrakis, C. (2008). A comparison of high-level full system power models. *Proceedings of the 2008 Conference on Power Aware Computing and Systems* (*Vol. 8*, p. 3). Academic Press.

Roberti, M. (2005). Financing for RFID Prison System. *RFID Journal*, December 31.

Rockwell Automation. (n.d.). Manufacturing and assembly Machine Automation And Control.

Roghanian, E., Rahimi, J., & Ansari, A. (2010). Comparison of first aggregation and last aggregation in fuzzy group TOPSIS. *Applied Mathematical Modelling*, *34*(12), 3754–3766. doi:10.1016/j.apm.2010.02.039

Rohatgi, P. (1992). Friction and wear of metal-matrix composites. In ASM Handbook (Vol. 18, p. 801). ASM.

Rojas, R. J., Solano-Charris, E., & Montoya-Torres, J. R. (2019). The storage location assignment problem: A literature review. *International Journal of Industrial Engineering Computations*, 199-224.

Rosensweig, R. E. (1997). *Ferrohydrodynamics*. Mineola, NY: Dover Publications, Inc.

Rosensweig, R. E. (2002). Heating magnetic fluid with alternating magnetic field. *Journal of Magnetism and Magnetic Materials*, *252*, 370–374. doi:10.1016/S0304-8853(02)00706-0

Rouwenhorst, B., Reuter, B., Stockrahm, V., Van Houtum, G., Mantel, R., & Zijm, W. (2000). Warehouse design and control: Framework and literature review. *European Journal of Operational Research*, *122*(122), 515–533. doi:10.1016/S0377-2217(99)00020-X

Roventi, G., Bellezze, T., Barbaresi, E., & Fratesi, R. (2013). Effect of carbonation process on the passivating products of zinc in $Ca(OH)_2$ saturated solution. *Materials and Corrosion*, *64*(11), 1007–1014. doi:10.1002/maco.201206868

Roy, S., & Bose, T. (2019). Efficient determination of local defect resonance frequencies from bicoherence plots using double excitations. *Mechanical Systems and Signal Processing*, *127*, 595–609. doi:10.1016/j.ymssp.2019.03.017

Roy, S., Bose, T., & Debnath, K. (2019). Detection of Local Defect Resonance Frequencies usingBicoherence Analysis. *Journal of Sound and Vibration*, *443*, 703–716. doi:10.1016/j.jsv.2018.12.006

Roy, S., & Sahoo, P. (2012). Potentiodynamic Polarization Behaviour of Electroless Ni-P-W Coatings. *ISRN Corrosion*, *2012*, 914867. doi:10.5402/2012/914867

Rubio, G. (2017). Multivariant perspective of forecasts in industrial SMEs of Ibague (Colombia). *Revista Facultad de Ciencias Económicas: Investigación y Reflexión*, *15*(2), 25–40.

Rush, C., & Roy, R. (2000). Analysis of cost estimating processes used within a concurrent engineering environment throughout a product life cycle. *Proceedings of the 7th ISPE International Conference on Concurrent Engineering: Research and Applications*, Lyon, France (pp. 58-67). Technomic Inc.

Ruth, M., & Hannon, B. (2012). *Modeling Dynamic Economic Systems* (2nd ed.). Springer. doi:10.1007/978-1-4614-2209-9

Saaty, T. L. (1980). *The Analytic Hierarchy Process*. New York: McGraw-Hill Company.

Saaty, T. L. (2008). Decision making with the analytic hierarchy process. *International Journal of Services Sciences*, *1*(1), 83–98. doi:10.1504/IJSSCI.2008.017590

Sabbaghi, A., & Ganesh, V. (2008). Effectiveness and Efficiency of RFID Technology in Supply Chain Management: Strategic Values and Challenges. *Journal of Theoretical and Applied Electronic Commerce Research*, *3*(2), 71–81. doi:10.4067/S0718-18762008000100007

Sadettin, O. (2007a). Analysis of free and forced vibration of a cracked cantilever beam. *NDT & E International*, *40*(6), 443–450. doi:10.1016/j.ndteint.2007.01.010

Sadoyan, H., Zakarian, A., & Mohanty, P. (2006). Data mining algorithm for manufacturing process control. *International Journal of Advanced Manufacturing Technology*, *28*(3-4), 342–350.

Safari, M., & Farzin, M. (2015). Experimental investigation of laser forming of a saddle shape with spiral irradiating scheme. *Optics & Laser Technology*, *66*, 146–150. doi:10.1016/j.optlastec.2014.09.003

Sagüés, A. A., Pech-Canul, M. A., & Al-Mansur, A. S. (2003). Corrosion macrocell behavior of reinforcing steel in partially submerged concrete columns. *Corrosion Science*, *45*(1), 7–32. doi:10.1016/S0010-938X(02)00087-2

Sahin, Y., & Murphy, S. (1996). Wear performance of aluminium alloy composites containing unidirectionally-oriented silicon carbide coated boron fibres. *Wear*, *197*(1-2), 248–254. doi:10.1016/0043-1648(96)06959-1

Sahoo, P., & Das, S. K. (2011). Tribology of electroless nickel coatings – a review. *Materials & Design*, *32*(4), 1760–1775. doi:10.1016/j.matdes.2010.11.013

Salazar, E., & Martínez, J. (2017). Storage Location Allocation Model for a Just In Sequence (JIS) production System. *Proceedings of the International Conference on Industrial Engineering and Operations Management,* Bogota, Colombia (pp. 277-288). Academic Press.

Salazar, M., & Cabrera, M. (2007). Pronóstico de demanda por medio de redes neuronales artificiales. *Ingenierías. Universidad Autónoma de Nuevo León UANL*, *10*(35), 6–12.

Salles, M. (2006). Decision making in SMEs and information requirements for competitive intelligence. *Production Planning & Control: The Management of Operations*, *17*(3), 229–237. doi:10.1080/09537280500285367

Sangüesa, M., Mateo, R., & Ilzarbe, L. (2006). *Teoría y practica de la calidad*. Madrid: Paraninfo.

Saravanan, K., Sathiyanarayanan, S., Muralidharan, S., Azim, S. S., & Venkatachari, G. (2007). Performance evaluation of polyaniline pigmented epoxy coating for corrosion protection of steel in concrete environment. *Progress in Organic Coatings*, *59*(2), 160–167. doi:10.1016/j.porgcoat.2007.03.002

Sarret, M., Müller, C., & Amell, A. (2006). Electroless Ni-P micro and nano-composite coatings. *Surface and Coatings Technology*, *201*(1-2), 389–395. doi:10.1016/j.surfcoat.2005.11.127

Sasikala, P. (2013). Energy efficiency in cloud computing: Way towards green computing. *International Journal of Cloud Computing*, *2*(4), 305–324. doi:10.1504/IJCC.2013.058095

Sathish, S., & Rathnaraj, J. D. (2015). Experimental studies of the mechanical and microstructure properties of lm6 metal matrix composite with borosilicate glass reinforcement. *International Journal of Applied Engineering Research, 10*(11), 10063–10069.

Satishkumar, T. P., Satheeshkumar, S., & Naveen, J. (2014). Glass fiber-reinforced polymer composites – a review. *Journal of Reinforced Plastics and Composites, 33*(13), 1258–1275. doi:10.1177/0731684414530790

Sau, D. C., Mohanty, S., & Biswal, K. C. (2010). Experimental studies and empirical models for the prediction of bed expansion in gas-solid tapered fluidized beds. *Chemical Engineering and Processing: Process Intensification.* doi:10.1016/j.cep.2010.02.010

Sau, D. C., & Biswal, K. C. (2011). Computational fluid dynamics and experimental study of the hydrodynamics of a gas–solid tapered fluidized bed. *Applied Mathematical Modelling, 35*(5), 2265–2278. doi:10.1016/j.apm.2010.11.037

Schaeffer, D. G. (1987). Instability in the evolution equations describing incompressible granular flow. *Journal of Differential Equations, 66*(1), 19–50. doi:10.1016/0022-0396(87)90038-6

Schreier, C., Udomkit, N., & Macchi, L. (2016). Cultural and Intuitive Intelligence of SME's Leaders. [GBR]. *GSTF Journal on Business Review, 4*(3), 15–20.

Scott, C. D., & Hancher, C. W. (1976). Use of a tapered fluidized bed as a Continuous bioreactor. *Biotechnology and Bioengineering, 18*(10), 1393–1403. doi:10.1002/bit.260181006 PMID:963280

Scott, J., Ho, W., Dey, P. K., & Talluri, S. (2015). A decision support system for supplier selection and order allocation in stochastic, multi-stakeholder and multi-criteria environments. *International Journal of Production Economics, 166*, 226–237. doi:10.1016/j.ijpe.2014.11.008

Segers, J., Hedayatrasa, S., Verboven, E., Poelman, G., Paepegem, W. V., & Kersemans, M. (2019). In-plane local defect resonances for efficient vibrothermography of impacted carbon fiber-reinforced polymers (CFRP). *NDT & E International, 102*, 218–225. doi:10.1016/j.ndteint.2018.12.005

Segers, J., Kersemans, M., Hedayatrasa, S., Calderon, J., & Paepegem, W. V. (2018). Towards in-plane local defect resonance for non-destructive testing of polymers and composites. *NDT & E International, 98*, 130–133. doi:10.1016/j.ndteint.2018.05.007

Serdar, M., Žulj, L. V., & Bjegović, D. (2013). Long-term corrosion behaviour of stainless reinforcing steel in mortar exposed to chloride environment. *Corrosion Science, 69*, 149–157. doi:10.1016/j.corsci.2012.11.035

Serrat, O. (2017). The five whys technique. In *Knowledge solutions* (pp. 307–310). Singapore: Springer.

Sezen, H. K., & Günal, M. (2009). *Yöneylem Araştırmasında Benzetim*. Bursa: Ekin Yayınevi.

Sharma, J. R., Rawani, A. M., & Barahate, M. (2008). Quality function deployment: A comprehensive literature review. *International Journal of Data Analysis Techniques and Strategies, 1*(1), 78–103. doi:10.1504/IJDATS.2008.020024

Sharma, S., Satish, B., Girish, B., Kamath, R., & Asanuma, H. (1998). Dry sliding wear of short glass fibre reinforced zinc-aluminum composites. *Tribology International, 31*(4), 183–188. doi:10.1016/S0301-679X(98)00020-6

Shen, J., Gui, Z., Ji, S., Shen, J., Tan, H., & Tang, Y. (2018). Cloud-aided lightweight certificateless authentication protocol with anonymity for wireless body area networks. *Journal of Network and Computer Applications, 106*, 117–123. doi:10.1016/j.jnca.2018.01.003

Shen, L., Johnsson, F., & Leckner, B. (2004). Digital image analysis of hydrodynamics two-dimensional bubbling fluidized beds. *Chemical Engineering Science, 59*(13), 2607–2617. doi:10.1016/j.ces.2004.01.063

Sheth, A. P., Gala, S. K., & Navathe, S. B. (1993). On Automatic Reasoning for Shema Integration. *International Journal of Intelligent and Cooperative Information Systems*, *2*(1), 23–50. doi:10.1142/S0218215793000034

Sheth, A. P., & Larson, J. A. (1990). Federated Database Systems for Managing Distributed, Heterogenous, and Autonomous Databases. *ACM Computing Surveys*, *22*(3), 183–236. doi:10.1145/96602.96604

Shih, H. S., Shyur, H. J., & Lee, E. S. (2007). An extension of TOPSIS for group decision making. *Mathematical and Computer Modelling*, *45*(7-8), 801–813. doi:10.1016/j.mcm.2006.03.023

Shi, J., & Ming, J. (2017). Influence of defects at the steel-mortar interface on the corrosion behavior of steel. *Construction & Building Materials*, *136*, 118–125. doi:10.1016/j.conbuildmat.2017.01.007

Shi, J., Ming, J., Sun, W., & Zhang, Y. (2017). Corrosion performance of reinforcing steel in concrete under simultaneous flexural load and chlorides attack. *Construction & Building Materials*, *149*, 315–326. doi:10.1016/j.conbuildmat.2017.05.092

Shi, Y. F., Yu, Y. S., & Fan, L. T. (1984). Incipient Fluidization Condition for a Tapered Fluidized Bed. *Industrial & Engineering Chemistry Fundamentals*, *23*(4), 484–489. doi:10.1021/i100016a018

Shliomis, M. I. (1996). On "Maxwell's equations and vorticity: A note on the viscosity of magnetic fluids". *Journal of Magnetism and Magnetic Materials*, *159*(1-2), 236–237. doi:10.1016/0304-8853(96)00260-0

Shu, Z., Wang, J., Zhou, Q., Fan, C., & Li, S. (2015). Evaluation of multifluid model for heat transfer behavior of binary gas-solid flow in a downer reactor. *Powder Technology*, *281*, 34–48. doi:10.1016/j.powtec.2015.04.055

Shu, Z., Wan, J., Zhang, D., & Li, D. (2016). Cloud-integrated cyber-physical systems for complex industrial applications. *Mobile Networks and Applications*, *21*(5), 865–878. doi:10.100711036-015-0664-6

Silberschatz, A., Korth, H. F., & Sudarshan, S. (1997). *Database System Concepts*. McGraw-Hill.

Silva, J. M. M., & Gomez, A. J. M. A. (1990b). Experimental dynamic analysis of cracked free-free beams. *Experimental Mechanics*, *30*(1), 20–25. doi:10.1007/BF02322697

Singh, D. D. N., & Ghosh, R. (2006). Electroless nickel–phosphorus coatings to protect steel reinforcement bars from chloride induced corrosion. *Surface and Coatings Technology*, *201*(1), 90–101. doi:10.1016/j.surfcoat.2005.10.045

Singh, N. (2003). Emerging technologies to support supply chain management. *Communications of the ACM*, *46*(9), 243–247. doi:10.1145/903893.903943

Singh, R. I., Brink, A., & Hupa, M. (2013). CFD modeling to study fluidized bed combustion and gasification. *Applied Thermal Engineering*, *52*(2), 585–614. doi:10.1016/j.applthermaleng.2012.12.017

Singh, R. K., Suryanarayana, A., & Roy, G. K. (1992). Prediction of minimum velocity and minimum bed pressure drop for gas-solid fluidization in conical conduits. *Canadian Journal of Chemical Engineering*, *70*(1), 185–189. doi:10.1002/cjce.5450700127

Sinha, J. K., Friswell, M. I., & Edwards, S. (2002). Simplified models for the location of cracks in beam structures using measured vibration data. *Journal of Sound and Vibration*, *251*(1), 13–38. doi:10.1006/jsvi.2001.3978

Sirisomboon, K., & Charernporn, P. (2017). Effects of air staging on emission characteristics in a conical fluidized-bed combustor firing with sunflower shells. *Journal of the Energy Institute*. doi:10.1016/j.joei.2015.12.001

Sirisomboon, K., & Laowthong, P. (2019). Experimental investigation and prediction of heat transfer in a swirling fluidized-bed combustor. *Applied Thermal Engineering*, *147*, 718–727. doi:10.1016/j.applthermaleng.2018.10.097

Sivathanu, S., Liu, L., Yiduo, M., & Pu, X. (2010). Storage management in virtualized cloud environment. *Proceedings of the IEEE 3rd International Conference on Cloud Computing (CLOUD)* (pp. 204-211). 10.1109/CLOUD.2010.27

Slimani, I., Farissi, I., & Achchab, S. (2015). Artificial Neural Networks/or Demand Forecasting: Application Using Moroccan Supermarket Data. *Proceedings of the 15th International Conference on Intelligent Systems Design and Applications (ISDA)* (pp. 266-271). Academic Press. 10.1109/ISDA.2015.7489236

Smith, J. L., & Virmani, Y. P. (1996). Performance of epoxy coated rebars in bridge decks.

Smith, H. III, Herbig, P., Milewicz, J., & Golden, J. (1996). Differences in forecasting behavior between large and small firms. *Journal of Marketing Practice: Applied Marketing Science*, *2*(1), 35–51. doi:10.1108/EUM0000000000006

Sobhanallahi, M. A., Zendehdel Nobari, N., & Pasandideh, S. H. R. (2019). An Aggregated Supplier Selection Method Based on QFD and TOPSIS (Case Study: A Financial Institution). *Journal of Optimization in Industrial Engineering*, *12*(1), 31–40.

Soderquist, C., & Overakkers, S. (2010). Education for Sustainable Development: A Systems Thinking Approach. *Global Environmental Research*, *14*, 193–202.

Sofia, A. S., & Kumar, P. G. (2015). Implementation of energy efficient green computing in cloud computing. *International Journal of Enterprise Network Management*, *6*(3), 222–237. doi:10.1504/IJENM.2015.071135

Solodov, I. (2014). Resonant Acoustic Nonlinearity of Defects for Highly-Efficient Nonlinear NDE. *Journal of Nondestructive Evaluation*, *33*(2), 252–262. doi:10.100710921-014-0229-9

Solodov, I., Bai, J., Bekgulyan, S., & Busse, G. (2011). A local defect resonance to enhance acoustic wave-defect interaction in ultrasonic nondestructive evaluation. *Applied Physics Letters*, *99*(21), 211911. doi:10.1063/1.3663872

Solodov, I., Bai, J., & Busse, G. (2013). Resonant ultrasound spectroscopy of defects: A Case study of flat-bottomed holes. *Journal of Applied Physics*, *113*(22), 223512. doi:10.1063/1.4810926

Solodov, I., Derusova, D., & Rahammer, M. (2015). Thermosonic Chladni figures for defect-selective imaging. *Ultrasonics*, *60*, 1–5. doi:10.1016/j.ultras.2015.02.007 PMID:25773969

Solodov, I., Rahammer, M., Gulnizkij, N., & Kreutzbruck, M. (2016). Noncontact Sonic NDE and Defect Imaging Via Local Defect Resonance. *Journal of Nondestructive Evaluation*, *35*(3), 48. doi:10.100710921-016-0364-6

Solodov, I., Rahammer, M., & Kreutzbruck, M. (2019). Analytical evaluation of resonance frequencies for planar defects: Effect of a defect shape. *NDT & E International*, *102*, 274–280. doi:10.1016/j.ndteint.2018.12.008

Soloman, S. (2011). *Industrial sensors and control--The basics--Part I.* EE Times.

Song, Y., Zhang, Y., Sun, Y., & Shi, W. (2009). Utility analysis for internet-oriented server consolidation in VM-based data centers. *Proceedings of IEEE International Conference on Cluster Computing* (pp. 1-10). IEEE Press. 10.1109/CLUSTR.2009.5289190

Soto, P., & Dolan, S. (2004). *Las PYMES ante el desafío del siglo XXI: los nuevos mercados globales*. Mexico: Thomson.

Söylev, T. A., & Richardson, M. G. (2008). Corrosion inhibitors for steel in concrete: State-of-the-art report. *Construction & Building Materials*, *22*(4), 609–622. doi:10.1016/j.conbuildmat.2006.10.013

Sproston, J. L., Yanyo, L. C., Carlson, J. D., & El Wahed, A. K. (2002). Controllable Fluids in 2002-Status of ER and MR Fluid Technology. *Proceedings of 8th International Conference on New Actuators* (pp. 333–338). Academic Press.

Srikantaiah, S., Kansal, A., & Zhao, F. (2008). Energy Aware consolidation for cloud computing. *Proceedings of the 2008 Conference on Power Aware Computing and Systems* (*Vol. 10*). Academic Press.

Srinivasan, K. N., Meenakshi, R., Santhi, A., Thangavelu, P. R., & John, S. (2010). Studies on development of electroless Ni-B bath for corrosion resistance and wear resistance applications. *Surface Engineering*, *26*(3), 153–158. doi:10.1179/174329409X409468

Srivyas, P. D., & Charoo, M. (2018). Role of reinforcements on the mechanical and tribological behavior of aluminum metal matrix composites-a review. *Materials Today: Proceedings*, *5*(9), 20041–20053.

Staudt, F. H., Mascolo, M. D., Alpan, G., & Rodriguez, C. M. (2014). Warehouse performance measurement: classification and mathematical expressions of indicators. *Proceedings of the 5th International Conference: Information Systems, Logistics and Supply Chain ILS 2014* (pp. 1-9). Breda: Dinalog BETA.

Staudt, F. H., Alpan, G., Mascolo, M. D., & Rodriguez, C. M. (2015). Warehouse performance measurement: A literature review. *International Journal of Production Research*, *53*(18), 5524–5544. doi:10.1080/00207543.2015.1030466

Sterman, J. D. (1991). *A Skeptic's Guide to Computer Models*. Cambridge, USA: MIT System Dynamics In Education Project.

Sterman, J. D. (2000). Business Dynamics Systems Thinking And Modelling. In *A Complex World*. New York: McGraw-Hill.

Stierstadt, K., & Liu, M. (2015). Maxwell's stress tensor and the forces in magnetic liquids. *Zeitschrift für Angewandte Mathematik und Mechanik*, *95*(1), 4–37. doi:10.1002/zamm.201300192

Stockman, H. (1948). Communication by Means of Reflected Power. *Proceedings of the Institute of Radio Engineers*. Academic Press.

Suárez, C., & Hualpa, A. M. (2018). Warehouse Sizing based on Material Requirements Planning for a Polyurethane Coating Factory. *Ingeniería*, *3*(1), 48–69.

Subbaiah, K. V., Durga Prasad, K. G., & Narayana Rao, K. (2011). Customer-driven product planning using conjoint analysis and QFD-ANP methodology. *International Journal of Productivity and Quality Management*, *7*(3), 374–394. doi:10.1504/IJPQM.2011.039353

Sudagar, J., Lian, J., & Sha, W. (2013). Electroless nickel, alloy, composite and nano coatings–A critical review. *Journal of Alloys and Compounds*, *571*, 183–204. doi:10.1016/j.jallcom.2013.03.107

Suemitsu, I., Izui, K., Yamada, T., Nishiwaki, S., Noda, A., & Nagatani, T. (2016). Simultaneous optimization of layout and task schedule for robotic cellular manufacturing systems. *Computers & Industrial Engineering*, *102*, 396–407. doi:10.1016/j.cie.2016.05.036

Sun, C., Li, J., Shuang, S., Zeng, H., & Luo, J. L. (2018). Effect of defect on corrosion behavior of electroless Ni-P coating in CO_2-saturated NaCl solution. *Corrosion Science*, *134*, 23–37. doi:10.1016/j.corsci.2018.01.041

Sun, J., Zhou, Y., Ren, C., Wang, J., & Yang, Y. (2011). CFD simulation and experiments of dynamic parameters in gas–solid fluidized bed. *Chemical Engineering Science*, *66*(21), 4972–4982. doi:10.1016/j.ces.2011.06.035

Sutar, H., & Das, C. K. (2012). Mixing and Segregation Characteristics of Binary Granular Material in Tapered Fluidized Bed: A CFD Study. *Engineering*, *4*(4), 215–227. doi:10.4236/eng.2012.44029

Syamlal, M., & Gidaspow, D. (1985). Hydrodynamics of fluidization: Prediction of wall to bed heat transfer coefficients. *AIChE Journal. American Institute of Chemical Engineers*, *31*(1), 127–135. doi:10.1002/aic.690310115

Syamlal, M., & O'Brien, T. J. (1988). Simulation of granular layer inversion in liquid fluidized beds. *International Journal of Multiphase Flow*, *14*(4), 473–481. doi:10.1016/0301-9322(88)90023-7

Tada, H., Paris, P. C., & Irwin, G. R. (1985). *The Stress Analysis of Cracks Handbook*. Hellertown PA: Del Research Corporation.

Taiwo, A., Ayodeji, M., & Yusuf, A. (2012). Impact of Small and Medium Enterprises on Economic Growth and Development. *American Journal of Business and Management*, *1*(1), 18–22. doi:10.11634/21679606170644

Takagi, H. (1993). Queueing Analysis—a foundation of performance evaluation. *Discrete time Systems*, *3*(1), 141-158.

Takahashi, I. (1999). Vibration and stability of non-uniform cracked Timoshenko beam subjected to follower force. *Composite Structures*, *71*(5), 585–591. doi:10.1016/S0045-7949(98)00233-8

Tang, F., & Chen, G. (2015). Chemically reactive enamel coating of steel rebar for enhanced durability of reinforced concrete structures. *International Journal of Structural Engineering*, *6*(1), 56–72. doi:10.1504/IJSTRUCTE.2015.067992

Tang, F., Chen, G., Brow, R. K., Volz, J. S., & Koenigstein, M. L. (2012b). Corrosion resistance and mechanism of steel rebar coated with three types of enamel. *Corrosion Science*, *59*, 157–168. doi:10.1016/j.corsci.2012.02.024

Tang, F., Chen, G., Volz, J. S., Brow, R. K., & Koenigstein, M. L. (2012a). Microstructure and corrosion resistance of enamel coatings applied to smooth reinforcing steel. *Construction & Building Materials*, *35*, 376–384. doi:10.1016/j.conbuildmat.2012.04.059

Tang, F., Chen, G., Volz, J. S., Brow, R. K., & Koenigstein, M. L. (2013). Cement-modified enamel coating for enhanced corrosion resistance of steel reinforcing bars. *Cement and Concrete Composites*, *35*(1), 171–180. doi:10.1016/j.cemconcomp.2012.08.009

Tan, Z. Q., & Hansson, C. M. (2008). Effect of surface condition on the initial corrosion of galvanized reinforcing steel embedded in concrete. *Corrosion Science*, *50*(9), 2512–2522. doi:10.1016/j.corsci.2008.06.035

Teghem, J. Jr. (1986). Control of the service process in a queueing system. *European Journal of Operational Research*, *23*(2), 141–158.

Tenek, L. H., Henneke, E. G. II, & Gunzburger, M. D. (1993). Vibration of delaminated composite plates and some applications to non-destructive testing. *Composite Structures*, *23*(3), 253–262. doi:10.1016/0263-8223(93)90226-G

The green grid consortium. (2011). Trade and Industry, Establishment of the Japan data center council. Retrieved from http://www.thegreengrid.org

Thisse, L. C. (1998). Advanced quality planning: A guide for any organization. *Quality Progress*, *31*(2), 73–77.

Thomassey, S., Happiette, M., & Castelain, J. M. (2002). An automic textile sales forecast using fuzzy treatment of explanatory variables. *Journal of Textile and Apparel. Technology and Management*, *2*(4), 1–15.

Tian, G., Zhang, H., Feng, Y., Wang, D., Peng, Y., & Jia, H. (2018). Green decoration materials selection under interior environment characteristics: A grey-correlation based hybrid MCDM method. *Renewable & Sustainable Energy Reviews*, *81*, 682–692. doi:10.1016/j.rser.2017.08.050

Tian, N., & Zhang, Z. G. (2006). Synchronous working vacation policy for finite-buffer multiserver queueing system. *Applied Mathematics and Computation*, *217*(24), 9916–9932.

Tian, N., & Zhang, Z. G. (2006). *Vacation queueing models: theory and applications*. Springer. doi:10.1007/978-0-387-33723-4

Tian, Y., Liu, M., Cheng, X., Dong, C., Wang, G., & Li, X. (2019). Cr-modified low alloy steel reinforcement embedded in mortar for two years: Corrosion result of marine field test. *Cement and Concrete Composites*, *97*, 190–201. doi:10.1016/j.cemconcomp.2018.12.019

Tompkins, J. A., White, J. A., Bozer, Y. A., & Tanchoco, J. M. (2010). *Facilities Planning*. New York: John Wiley & Sons.

Torbati-Sarraf, H., & Poursaee, A. (2018). Corrosion of coupled steels with different microstructures in concrete environment. *Construction & Building Materials*, *167*, 680–687. doi:10.1016/j.conbuildmat.2018.02.083

Torres, J., Carrera, D., Hogan, K., Gavaldà, R., Beltran, V., & Poggi, N. (2008). Reducing wasted resources to help achieve green data centers. *Proceedings of IEEE International Symposium on Parallel and Distributed Processing* (pp. 1-8). IEEE Press. 10.1109/IPDPS.2008.4536219

Towil, D.R. (1996). Industrial Dynamics Modelling of Supply Chains. *Logistic Information Management*, *9*(4), 43-56.

Trahms, L. (2009). Biomedical applications of magnetic nanoparticles. In Colloidal Magnetic Fluids (pp. 327–358). Springer.

Treadaway, K. W. J., Cox, R. N., & Brown, B. L. (1989). Durability of corrosion resisting steels in concrete. *Proceedings - Institution of Civil Engineers*, *86*(2), 305–331. doi:10.1680/iicep.1989.1628

Tsai, J., Fang, J., & Chou, J. (2013). Optimized task scheduling and resource allocation on cloud computing environment using improved differential evolution algorithm. *Computer Operation Research Elsevier*, *40*(12), 3045–3055. doi:10.1016/j.cor.2013.06.012

Tsinidou, M., Gerogiannis, V., & Fitsilis, P. (2010). Evaluation of the factors that determine quality in higher education: An empirical study. *Quality Assurance in Education*, *18*(3), 227–244. doi:10.1108/09684881011058669

Ucler, C. (2017). Brainstorming the cryoplane layout by using the iterative AHP-QFD-AHP approach. *Aviation*, *21*(2), 55–63. doi:10.3846/16487788.2017.1344138

Uhlmann, E., Spur, G., Patzwald, R., & Bayat, N. (2002). Application of magnetic fluids in tribo-technical systems. *Journal of Magnetism and Magnetic Materials*, *252*, 337–340. doi:10.1016/S0304-8853(02)00724-2

Ullman, J. D. (1997). Information integration using logical views. *Proceedings of International Conference on Database Theory*, Delphi, Greece (pp. 19-40). Academic Press.

Umanath, K., Palanikumar, K., & Selvamani, S. (2013). Analysis of dry sliding wear behaviour of Al6061/SiC/Al2O3 hybrid metal matrix composites. *Composites. Part B, Engineering*, *53*, 159–168. doi:10.1016/j.compositesb.2013.04.051

van der Hoef, M. A., van Sint Annaland, M., Deen, N. G., & Kuipers, J. A. M. (2008). Numerical Simulation of Dense Gas-Solid Fluidized Beds: A Multiscale Modeling Strategy. *Annual Review of Fluid Mechanics*, *40*(1), 47–70. doi:10.1146/annurev.fluid.40.111406.102130

van Kampen, T., Akkerman, R., & van Donk, D. (2012). SKU classification: A literature review and conceptual framework. *International Journal of Operations & Production Management*, *32*(7), 850–876. doi:10.1108/01443571211250112

Varshney, K., Jain, M., & Sharma, G. C. (1989). A multi-server queueing model with balking and reneging via diffusion approximation. *Journal of Physical and Natural Science*, *10*, 10–15.

Velasco, S. J., & Campins, M. J. (2013). *Gestión de la producción en la empresa: Planificación, programación y control*. Madrid: Ediciones Pirámide.

Velte, A., Velte, T., & Elsenpeter, R. (2010). Your organization and cloud computing. In *Cloud Computing-A practical Approach*. McGraw-Hill.

Venkadeshwaran, K., Das, S., & Misra, D. (2010). Finite element simulation of 3-D laser forming by discrete section circle line heating. *International Journal of Engineering Science and Technology*, *2*(4), 163–175. doi:10.4314/ijest.v2i4.59284

Verbruggen, H., Baert, K., Terryn, H., & De Graeve, I. (2019). Molybdate-phosphate conversion coatings to protect steel in a simulated concrete pore solution. *Surface and Coatings Technology*, *361*, 280–291. doi:10.1016/j.surfcoat.2018.09.056

Victor, T., & Samutra, V. (2001). System Dynamics Origins, Development And Future Prospects of a Method. In Research Seminar in Engineering Systems. Academic Press.

Vinay, H. B., Govindaraju, H. K., & Banakar, P. (2015). Processing and characterization of glass fiber and carbon fiber reinforced vinyl ester based composites. *International Journal of Research in Engineering and Technology*, *4*(5), 401–406. doi:10.15623/ijret.2015.0405074

Vlot, A., & Gunnink, J. W. (2011). *Fibre metal laminates: an introduction.* Springer Science & Business Media.

Vogelesang, L. B., & Vlot, A. (2000). Development of fibre metal laminates for advanced aerospace structures. *Journal of Materials Processing Technology*, *103*(1), 1–5. doi:10.1016/S0924-0136(00)00411-8

Vogels, W. (2008). Beyond server consolidation. *Queue*, *6*(1), 20–26. doi:10.1145/1348583.1348590

Voila, E., Ricci, P., & Aliabadi, M. H. (2007d). Free vibration analysis of axially loaded cracked Timoshenko beam structures using the dynamic stiffness method. *Journal of Sound and Vibration*, *304*(1-2), 124–153. doi:10.1016/j.jsv.2007.02.013

Vollertsen, F. (1998). *Forming, sintering and rapid prototyping, Handbook of the Eurolaser Academy* (pp. 357–453). London: Chapman & Hall. doi:10.1007/978-1-4615-5297-0_6

Vollertsen, F., Komel, I., & Kals, R. (1995). The laser bending of steel foils for micro-parts by the buckling mechanism - A model. *Modelling and Simulation in Materials Science and Engineering*, *3*(1), 107–119. doi:10.1088/0965-0393/3/1/009

Wache, H., Vogele, T., Visser, U., Struckenschmidt, H., Schuster, G., Neumann, H., & Hubner, S. (2001). Ontology-Based Integration of Information – A Survey of Existing Approaches. In *Workshop on Ontologies and Information Sharing IJCAI-2001*, Seattle, WA, April 4-5 (pp. 108-117). Academic Press.

Wagner, G., & Choset, H. (2015). Subdimensional expansion for multirobot path planning. *Artificial Intelligence*, *219*, 1–24. doi:10.1016/j.artint.2014.11.001

Wang, K. & Li, Z. (2012). Power and performance analysis for web server cluster. *Advances in information Sciences and Service Sciences*, *4*(12), 285-293.

Wang, L., Zhang, F., Zheng, K., Vasilakos, A. V., Ren, S., & Liu, Z. (2014). Energy-efficient flow scheduling and routing with hard deadlines in data center networks. *Proceedings of the 2014 IEEE 34th International Conference on Distributed Computing Systems* (pp. 248-257). IEEE. 10.1109/ICDCS.2014.33

Wang, J., Ge, W., & Li, J. (2008). Eulerian simulation of heterogeneous gas-solid flows in CFB risers: EMMS-based sub-grid scale model with a revised cluster description. *Chemical Engineering Science*, *63*(6), 1553–1571. doi:10.1016/j.ces.2007.11.023

Wang, J., Gu, Y., & Li, X. (2012). Multi-robot task allocation based on ant colony algorithm. *Journal of Computers*, *7*(9), 2160–2167. doi:10.4304/jcp.7.9.2160-2167

Wang, J., & Liu, Y. (2010). EMMS-based Eulerian simulation on the hydrodynamics of a bubbling fluidized bed with FCC particles. *Powder Technology*. doi:10.1016/j.powtec.2009.09.022

Wang, K., & Wand, B. (2015). Timoshenko beam model for the vibration analysis of a cracked nano- beam with surface energy. *Journal of Vibration and Control*, *21*(12), 2452–2464. doi:10.1177/1077546313513054

Wang, L. Sun, J., Duan, S., & Xu, L. (2019). Simulation of cylindrical fluidized bed between granular flow and annulus surface. *International Journal of Thermal Sciences*. doi:10.1016/j.ijthermalsci.2018.12.026

Wang, L. Yang, D., Shen, Z., Mao, K., & Long, J. (2016). Thermal-hydraulic calculation and analysis of a 600 MW supercritical circulating fluidized bed boiler with annular furnace. *Applied Thermal Engineering*. doi:10.1016/j.applthermaleng.2015.11.014

Wang, L., Zhang, F., Aroca, J. A., Vasilakos, A. V., Zheng, K., Hou, C., & Liu, Z. (2013). GreenDCN: A general framework for achieving energy efficiency in data center networks. *IEEE Journal on Selected Areas in Communications*, *32*(1), 4–15. doi:10.1109/JSAC.2014.140102

Wang, S., Zhou, A., Bao, R., Chou, W., & Yau, S. S. (2018). Towards green service composition approach in the cloud. *IEEE Transactions on Services Computing*, 1. doi:10.1109/TSC.2018.2868356

Wang, W., & Li, J. (2007). Simulation of gas–solid two-phase flow by a multi-scale CFD approach—Of the EMMS model to the sub-grid level. *Chemical Engineering Science*, *62*(1-2), 208–231. doi:10.1016/j.ces.2006.08.017

Wang, Y. Q., Kong, G., Che, C. S., & Zhang, B. (2018). Inhibitive effect of sodium molybdate on the corrosion behavior of galvanized steel in simulated concrete pore solution. *Construction & Building Materials*, *162*, 383–392. doi:10.1016/j.conbuildmat.2017.12.035

Wang, Z. (2006). *Experimental studies and CFD simulations of conical spouted bed hydrodynamics*. University of British Columbia.

Wauer, J. (1990a). Cracked rotor Dynamics: A state of the art survey. *Applied Mechanics Reviews*, *43*, 13–17. doi:10.1115/1.3119157

Wei, S., Zhang, J., & Li, Z. (1997). A supplier-selecting system using a neural network. In *IEEE international conference on intelligent processing systems* (pp. 468–471). IEEE.

Wei, D., Liu, Y., & Xiang, Z. (2012b). An analytical method for free vibration analysis of functionally graded beams with edge cracks. *Journal of Sound and Vibration*, *33*(7), 1686–1700. doi:10.1016/j.jsv.2011.11.020

Weishaar, A., Carpenter, M., Loucks, R., Sakulich, A., & Peterson, A. M. (2018). Evaluation of self-healing epoxy coatings for steel reinforcement. *Construction & Building Materials*, *191*, 125–135. doi:10.1016/j.conbuildmat.2018.09.197

Wheeler Frederick, P. (1994). How to Get Managers To Use System Dynamics. *Proceedings of the International System Dynamics Conference*, Stirling, Scotland. Academic Press.

Wiederhold, G. (1992). Mediators in the architecture of future information systems. *Computer*, *25*(3), 38–49.

Wikipedia. (n.d.). Wiki. Retrieved from http://en.wikipedia.org/wiki/

Wilson, S., & Alpas, A. (1997). Wear mechanism maps for metal matrix composites. *Wear*, *212*(1), 41–49. doi:10.1016/S0043-1648(97)00142-7

Wind, Y., & Saaty, T. L. (1980). Marketing applications of the analytic hierarchy process. *Management Science*, *26*(7), 641–658. doi:10.1287/mnsc.26.7.641

Winston, W. (2005). *Investigación de operaciones. Aplicaciones y algoritmos*. México: Thomson.

Wiratmadja, I., Rumanti, A., Reynaldo, R., & Kurniawati, A. (2016). Intuition Concept in Small and Medium Enterprises through Strategic Decision-Making. *International Review of Management and Marketing*, *6*(8), 322–326.

Woizeschke, P. (2019). Approach for bidirectional laser bending of sheet metal with one-sided accessibility. *CIRP Annals*, *68*(1), 277–280. doi:10.1016/j.cirp.2019.04.098

World Trade Organization [WTO]. (2016). *Informe sobre el comercio mundial 2016*. Retrieved from https://www.wto.org/spanish/res_s/booksp_s/world_trade_report16_s.pdf

Wronkowska, A. A. (1993). In Situ and Ex Situ Characterization of Passive Layers on $Ni_{1-x}P_x$ in Alkaline Solution. *Journal of the Electrochemical Society*, *140*(4), 995–1005. doi:10.1149/1.2056242

Wu, C., Chen, G., Volz, J. S., Brow, R. K., & Koenigstein, M. L. (2012). Local bond strength of vitreous enamel coated rebar to concrete. *Construction & Building Materials*, *35*, 428–439. doi:10.1016/j.conbuildmat.2012.04.067

Wu, G. Y., & Shih, Y. S. (2005c). Dynamic instability of rectangular plate with an edge crack. *Composite Structures*, *84*(1-2), 1–10. doi:10.1016/j.compstruc.2005.09.003

Wu, Z. (1993). *Coal, Lignite, and Peat; Fluidized-bed combustion technology assessment, chemical Reactions; Fluidization*, Pollutants.

Xiaoyan, C., & Zhenliang, L. (2007, August). Integrated Computer-aided Control and Optimization of Béchamel Production Processes. *Proceedings of the 2007 IEEE International Conference on Automation and Logistics* (pp. 1646-1651). IEEE.

Xie, J., Mei, Y. T., Ernst, A., Li, X., & Song, A. (2015). *A Restricted Neighbourhood Tabu Search for Storage Location Assignment Problem. Proceedings of the Congress on Evolutionary Computation (CEC)*, Sendai, Japan *(pp. 2805-2812)*. IEEE.

Xiong, C., Li, W., Jin, Z., Gao, X., Wang, W., Tian, H., & Jiang, L. (2018). Preparation of phytic acid conversion coating and corrosion protection performances for steel in chlorinated simulated concrete pore solution. *Corrosion Science*, *139*, 275–288. doi:10.1016/j.corsci.2018.05.018

Yadav, S., Pathak, V. K., & Gangwar, S. (2019). A novel hybrid TOPSIS-PSI approach for material selection in marine applications. *Sadhana*, *44*(3), 58. doi:10.100712046-018-1020-x

Yan, D., Reis, S., Tao, X., Chen, G., Brow, R. K., & Koenigstein, M. (2012). Effect of chemically reactive enamel coating on bonding strength at steel/mortar interface. *Construction & Building Materials*, *28*(1), 512–518. doi:10.1016/j.conbuildmat.2011.08.075

Yang, C. L., Huang, R. H., & Ke, W. C. (2012). Applying QFD to build green manufacturing system. *Production Planning and Control*, *23*(2-3), 145–159. doi:10.1080/09537287.2011.591632

Yang, E. C., Zhao, X., & Li, Y. H. (2015b). Free Vibration Analysis for Cracked FGM Beams by Means of a Continuous Beam. *Shock and Vibration*, *2015*, 1–13.

Yang, J., Chen, Y., Xiang, Y., & Jia, X. L. (2008b). Free and forced vibration of cracked inhomogeneous beams under an axial force and a moving load. *Journal of Sound and Vibration*, *312*(1-2), 166–181. doi:10.1016/j.jsv.2007.10.034

Yang, N., Wang, W., Ge, W., & Li, J. (2003). CFD simulation of concurrent-up gas-solid flow in circulating fluidized beds with structure-dependent drag coefficient. *Chemical Engineering Journal*, *96*(1-3), 71–80. doi:10.1016/j.cej.2003.08.006

Yang, N., Wang, W., Ge, W., Wang, L., & Li, J. (2004). Simulation of heterogeneous structure in a circulating fluidized-bed riser by combining the two-fluid model with the EMMS approach. *Industrial & Engineering Chemistry Research*, *43*(18), 5548–5561. doi:10.1021/ie049773c

Yang, S., Wang, S., Luo, K., Fan, J., & Chew, J. W. (2019). Numerical investigation of the back-mixing and non-uniform characteristics in the three-dimensional full-loop circulating fluidized bed combustor with six parallel cyclones. *Applied Thermal Engineering*. doi:10.1016/j.applthermaleng.2019.03.032

Yan, T., & Yang, J. (2011b). Forced vibration of Edged – Cracked functionally graded beam due to a transverse moving load. *Procedia Engineering*, *14*, 3293–3300. doi:10.1016/j.proeng.2011.07.416

Yao, D. Q. (2012). Application of the Theory of Constraints (TOC) to batch scheduling in process industry. *International Journal of Applied Industrial Engineering*, *1*(1), 10–22. doi:10.4018/ijaie.2012010102

Ye, C., Wang, Q., Luo, Z., Xie, G., Jin, K., Siyil, M., & Cen, K. (2017). Characteristics of Coal Partial Gasification on a Circulating Fluidized Bed Reactor. *Energy & Fuels*, *31*(3), 2557–2564. doi:10.1021/acs.energyfuels.6b02889

Ye, C., Wang, Q., Yu, L., Luo, Z., & Cen, K. (2018). Characteristics of coal partial gasification experiments on a circulating fluidized bed reactor under CO2–O2 atmosphere. *Applied Thermal Engineering*, *130*, 814–821. doi:10.1016/j.applthermaleng.2017.11.071

Yeomans, S. R. (1994). Performance of black, galvanized, and epoxy-coated reinforcing steels in chloride-contaminated concrete. *Corrosion*, *50*(1), 72–81. doi:10.5006/1.3293496

Yeomans, S. R. (2004). Galvanized steel in concrete: an overview. In S. R. Yeomans (Ed.), *Galvanized Steel Reinforcement in Concrete* (pp. 1–30). Oxford, UK: Elsevier Science Ltd.

Yokoyama, T., & Chen, M. C. (1998c). Vibration analysis of edge-cracked beams using a line–spring model. *Engineering Fracture Mechanics*, *59*(3), 403–409. doi:10.1016/S0013-7944(97)80283-4

Yoon, S., & Kim, J. (2017). Efficient multi-agent task allocation for collaborative route planning with multiple unmanned vehicles. *IFAC-PapersOnLine*, *50*(1), 3580–3585. doi:10.1016/j.ifacol.2017.08.686

Yu, Y., & Wen, C. (1966). Mechanics of fluidization. *Chemical Engineering Progress*. doi:10.1016/S0032-0633(98)00014-2

Zamzam, M. (1991). Abrasive wear of aluminum-matrix composites. *Metal*, *45*(3), 250–254.

Zavadskas, E. K., Mardani, A., Turskis, Z., Jusoh, A., & Nor, K. M. (2016). Development of TOPSIS method to solve complicated decision-making problems-An overview on developments from 2000 to 2015. *International Journal of Information Technology & Decision Making*, *15*(3), 645–682. doi:10.1142/S0219622016300019

Zeballos, E. (2003). Micro, Small and Medium-sized Enterprises in Latin America. *CEPAL Review*, *79*(79), 51–68. doi:10.18356/def1ede4-en

Zehner, P., & Schlünder, E. U. (1970). Wärmeleitfähigkeit von Schüttungen bei mäßigen Temperaturen. *Chemieingenieurtechnik (Weinheim)*, *42*(14), 933–941. doi:10.1002/cite.330421408

Zehra, B. (1997). *Sistem Dinamiğinin İşletmelerde Uygulanması*. Bursa: Ekin Kitabevi Yayınları.

Zhang, R., & Liu, L. (2010). Security models and requirements for healthcare application clouds. *Proceedings of the IEEE 3rd International Conference on Cloud Computing (CLOUD)* (pp. 268-275). IEEE Press. 10.1109/CLOUD.2010.62

Zhang, G., Nishi, T., Turner, S. D., Oga, K., & Li, X. (2016). An Integrated Strategy for a Production Planning and Warehouse Layout Problem: Modeling and Solution Approaches. *Omega*, *68*, 85–94. doi:10.1016/j.omega.2016.06.005

Zhang, G., Patuwo, B., & Hu, M. (1998). Forecasting with artificial neural networks: The state of art. *International Journal of Forecasting*, *14*(1), 35–62. doi:10.1016/S0169-2070(97)00044-7

Zhang, H., Peng, Y., Tian, G., Wang, D., & Xie, P. (2017). Green material selection for sustainability: A hybrid MCDM approach. *PLoS One*, *12*(5), 1–26. doi:10.1371/journal.pone.0177578 PMID:28498864

Zhang, X., Ghamari, M., & Ratner, A. (2013). Numerical modeling of co-firing a light density biomass, oat (Avena sativa) hulls, and chunk coal in fluidized bed boiler. *Biomass and Bioenergy*, *56*, 239–246. doi:10.1016/j.biombioe.2013.05.006

Zheng, H., Dai, J. G., Poon, C. S., & Li, W. (2018). Influence of calcium ion in concrete pore solution on the passivation of galvanized steel bars. *Cement and Concrete Research*, *108*, 46–58. doi:10.1016/j.cemconres.2018.03.001

Zheng, T., & Ji, T. (2012a). An approximate method for determining the static deflection and natural frequency of a cracked beam. *Journal of Sound and Vibration*, *331*(11), 2654–2670. doi:10.1016/j.jsv.2012.01.021

Zhou, X., Wang, H., Ding, B., Hu, T., & Shang, S. (2019). Balanced connected task allocations for multi-robot systems: An exact flow-based integer program and an approximate tree-based genetic algorithm. *Expert Systems with Applications*, *116*, 10–20. doi:10.1016/j.eswa.2018.09.001

Zhu, Z., Tang, B., & Yuan, J. (2017). Multirobot task allocation based on an improved particle swarm optimization approach. *International Journal of Advanced Robotic Systems*, *14*(3). doi:10.1177/1729881417710312

Ziegler, P., & Dittrich, K. R. (2007). *Data Integration – Problems, Approaches, and Perspectives*. University of Zurich.

Zimmermann, K., Zeidis, I., Naletova, V. A., & Turkov, V. A. (2004a). Waves on the surface of a magnetic fluid layer in a traveling magnetic field. *Journal of Magnetism and Magnetic Materials*, *268*(1-2), 227–231. doi:10.1016/S0304-8853(03)00503-1

Zimmermann, K., Zeidis, I., Naletova, V. A., & Turkov, V. A. (2004b). Traveling waves on a free surface of a magnetic fluid layer. *Journal of Magnetism and Magnetic Materials*, *272–276*, 2343–2344. doi:10.1016/j.jmmm.2003.12.966

Zou, Y., Cheng, Y., Cheng, L., & Liu, W. (2010). Effect of tin addition on the properties of electroless Ni-P-Sn ternary deposits. *Materials Transactions*, *51*(2), 277–281. doi:10.2320/matertrans.MC200917

Zuluaga, M., Molina, P., & Guisao, G. (2011). Supply chain management strategies based on demand planning in Colombia. *Revista Politecnica*, *12*, 1900–23510.

Zweschper, T., Dillenz, A., Riegert, G., Scherling, D., & Busse, G. (2003). Ultrasound excited thermography using frequency modulated elastic waves. *Insight (American Society of Ophthalmic Registered Nurses)*, *45*, 178–182.

About the Contributors

Prasanta Sahoo received a Ph.D. degree in Mechanical Engineering in 2002 and M.Tech. degree in Mechanical Engineering (Machine Design) in 1994 both from Indian Institute of Technology, Kharagpur, Bachelor of Mechanical Engineering degree in 1991, from Jadavpur University, Kolkata. Currently, he is Professor at the Department of Mechanical Engineering of Jadavpur University, Kolkata, India. He has more than 25 years of teaching and research experience in manufacturing, materials and mechanical engineering, with special emphasis in tribology of materials. He has also interest in structural mechanics, contact mechanics, fractal modeling, composites, and coatings. He has guided large numbers of Ph.D. and master's students as well as coordinated and participated in several financed research projects. He has worked as an evaluator of projects for international research agencies as well as examiner of Ph.D. thesis for many universities and institutes. He is on the editorial board of several international journals and conferences. Presently, he acts as reviewer for more than 40 prestigious web of science journals. In addition, he has also authored 2 books, over 20 book chapters and more than 400 articles in peer-reviewed journals and conference proceedings (h-index 25+/3000+ citations).

* * *

Sumanta Banerjee obtained his bachelor's degree in mechanical engineering from BIT Mesra. He completed his post-graduate studies in the Mechanical Engineering Department at Jadavpur University, finally obtaining the coveted PhD (Engg.) Degree (from Jadavpur University) in 2012. Prior to Doctoral program, he has held Junior Research Fellowship position in Chemical Engineering and has worked in the field of Compressible Flows in High Mach number regimes. Dr. Sumanta Banerjee has joined the Mechanical Engineering Department at the Heritage Institute of Technology in 2013 as an assistant professor, and has been serving the institute since then. His research interests are Convective heat and mass transfer, transport in porous media, thermal management in micro-scaled devices, and dynamics of magneto-rheological fluids.

Tanmoy Bose is working as a full-time assistant professor at NIT Meghalaya. He has obtained a BTech degree from Jalpaiguri Government Engineering College (under WBUT) in 2008, master's in engineering from Jadavpur University in 2010, and a PhD from IIT Kharagpur 2015. Research interests of Dr. Bose are nonlinear ultrasonic, non-destructive testing, and acoustics.

Santiago-Omar Caballero-Morales is a full-time professor-researcher at the Postgraduate Department of Logistics and Supply Chain Management in Universidad Popular Autonoma del Estado de

Puebla (UPAEP) A.C. His main research topics are simulation of industrial processes, combinatorial optimization, artificial intelligence, pattern recognition, statistical quality control, design of distribution networks, inventory control, service robotics.

Patricia Cano-Olivos is a full-time professor of Logistics and Supply Chain Management Doctorate at Centro Interdisciplinario de Posgrados (CIP) at Universidad Popular Autónoma del Estado de Puebla (UPAEP).

Sandip Das is a faculty member in the Department of Mechanical Engineering, Jadavpur University. He has 31 years of teaching experience. His fields of interest are thermal science, instrumentation and measurement and thermal-hydraulic studies in two-phase flow. Presently, he heads a couple of significant projects of national importance in the field of simulation of accident conditions in nuclear power plants.

Shubhajit Das received his B.Tech degree in mechanical engineering from Tezpur University, Assam, India in 2010, M.Tech degree in mechanical engineering (design and manufacturing) from NIT Silchar, Assam, India in 2012 and pursuing his Ph.D. in mechanical engineering from NERIST, Deemed University, Arunachal Pradesh, India. Currently, he is working as an Assistant Professor in the Department of Mechanical Engineering, NIT Arunachal Pradesh. He has seven years of teaching and research experience. He has published 10 research papers in various national or international journals or conferences. He has guided 9 B.Tech projects, 6 M.Tech projects and presently guiding 3 B.Tech and 2 M.Tech students. His current research interest includes conventional and non-conventional machining of metal matrix composites, material characterization and optimization, tribology. He is a member of TSI and IAENG.

K. G. Durga Prasad received a PhD degree in Mechanical Engineering, a M.E. degree in Mechanical Engineering (Industrial Engineering) and a Bachelor of Mechanical Engineering from Andhra University, Visakhapatnam, India. Currently, he is working as a Professor of Mechanical Engineering, Engineering & Technology Program, Gayatri Vidya Parishad College for Degree and P.G. Courses (A), Visakhapatnam, India. He published 26 papers in various international and national journals and 15 papers presented at various conferences. He also published 3 book chapters, one monograph and two textbooks. He is a charted engineer and a life member of Institution of Engineers (India). His research areas of interests are product design and development, supply chain management, multi criteria decision making, and composite materials.

Maidely Fernández-Rodríguez is a student of the Master in Logistics and Supply Chain Management at Universidad Popular Autonoma del Estado de Puebla (UPAEP) in Mexico.

Rajib Ghosh did his Btech at the BPUT University in Mechanical Engineering, qualified GATE in 2017 and received his Mtech in 2019 from NIT Agartala in machine design.

Veena Goswami is currently a Professor in the School of Computer Application, Kalinga Institute of Industrial Technology, Bhubaneswar, India. She received her PhD from Sambalpur University, India, and then worked as a research associate at the Indian Institute of Technology, Kharagpur for two years. Her research interests include continuous- and discrete-time queues. She has published research articles in INFORMS Journal on Computers and Mathematics with Applications, Applied Mathematical Modelling, Computers & Industrial Engineering, Applied Mathematics and Computation, etc. She is a life member of Indian Society for Technical Education and Orissa Information Technology Society, Orissa Mathematical Society, Indian Science Congress Association and Indian Statistical Association.

N. S. V. N. Hanuman received an M.Tech degree in mechanical engineering from Jawaharlal Nehru Technological University Hyderabad, India in 2014. He has worked on finite element analysis to study the free vibration characteristics of laminated composite beams using a higher-order shear deformation theory and the conventional first-order shear deformation theory. Since 2018, he is pursuing a Ph.D. degree at the National Institute of Technology Meghalaya, Meghalaya, India, where he is working on the non-destructive testing of composite materials using a nonlinear ultrasonic technique. He is currently working on detection of delamination in composites by determining the local defect resonance of the defect using fully acoustic based NDT.

Paramasivan Kalvettukaran is currently a research associate in School of Laser Science & Engineering, Jadavpur University, Kolkata, India. He completed his graduation in Mechanical Engineering from Government College of Engineering, Tirunelveli, India in 2009 and his post graduation in Computer Aided Design at Alagappa Chettiyar College of Engineering and Technology, Karaikkudi, India, in 2011. His research interest includes finite element modeling and laser processing of materials.

P. Krishna Murthy is an Assistant Professor of Mechanical Department, Gayatri Vidya Parishad College for degree and P.G Courses (A), Visakhapatnam, Andhra Pradesh, India. He received his M.Tech Degree in mechanical engineering (machine design) from Gitam University, Visakhapatnam, Andhra Pradesh, India. His areas of interest are product design, finite elements analysis, and composite materials.

Pinakeswar Mahanta is a Professor in the Department of Mechanical Engineering, IIT Guwahati. His major area of research is on thermodynamics, heat transfer and renewable energy. He did his BSc Engg from Regional Engineering College, Rourkela, MTech from IIT Kharagpur and PhD from IIT Guwahati. He is engaged in collaborative research in the area of bio-energy and fluidization technologies with University of Nottingham, University of Birmingham, and Loughborough University, UK, University of Cranfield, UK and UCL, Belguim. He has published more than 60 papers in internationally reputed peer reviewed journals and supervised 9 PhDs. Currently, he holds the position of the Director of NIT Arunachal Pradesh.

Arindam Majumer received his B.E. Degree in Mechanical Engineering from the North Maharashtra University, Jalgoun, and M.E. degree in Mechanical Engineering (Machine Design) from the Bengal Engineering and Science University, Shibpur in 2007 and 2009, respectively. He is currently working as an Assistant Professor in the Department of Mechanical Engineering, National Institute of Technology, Agartala, India. His areas of interest are the effect of process parameters on the material properties of various materials during welding and machining, as well as nonlinear optimization.

B. L. Manasa is working as Assistant Professor in Mechanical Engineering Department, University College of Engineering, JNTUK, Kakinada, Andhra Pradesh, India. She received B.Tech and M.Tech Degrees from JNT University, Kakinada, India. She has published 8 papers in International / National journals and 2 conference papers. Her teaching and research interests are robotics, machine design, finite element analysis, product design and development, and composite materials.

Natalie-Nicole Mar-Hernández is a student of Logistics and Supply Chain Management Master at Centro Interdisciplinario de Posgrados (CIP) at Universidad Popular Autónoma del Estado de Puebla (UPAEP).

Sundar Marimuthu specialises in laser/material interaction and theoretical modelling of thermal interaction and fluid flow. He undertook post-doctoral research on innovative laser processing of materials including laser cleaning and micro-machining of metals, ceramics and composites using short and ultra-short pulsed UV laser. At present, he is working as a senior research engineer at the Manufacturing Technology Centre, United Kingdom.

José Luis Martínez-Flores is a full time professor and Coordinator of Logistics and Supply Chain Management Doctorate at Centro Interdisciplinario de Posgrados (CIP) at Universidad Popular Autónoma del Estado de Puebla (UPAEP).

Jose Luis Martínez Flores is a researcher and Academic Director of the Postgraduate in Logistics and Supply Chain Management at Universidad Popular Autonoma del Estado de Puebla (UPAEP) in Mexico. He received a Ph.D. in mathematics. His objectives in research are the optimization and implementation of models for problems in the field of logistics through the use of information technologies.

Dipten Misra is the Director of School of Laser Science & Engineering and a faculty in the department of Mechanical Engineering, Jadavpur University. He has 31 years of teaching experience. His research interests are heat and mass transfer, computational fluid dynamics and laser processing of materials.

Arkadeb Mukhopadhyay received a Bachelor of Technology in Mechanical Engineering from Asansol Engineering College under West Bengal University of Technology. He received Master of Engineering in Mechanical Engineering from Jadavpur University. He pursued his doctoral work at Jadavpur University under DST PURSE II scheme of Govt. of India. Presently, he is an Assistant Professor in the Department of Mechanical Engineering of Birla Institute of Technology Mesra, Ranchi. His area of research interest is tribology.

G. B. Mund is currently a Professor in the School of Computer Engineering, Kalinga Institute of Industrial Technology, Bhubaneswar, India. He received his Master and Ph.D. in Computer Science from Indian Institute of Technology, Kharagpur, India. His areas of research interest include program analysis, algorithms, graph theory and discrete-time queues. He has published research articles in information and software technology, journal of systems and software, computers and mathematics with applications, computers and industrial engineering, etc. He is a life member of the Indian Society for Technical Education, Orissa Mathematical Society and Orissa Information Technology Society.

Kamalendu Pal is with the Department of Computer Science, School of Mathematics, Computer Science and Engineering, City, University of London. Kamalendu received his BSc (Hons) degree in physics from Calcutta University, India, Postgraduate Diploma in Computer Science from Pune, India;

MSc degree in Software Systems Technology from Sheffield University, a Postgraduate Diploma in Artificial Intelligence from Kingston University, a MPhil degree in Computer Science from University College London, and a MBA degree from University of Hull, United Kingdom. He has published dozens of research papers in international journals and conferences. His research interests include knowledge-based systems, decision support systems, computer integrated design, software engineering, and service-oriented computing. He is a member of the British Computer Society, the Institution of Engineering and Technology, and the IEEE Computer Society.

B. Panigrahi is working as anAssistant Professor in the Department of Mechanical Engineering, BITS Pilani, Hyderabad campus.

Sudhansu Shekhat Patra is currently an Associate Professor in the School of Computer Application, KIIT University, Bhubaneswar, India. He received his Master degree in Computer Application from Motilal Nehru National Institute of Technology, Allahabad, India, his M.Tech (Computer Science & Engg) from Utkal University, Bhubaneswar, India and his Ph.D. in Computer Science from Kalinga Institute of Industrial Technology, Deemed to be University, Bhubaneswar, India. His research interests include grid computing, Cloud computing, and algorithms. He is a life member of Indian Society for Technical Education.

Goutam Pohit is working as a Professor in the Department of Mechanical Engineering, Jadavpur University, Kolkata.

Mohan Kumar Pradhan is Assistant Professor in the Department of Mechanical Engineering and Head of Metrology & Adv. and CAM lab. of the Maulana Azad National Institute of Technology, Bhopal, India (An Institute of National Importance, under MHRD, Govt. of India). He received his M. Tech and Ph.D. from National Institute of Technology, Rourkela, India. He has over 20 years of teaching and research experience in manufacturing and 10 years of postdoctoral research experience. He has advised over 80 graduates, 40 Postgraduate and supervised 3 Ph.D. and currently supervising 5 Ph.D. students. Dr. Pradhan's research interests include manufacturing, non-traditional machining, metrology, micro-machining, hybrid machining, and process modeling and optimization. Dr. Pradhan has more than 60 refereed publications and nearly 60 technically edited papers, which were published in conference proceedings, edited two books, five conference proceedings and five journals as guest editor, authored seven book chapters. Dr. Pradhan is a Charted Engineer, a life fellow of IIPE and a life member of ISTE, IACSIT, IAENG, and MIE (I).

Subhankar Roy received an M.Tech degree in Mechanical engineering from the National Institute of Technology Silchar, Assam, India, in 2016. He has worked on the design of ultrasonic horns for improving the magnification factor in ultrasonic machining (USM) using finite element analysis. His work proposed two new designs of ultrasonic horns which can be used to obtain better magnification and so a better machining rate in USM. Since 2017, he is pursuing Ph.D. degree at the National Institute of Technology Meghalaya, Meghalaya, India where he is working on the structural health monitoring of composite materials using nonlinear wave spectroscopy. He is currently working on detection of delamination in composites by determining the local defect resonance of the defect. He is also working on vibro-thermography technique to perform defect selective thermal imaging of defects in composite structures.

Sarmila Sahoo received a Bachelor of Engineeing (Civil Engineering) from Bengal Engineering and Science University, Shibpur, Master of Civil Engineering (Structural Engineering) from Jadavpur University and Doctor of Philosophy (Engineering) from Jadavpur University. She was the recipient of Senior Research Fellowship (Council of Scientific and Industrial Research) of Govt. of India. Presently she is working as Associate Professor of Civil Engineering at the Heritage Institute of Technology, Kolkata. Her research interests are finite elements method, vibration of plates and shells, composite materials. She has published over 75 research articles in international journals and conference proceedings. She has authored a book on design aids for stiffened composite shells with cutout (Springer). Dr. Sahoo is on the editorial board of two international journals.

Rituraj Saikia is a research scholar in the department of Mechanical Engineering, IIT Guwahati. He is also serving as a Technical Officer Gr-I in the same department. His research area is Thermal Engineering and fluidization. He did his BE in Mechanical Engineering from Assam Engineering College, Guwahati, Assam and MTech from IIT Guwahati. Currently, he is doing his PhD in the department of Mechanical Engineering, IIT Guwahati.

Diana Sánchez-Partida is a Doctor in Logistics and Supply Chain Management. Currently, she is Professor-Researcher of the Postgraduate in Logistics and Supply Chain Management at UPAEP University. She has been granted a doctorate and post-doctorate scholarship by CONACyT. She is the leader of the Humanitarian Logistics Group in the same institution. Her areas of interest are disaster risk reduction, humanitarian logistics, resilience in economic activities and logistics operations, among others. She has participated in applied projects such as programming schedules for schools, levels, and control of inventories, planning, and control production. She is a member of the National System of Researchers in the category (Level 1) from January 2018 to December 2021.

Onur Mesut Şenaras is a PhD student in Automotive Engineering at Bursa Uludağ University. He also works as an engineer at an automotive plant.

K. D. S. Sravani is a graduate student of Mechanical Engineering, A.U. College of Engineering for Women, Andhra University, India. She has published one book chapter and two conference papers. Her areas of research interests are product development and optimization.

Index

T

U

V

W

X

www.ingramcontent.com/pod-product-compliance
Lightning Source LLC
LaVergne TN
LVHW080620300926
849162LV00005B/1221